MATHEMATICAL MODELING IN NUTRITION AND THE HEALTH SCIENCES

Edited by

Janet A. Novotny

Beltsville Human Nutrition Research Center
U.S. Department of Agriculture
Beltsville, Maryland

Michael H. Green

Department of Nutrition
Pennsylvania State University
State College, Pennsylvania

and

Ray C. Boston

School of Veterinary Medicine
University of Pennsylvania
Kennett Square, Pennsylvania

Kluwer Academic / Plenum Publishers
New York, Boston, Dordrecht, London, Moscow

Library of Congress Cataloging-in-Publication Data

ISBN 0-306-47881-1

©2003 Kluwer Academic / Plenum Publishers, New York
233 Spring Street, New York, New York 10013

http://www.kluweronline.com

10 9 8 7 6 5 4 3 2 1

A C.I.P. record for this book is available from the Library of Congress

Permissions for books published in Europe: *permissions@wkap.nl*
Permissions for books published in the United States of America: *permissions@wkap.com*

Printed in United Kingdom by Biddles/IBT Global

DEDICATION

This book is dedicated to the memory of Dr. Loren A. Zech Sr. (1943-1997). Dr. Zech was a senior investigator at the National Institutes of Health, where he was a clinician, an active researcher of physiology and metabolism using compartmental modeling, and a guide for the development and use of the SAAM kinetic modeling software. Dr. Zech was a mentor and friend to many of the authors in this book, and his knowledge and enthusiasm were an inspiration to all who worked with him. In spite of his many other responsibilities, he always made seemingly unlimited time and attention available to help both friends and strangers with modeling endeavors, with absolutely no expectations in return. His generosity in sharing his knowledge and love of modeling was unsurpassed. It was an honor to have known him and an honor to have worked with him.

CONTRIBUTORS

Steven A. Abrams
USDA/ARS Children's Nutrition Research Center and Section of Neonatology
Department of Pediatrics
Baylor College of Medicine 77030
Houston, TX

John A. Adam
Department of Mathematics and Statistics
Old Dominion University
Norfolk, VA 23529

Assad Al-Ammar
Department of Chemistry
University of Massachusetts
Amherst, MA 01003

Ramon M. Barnes
University Research Institute for Analytical Chemistry
Amherst, MA 01002

Gary Beecher
U. S. Department of Agriculture
Beltsville Human Nutrition Research Center
Beltsville, MD 20705

Richard N. Bergman
Department of Physiology and Biophysics and USC Diabetes Research
Center
University of Southern California
Los Angeles, CA 90033

Ray Boston
Biomathematics Unit
School of Veterinary Medicine
University of Pennsylvania
Kennett Square, PA 19348

Steven Britz
U. S. Department of Agriculture
Beltsville Human Nutrition Research Center
Beltsville, MD 20705

Rebecca J. Bryant
Department of Food and Nutrition
Purdue University
West Lafayette, IN 47907

Chris C. Calvert
Department of Animal Science
University of California
Davis, CA   95616

John Cant
Department of Animal and Poultry Science
University of Guelph
Guelph
Ontario, Canada N1G 2W1

Raymond J. Carroll
Department of Biostatistics
The University of Texas M.D. Anderson Cancer Center
Houston, TX 77030

Francis Caulfield
U. S. Department of Agriculture
Beltsville Human Nutrition Research Center
Beltsville, MD 20705

Robert S. Chapkin
Faculty of Nutrition
Texas A & M University
College Station, TX 77843

Beverly Clevidence
U. S. Department of Agriculture
Beltsville Human Nutrition Research Center
Beltsville, MD 20705

Stephen P. Coburn
Fort Wayne State Developmental Center
Fort Wayne, IN 46835

Patricia S. Cooper
Department of Medicine
Medical College of Virginia
Virginia Commonwealth University
Richmond, VA 23298

David L. Costill
Ball State University
Muncie, IN 47306

Karen L. Ericson
Indiana University-Purdue University Fort Wayne
Fort Wayne, IN 46805

William J. Fink
Ball State University
Muncie, IN 47306

Danny G. Fox
Department of Animal Science
Cornell University
Ithaca, NY 14853

Joanne Balmer Green
Nutrition Department
The Pennsylvania State University
University Park, PA 16802

Michael H. Green
Nutrition Department
The Pennsylvania State University
University Park, PA 16802

Peter Greif
Laboratory of Computational and Experimental Biology
Division of Cancer Biology and Diagnosis
National Cancer Institute
Bestheda, MD 20892

Ian J. Griffin
USDA/ARS Children's Nutrition Research Center and Section of Neonatology
Department of Pediatrics
Baylor College of Medicine 77030
Houston, TX

David L. Hachey
Vanderbilt University
Nashville, TN 37232

James L. Hargrove
Department of Foods and Nutrition
The University of Georgia
Athens, GA 30602

Diane K. Hartle
Department of Pharmaceutical and Biomedical Sciences
College of Pharmacy
The University of Georgia
Athens, GA 30602

James F. Hochadel
Intramural Research Support Program-Science Applications International Corp.
Frederick, MD 21702

Mee Young Hong
Faculty of Nutrition
Texas A & M University
College Station, TX 77843

Jinah Hwang
Department of Foods and Nutrition
The University of Georgia
Athens, GA 30602

Heidi A. Johnson
Department of Nutrition
University of California
Davis, CA   95616

Daniel Kerner
Civilized Software Inc.
Silver Spring, MD 20906

Thomas R. Kiffe
Department of Mathematics
Texas A&M University
College Station, TX 77843

Kirk C. Klasing
Department of Animal Science
University of California
Davis, CA   95616

Gary Knott
Civilized Software Inc.
Silver Spring, MD 20906

Richard A. Kohn
Department of Animal and Avian Sciences
University of Maryland
College Park, MD 20742

Kevin C. Lewis
Drug-Nutrient Interactions Group
Basic Research Laboratory
National Cancer Institute
National Institutes of Health
Frederick, MD 21702

Oscar Linares
University of Michigan Geriatrics Clinic
Ann Arbor, MI 48105

Joanne R. Lupton
Faculty of Nutrition
Texas A & M University
College Station, TX 77843

J. Dennis Mahuren
Fort Wayne State Developmental Center
Fort Wayne, IN 46835

Berdine R. Martin
Department of Food and Nutrition
Purdue University
West Lafayette, IN 47907

James H. Matis
Department of Statistics
Texas A&M University
College Station, TX 77843

Mary C. McKenna
Department of Pediatrics
University of Maryland School of Medicine
Baltimore, MD 21201

Peter Moate
Biomathematics Unit
School of Veterinary Medicine
University of Pennsylvania
Kennett Square, PA 19348

Jeffrey S. Morris
Department of Biostatistics
The University of Texas M.D. Anderson Cancer Center
Houston, TX 77030

Edward B. Neufeld
National Institutes of Neurological Disorders and Stroke
National Institutes of Health
Bethesda, MD 20891

Janet A. Novotny
U. S. Department of Agriculture
Beltsville Human Nutrition Research Center
Beltsville, MD 20705

Thomas A. Pauly
Fort Wayne State Developmental Center
Fort Wayne, IN 46835

David R. Pearson
Ball State University
Muncie, IN 47306

Peter G. Pentchev
National Institutes of Neurological Disorders and Stroke
National Institutes of Health
Bethesda, MD 20891

Robert D. Reynolds
University of Illinois at Chicago
Chicago, IL 60612

Judah Rosenblatt
University of Texas Medical Branch
Galveston, TX 77555

Wayne E. Schaltenbrand
Fort Wayne State Developmental Center
Fort Wayne, IN 46835

Charles C. Schwartz
Department of Medicine
Medical College of Virginia
Virginia Commonwealth University
Richmond, VA 23298

Darko Stefanovski
Biomathematics Unit
School of Veterinary Medicine
University of Pennsylvania
Kennett Square, PA 19348

Anne Sumner
Diabetes Branch
National Institute of Digestive Disease and K
Bethesda, MD 20892

Douglas W. Townsend
Indiana University-Purdue University Fort Wayne
Fort Wayne, IN 46805

Nancy D. Turner
Faculty of Nutrition
Texas A & M University
College Station, TX 77843

Michael E. Van Amburgh
Department of Animal Science
Cornell University
Ithaca, NY 14853

Julie M. VandenBroek
Department of Medicine
Medical College of Virginia
Virginia Commonwealth University
Richmond, VA 23298

Naisyin Wang
Department of Statistics
Texas A & M University
College Station, TX 77843

Yao Wang
University of Illinois at Chicago
Chicago, IL 60612

Meryl E. Wastney
Metabolic Modeling Services
Dalesford
Hamilton, New Zealand

Connie M. Weaver
Department of Food and Nutrition
Purdue University
West Lafayette, IN 47907

E. Paul Wileyto
Section of Epidemiology and Public Health
School of Veterinary Medicine
University of Pennsylvania
New Bolton Center
Kennett Square, PA 19348

Loren A. Zech Jr.
Drug-Nutrient Interactions Group
Basic Research Laboratory
National Cancer Institute
National Institutes of Health
Frederick, MD 21702

Paula J. Ziegler
Ball State University
Muncie, IN 47306

PREFACE

Over the years, research in the life sciences has benefited greatly from the quantitative tools of mathematics and modeling. Many aspects of complex biological systems can be more deeply understood when mathematical techniques are incorporated into a scientific investigation. Modeling can be fruitfully applied in many types of biological research, from studies on the molecular, cellular, and organ level, to experiments in whole animals and in populations.

Using the field of nutrition as an example, one can find many cases of recent advances in knowledge and understanding that were facilitated by the application of mathematical modeling to kinetic data. The availability of biologically important stable isotope-labeled compounds, developments in sensitive mass spectrometry and other analytical techniques, and advances in the powerful modeling software applied to data have each contributed to our ability to carry out ever more sophisticated kinetic studies that are relevant to nutrition and the health sciences at many levels of organization. Furthermore, we anticipate that modeling is on the brink of another major advance: the application of kinetic modeling to clinical practice. With advances in the ability of models to access large databases (e.g., a population of individual patient records) and the development of user interfaces that are "friendly" enough to be used by clinicians who are not modelers, we predict that health applications modeling will be an important new direction for modeling in the 21st century.

This book contains manuscripts that are based on presentations at the seventh conference in a series focused on advancing nutrition and health research by fostering exchange among scientists from such disciplines as nutrition, biology, mathematics, statistics, kinetics, and computing. The themes of the six previous conferences included general nutrition modeling (Canolty and Cain, 1985; Hoover-Plow and Chandra, 1988), amino acids and carbohydrates (Abumrad, 1991), minerals (Siva Subramanian and Wastney, 1995), vitamins, proteins, and modeling theory (Coburn and Townsend, 1996), and physiological compartmental modeling (Clifford and Müller, 1998). The seventh conference in the series was held at The Pennsylvania State University from July 29 through August 1, 2000. The meeting began with an instructive and entertaining keynote address by Professor Britton Chance, Eldridge Reeves Johnson University Professor Emeritus of Biophysics, Physical Chemistry, and Radiologic Physics, University of Pennsylvania. Dr. Chance, a pioneer in the field of mathematical modeling, presented a personal perspective on the history of enzyme kinetics and biological oscillators. His many important contributions to a wide variety of projects were highlighted in his remarks.

The papers in this text are grouped into several sections that reflect the topics covered at the conference. All manuscripts were reviewed and refereed by experts in the field. The first group of papers focuses on historical perspectives and future directions in modeling, and includes important new example of health applications modeling. The second part includes three didactic presentations on the development and use of mathematical models. Papers in the third and fourth sections cover theoretical issues in modeling and experimental aspects of kinetic data. In the next two parts, modeling of the metabolism of selected vitamins, minerals, lipids, and protein, as well as growth, digestion, fermentation, and lactation, are discussed. Then several manuscripts on modeling in disease states are presented. The final group of papers describes state-of-the-art modeling software packages.

It is our hope that the papers in this text will be useful to beginning modelers as well as experienced scientists. As we look to the future of research in nutrition and the life sciences, we remain firmly convinced that mathematical modeling will be a powerful tool for solving complex problems and advancing both practical and theoretical knowledge.

REFERENCES

Abumrad, N., ed., 1991, Mathematical models in experimental nutrition, *J. Parenter. Enteral. Nutr.* 15:44s-98s.

Canolty, N., and Cain, T.P., eds, 1985, *Mathematical Models in Experimental Nutrition,* University of Georgia, Athens.

Clifford, A.J., and Müller, H.-G., eds., 1998, *Mathematical Modeling in Experimental Nutrition,* Plenum Press, New York.

Coburn, S.P., and Townsend, D. W., eds., 1996, Mathematical modeling in experimental nutrition, *Adv. Food Nutr. Res.* 40:1-362.

Hoover-Plow, J., and Chandra, R.K., eds., 1988, Mathematical modeling in experimental nutrition, *Prog. Food Nutr. Sci.* 12:211-338.

Siva Subramanian, K.N., and Wastney, M.E., eds., 1995, *Kinetic Models of Trace Element and Mineral Metabolism during Development,* CRC Press, New York.

ACKNOWLEDGMENTS

The 7th Conference on Mathematical Modeling in Nutrition and the Health Sciences held at The Pennsylvania State University from July 29 to August 1, 2000 was an outreach program of the College of Health and Human Development and its Department of Nutrition. The conference was supported by generous contributions from the National Institutes of Health (NIH R13 DK 56755), the United States Department of Agriculture, Penn State University, the University of Pennsylvania, Agribrands International, Hoffmann-LaRoche, Purina Mills, Inc., and Isotec, Inc.

We thank the staff, and especially Janet Patterson, at Penn State's Conferences and Institutes for scheduling excellent facilities at the Penn Stater Conference Center Hotel. We gratefully acknowledge the contributions of numerous expert referees who provided scientific review of the papers in this volume.

We are sincerely grateful to Joanne Balmer Green (Department of Nutrition, Penn State University) for performing technical and scientific editing of the papers with great attention to quality and consistency, as well as to Darko Stefanovski (Department of Clinical Studies, School of Veterinary Medicine, University of Pennsylvania) for creating a unified camera-ready document from the variety of formats submitted by authors. We also thank Lisa Moore Shelton (Department of Clinical Studies, School of Veterinary Medicine, University of Pennsylvania) for early editorial assistance.

CONTENTS

HISTORICAL PERSPECTIVES AND FUTURE DIRECTIONS

THE MINIMAL MODEL OF GLUCOSE REGULATION: A BIOGRAPHY . 1
Richard N. Bergman

CORNERSTONES TO SHAPE MODELING FOR THE 21ST CENTURY: INTRODUCING THE AKA-GLUCOSE PROJECT 21
Raymond C. Boston, Darko Stefanovski, Peter Moate, Oscar Linares, and Peter Greif

DEVELOPMENT AND USE OF MATHEMATICAL MODELS

ASPECTS OF EFFECTIVE MODELING 43
Judah Rosenblatt

FITTING A MATHEMATICAL MODEL TO BIOLOGICAL DATA: INTRACELLULAR TRAFFICKING IN NIEMANN-PICK C DISEASE 63
Meryl E. Wastney, Peter G. Pentchev, and Edward B. Neufeld

A SATURATION KINETIC MODEL TO TEACH BALANCE OF ESSENTIAL FATTY ACIDS IN NUTRITION EDUCATION 77
James L. Hargrove, Jinah Hwang, and Diane K. Hartle

THEORETICAL MODELING ISSUES

MODELING PROCESSES FROM PROBABILITIES 87
James H. Matis and Thomas R. Kiffe

UNDERSTANDING THE RELATIONSHIP BETWEEN CARCINOGEN-INDUCED DNA ADDUCT LEVELS IN DISTAL AND PROXIMAL REGIONS OF THE COLON 105
Jeffrey S. Morris, Naisyin Wang, Joanne R. Lupton, Robert S. Chapkin, Nancy D. Turner, Mee Young Hong, and Raymond J. Carroll

EXPERIMENTAL ASPECTS OF KINETIC DATA

METHODOLOGICAL ISSUES IN STABLE ISOTOPE- BASED KINETIC STUDIES IN CHILDREN 117
Ian J. Griffin and Steven A. Abrams

INTRINSIC LABELING OF PLANTS FOR BIOAVAILABILITY STUDIES 131
Janet A. Novotny, Steven Britz, Frances Caulfield, Gary Beecher, and Beverly Clevidence

ELEMENTAL MASS SPECTROMETRY FOR COMPARTMENTAL BIOLOGICAL MODELING 141
Assad Al-Ammar and Ramon M. Barnes

MODELING OF VITAMINS, MINERALS, AND CHOLESTEROL

THE USE OF MODEL-BASED COMPARTMENTAL ANALYSIS TO STUDY VITAMIN A METABOLISM IN A NON-STEADY STATE . 159
Michael H. Green and Joanne Balmer Green

MODELING SHORT (7 HOUR)- AND LONG (6 WEEK)-TERM KINETICS OF VITAMIN B-6 METABOLISM WITH STABLE ISOTOPES IN HUMANS 173
Stephen P. Coburn, Douglas W. Townsend, Karen L. Ericson, Robert D. Reynolds, Paula J. Ziegler, David L. Costill, J. Dennis Mahuren, Wayne E. Schaltenbrand, Thomas A. Pauly, Yao Wang, William J. Fink, David R. Pearson, and David L. Hachey

CALCIUM UTILIZATION IN YOUNG WOMEN: NEW INSIGHTS FROM MODELING 193
Meryl E. Wastney, Berdine R. Martin, Rebecca J. Bryant, and Connie M. Weaver

MODELING CHOLESTEROL IN HUMANS: UPDATE AND DEALING WITH THE PROBLEM OF EXCHANGE *IN VIVO* USING THE BLOOD CELL-LIPOPROTEIN PARADIGM 207
Charles C. Schwartz, Julie M. VandenBroek, and Patricia S. Cooper

MODELING OF PROTEIN METABOLISM, ENERGY, AND GROWTH

CHALLENGING THE ASSUMPTIONS IN ESTIMATING PROTEIN FRACTIONAL SYNTHESIS RATE USING A MODEL OF RODENT PROTEIN TURNOVER 221
Heidi A. Johnson, Chris C. Calvert, and Kirk C. Klasing

SIMULATING PATTERNS OF CHANGE IN RATES OF SECRETION OF PROTEIN INTO MILK 239
John Cant

MECHANISTIC EQUATIONS TO REPRESENT DIGESTION AND FERMENTATION 253
Richard A. Kohn

MODELING GROWTH OF CATTLE FOR APPLICATION WITHIN THE STRUCTURE OF THE CORNELL NET CARBOHYDRATE AND PROTEIN SYSTEM 267
Danny G. Fox and Michael E. Van Amburgh

MODELING TO EXPLORE DISEASE

MATHEMATICAL MODELS OF TUMOR GROWTH: FROM EMPIRICAL DESCRIPTION TO BIOLOGICAL MECHANISM 287
John A. Adam

A SYSTEMS MODELING APPROACH TO THE STUDY OF RETINOID FUNCTION: IMPLICATIONS FOR EVALUATION OF RETINOIDS IN CANCER CHEMOPREVENTION AND/OR CHEMOTHERAPY 301
Kevin C. Lewis, James F. Hochadel, and Loren A. Zech Jr.

GLUTAMATE METABOLISM IN PRIMARY CULTURES OF RAT BRAIN ASTROCYTES: RATIONALE AND INITIAL EFFORTS TOWARD DEVELOPING A COMPARTMENTAL MODEL 317
Mary C. McKenna

STATE-OF-THE-ART MODELING SOFTWARE

WinSAAM: APPLICATION AND EXPLANATION 343
Janet A. Novotny, Peter Greif, and Raymond C. Boston

**STATA: A STATISTICAL ANALYSIS SYSTEM FOR EXAMINING
    BIOMEDICAL DATA** 353
Raymond C. Boston and Anne E. Sumner

**USING ADVANCED CONTINUOUS SIMULATION LANGUAGE (ACSL)
    TO SIMULATE, SOLVE, AND FIT MATHEMATICAL
    MODELS IN NUTRITION** 371
Heidi A. Johnson

SOLVING AND FITTING FRANCE'S RUMEN MODEL WITH MLAB 389
Gary Knott and Daniel Kerner

**STELLA® RESEARCH SOFTWARE FOR TEACHING CONCEPTS OF
    NUTRIENT DYNAMICS IN THE UNDERGRADUATE
    CLASSROOM** 401
James L. Hargrove

A REVIEW OF 'SCIENTIST' SOFTWARE 407
E. Paul Wileyto

INDEX 415

PART I
HISTORICAL PERSPECTIVES AND FUTURE DIRECTIONS

THE MINIMAL MODEL OF GLUCOSE REGULATION: A BIOGRAPHY

Richard N. Bergman[*]

INTRODUCTION AND HISTORICAL CONTEXT

The life history of the so-called "minimal model" of glucose regulation can be traced back over two decades. During the decades after World War II, the science of cybernetics was introduced by Norbert Weiner at MIT (Weiner, 1965). Weiner recognized the importance of the rapidly developing fields of control theory and systems analysis to problems in biology and medicine. During the ensuing decade, much was written about the potential benefits of applying mathematical analysis to biology (Yates et al., 1972).

Ironically, the biological research community was resistant to the application of modeling, especially in systems physiology. While giants in the fields of cardiovascular and respiratory function emerged (Manning and Guyton, 1982), little progress was made in endocrinology. Traditional physiologists were less than impressed with the early work of Rashevsky (1940) and others, perhaps because it appeared to be somewhat aloof from experimental work. Also, very few endocrinologists, then and now, were selected for or trained in mathematics and physics. Most endocrine scientists were less than comfortable with the paradigms of physical science, in which mathematical constructs are a necessary component of the research process. In fact, in the 1970s, it was difficult to publish mathematical models in traditional endocrine journals. This difficulty was eased by the eventual introduction of special modeling subsections in the American Journal of Physiology.

It was during this skeptical period that my co-workers and I entered the field of endocrine modeling. We reasoned that several precepts should be kept in mind if modeling were to be effectively applied to endocrine systems. First, there must be a close interaction between experiments and models. Second, modeling should have explicit goals and lead to understanding even for non-modelers, and third, if possible, modeling should lead to improvements in clinical applications or research. We hope our efforts over the last two decades have been consistent with these precepts. The apparent

[*] Richard N. Bergman, Department of Physiology and Biophysics and USC Diabetes Research Center, University of Southern California, Los Angeles, CA 90033.

success of our efforts is supported by the precipitous rise in citations of the minimal model (Figure 1). It is of interest to note that our modeling papers went virtually uncited for the first 5 years after the model was published. Anguish associated with our lack of early citation is assuaged by the fact that Turing's "On computable numbers..." (1936), the basis of the digital computer, was cited only 3 times in its first 13 years of existence. (Promotion committees take note!)

THE STUDY OF DIABETES

Diabetes, and regulation of the level of glucose in the bloodstream, was an excellent candidate for systems analysis. Diabetes is an important disease, and it has become a major factor in overall costs of healthcare in the U.S. (c.f., Newsweek, September 4, 2000). It is estimated that 16,000,000 Americans have this disease, and it is a primary cause of blindness, kidney failure, neuropathy, and heart disease. The recent rise in diabetes is primarily in the so-called "adult type" (Type 2) and is related to increasing obesity and poor diet. Diabetes is prevalent in minority groups including Native Americans, Mexican Americans, and African Americans. As the minority population in the U.S. has swelled, so has the prevalence of diabetes. Thus, there is an overwhelming justification for studying diabetes, so that we may investigate prediction, prevention, therapy, and cure.

Type 2 diabetes is an inter-organ disease (Figure 2). While failure of the β-cells of the pancreas to secrete insulin is the sole cause of Type 1 ("juvenile") diabetes, Type 2 diabetes is related to a failure of the *system* regulating blood glucose. That is, failure of muscle tissue, fat tissue, liver tissue, and pancreas are *all* involved. Thus, to explain diabetes, it is potentially of benefit to understand the dynamic interrelationships among the metabolic organs of the body. The importance of this disease, as well as the intellectual challenge of the interactive relation among the different organs, has provided the justification for devoting our efforts to this problem.

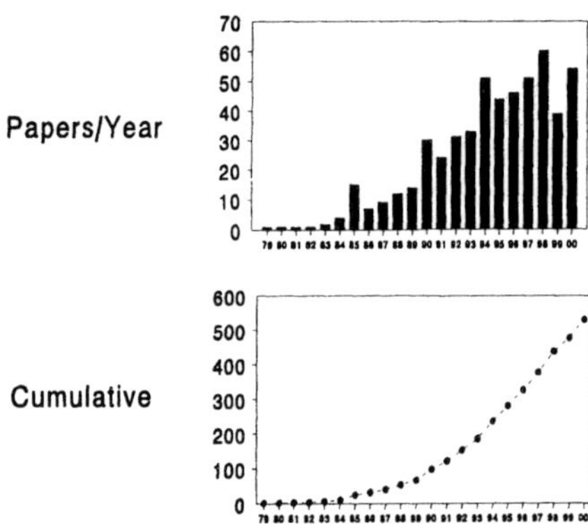

Figure 1. Full manuscripts exploiting the minimal model technology; 1979-2000. Top panel: estimate of publications/year; lower panel, cumulative publications. Note the very few publications before 1985.

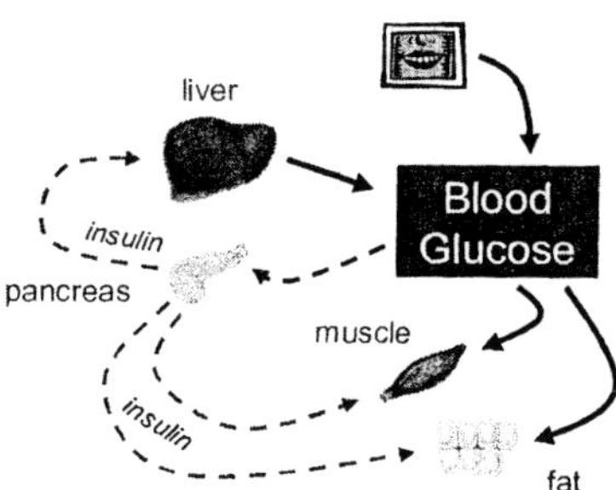

Figure 2. Simplified diagram of the regulation of glucose concentration in blood. Sources of glucose in blood include dietary intake of carbohydrate and *de novo* production by the liver and kidney (latter not shown). Glucose is produced primarily by the liver and used primarily by the brain and other insulin-independent tissues (e.g., the gut) under fasting conditions. Carbohydrate intake elicits a rise in glucose which elicits insulin secretion by β-cells of the pancreas. Plasma insulin suppresses liver glucose output, suppresses release of lipid by fat tissues, and stimulates glucose uptake by muscle. The effect of insulin on the liver can be direct but is also mediated by other signals (free fatty acids, not indicated).

ORIGIN OF THE MINIMAL MODEL: PARTITION ANALYSIS

In normal individuals, blood glucose concentration is maintained within exquisitely narrow limits, given the vicissitudes of environment (e.g., meal composition, exercise needs, temperature, pregnancy). Because glucose is the preeminent energy source of the central nervous system, a redundant mechanism prevents blood glucose from falling much below 4.5 mM, even after months of starvation (Cahill et al., 1966). Signals that support blood glucose when an external energy source is unavailable include low insulin, glucagon, catecholamines, growth hormone, steroids, and free fatty acids. Additionally, because elevated glucose can degrade activity of important proteins ("glycation;" see Brownlee, 1994), insulin is secreted when glucose increases, preventing prolonged hyperglycemia. Insulin has multiple effects, including restriction of liver glucose production, restriction of release of fat by adipose cells, and acceleration of glucose use by muscle (muscle comprises 50% of body weight). Thus, the blood glucose level is maintained in a narrow range in normal individuals by a sophisticated negative feedback system involving many different blood-borne signals. The brain is also involved in sensing glucose to prevent hypoglycemia.

It is daunting to develop a model of a feedback system with such a large number of components. Such models have large numbers of nonlinear relationships and parameters, and they prove difficult to test experimentally or exploit profitably. In fact, the insulin-secreting cells, the β-cells of the pancreas, demonstrate particularly nonlinear behavior and modeling challenges (Grodsky et al., 1972).

To cope with the modeling challenge, we reasoned that model development must be based upon real experimental data. We exploited the simplest perturbation test: the injection of glucose into the bloodstream, approximating a delta function. While this test was a classic one, we were the first to sample blood at 1-min intervals. Frequent sampling revealed rich system dynamics as shown in Figure 3. We observed sequential phases of glucose response: an insulin-independent glucose decline, an acceleration of the decline by plasma hyperinsulinemia, baseline crossing and a period of glucose below

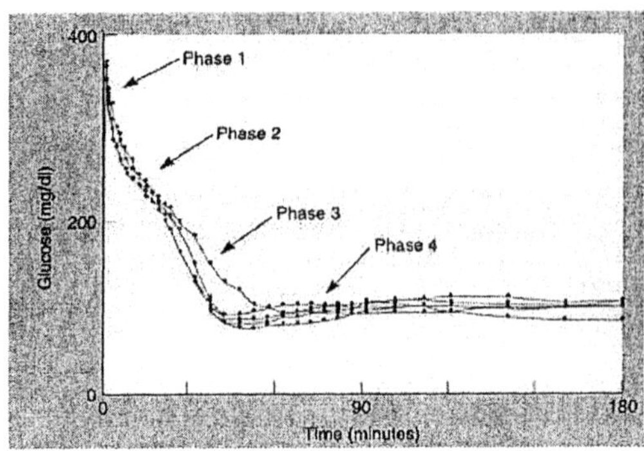

Figure 3. Phases of glucose renormalization after injection. Phase 1: intravenous glucose injection (at t=0) elevates the concentration due to distribution in extracellular space. Phase 2: the early decline (<20 min) is primarily uptake by insulin-independent tissues including brain, gut, liver, and red blood cells. Phase 3: the rate of decline is accelerated (at 20 min in this example) due to glucose-stimulated elevation in plasma insulin. Glucose may fall below basal value (in insulin sensitive individuals), with a nadir at 40-45 min. Phase 4: slow return of glucose concentration, which finally returns to its pre-injection value by 180 min. Reprinted with permission from Bergman (1995).

baseline, and the eventual (90 min or later) return to glucose baseline. These observations debunked previous assumptions that the post-injection glucose pattern was simply exponential. Clearly, important information regarding blood glucose regulation could be extracted from the intriguing dynamic relationship between plasma insulin and glucose.

It was naïve to expect that multiple parameters of a complex model could be estimated from the admittedly interesting post-injection glucose/insulin dynamics. Therefore, we attempted to match the model with the easily obtainable post-injection data: we asked what was the simplest, physiologically based (i.e., isomorphic) model which could account for the known (measured) plasma dynamics. Secondly, because of the known difficulty of modeling β-cells, we treated the (measured) plasma insulin as "input" and the resulting glucose as "output," thereby simplifying the challenge to modeling the effects of insulin *per se* on glucose. We called this "partition analysis" (Bergman and Cobelli, 1980).

It is possible to imagine systematic simplification of a complex model structure. Alternatively, one could build increasingly more complex models, beginning with the simplest conceptual structure. We took the latter approach and settled on a fairly simple model (Figure 4). There are substantial simplifications in this model. First, we assumed one-compartment distribution kinetics for glucose. Later research by Cobelli and colleagues (Vicini et al., 1997) has shown this to be an oversimplification. Second, we modeled only the effect of plasma insulin on glucose disappearance, and we assumed that effects of other hormones were inconsequential or did not affect glucose dynamics in a predictable fashion. This assumption may need further investigation. Further, we expressed the sum effects of insulin to both *reduce glucose production* and *enhance glucose output* as overall insulin action. Obviously, it would have been better to separate out these components, but that can only be done if tracer glucose is included, as

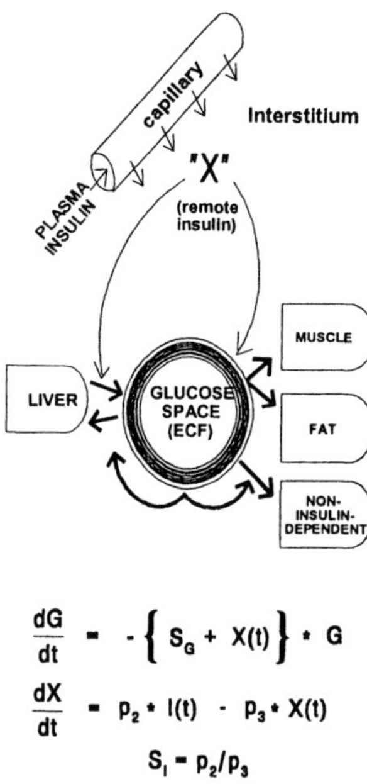

$$\frac{dG}{dt} = -\left\{ S_G + X(t) \right\} * G$$

$$\frac{dX}{dt} = p_2 * I(t) - p_3 * X(t)$$

$$S_I = p_2/p_3$$

Figure 4. The minimal model. Plasma insulin crosses the capillary endothelium sluggishly to enter the interstitial compartment. Glucose is produced by or is taken up by the liver in proportion to interstitial insulin; similarly, interstitial insulin is the signal for glucose utilization by peripheral tissues (muscle and fat). Glucose can also suppress liver glucose production or stimulate glucose uptake by insulin-independent tissues. The first equation describes the mechanisms which account for return of glucose to basal after deflection due to glucose injection (c.f., Figure 3): the effect of glucose itself to enhance net glucose uptake ($S_G * G$) as well as the effect of interstitial insulin which also depends upon plasma glucose level [$X(t) * G$]. The second equation describes first-order movement of insulin across capillary endothelial barriers in proportion to plasma insulin [$p_2 * I(t)$]; interstitial insulin is cleared by a first order process [$p_3 * X(t)$]. Insulin sensitivity may be expressed as the insulin sensitivity index (S_I) which is equal to a ratio of parameters (p_2/p_3). See text description of S_I. Adapted from Bergman and Ader (1993).

has been done in later studies (Caumo et al., 1996; Cobelli et al., 1997, 1998). Finally, we assumed that the time course of insulin's effects to increase utilization and suppress production were similar. This assumption has been tested in detail (Bradley et al., 1992).

This minimal model was able to account for the phasic renormalization of glucose post-injection. There were several assumptions that were implemented as being absolutely required to account for the data. First, glucose influences its own net disappearance from plasma. We termed this "glucose effectiveness," a somewhat awkward term which has now entered the lexicon (Ader et al., 1985; Best et al., 1996; Basu et al., 1997). Second, insulin acts *interactively* with glucose to inhibit production and accelerate glucose uptake. Third, the effect of insulin on glucose economy is *delayed in time*. It was assumed that the slowness of insulin's effect was due to slow passage from the plasma to the interstitial compartment. This assumption has likewise come under extensive investigation (Yang et al., 1989; Jansson et al., 1993; Miles et al., 1995). These several assumptions will be discussed later.

LONGEVITY OF THE MINIMAL MODEL

With retrospect, it is astounding to us that the minimal model is still being used and studied after lo these many years. In this paper, I will review why this may be true. Proposed answers may not be totally correct, but it is hoped that this saga may provide useful information to others regarding the longevity of their own equally excellent modeling efforts.

Insulin Sensitivity: Model Validation

Factors that determine the longevity of a model include whether it is validated, whether it is useful to other investigators (or clinicians) than those who developed it, and whether it can be easily implemented and exploited by multiple groups.

It was becoming clear in the early 1980s that the concept of insulin resistance was an important factor in the pathogenesis of Type 2 diabetes. The question arose as to whether insulin sensitivity could be calculated from the minimal model. If so, the utility of the model could increase, as many diabetes investigators were interested in measuring insulin resistance (the inverse of sensitivity) in various groups potentially at risk for Type 2 diabetes. Examples include obese individuals (Wing et al., 1994), individuals taking various therapeutic agents, pregnant women (Buchanan et al., 1990), elderly subjects (Garcia et al., 1997), and subjects with impaired glucose tolerance (Taniguchi et al., 1994). Independent methods for measuring sensitivity were developed (Bergman et al., 1985; Ader and Bergman, 1987), and comparing model-based insulin sensitivity with analogous sensitivity parameters obtained independently could test the veracity of the model.

Andres and colleagues (1966) introduced the so-called "glucose clamp" methodology for assessment of insulin sensitivity in 1966. This method was applied in the 1970s by his group and later standardized by DeFronzo et al. (1979). The clamp method depends upon a challenging protocol in which insulin is infused intravenously to raise the plasma insulin concentration. While this would normally cause glucose to decline, the latter is prevented by infusing glucose. The rate of glucose infusion *required* to prevent the decline is a measure of insulin sensitivity.

We set out to compare insulin sensitivity from the model with that from the clamp. This required us to calculate sensitivity from the model. Parameters of insulin sensitivity were defined for the minimal model $(S_I)^*$ and for the glucose clamp $(S_{I(clamp)})$.[†] These parameters were found to be strongly correlated in a comparative study done in dogs

---

[*] $S_I = \{\delta^2[r(t)]/[\delta G(t)\delta I(t)]\}$, where r(t) is the rate of glucose decline, and G(t) and I(t) are glucose and insulin concentrations, respectively.

[†] $S_{I(clamp)} = GINF/[G \times \Delta I]$, where GINF is the steady state rate of glucose infusion during clamps, G is the prevailing glucose concentration, and ΔI is the increment of insulin in plasma. For further explanation, see Bergman et al. (1987).

(Finegood et al., 1984). This correlation was viewed as a powerful validation of the minimal model as an accurate representation of glucose metabolism *in vivo*. It appeared to validate the approach of simplifying models to allow for identification from a simple clinically useful test. A strength of this validation was that the model enabled calculation of metabolic parameters from a totally independent and dissimilar experimental protocol. Limitations of the validation were that it might not be possible to extrapolate results from the normal dog model to other species (e.g., humans) and other conditions (e.g., diabetes).

Controversy: Validation of S_I

Donner et al. (1985) compared insulin sensitivity measurements in human subjects in differing metabolic states (including Type 2 diabetes) and reported lack of significance between clamp and minimal model-based insulin sensitivity indices. It was not clear whether this lack of correlation was due to structural inadequacy of the model itself, to lack of application of the model to human subjects, and/or to inadequate design of test protocols of the model.

The failure of the model in the hands of Donner and colleagues energized our group to examine the model's limitations. In human subjects, the simple injection of glucose was inadequate to elicit dynamics with sufficient richness to yield precise estimates of model parameters. Compared to smaller animals, humans are remarkably insulin resistant, and this resistance is considerably worsened in those at risk for diabetes (Martin et al., 1992). The insulin provoked by glucose injection has a proportionately lesser effect on the pattern of glucose decline in man than in the dog, for example. Simulation studies led us to include a second injection in the test protocol: glucose followed 20 min later by tolbutamide or insulin itself. When this protocol was used, S_I from the model and $S_{IP\ (clamp)}$ were indeed well correlated (Korytkowski et al., 1995; Dunaif and Finegood, 1996). In addition, the controversy surrounding S_I estimates led us to define insulin sensitivity measurements in similar units. Then the sensitivity measurements were not only correlated but were also *equivalent* overall (Figure 5; see Bergman et al., 1987). The equivalence of sensitivity measurements from these two entirely independent methodologies established the validity of the minimal model as a useful and correct representation of insulin-dependent glucose utilization in animals and humans.

Ease of Application

Having established a protocol which was applicable in human patients, S_I was measured in a plethora of human conditions. One factor which apparently catalyzed the devolution was making available a relatively "friendly" piece of software – "MINMOD" – which could be obtained from the University of Southern California.

Use of this software did not require mathematical skill, although it has been necessary for our laboratory to maintain an informal information agency to answer questions about MINMOD application. Fortunately, additional software packages were available to compute model parameter values. These include various generations of the SAAM program [e.g., WinSAAM, now available from Dr. Ray Boston and colleagues (Grief et al., 1998)] and the MLAB program (Knott, 1979) which has a minimal model application. In fact, in IRAS, a large epidemiological study in which 1524 separate minimal model protocols were performed on human subjects, parameter values were calculated in each individual using MINMOD and MLAB (Howard et al., 1996).

Validation and availability of software resulted in insulin sensitivity being assessed for many different conditions in many different populations (c.f., Bergman, 1989). One of the most significant applications was the examination of risk factors leading to diabetes. The Joslin Research Center had performed intravenous glucose tests over many years, beginning 40 years ago (Martin et al., 1992). In one study, they performed these tests on offspring of two Type 2 diabetics. Because diabetes is genetically transmitted, these individuals had a much greater risk for diabetes than the population in general. In fact, over a 30 year observation period, 25 offspring developed diabetes among the 155 subjects studied. By MINMOD analysis of the probands, the factors that conferred diabetes risk could be identified. In fact, the most insulin-resistant subjects (i.e., those with the lowest S_I values) had a much higher risk of diabetes than the population in general. Thus, by *a posteriori* analysis of the data, it was shown from this study that insulin resistance *per se* is a powerful risk factor for the disease.

PHYSIOLOGICAL CONCEPTS EMERGING FROM THE MINIMAL MODEL

Unlike empirical models, isomorphic models require implicit physiological assumptions. Success of the model suggests, but does not prove, that these assumptions are true. It behooves the proponent and utilizer of models to examine these assumptions experimentally to lend credence to or question the validity of the model itself. Several

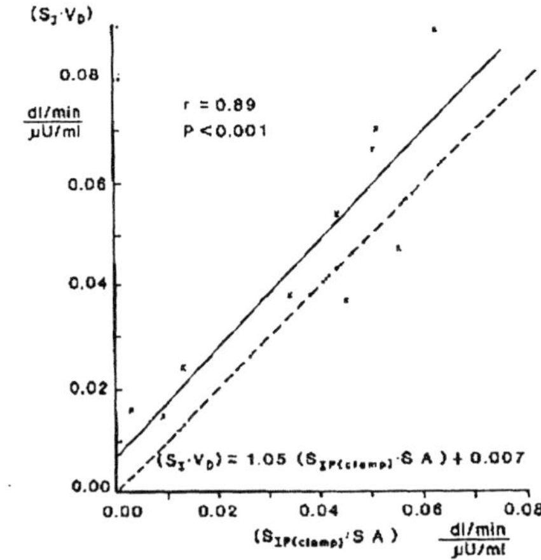

Figure 5. Validation of the minimal model in human subjects. The insulin sensitivity parameter S_I was calculated from the minimal model as well as from glucose clamp experiments. Parameters were corrected to similar units, and the correlation between them exhibited a slope not different from 1.0. The relationship was shifted parallel to the ordinate due to inclusion of liver suppression in S_I from minimal model but not from clamp experiments. Adapted from Bergman et al. (1987).

of the assumptions implicit in the minimal model have been tested, and while some have been confirmed, others have not. In addition, we believe that considerable physiological understanding has resulted from this model-testing process.

Assumption 1: Net Glucose Utilization (at Basal Insulin) is Proportional to Glucose *Itself* (Glucose Effectiveness)

Ader et al. (1985, 1997) tested this assumption in dogs. Insulin release was suppressed with somatostatin; basal insulin was replaced with intraportal insulin infusion. After glucose injection (Figure 6), glucose was in fact renormalized, demonstrating that an acute insulin response from the β-cells was not necessary for re-regulation of the plasma glucose concentration. The fractional rate of decline of post-injection glucose was equal to parameter S_G from the model (Ader et al., 1985).[*] Additional simulation studies showed that S_G was an important factor in glucose regulation, particularly in states of impaired glucose tolerance (IGT) and diabetes. While glucose effectiveness accounted for 40% of glucose normalization in normal individuals, this proportion was 70% in IGT subjects and almost 100% in Type 2 diabetics. The large contribution in diabetes was due to a sinister combination of insulin resistance and β-cell dysfunction (Beard et al., 1987). These studies demonstrated that glucose effectiveness is an important contributor to diabetes risk.

Controversy: S_G Underestimation in Type 1 Diabetes

Quon and colleagues (1994) raised a crucial issue related to the assessment of glucose effectiveness by the minimal model. They performed studies in Type 1 diabetics (total β-cell destruction with exogenous insulin replacement) and showed that S_G was overestimated by the minimal model in such subjects. They proposed that the model overestimated glucose effectiveness and concomitantly underestimated the contribution of insulin *per se* to glucose normalization post-injection. They suggested that this apparent artifact may be due to the model's assumption that glucose distribution follows one-compartment distribution kinetics.

Quon and colleagues were correct. Ni et al. (1997) examined the significance of the single compartment assumption to the estimate of S_G. The single compartment assumption results in overestimation of S_G at low values and underestimation at high values; the estimated value at 0.02 min^{-1} (the mean value in humans) is correct (Figure 7). Thus, the quandary still exists: a two-compartment model is not identifiable without the concomitant injection of glucose and "tracer" (i.e., labeled glucose). Fortunately this "undermodeling" does not invalidate the estimate of S_I (McDonald et al., 2000). Further work is necessary to obtain accurate and precise estimates of S_G from the injection protocol. The Quon "paradox" represents a second controversy that resulted when model assumptions were investigated. This example yields increased insight into how the modeling process may be improved.

---

[*] S_G equals model parameter p1.

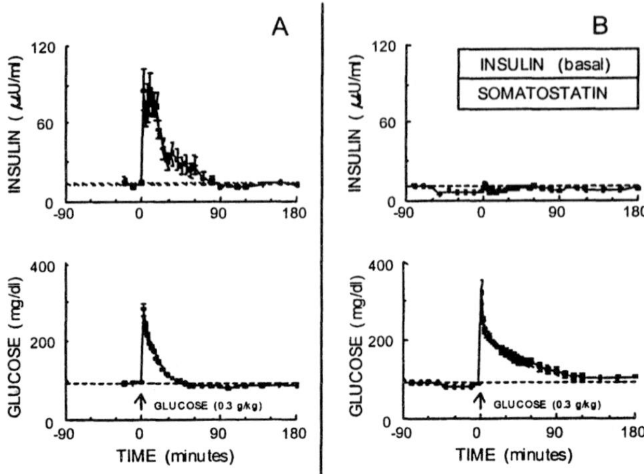

Figure 6. Examination of the effect of glucose *per se*, independent of a change in insulin on glucose disappearance at basal insulin. (A) Standard intravenous glucose tolerance test in conscious dogs. Glucose is injected at t=0, eliciting a prompt insulin response. (B) Dynamic insulin response is blocked by somatostatin; basal insulin is maintained by exogenous insulin infusion. Despite the absence of a dynamic insulin response, glucose declines to near basal value, albeit at a slower rate (B compared to A). These data prove that the effect of glucose on its own disappearance ("glucose effectiveness") is critical to reattainment of basal glucose after glucose injection. Adapted from Ader et al. (1985).

Assumption 2: Insulin Acts Slowly to Accelerate Glucose Utilization and Suppress Endogenous Glucose Output

The assumption of sluggish insulin action could not be avoided in modeling glucose kinetics. Two explanations seemed possible: there might be slow movement of insulin

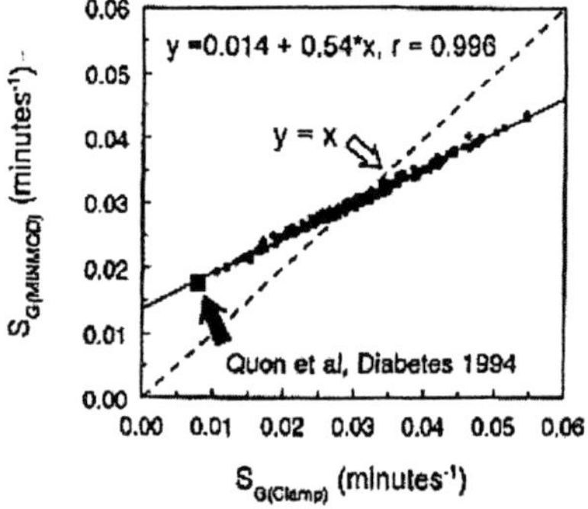

Figure 7. The Quon "paradox" explained. The calculation of glucose effectiveness (S_G) from the minimal model is only equal to the "actual" value at an approximate value of 0.028 min^{-1}. Below that value, effectiveness is overestimated, and it is underestimated at a value above 0.028 min^{-1}. These data explain the report of Quon et al. (1994) that effectiveness is overestimated in Type 1 diabetics, in whom S_G is below the normal value. Reprinted with permission from Ni et al. (1997).

across the capillary endothelial tissues to gain cellular access from interstitial fluid and/or the action of insulin to accelerate glucose uptake after receptor binding might be slow. The latter was inconsistent with measurements demonstrating rapid effects of insulin on glucose utilization *in vitro*. That the delay was due to transendothelial transport of insulin (TET) was confirmed in experiments with dogs in which we compared the time course of insulin concentration in blood and insulin in interstitial fluid *vs.* the rate of glucose utilization. These comparisons (Yang et al., 1989; Poulin et al., 1994; Hamilton-Wessler et al., 1999) revealed that there was a substantial delay between plasma insulin and interstitial insulin and that the rate of glucose utilization was exactly proportional to hormone concentrations in the interstitium but not in plasma. Our data were consistent with the concept that the transport of insulin across the interstitium determined the time course of insulin on glucose utilization *in vivo* (Figure 8).

The mechanism of the transendothelial transport of insulin is controversial. King and Johnson (1985) have suggested that insulin movement is mediated by insulin receptors on endothelial cells: i.e., that insulin binds and is carried across endothelial cells and deposited in the interstitial space. Our data argue otherwise: we found no evidence for saturation of TET even at pharmacological insulin concentrations (Steil et al., 1996), and hyperinsulinemia in blood has no effect on insulin transport of insulin-analog molecules which also bind to insulin receptors (Dea et al., 1997). It is apparent that insulin transport is mediated by a nonsaturable mechanism, possibly transcytosis or paracellular diffusion.

Significance of TET

Regardless of mechanism, the sluggish action of insulin predicted by modeling and confirmed by direct measurement is physiologically significant. It appears inevitable that a protein hormone of molecular weight over 5000 daltons would exit the plasma slowly. One is tempted to speculate that the biphasic secretory response characteristic of β-cell insulin secretion may have evolved to compensate for the slow TET. We measured interstitial insulin concentration in the face of biphasic plasma insulin (Getty et al., 1998). The two phases of secretion into blood act like a "pulse-infusion" pattern

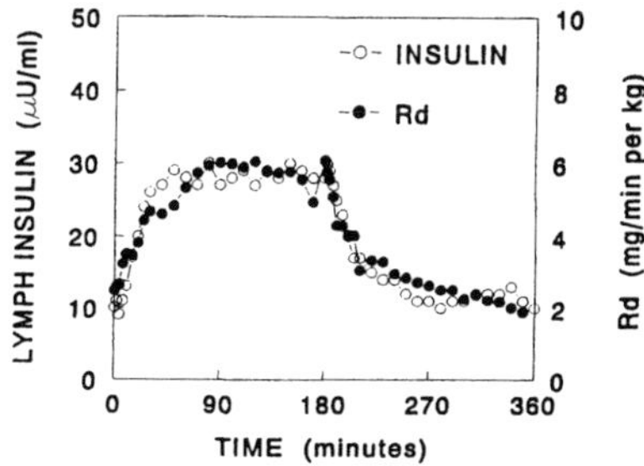

Figure 8. Striking dynamic similarity between insulin in lymph (interstitial insulin) and glucose utilization (Rd). Data are from euglycemic clamps. The two measurements are entirely independent and yet have almost exactly the same time course, demonstrating that glucose uptake by muscle tissue is proportional to interstitial insulin concentration. Adapted from Yang et al. (1989).

and result in an apparent step-like increase in interstitial insulin (Figure 9). Also, there is some evidence that a reduction in transendothelial transport may be responsible for insulin resistance secondary to central adiposity. After fat feeding in the dog, insulin sensitivity is reduced 60%. We modeled plasma and interstitial insulin kinetics, and we estimated TET before and after the elevated fat diet. We found that reduced TET was responsible for about 30% of the reduction in insulin sensitivity (Hamilton-Wessler et al., 2000). The mechanism by which increased dietary fat may reduce the rate of insulin transport across the endothelium is unclear, but one possibility is that increased lipid in blood may alter the fluidity of the membrane. Appreciation of the importance of the movement of insulin across the endothelial membrane was a direct outcome of the development of the minimal model. Several groups have become interested in interstitial insulin, reconfirming the impact that quantitative modeling can have on experimental investigation.

Assumption 3: Similar Dynamics of Insulin Activation of Glucose Uptake and Suppression of Glucose Output

Traditionally it has been believed that insulin suppressed liver glucose production rapidly, while activation of glucose uptake was slow. However, it appeared unnecessary to make such an assumption in the minimal model. Was it possible that suppression of endogenous glucose production (EGP) was as sluggish as activation of glucose utilization?

Availability of accurate methods confirmed that insulin effects on glucose disposal and production had similar dynamics – in fact, they were mirror images (Bradley et al.,

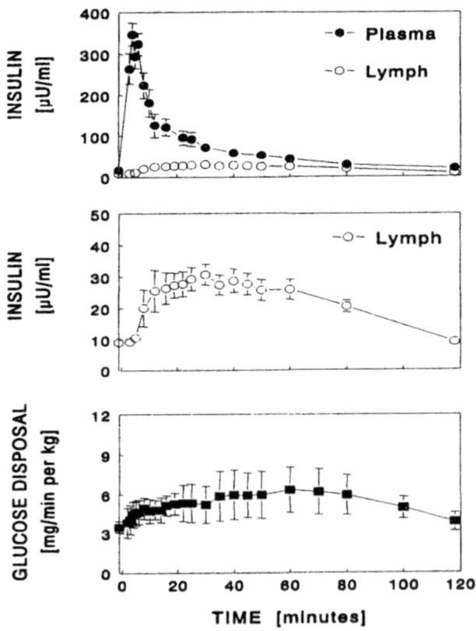

Figure 9. Plasma and interstitial insulin and peripheral glucose disposal. Biphasic plasma insulin results in approximately "square-wave" time course of interstitial insuin. These results may explain why a biphasic insulin time course evolved. Adapted from Getty et al. (1998).

1992). This raised the possibility that these processes were mechanistically linked. Independent evidence of an extrahepatic "signal" controlling EGP (Rebrin et al., 1995; Lewis et al., 1997) led to the concept that insulin regulated liver glucose output via plasma free fatty acid (FFA) concentration. The similarity of insulin sensitivity of adipocytes and liver supported this contention, as did additional evidence. First, there is a strong correlation between plasma FFA and EGP under a variety of conditions (Hamilton-Wessler et al., 1999). Second, the effect of insulin on EGP is obliterated if plasma FFA do not decline (Rebrin et al., 1996). Finally, suppression of FFA at constant plasma insulin results in suppression of EGP (Mittelman and Bergman, 2000).

Our data led us to propose the "single gateway hypothesis" (Figure 10). There are two primary postulates. First, the time courses of insulin's actions to increase glucose disposal and to suppress EGP are both limited by TET. Second, insulin action on the liver is controlled by plasma FFA concentration. According to this concept, reduction in EGP after insulin administration is mediated by suppression of lipolysis, lowering FFA and reducing liver glucose production.

Significance of the Single Gateway Concept

The putative regulation of EGP by FFA has important therapeutic as well as pathogenic implications. Insulin sensitizing drugs – thiazolidinediones – appear to function by reducing plasma FFA and thus enhancing peripheral insulin action. Assuming that FFA regulate the liver, thiazolidinediones would similarly enhance liver sensitivity by reducing plasma FFA.

A possibly more important implication of FFA control of the liver is due to elevated FFA levels in the portal vein leading to the liver. Visceral obesity – the storage of lipid in the omental adipose depot – is a risk factor for several chronic diseases including diabetes, hypertension, heart disease, and cancer. Central fat cells are resistant to insulin. Therefore, it is reasonable to assume that visceral obesity is associated with an increased flux of FFA to the liver. Given the evidence for FFA stimulation of EGP, overproduction of glucose by the liver would be expected. Thus, the concept of the single gateway may explain the insulin resistance of the liver that is associated with visceral obesity.

APPLICATION OF THE MINIMAL MODEL TO HOLISTIC GLUCOSE REGULATION

As discussed, the minimal model only represents one half of the totality of glucose regulation: the effect of insulin on glucose production/disposal. The other half of the regulatory duo is insulin secretion. By combining these components, it is possible to gain an integrated picture of the capacity of an organism to regulate his/her glucose concentration by altering insulin secretion and/or insulin sensitivity.

We may think about the *product* of insulin secretion and insulin sensitivity as the "disposition index" (DI), a parameter analogous to the "closed loop gain" that is a measure of the quality of regulation of a negative feedback system. Then

$$\text{Disposition Index} = \text{insulin sensitivity} * \text{insulin secretion} \tag{1}$$

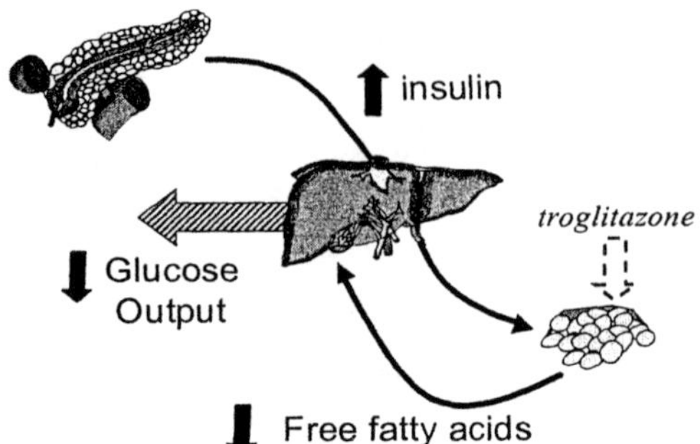

Figure 10. The single gateway hypothesis. Data indicate that control of liver glucose production is more dependent upon FFA than portal insulin itself. By this concept, pancreatic effluent insulin traverses the liver, enters the systemic circulation, and then suppresses adipocyte lipolysis. The latter results in reduced plasma free fatty acid concentration, which reduces FFA stimulation of glucose output by the liver.

or

$$DI = S_I * AIR_{glucose} \qquad (2)$$

where S_I is the sensitivity index from the minimal model and $AIR_{glucose}$ is a measure of insulin secretion (acute insulin response), defined as the integrated insulin concentration during the first phase (from 0-10 min after glucose injection, above basal). Equation 2 represents the hyperbolic relationship between insulin sensitivity and insulin secretion. Physiologically, this relationship implies that environmental changes in insulin sensitivity (for example, during pregnancy, changes in body weight, changes in fitness) will be compensated by changes in insulin secretion. Considerable data support this hyperbolic relationship (Bergman et al., 1981; Kahn et al., 1993; Buchanan et al., 2000). For example, in normal pregnancy, severe insulin resistance (decrease in S_I) is compensated by an approximately equal and opposite increase in $AIR_{glucose}$, such that glucose tolerance remains nearly constant (Figure 11).[*]

The disposition index can be a very useful concept. For example, Bogardus and colleagues (Weyer et al., 1999) have demonstrated that DI is an excellent predictor of which individuals will develop Type 2 diabetes in the Pima Native American population. The Pimas have the highest risk of diabetes of any population in the world.

The hyperbola in Figure 11 is rectangular: from this description, the *causality* of the relationship between S_I and AIR cannot be inferred. However, from experimental data, we know that the overall sensitivity of the pancreatic β-cells is highly adaptable (Kahn et al., 1989). Therefore, we expect that DI is primarily a measure of the capacity of the β-cells to compensate for changes in insulin sensitivity. In fact, it may be this interpretation of DI that makes it a powerful predictor of diabetes. Low DI implies that β-cells have limited compensating capacity. A low DI may well be a signal that the β-

---

[*] In fact, as Bogardus (personal communication) pointed out, changes in S_I are not totally compensated by $AIR_{glucose}$, as the system retains an "error signal," so that when S_I returns to normal, there is a tendency for $AIR_{glucose}$ to return to normal as well.

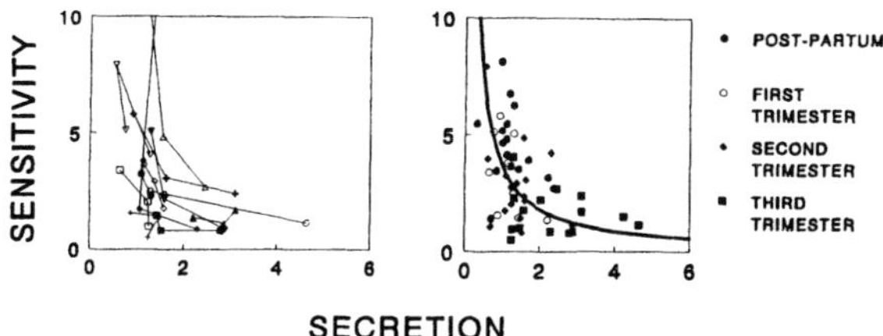

Figure 11. The hyperbolic relationship between insulin sensitivity and insulin secretion which is observed *in vivo*. During pregnancy, reproductive hormones lead to severe insulin resistance, which elicits a predictable enhancement of β - cell insulin secretion. Data from individual human subjects are connected in the left panel.

cells are injured or inherently unable to compensate. DI has been a powerful tool in the search for the genes for Type 2 diabetes (Valle et al., 1998).

RETROSPECTIVE

What is the life span of a mathematical model? Clearly, there is a good deal of "infant mortality" or models that appear in the literature one time never to resurface. Such models may not be data based, or the utility of such models may be unclear. Other models may appear and engender experimental tests which may not confirm or be supplanted by more appropriate representations, leading to the "early model death syndrome." Other models have existed for decades and have either entered the lexicon from a methodological perspective [e.g., the SAAM programs for distribution and metabolism (Foster et al., 1989)] or have greatly enhanced our understanding of a fundamental biological process [best example: the Hodgkin-Huxley model of nerve excitability (1951)].

Where in this landscape of models does the "minimal model" of glucose regulation lie? Clearly, the model has survived as a point of discussion for several decades. The model still has problems related to identifiability of glucose effectiveness. This results from the one-compartment assumption. Also, there has been extensive debate as to whether S_I is the "best" measure of insulin sensitivity (Donner et al., 1985). However, the model has been and continues to be widely applied. Why is this?

I believe that longevity of the minimal model can be traced to the following:

- simplicity of the concept: nonmathematical observers can comprehend the concepts underlying the model;
- validation of the model by many independent investigators;
- utility of the model: many investigators in a variety of fields want to measure insulin sensitivity;
- availability of friendly software to calculate model parameters;
- proof of the pudding: the model has provided a powerful tool to predict which individuals will develop Type 2 diabetes before the onset of the disease; and
- interesting and important physiological and pathophysiological concepts have emerged from testing the model.

I will be deeply gratified if the lessons my colleagues and I have learned from our experience with the minimal model provide a learning experience for other scientists embarking on the difficult but challenging journey of integrating mathematical modeling with their research programs.

It must be understood that the journey outlined represents a collective biography not only of the minimal model but of the many investigators who have toiled in its presence.

ACKNOWLEDGEMENTS

The studies described in this review were supported by the National Institutes of Health (DK 29867).

CORRESPONDENCE

Please address all correspondence to:
Richard N. Bergman
Keck-USC School of Medicine
Department of Physiology and Biophysics
1333 San Pablo Street
Los Angeles, CA 90033
rbergman@usc.edu

REFERENCES

Ader, M., Pacini, G., Yang, Y.J., and Bergman, R.N., 1985, Importance of glucose per se to intravenous glucose tolerance: comparison of the minimal model prediction with direct measurements, *Diabetes* 34:1092-1103.

Ader, M., and Bergman, R.N., 1987, Insulin sensitivity in the intact organism, in: *Bailliere's Clinics in Endocrinology and Metabolism,* K.G.M.M. Alberti, P. D. Home, and R. Taylor, eds., Bailliere Tindall, London.

Ader, M., Ni, T.-C., and Bergman, R.N., 1997, Glucose effectiveness assessed under dynamic and steady state conditions: comparability of uptake versus production components, *J. Clin. Invest.* 99:1187-1199.

Andres, R., Swerdloff, R., Pozefsky, T., and Coleman, D., 1966, Manual feedback technique for the control of blood glucose concentration, in: *Automation in Analytical Chemistry,* J. Skeggs, ed., Mediad, Inc., New York.

Basu, A., Caumo, A., Bettini, F., Gelisio, A., Alzaid, A., Cobelli, C., and Rizza, R.A., 1997, Impaired basal glucose effectiveness in NIDDM: contribution of defects in glucose disappearance and production, measured using an optimized minimal model independent protocol, *Diabetes* 46:421-432.

Beard, J. C., Ward, W.K., Halter, J.B., Wallum, B.J., and Porte, D., Jr., 1987, Relationship of islet function to insulin action in human obesity, *J. Clin. Endocrinol. Metab.* 65:59-64.

Bergman, R.N., 1989, Toward physiological understanding of glucose tolerance: minimal-model approach (Lilly Lecture), *Diabetes* 38:1512-1527.

Bergman, R.N., 1995, Insulin sensitivity from the minimal model, in: *Research Methodologies in Human Diabetes -- Part 2*, Walter de Gruyter, Berlin.

Bergman, R.N., and Ader, M., 1993, Concepts emerging from the minimal model approach, in: *Current Topics in Diabetes Research*, F. Belfiore, R.N. Bergman, and G.M. Molinatti, eds., Karger, Basel.

Bergman, R.N., and Cobelli, C., 1980, Minimal modeling, partition analysis and the estimation of insulin sensitivity, *Fed. Proc.* 39:110-115.

Bergman, R.N., Finegood, D.T., and Ader, M., 1985, Assessment of insulin sensitivity in vivo, *Endocr. Rev.* 6:45-86.

Bergman, R.N., Phillips, L.S., and Cobelli, C., 1981, Physiologic evaluation of factors controlling glucose tolerance in man: measurement of insulin sensitivity and B-cell glucose sensitivity from the response to intravenous glucose, *J. Clin. Invest.* 68:1456-1467.

Bergman, R.N., Prager, R., Volund, A., and Olefsky, J.M., 1987, Equivalence of the insulin sensitivity index in man derived by the minimal model method and the euglycemic glucose clamp, *J. Clin. Invest.* 79:790-800.

Best, J.D., Kahn, S.E., Ader, M., Watanabe, R.M., Ni, T.-C., and Bergman, R.N., 1996, Role of glucose effectiveness in the determination of glucose tolerance, *Diab. Care* 19:1018-1030.

Bradley, D.C., Poulin, R.A., and Bergman, R.N., 1992, Dynamics of hepatic and peripheral insulin effects suggest common rate-limiting step in vivo, *Diabetes* 42:296-306.

Brownlee, M., 1994, Glycation and diabetic complications, *Diabetes* 43:836-841.

Buchanan, T.A., Metzger, B.E., Freinkel, N., and Bergman, R.N., 1990, Insulin sensitivity and B-cell responsiveness to glucose during late pregnancy in lean and moderately obese women with normal glucose tolerance or mild gestational diabetes, *Am. J. Obstet. Gynecol.* 162:1008-1014.

Buchanan, T.A., Xiang, A.H., Peters, R.K., Kjos, S.L., Berkowitz, K., Marroquin, A., Goico, J., Ochoa, C., and Azen, S.P., 2000, Response of pancreatic b-cells to improved insulin sensitivity in women at high risk for type 2 diabetes, *Diabetes* 49:782-788.

Cahill, G.F., Jr., Herrera, M.G., Morgan, A.P., Soeldner, J.S., Steinke, J., Levy, P.L., Reichard, G.A., and Kipnis, D.M., 1966, Hormone-fuel interrelationships during fasting, *J. Clin. Invest.* 45:1751-1769.

Caumo, A., Vicini, P., and Cobelli, C., 1996, Is the minimal model too minimal? *Diabetologia* 39:997-1000.

Cobelli, C., Vicini, P., and Caumo, A., 1997, If the minimal model is too minimal, who suffers more: S_G or S_I? *Diabetologia* 40:362-364.

Cobelli, C., Bettini, F., Caumo, A., and Quon, M.J., 1998, Overestimation of minimal model glucose effectiveness in presence of insulin response is due to undermodeling, *Am. J. Physiol.* 275:E1031-E1036.

Dea, M., Hamilton-Wessler, M., Ader, M., Poulin, R.A., Moore, D., and Markussen, J., 1997, Long-acting insulin analogue NN304 has similar transendothelial transport to porcine insulin, *Diabetes* 46:164A.

DeFronzo, R.A., Tobin, J.D., and Andres, R., 1979, Glucose clamp technique: a method for quantifying insulin secretion and resistance, *Am. J. Physiol.* 237:E214-E223.

Donner, C.C., Fraze, E., Chen, Y.D.I., Hollenbeck, C.B., Foley, J.E., and Reaven, G.M., 1985, Presentation of a new method for specific measurement of in vivo insulin-stimulated glucose disposal in humans: comparison of this approach with the insulin clamp and minimal model techniques, *J. Clin. Endocrinol. Metab.* 60:723-726.

Dunaif, A., and Finegood, D.T., 1996, b-Cell dysfunction independent of obesity and glucose intolerance in the polycystic ovary syndrome, *J. Clin. Endocrinol. Metab.* 81:942-947.

Finegood, D.T., Pacini, G., and Bergman, R.N., 1984, The insulin sensitivity index: correlation in dogs between values determined from the intravenous glucose tolerance test and the euglycemic glucose clamp, *Diabetes* 33:362-368.

Foster, D.M., Boston, R.C., Jacquez, J.A., and Zech, L., 1989, A resource facility for kinetic analysis: modeling using the SAAM computer programs, *Health Phys.* 57 (Suppl 1):457-466.

Garcia, G.V., Freeman, R.V., Supiano, M.A., Smith, M.J., Galecki, A.T., and Halter, J.B., 1997, Glucose metabolism in older adults: a study including subjects more than 80 years of age, *J. Am. Geriatr. Soc.* 45:813-817.

Getty, L., Hamilton-Wessler, M., Ader, M., Dea, M.K., and Bergman, R.N., 1998, Biphasic insulin secretion during intravenous glucose tolerance test promotes optimal interstitial insulin profile, *Diabetes* 47:1941-1947.

Greif, P., Wastney, M., Linares, O., and Boston, R., 1998, Balancing needs, efficiency, and functionality in the provision of modeling software: a perspective of the NIH WinSAAM project, in: *Mathematical Modeling in Experimental Nutrition*, A.J. Clifford and H.-G. Müller, eds., Plenum Press, New York.

Grodsky, G.M., 1972, A threshold distribution hypothesis for packet storage of insulin and its mathematical modeling, *J. Clin. Invest.* 51:2047-2059.

Hamilton-Wessler, M., Ader, M., Dea, M., Moore, D., Jorgensen, P.N., Markussen, J., and Bergman, R.N., 1999, Mechanism of protracted metabolic effects of fatty acid acylated insulin, NN304, in dogs: retention of NN304 by albumin, *Diabetologia* 42:1254-1263.

Hamilton-Wessler, M., Ellmerer, M., Dea, M.K., Mittelman, S.D., van Citters, G.W., and Kim, S.P., 2000, Insulin resistance in dogs with transendothelial transport defect, *Diabetes* 49 (Suppl 1), A59.

Hodgkin, A.L., 1951, The ionic basis of electrical activity in nerve and muscle, *Biol. Rev.* 26: 339-401.

Howard, G., O'Leary, D.H., Zaccaro, D., Haffner, S., Rewers, M., Hamman, R., Selby, J.V., Saad, M.F., Savage, P., and Bergman, R. for the IRAS Investigators, 1996, Insulin sensitivity and atherosclerosis: the Insulin Resistance Atherosclerosis Study (IRAS), *Circulation* 93:1809-1817.

Jansson, P.A., Fowelin, J.P., von Schenck, H.P., Smith, U.P., and Lonnroth, P.N., 1993, Measurement by microdialysis of the insulin concentration in subcutaneous interstitial fluid: Importance of the endothelial barrier for insulin, *Diabetes* 42:1469-1473.

Kahn, S. E., Beard, J.C., Schwartz, M.W., Ward, W.K., Ding, H.L., Bergman, R.N., Taborsky, G.J., Jr., and Porte, D., Jr., 1989, Increased B-cell secretory capacity as mechanism for islet adaptation to nicotinic-acid-induced insulin resistance, *Diabetes* 38:562-568.

Kahn, S.E., Prigeon, R.L., McCulloch, D.K., Boyko, E.J., Bergman, R.N., Schwartz, M.W., Neifing, J.L., Ward, W.K., Beard, J.C., Palmer, J.P., and Porte, D., Jr., 1993, Quantification of the relationship between insulin sensitivity and B-cell function in human subjects: evidence for a hyperbolic function, *Diabetes* 42:1663-1672.

King, G.L., and Johnson, S.M., 1985, Receptor-mediated transport of insulin across endothelial cells, *Science* 227:1583-1586.

Knott, G.D., 1979, MLAB -- a mathematical modeling tool, *Comput. Progr. Biomed.* 10:271-280.

Korytkowski, M.T., Berga, S.L., and Horwitz, M.J., 1995, Comparison of the minimal model and the hyperglycemic clamp for measuring insulin sensitivity and acute insulin response to glucose, *Metabolism* 44:1121-1125.

Lewis, G.F., Vranic, M., Harley, P., and Giacca, A., 1997, Fatty acids mediate the acute extrahepatic effects of insulin on hepatic glucose production in humans, *Diabetes* 46:1111-1119.

Manning, R.D., Jr., and Guyton, A.C., 1982, Control of blood volume, *Rev. Physiol. Biochem. Pharmacol.* 93:70-114.

Martin, B.C., Warram, J.H., Krolewski, A.S., Bergman, R.N., Soeldner, J.S., and Kahn, C.R., 1992, Role of glucose and insulin resistance in development of type 2 diabetes mellitus: results of a 25-year follow-up study, *Lancet* 340:925-929.

McDonald, C., Dunaif, A., and Finegood, D.T., 2000, Minimal-model estimates of insulin sensitivity are insensitive to errors in glucose effectiveness, *J. Clin. Endocrinol. Metab.* 85: 2504-2508.

Miles, P.D.G., Levisetti, M., Reichart, D., Khoursheed, M., Moossa, A.R., and Olefsky, J.M., 1995, Kinetics of insulin action in vivo: identification of rate-limiting steps, *Diabetes* 44:947-953.

Mittelman, S.D., and Bergman, R.N., 2000, Inhibition of lipolysis causes suppression of endogenous glucose production independent of changes in insulin, *Am. J. Physiol.* 279: E630-E637.

Ni, T.-C., Ader, M., and Bergman, R.N., 1997, Reassessment of glucose effectiveness and insulin sensitivity from minimal model analysis: a theoretical evaluation of the single-compartment glucose distribution assumption, *Diabetes* 43:1813-1821.

Poulin, R.A., Steil, G.M., Moore, D.M., Ader, M., and Bergman, R.N., 1994, Dynamics of glucose production and uptake are more closely related to insulin in hindlimb lymph than in thoracic duct lymph, *Diabetes* 43:180-190.

Quon, M.J., Cochran, C., Taylor, S.I., and Eastman, R.C., 1994, Non-insulin-mediated glucose disappearance in subjects with IDDM: discordance between experimental results and minimal model analysis, *Diabetes* 43:890-896.

Rashevsky, N., 1940, *Advances and Applications of Mathematical Biology,* University of Chicago Press, Chicago.

Rebrin, K., Steil, G.M., Getty, L., and Bergman, R.N., 1995, Free fatty acid as a link in the regulation of hepatic glucose output by peripheral insulin, *Diabetes* 44:1038-1045.

Rebrin, K., Steil, G.M., Mittelman, S., and Bergman, R.N., 1996, Causal linkage between insulin regulation of lipolysis and liver glucose output, *J. Clin. Invest.* 98:741-749.

Turing, A.M., 1936, On computable numbers, with an application to the Entscheidungsproblem, *Proc. London Math. Soc.* 2:230-265.

Steil, G. M., Ader, M., Moore, D.M., Rebrin, K., and Bergman, R.N., 1996, Transendothelial insulin transport is not saturable in vivo: no evidence for a receptor-mediated process, *J. Clin. Invest.* 97:1497-1503.

Taniguchi, A., Nakai, Y., Fukushima, M., Imura, H., Kawamura, H., Nagata, I., Florant, G.L., and Tokuyama, K., 1994, Insulin sensitivity, insulin secretion, and glucose effectiveness in subjects with impaired glucose tolerance: a minimal model analysis, *Metabolism* 43:714-718.

Valle, T., Tuomilehto, J., Bergman, R.N., Ghosh, S., Hauser, E.R., Eriksson, J., Nylund, S.J., Kohtamaki, K., Toivanen, L., Vidgren, G., Tuomilehto-Wolf, E., Ehnholm, C., Blaschak, J., Langefeld, C.D., Watanabe, R.M., Magnuson, V., Ally, D.S., Hagopian, W.A., Ross, E., Buchanan, T.A., Collins, F., and Boenke, M., 1998, Mapping genes for NIDDM: design of the Finland-United States investigation of NIDDM genetics (FUSION) study, *Diab. Care* 21: 949-958.

Vicini, P., Caumo, A., and Cobelli, C., 1997. The hot IVGTT two-compartment minimal model: indexes of glucose effectiveness and insulin sensitivity, *Am. J. Physiol.* 273:E1024-E1032.

Weiner, N., 1965, *Cybernetics or Control and Communication in the Animal,* MIT Press, Boston.

Weyer, C., Bogardus, C., Mott, D.M., and Pratley, R.E., 1999, The natural history of insulin secretory dysfunction and insulin resistance in the pathogenesis of type 2 diabetes mellitus, *J. Clin. Invest.* 104:787-794.

Wing, R.R., Blair, E.H., Bononi, P., Marcus, M.D., Watanabe, R., and Bergman, R.N., 1994, Caloric restriction per se is a significant factor in improvements in glycemic control and insulin sensitivity during weight loss in obese NIDDM patients, *Diab. Care* 17:30-36.

Yang, Y.J., Hope, I.D., Ader, M., and Bergman, R.N., 1989, Insulin transport across capillaries is rate limiting for insulin action in dogs, *J. Clin. Invest.* 84:1620-1628.

Yates, F.E., Marsh, D.J., and Iberall, A.S., 1972, Integration of the whole organism: a foundation for a theoretical biology, in: *Challenging Biological Problems: Directions Towards Their Solution,* J.A. Behnke, ed., Oxford University Press, New York.

CORNERSTONES TO SHAPE MODELING FOR THE 21ST CENTURY: INTRODUCING THE AKA-GLUCOSE PROJECT

Ray Boston, Darko Stefanovski, Peter Moate, Oscar Linares, and Peter Greif [*]

INTRODUCTION

For some reason, the start of a new century, as though magically demarcating history, seems to confer the right to reflect on progress as we embrace a 'new' period or 'new' phase of endeavor. Accordingly, we begin this paper by reviewing what computer-based mathematical modeling in biology (CMMB) has brought us thus far, as we begin the 21st century (or could we say, 'the third millennium'). We will decompose the stages of CMMB evolution in the last century into three generations, noting the hallmark accomplishments in each.

After considering historical developments in CMMB, we address two important questions which arise as we leave one epoch and begin another: what have we learned from our work effort in CMMB so far, and what are the areas which will have the highest returns per unit of effort in the new epoch? It is against this backdrop that we present the sections on 'cornerstones' that will shape modeling in the 21st century. We believe that the first cornerstone is the model development environment (which embraces modeling philosophy and software). This is exemplified by the modeling program 'WinSAAM' which has been significantly enhanced by the work of co-authors R.C. Boston and P. Greif. In the discussion of this cornerstone, we describe some of the features of WinSAAM and focus on a new tool called 'the project manager' which enables population analysis of kinetic models. As the second cornerstone, we chose the addition of database technology to modeling. This cornerstone is built on work by R.C. Boston, D. Stefanovski, E. Janczewski, M. Petrova and P.J. Moate (Biomathematics Unit, School of Veterinary Medicine, University of Pennsylvania). We focus on an entirely new approach to model dissemination, the AKA-Glucose project. We consider the third

---

[*] Ray Boston, Darko Stefanovski, and Peter Moate, Biomathematics Unit, School of Veterinary Medicine, University of Pennsylvania, Kennett Square, PA 19348. Oscar Linares, University of Michigan Geriatrics Clinic, Ann Arbor, MI 48105. Peter Greif, Laboratory of Computational and Experimental Biology, Division of Cancer Biology and Diagnosis, National Cancer Institute, Bethesda MD 20892.

cornerstone of modeling to be data exchange. Contributions by co-author R.C. Boston and A.E. Sumner (Diabetes Branch, NIDDK) are highlighted. We demonstrate how both project management and database technology in the CMMB setting can be conducive to data exchange. Finally, we believe that a fourth cornerstone of modeling for the 21st century is the establishment of a structured modeling community working together to promote and apply modeling methodology in the biomedical sciences. This effort has been spear-headed by co-author R.C. Boston and N. Canolty (Department of Foods and Nutrition, University of Georgia). We present the case for such a community and propose a plan for its implementation.

HISTORY OF CMMB: GENERATION 1

In its first generation, CMMB used scientific subroutines tightly linked into a batch-processing environment (program) to enable program users to perform a limited number of (pre-scheduled) processing steps to explore systems. The very limited power of computers available in this generation (conceivably spanning 1960 to 1975) meant that a rigid input, output, and processing structure was the best that could be provided. Indeed, when we recall that computers of this period were one thousandth the size, one thousandth the speed, and one thousand times the cost of today's computers (a 10^9 deficiency), it is remarkable that anything was achieved at all. And yet remarkable were the feats of the CMMB pioneers of this generation. Necessity gave birth to modeling constructs and operational units, to accurate and extremely fast numerical integrators, and to robust and efficient data fitting procedures. Of course, the most important outgrowth of this era was the unequivocal emergence of a role for mathematical modeling in biology, specifically in nutrition, in clinical science, and in agriculture and the environment.

The tools and techniques developed and refined in this generation were of such high quality and so germane to the CMMB setting that, in many instances, they endure even today. Specifically, a recent re-assessment (Boston et al., 1996) of a specific numerical integrator of this generation has shown it to be both faster and more accurate than other more modern integrators when used in its preferred setting. Furthermore, an evaluation of optimizers (Boston and McNabb, 1990) has revealed that optimization performance of software of this generation [specifically the SAAM (Simulation, Analysis, and Modeling) optimizer] is approximately 10 times faster than other competing systems with no loss of precision.

The SAAM software was a prime example of CMMB software of this generation. Areas to which it was applied include iodine metabolism (Lewallen et al., 1959; Berman et al., 1968), glucose-insulin interaction (Insel et al., 1974; Sherwin et al., 1974), lithium (Temple et al., 1972), magnesium (Avioli and Berman, 1966, 1968), phosphorus (Brauer et al., 1960), calcium (Neer et al., 1967; Birge et al., 1969), and anticancer drug action (Leme et al., 1975).

HISTORY OF CMMB: GENERATION 2

A severe, though entirely unintentional, shortcoming of the first generation of CMMB was the virtual exclusion of all but the mathematically and computationally competent from the modeling process. Investigators wanting to embrace modeling as an investigatory step needed to bring their data to a modeling center (such as Berman's 'Laboratory of Theoretical Biology' at the National Cancer Institute) and work there with an expert to explore ideas and craft scientific explanations. Indeed, it was principally this issue that gave rise to the second generation of CMMB software, the CLI (command line interface) interactive modeling software. Typified by CONSAM (Boston et al., 1981, 1982), the CLI version of the SAAM program, this generation spanned the period 1976 to 1990, and its introduction witnessed a dramatic popularization of modeling as the tool of choice among investigators who wanted to explore kinetic and dynamic biomedical systems.

The second generation brought entirely new problems for CMMB software development: the tools, solvers, integrators, and optimizers were in place, the modeling constructs and operational units were tried and tested, but now the problems of task scheduling and mode of software use needed to be addressed. Breaking apart the formerly strictly scheduled processing and allowing users, via command line, to execute any reasonable processing step at any time forced careful evaluation of exactly what processing was reasonable at particular times (or software states). Here, the CONSAM designers relied very heavily on the state transition paradigm of software development to solve this scheduling problem, and they in fact implemented an early version of the modeless processing procedure available today in successful modeling software.

As we have already noted, modeling flourished in this period and, while there were few new tools actually developed, exposure to their power via the CLI and, more importantly, via modeless CLI operation, led to new applications of CMMB in the metabolism of zinc (Foster et al., 1979), lipoproteins (Schwartz et al., 1982), and amino acids (Hall et al., 1977), and in metabolic actions (France and Thornley, 1984). Indeed, a full model of carbon atom movement in association with the TCA cycle was developed (Goebel et al., 1982).

Between the first and second generation of CMMB software, computers became a little 'friendlier.' Users didn't (yet) own their own computers, but they could at least communicate with them in a more straightforward fashion. Actually, users began to own personal computers in the latter part of the second generation; however, the style of computer use didn't change all that much.

HISTORY OF CMMB: GENERATION 3

The third generation witnessed dramatic changes in ownership, style of computer use, and processing power of computers used by CMMB investigators. Most significantly, following the 'Apple' innovation, computers became graphically driven with windows, dialog boxes, and menus harnessing actions such as selecting, grouping, dragging, and activating (opening). The new problem posed for CMMB software developers was how far to go with graphical management. Fully graphically managed meant moded or controlled processing, whereas less graphically managed meant less moded or more user-directed processing flow. It was here that we saw a split in the SAAM community of investigators, with the moded or graphical modeling paradigm advanced in the SAAM II stream, and the modeless or user driven paradigm advanced in the WinSAAM stream. In the moded stream, along with SAAM II, we also saw Stella and JMP, and in the modeless stream, along with WinSAAM, we also saw Scientist and STATA.

The feature of moded software was that all actions were driven by the activation of graphical objects, from creating the model to advancing data analysis using it. And, most particularly, the flow of analysis using this paradigm was largely dictated by the 'current state' of the software. That is, while the software was in a particular state (e.g., creating the model graphic), concurrent or asynchronous modeling activities relating to model processing were suspended until the software services supporting the 'current state' were exited. Modeless software, on the other hand, incorporated lexical specification of models as well as data, and here processing was advanced using commands. The modeless software user was able to dictate the processing flow to a much greater degree than the moded software user.

Of course, just as no software can be fully moded (except for trivial tasks), neither can software be fully modeless. The degree to which either agenda was advanced in the respective software schools was made possible by the innovative steps in the second generation to understand the concept of moded processing and to develop software from a state transition perspective.

The application yield of the third generation of modeling software was to enable modeling to be incorporated into university lecture and laboratory programs and to dramatically diversify the application domains in biology and medicine. New applications included, for example, spread of diseases (Anderson et al., 1986) and DNA transcription processes (Phillips et al., 1994).

We are now ready to look to the future by reviewing selected 'cornerstones' that we believe will shape modeling in the next generation.

CORNERSTONE 1: THE MODEL DEVELOPMENT ENVIRONMENT

The model development environment is the combination of the modeling philosophy and the suite of modeling programs that modelers use to investigate systems employing the CMMB paradigm. Typically, confronted by a scientific problem and with the relevant data and information at hand, the investigator (modeler) sets about assessing plausible scientific explanations of the problem based largely upon their consistency with the data and information. With this agenda in mind, the model development environment needs to support an array of tools and facilities for the fabrication of models and data sets, for the exploration of models, and for the evaluation of models.

In a paper presented at the 6th Conference on Mathematical Modeling in Experimental Nutrition in 1997 (Greif et al., 1998), we introduced the Windows version ('WinSAAM') of the NIH SAAM/CONSAM modeling software. WinSAAM incorporated both the functionality of SAAM and the operational control of CONSAM with the added feature of being a fully Windows integrated system. From the model and dataset fabrication perspective, WinSAAM supports a highly refined graphics editor (Figure 1). With field entry support, modeling objects and data can be easily entered directly in lexical style, or they can be incorporated in the same style as tab delimited text from the clipboard/paste buffer or from external files. Local font management ensures easy checking of alignment of field entries. Furthermore, to help guide users with field entry, we provide local 'entry tips' concerning the nature of entries anticipated in each field.

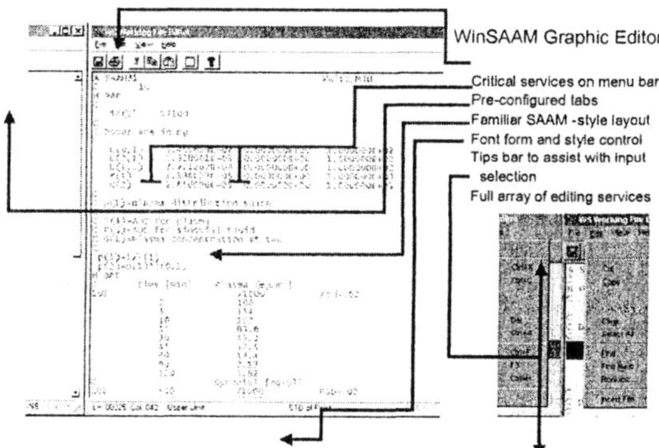

Figure 1. The WinSAAM graphic editor. Shown are special features of this environment that ease model fabrication.

Once the model and the data are entered, the WinSAAM user exits from the editor and returns to the 'terminal' (or 'console') window. It is here where much of the exploratory phase of the modeling exercise takes place (Figure 2). Equipped with a comprehensive array of (formerly CONSAM) commands, the user is able to compile, solve, and fit (candidate) models as well as display either graphically (Figure 3) or lexically the consequences of these actions. Building upon our experiences with CONSAM, we have taken care to ensure that sensible processing flow is promoted and that the valuable results derived by the user are never inadvertently overwritten.

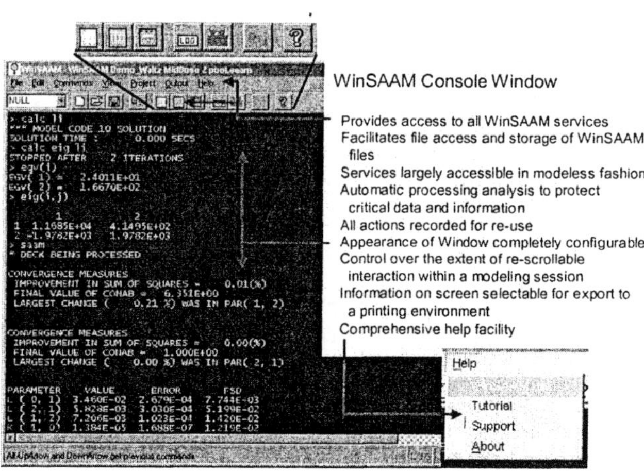

Figure 2. The WinSAAM console window. Shown are special features that facilitate model exploration.

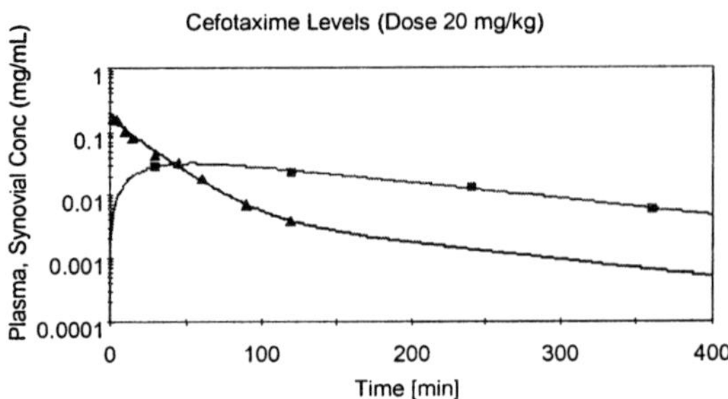

Figure 3. The WinSAAM charting system. The figure highlights special features that make charting flexible and easy. The final plots produced here are of publication quality.

The WinSAAM system supports an array of facilities to enable critical evaluation of modeling activities. In addition to the charting system, there are spreadsheet displays (Figure 4) and batch style output provisions. The strength of the latter is that it enables the encapsulation of a total overview of a model within a single output structure. Furthermore, modeling actions and responses can be easily tracked in WinSAAM using the history facility or the event-logging window.

File management in WinSAAM has been carefully refined to ensure the widest possible protection to the user. However, because some of the file control operations are accessible to the user whereas others take place in the 'background' (i.e., they are automatically managed by the WinSAAM file processing machinery), there may be some confusion concerning the 'data set' actually being used at any time. In Figure 5, we show the possible data states together with actions causing migration among these states.

Compt.	Cat.	Time	QC	QO	SD	FSD	Weight
1	F(1)	2	1.55E-01	1.68E-01	3.36E-03	9.34E-03	3.55E-03
1	F(1)	5	1.37E-01	1.54E-01	3.08E-03	8.11E-03	4.23E-03
1	F(1)	10	1.12E-01	1.07E-01	2.14E-03	6.25E-03	8.76E-03
1	F(1)	15	9.19E-02	8.36E-02	1.67E-03	4.80E-03	1.44E-02
1	F(1)	30	5.12E-02	4.53E-02	9.06E-04	5.12E-03	4.89E-02
1	F(1)	45	2.92E-02	3.25E-02	6.50E-04	8.55E-03	9.50E-02
1	F(1)	60	1.73E-02	1.84E-02	3.68E-04	1.06E-02	2.96E-01
1	F(1)	90	7.15E-03	7.17E-03	1.43E-04	9.83E-03	1.95E+00
1	F(1)	120	3.88E-03	3.82E-03	7.64E-05		.87E+00
2	F(2)	30	2.88E-02	2.94E-02	5.89E-04		.16E-01
2	F(2)	120	2.52E-02	2.38E-02	4.75E-04		.77E-01
2	F(2)	240	1.25E-02	1.34E-02	2.69E-04		.56E-01
2	F(2)	360	6.07E-03	5.93E-03	1.19E-04		.85E+00

Captures computed values and parameter values, together with salient solution
and fit indices
Provides key data exchange linkages between WinSAAM and other software
Synchronized with *solve* and *iterate* invocation from the terminal window

Param.	Form	Value	Minimum	Maximum	FSD	Exclude
K(1)	A	1.38E-05	0.00E-01	1.00E+03	1.02E-02	Included
K(2)	A	2.65E-05	0.00E-01	1.00E+03	4.77E-02	Included
L(0,1)	A	3.46E-02	0.00E-01	1.00E+02	7.74E-03	Included
L(1,2)	A	7.21E-03	0.00E-01	1.00E+02	1.42E-02	Included
L(2,1)	A	5.83E-03	0.00E-01	1.00E+02	5.20E-02	Included
P(1)	D	7.22E+04				
P(2)	D	2.50E+03				

Figure 4. The WinSAAM spreadsheet environment. This facilitates results summarization and the exportation of WinSAAM results to other spreadsheet-compatible systems.

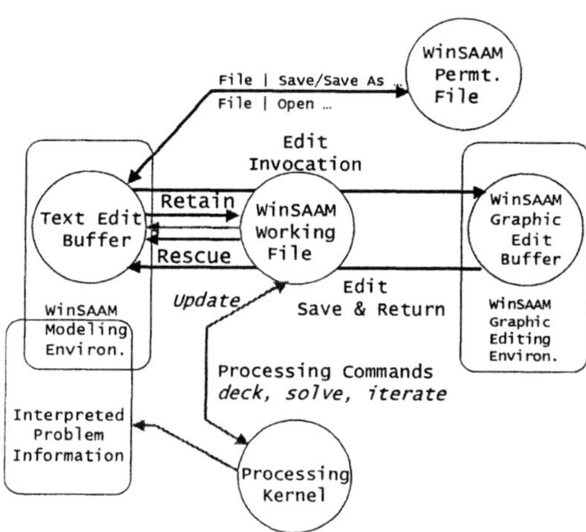

Figure 5. The states of WinSAAM datasets. Specifically included here are the text edit buffer, the working file, the graphic edit buffer, the permanent file, and the modeling environment. Shown are steps that may be taken either by WinSAAM (retain or rescue in an automatic style) or by the user to maneuver between these states to protect and preserve critical data and calculations.

Once the initial modeling foray is completed and a model which is consistent with representative samples of data from our 'study population' is found, a number of important questions emerge:

- How consistent are the model parameters and the critical model-based clinical indices across all subjects?
- How sensitive is our model to demographic and clinical traits of sub-populations from within our study?
- How well are all model parameters and model-based clinical indices resolved for each subject from this study population?
- What is the variability between and within demographic/clinical groups from the entire study population?
- What are the best estimates and uncertainties of parameters characterizing demographic/clinical groups from the entire study population?

To assist investigators with these issues, we have added an entirely new tool to the WinSAAM investigatory repertoire: the project manager. Underpinning the project manager is the concept of a 'study stream' or series of WinSAAM files assembled into a single project for some analytic purpose. The project manager supports the following activities:

- establishing a study stream (Figure 6A)
- editing a study stream (Figure 6B)
- saving a study stream
- retrieving a study stream
- invoking the independent processing of subjects in a study stream (Figure 6B)

- invoking population modeling of a study stream (Figure 6B)
- assembling the study stream processing results into a series of spreadsheet tables for direct incorporation into other analytic and productivity environments (Figure 6C)
- assembling the study stream processing results into a comprehensive text file

It is important for us to distinguish here between the two types of processing analyses to which the study stream may be submitted: independent processing and population modeling. As the name suggests, 'independent processing' implies the parsing of independently examined, subject-level WinSAAM files into a sequence of tables including a data (and solutions) table, a parameters table, a covariance table, a correlation table, and a partials matrix table. The contents of each of these tables can then be freely manipulated within WinSAAM using regular spreadsheet style instructions or possibly exported to other software for further processing.

The population modeling tool uses a global two-stage (GTS) iterative procedure (Lyne et al., 1992) to accommodate intra-subject variability in model parameters as it attempts to derive population parameter estimates using a mixed effects modeling approach. As compared with the single stage approach of SAAM II, the main advantage of the GTS scheme is that it requires that each individual's results be fitted. This approach reduces the risk of misclassification bias resulting from pooling subjects from heterogeneous populations. A disadvantage of this approach is that the control we assert over the model's error structure is somewhat limiting.

With the 'population modeling' processing stream, one acquires a set of spreadsheet tables characterizing the variability of subject parameters within each demographic/clinical group around the population means for that group and the variance of the population's parameters between demographic/clinical groups (see Figure 7). In addition, the entire set of subject level tables is also provided, as with the independent processing stream.

CORNERSTONE 2: MODEL DISSEMINATION AND DATABASE TECHNOLOGIES – THE AKA-GLUCOSE PROJECT

Once it became clear that models such as those discussed in this book may have a utility beyond their immediate development purpose, efforts have been expended to disseminate them so that other investigators may use them and build upon them. The key concerns associated with dissemination are the degree to which the development environment needs to be disseminated with the model, the ease with which the model can be used in its disseminated form, and the level of protection that intended users are provided with regard to inappropriate use of the model. (Recall that these models are often applied in clinical settings where the consequences of misuse could be quite profound.) Furthermore, the problem of a mechanism for data entry into the 'model' at a user site always has to be factored into the dissemination plan.

As part of this cornerstone of modeling, we discuss an approach to model dissemination that enables operational functionality, integrated and secure data entry into the model, and environment management for post-processing of model determinations.

A

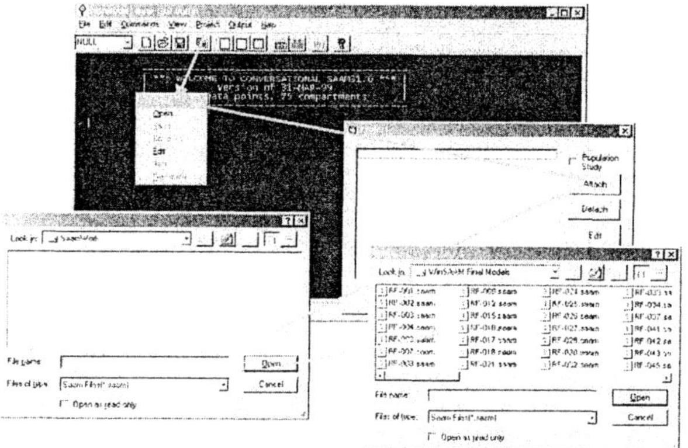

B

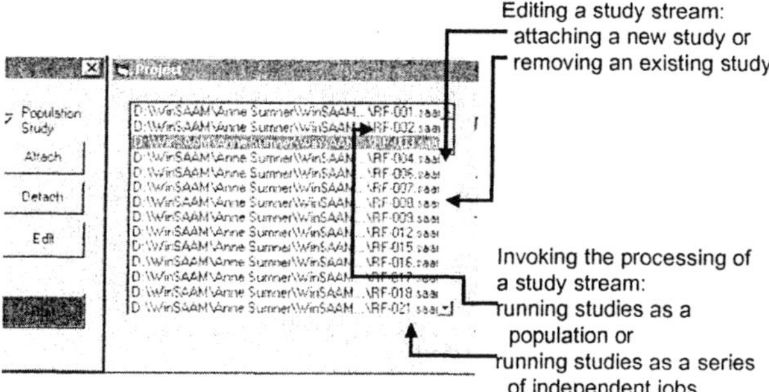

C

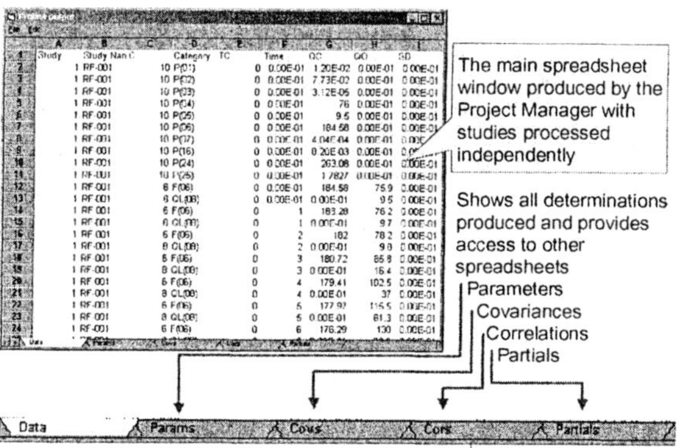

Figure 6. The WinSAAM project manager. (A) Creating a new project. (B) Editing a project and invoking the processing of a project stream. (C) Spreadsheet tables produced in conjunction with a sequential processing run, with jobs processed independently of one another.

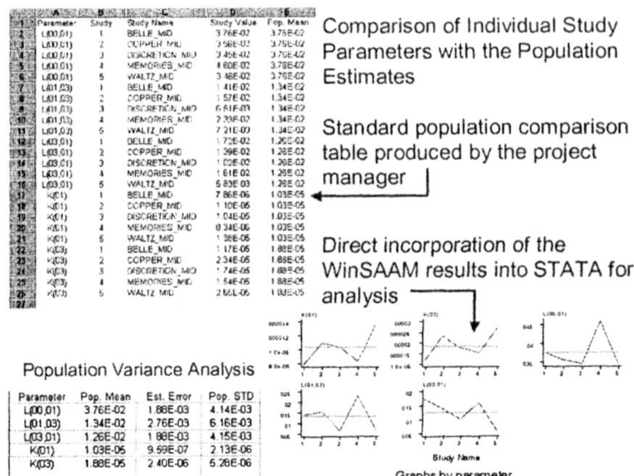

Figure 7. The WinSAAM project manager. This shows the assembly of population studies analysis into the WinSAAM designer spreadsheet (automatically) followed by the transport of the population parameter estimates to STATA for graphical and numerical summarization.

For many years, investigators have anticipated the need for database management of kinetic data. Indeed, Wastney's extremely successful 'model library' (Wastney et al., 1998) was initially conceived as a model database. However, for technical reasons, it advanced along a file-structured library path as opposed to the intended database route. A database is the most appropriate approach to managing the data, information, and models we work with because it yields integrated access to the contents. Specifically, a database approach to the management of kinetic data and models enables:

- direct access to each individual study within the database
- proper administrative control over alteration, removal, and addition of (new) objects within the database
- secure and consistent processing of encapsulated studies
- a reliable and well documented array of access procedures (queries)
- a high degree of portability and exchangeability of its contents with other productivity and statistical environments
- the capacity to 'stand alone' and to be sharable among participating individuals and groups (indeed the potential for network-based modeling becomes available)
- flexibility of perspective in that a database could be created around a central model theme, with database records comprising new studies based on the same model; or the theme could be a modeling application, and hence records would coincide with new or different features of the system of interest

Most importantly, with funding agencies (e.g., NIH, FDA, CDC, and USDA) stipulating that publicly funded experimental data be retained for up to 10 years from the conclusion of the funding, database technology provides the ideal information and research development repository.

We saw two major issues confronting us as we considered the incorporation of database technology into kinetic analysis: firstly, the normalization of kinetic information (models and data) and secondly, a means whereby powerful local processing and analysis could be available. *Normalization* in database terms means organizing the information and making it amenable to database management; *processing* here means performing model solutions and model parameter adjustments (data fitting) as needed and on a record-by-record basis.

For our first foray into such a technological endeavor, we elected to use a published model of high clinical utility. This would assure that, among other things, as our ideas developed, there might be wide interest in our results. Because of our interest in glucose metabolism and specifically, in insulin regulation of glucose, a natural model of choice was Bergman's Minimal Model (Pacini and Bergman, 1985). From this point on, our project became known as the A(utomated) K(inetic) A(nalysis)-Glucose project.

The Minimal Model describes the response of glucose and insulin to the intravenous glucose tolerance test (IVGTT). Specifically, with the IVGTT protocol, a large glucose bolus is injected into the circulation of a subject, and the time course of circulating glucose and insulin levels are recorded. Variations to the standard approach allow, for example, a short insulin infusion about 20 minutes after the glucose injection. The objective of the Minimal Model is to use the IVGTT data in a kinetically sensible way to provide the best possible estimates for parameters describing insulin sensitivity and glucose responsiveness.

In Figure 8, we show a schematic diagram of the Minimal Model depicted as a graphic within the AKA-Glucose protocol dialog window. The equations describing circulating glucose (G; mg/dL) and insulin action (X; min^{-1}) during an IVGTT are as follows:

$$G` = -G(X + S_G) + S_G.G_b \qquad G(0) = G0$$
$$X` = -p_2[X - S_I(I - I_b)] \qquad I - I_b = 0 \text{ if } I <= I_b$$

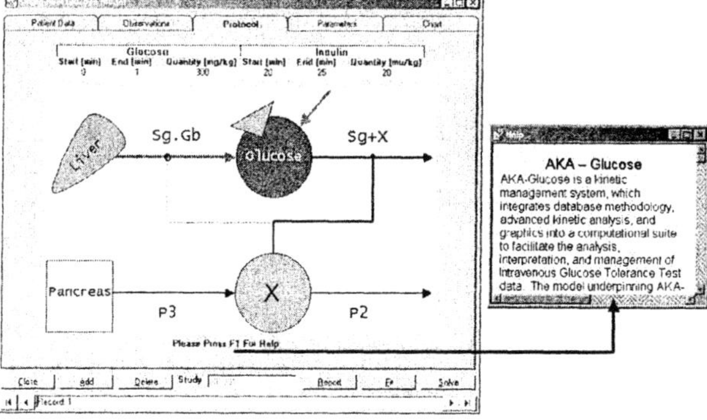

Figure 8. A schematic of the Minimal Model, shown in the AKA-Glucose protocol dialog window. The protocol window specifies the study protocol and provides access to AKA-Glucose help.

where

S_G = glucose effectiveness (min^{-1})
S_I = insulin sensitivity $(L\cdot mU^{-1}\cdot min^{-1})$
G_b = basal glucose (mg/dL)
I_b = basal insulin (mU/L)
I = circulating insulin (mU/L)
$G0$ = distributed glucose bolus (mg/dL)
p_2 = model parameter describing disposal of insulin (min^{-1})
G' and X' denote derivatives with respect to time

A key ingredient to solving the normalization problem was to develop a table-oriented data dictionary. On the one hand, this table needed to portray the calculations to be performed by the calculations engine and, on the other hand, it revealed where the database system (AKA-Glucose) could retrieve those calculations and how they were represented. In Figure 9, we present a tabulation of the AKA-Glucose data dictionary.

A by-product of earlier work (Boston et al., 2000) which related to the implementation of the project management tool in WinSAAM was the refinement of a Windows executable (stand alone) version of SAAM with the capability of producing comprehensive spreadsheet-style output tables (see earlier). Accordingly, it seemed highly appropriate to us in our evaluation of alternate model processing methods for AKA-Glucose that we explore the feasibility of the SAAM executable module as the processing base. This has subsequently proven to be a most effective implementation path, as the

SAAM Modeling Construct or Operational Unit	Clinical Index	Unit	Typical Value
P(1)	SG	min^{-1}	2.2e-2
P(2)	P2	min^{-1}	5.0e-2
P(3)	P3	$[mU/L]^{-1}.min^{-2}$	1.0e-5
P(4)	Gb	$mg.dL^{-1}$	80
P(5)	Ib	$mU.L^{-1}$	10
P(6)	G0	$mg.dL^{-1}$	200
P(7)	SI = P3/P2	$[mU/L]^{-1}.min^{-1}$	2.0e-4
P(8)	Apparent distribution space, Vg = 300.Weight/G0	dL	140
P(16)	GEZI = SG − SI*Ib	min^{-1}	1.8e-2
P(24)	β-cell function = 20.Ib/(Gb/18 − 3.5)	$mU.min^{-1}$	200
P(25)	Insulin resistance = Gb.Ib/405	$mm.mU.L^{-2}$	2.0
UF(6)	Glucose disposal = -(SG+X).F6	$mg.dL^{-1}.min^{-1}$	
UF(7)	P3.(FF(8) -Ib)	min^{-2}	
F(9)	AIRg	$mU.L^{-1}.min$	400
F(7)	Insulin Action, X	min^{-1}	

Figure 9. The AKA-Glucose data dictionary table.

entire processing strengths of the first generation of CMMB software were immediately available to AKA-Glucose.

The appearance of the AKA-Glucose database system is shown in Figures 8 and 10A-F. In essence, AKA-Glucose has the appearance of a 'tab driven' series of dialog windows. Each of these encapsulates its own array of tools and functions, and each is expandable to fill the entire screen using conventional window manipulation objects. Hence we refer to these as *dialog windows*.

In Figure 10A, the patient dialog window, patient demographics and clinical measurements can be entered, edited, and displayed. In addition, at any stage in the life of the database, new demographic or clinical information can be compiled into the database as availability and need dictate. This feature provides a unique opportunity to expand our population modeling facility, as development time permits, by admitting the incorporation of error variability associated with demographics into the population parameter estimation procedure.

Figure 10B shows the observation dialog window. This provides a structured data entry and data editing control for the glucose and insulin observations. We also specify the error structure of the data in this dialog box, and furthermore, we can control (toggle) the capacity of individual observations to influence the parameter estimation process.

The protocol dialog window (Figure 8) is where the user specifies the glucose and insulin administration protocols and can obtain 'help' with the use of AKA-Glucose. In the parameters dialog window (Figure 10C), all the model determinations and their uncertainties (fractional standard deviations or FSDs) are displayed. If, after extensive iterative or manual adjustment of the parameters, the quality of fit seems to be deteriorating, a 'good' (factory setting) set of parameter values can be retrieved. Alternatively, if after a single iteration, the quality of the fit seems poorer in some important sense, then the original adjustable parameter values can be restored. If we have estimates of the errors of (some of) the adjustable parameters from some other procedure, then they can be applied via this dialog box as well. At any stage of the analysis of a glucose response, the user can control the adjustability of each parameter (S_G, p_2, p_3, and G0) and independently set the 'current' value of any of these parameters to some desired value.

In the chart window (Figure 10D), we display color-coded representations of observations and calculations associated with glucose and insulin. Hollow objects are used to depict unweighted observations. The left and right vertical axes are labeled and assigned to glucose and insulin, respectively. To enable easy reading of the graphs, horizontal grid lines are provided. If desired, a profile of the insulin-attributable fraction of glucose disposal can be displayed by toggling on a display radio button within the chart window.

Figures 10E and F show the AKA-Glucose button controls. There are two processing buttons, *solve,* which invokes solutions of the Minimal Model with the current parameter values, and *fit,* which causes new, potentially better-fitting values of the adjustable parameters of the model to be found. There are three database management buttons which lead to the addition of a new subject, the removal of a selected subject, or the compression of the database and processing termination. We point out that new subjects can be added into the database either in spreadsheet (tab delimited) format or in WinSAAM file format. The sixth button, the report generation button, provides access to many of the important output facilities of AKA-Glucose:

A

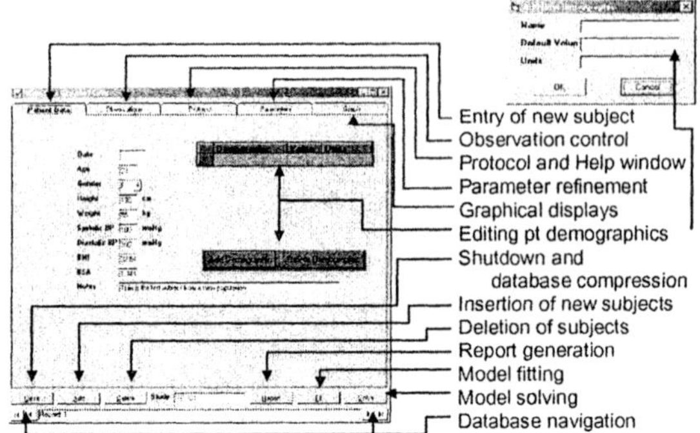

B

- Independent data weighting
- Independent entry of glucose and insulin
- Error assignment to glucose data
- Editorial control with data

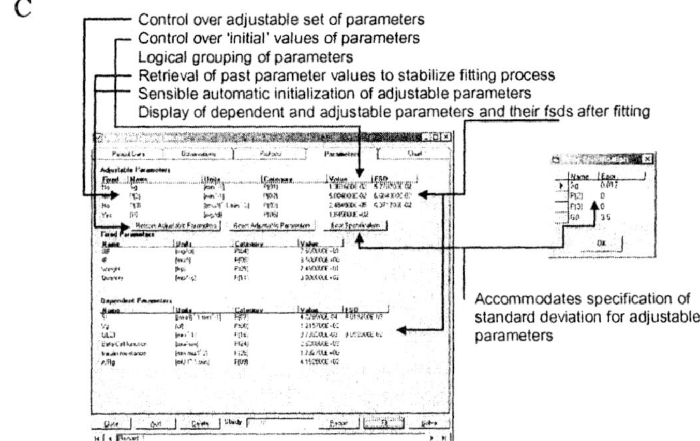

C

Control over adjustable set of parameters
Control over 'initial' values of parameters
Logical grouping of parameters
Retrieval of past parameter values to stabilize fitting process
Sensible automatic initialization of adjustable parameters
Display of dependent and adjustable parameters and their fsds after fitting

Accommodates specification of
standard deviation for adjustable
parameters

D

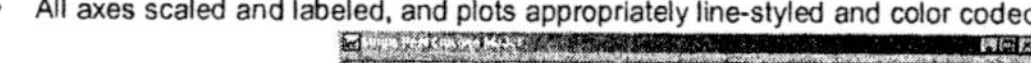

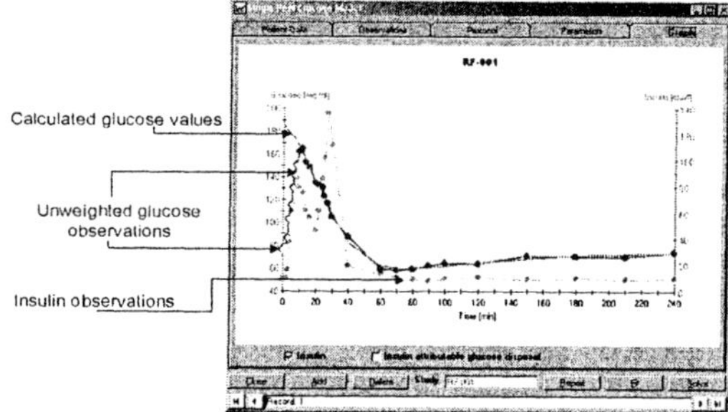

E

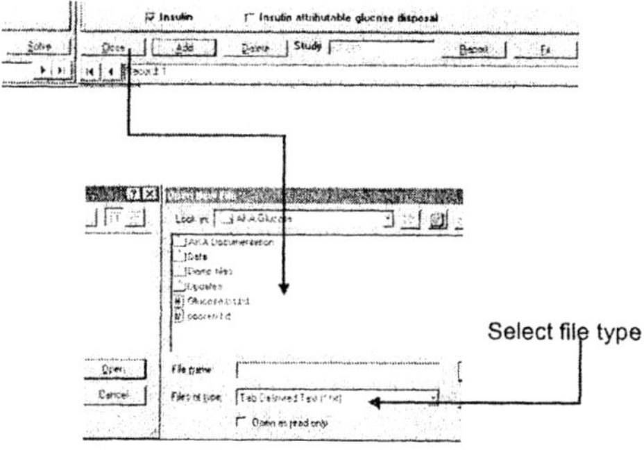

F

Activating the 'Report' button enables production of the following:

1. A printout of the IVGTT charts
2. A parameters report for the current study
3. A model predictions report for the current study
4. A tab delimited text file for the entire study for direct incorporation into statistical software
5. A WinSAAM compatible file for further detailed investigation

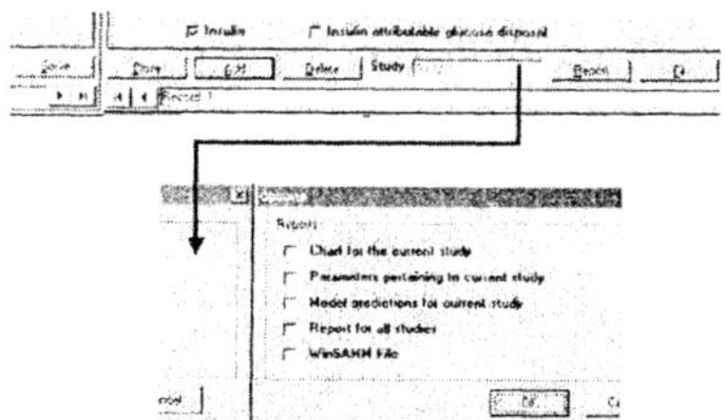

Figure 10. The AKA-Glucose kinetic database environment. (A) The patient dialog window. (B) The observations dialog window. (C) The parameters dialog window. (D) The chart window. (E) The AKA-Glucose button controls: adding a subject to the database. (F) The AKA-Glucose button controls: producing reports from the database.

- charts of observed and calculated values for the current study (printout)
- parameter values and their uncertainties for the current study (printout)
- model predictions and their uncertainties for the current study (printout)
- all database determinations for all subjects (tab delimited file format)
- WinSAAM processable file format for the current study

Lastly, AKA-Glucose incorporates database navigation tools enabling rapid direct retrieval of any database entry. In Figure 11, we present a 'data flow' overview of the AKA-Glucose environment.

CORNERSTONE 3: DATA EXCHANGE AND POST-FITTING ANALYSIS

The ease with which CMMB software may be refined and validated is in part due to the fact that its developers can specialize and keep abreast of a small array of critical computational tools pertinent to their needs. Once developers advance beyond their locally driven needs, fragmentation of focus and less than viable products emerge. The WinSAAM group has drawn the line at model solving and model fitting. Furthermore, the group has attempted to deliver an efficient set of procedures that enable users to progress to their chosen statistical and productivity settings in virtually seamless fashion. Indeed, if we glance back at Figure 11, the provision of such data exchange capabilities has been heavily used in the implementation of the AKA-Glucose initiative.

Whereas there are many statistical packages that might embrace the WinSAAM environment, it is difficult to imagine a more synergistic association than that which STATA offers us. STATA is a CLI system like WinSAAM, and it is fast, efficient, and accurate (to which we also aspire). Indeed, STATA provides the most natural setting for post processing of WinSAAM (and AKA-Glucose) results. Data can be exchanged between WinSAAM, STATA, and AKA-Glucose via the paste buffer or as export files as shown in Table 1.

A good example illustrating the synergy between STATA and WinSAAM is the use of STATA to provide good initial estimates for S_G and G0 in conjunction with fitting the Minimal Model to IVGTT data using WinSAAM. Studies are read into STATA and, after differentiating the glucose values with respect to time (leading to G` values), robust regression is used to fit the function

Table 1. Methods for data exchange between WinSAAM, STATA, and AKA-Glucose

Exchange method	Data source	Data destination	Trasfer vehicle
1	WinSAAM Spreadsheet	STATA	Paste Buffer
2	WinSAAM Program Manager	STATA	Paste Buffer
3	STATA	WinSAAM	Paste Buffer (special)
4	AKA-Glucose	WinSAAM	WinSAAM File Format
5	WinSAAM	AKA-Glucose	WinSAAM File Format
6	WinSAAM Spreadsheet	AKA-Glucose	Tab delimited Text File
7	AKA-Glucose	STATA	Tab delimited Text File
8	STATA	AKA-GLucose	Tab delimited Text File

$$Z = G` / G \qquad\qquad 10 <= t <= 20 \text{ mins}$$

to a constant. This yields estimates of S_G and its error. Then, using this S_G, the transformed values W, where

$$W = \ln [\, G\,] - S_G.t \qquad\qquad 10 <= t <= 20 \text{ mins}$$

are also robustly regressed on a constant, yielding estimates of G0 and its error. These values are then provided as initial guesses to their Minimal Model counterparts in WinSAAM, and the final values for p_2 and S_I and the refined S_G and G0 are found.

In Figure 12, we show a preliminary analysis in which STATA was used to derive S_G and G0 estimates and final fits to the studies yielding S_I, p_2, S_G, and G0. In Figure 13, we show the STATA commands leading to the S_G and G0 estimates, along with their precisions. Whereas efforts to locate precise values for the Minimal Model parameters seemed in the past to have been plagued with identification problems, this new approach has been found to yield precise parameter estimates in 100% of cases.

Of course, data exchange for post-processing analysis is not the only need we have for data exchange. In order to create presentations, publications, lecture and demonstration material, and grant sections about our modeling work, we need to be able to export the relevant material into so-called 'productivity settings.' The WinSAAM development group has always been mindful of this need and, to this end, we have maintained consistency of our graphic and text outputs with the Office productivity setting. Indeed, this paper was completely prepared in Word and PowerPoint, attesting to the ease with which productivity enhancement is available.

CORNERSTONE 4: MODELING IN A COMMUNITY – SHOULD WE CONTINUE TO BOWL ALONE?

After introspection and soul searching, we have to honestly admit that mathematical modeling is a poor cousin of science. Even after fifty years of convincing exploitation, it

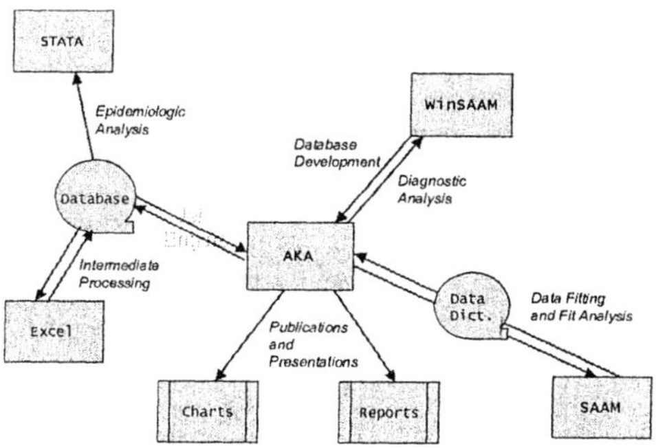

Figure 11. AKA-Glucose. Data exchanges managed by or supported by AKA-Glucose.

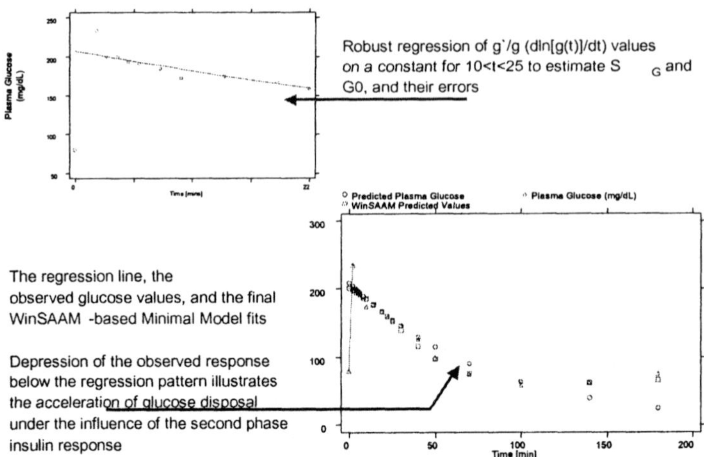

Figure 12. The analysis of IVGTT data using STATA. Shown here is the estimation of S_G and G0 values and their subsequent use to profile the first phase of glucose disposal. The upper plot shows the STATA fit to glucose observation between t=10 and 20 minutes. The lower plot shows the entire set of IVGTT glucose observation, the WinSAAM-predicted glucose profile (using the STATA S_G and G0), and the STATA fit to the indicated glucose data. Note the depression of the glucose values below the STATA predictions between t=25 and 100 minutes. This highlights response to the IVGTT insulin injection applied to the subject at t=20 minutes.

lacks formality and structure. Despite genuine efforts by journals to admit us in some formal way to their ranks, we fail to prosper. For example, the American Journal of Physiology Modeling Forum closed after about 8 years of operation. The reasons for this situation seem fairly clear: we don't freely and openly collaborate with one another, we don't promote any particular standards, and we don't really have any idea how our colleagues from other disciplines appreciate our efforts, or whether they even understand, or care, about those efforts. There is no structured educational program channeling participants to modeling, and there is no society promoting our work, championing our worth, or soliciting commitments from funding agencies.

The key point here is 'do we care?' For those who are succeeding, taking on these issues would divert them from their productive agendas. Indeed pure altruism would need to prevail. However, taking on these issues would be giving back some of the energy and excitement the journey into modeling methodology has brought.

If we do acknowledge the gravity of this problem and if we are prepared to accept the challenge, then we are well placed to deal with it. Some considerations might include:

- Establish an oversight group/committee
  - o Internationally based
  - o Funded secretariat
  - o Communications hub
- Establish an approach to software certification
  - o Specification of functions, standardized inputs and outputs

- o Layered approach to certification
- o Solving, statistics, fitting, inter-software exchange
- o Establishment of information repository of algorithms
- Establish an educational subcommittee
  - o Identify course structure and content for undergraduates/graduates
  - o Coordinate student exchange among centers (internships, practicums)
  - o Publicize the existence of educational programs
  - o Certify instructional material and texts
  - o Approve examinations and standards
  - o Solicit employment demands and opportunities
- Establish a meeting forum and annual agendas and locations
- Confirm the adequacy of existing journals for our publications
- Establish an information bulletin updating the community
- Network
  - o Promote one another's ideas and interests
  - o Collaborate and cooperate on large scale projects
  - o Exchange information and ideas
  - o Helpfully preview work efforts
  - o Visit sites of one another as opportunity permits

We can argue the point: what is at risk if we don't explore these ideas *vs.* what is at risk if we do? If we don't explore these ideas, we will almost surely remain in our current state of limbo. Young investigators will cease to express interest in our ideas, collaborators will remain in obscurity about what we do, and most importantly, no structured or disciplined basis for what we do will emerge. Of course, the risk of advancing these ideas is a potential sacrifice of one's identity and loss of momentum with one's work. The bottom line, though, is that the yield from a team is better in every respect than that from the sum of the individual players. In Putman's words (Putman, 2000), "let's abandon bowling alone."

CONCLUSIONS AND SUMMARY

In this paper, we have reflected on the historical development during the twentieth century of several major cornerstones on which the edifice of mathematical modeling in nutrition and the health sciences was built. When we consider the scope and magnitude of problems in nutrition and the health sciences that have been addressed and solved by mathematical modeling, we as a group can justifiably feel a certain amount of satisfaction. But we should not be complacent. So much more remains to be done. The increasing pace of developments in biology (e.g., the human genome project) places a whole new range of challenges before us.

We can also reflect on the mathematical basis of modeling and consider that we enter the twenty-first century with a solid foundation on which to build bigger and better models. At the same time, we have at our disposal computers of immense power. At the start of the twentieth century, no one could have foreseen where we are today. In our

```
use "stata demo_a -001", clear
        label var time "Time [mins]"
        label var glc "Plasma Glucose [mg/dl]"

dydx glc ti if ti<=30, gen(gp)
gen gd=gp/glc
rreg gd if time>10 & time<=25, nolog
        scalar define sg=_b[_cons]
        scalar sesg=_se[_cons]

gen z=ln(glc) -sg*time
rreg z if time>=10 & time<25, nolog
        scalar g0=exp(_b[_cons])
        scalar seg0=g0*_se[_cons]

gen g_pred=g0*exp(sg*time)
        label var g_pred "Predicted Plasma Glucose"

gra g_pred glc time if time<25, c(l.) s(io) ylabel border
scalar list sg sesg g0 seg0
gra g_pred glc time, c(LL) s(OT) xlabel ylabel border

merge using "winsaam demo_a -001
label var qc "WinSAAM Predicted Values"
gra g_pred glc qc time, c(LLL) s(OTS) xlabel ylabel border
```

Figure 13. Processing kinetic data using STATA. The STATA commands used to produce S_G, G0, and the predicted response displayed in Figure 12.

lifetimes, computers have been developed from lumbering behemoths with the calculating ability of an abacus to the present day machines with calculating capabilities that we are only beginning to appreciate. There is general consensus that developments in the field of computing will continue far into the twenty-first century. Therefore, we can confidently assert that developments in modeling will not be greatly limited by our present mathematical foundation or by the capabilities of computers.

In this paper, we have mainly dwelt on three cornerstones:

- the model development environment
- model dissemination and database technologies
- data exchange and post-fitting analysis

We have focused on SAAM, WinSAAM, and AKA-Glucose as illustrating these cornerstones, while acknowledging that there are many other modeling programs that also represent them. However, a building usually has four cornerstones. If one cornerstone is missing, then there may be a structural weakness in the entire building. We contend that, at the start of the twenty-first century, the edifice of modeling is missing the important fourth cornerstone: a modeling community. As modelers, we should cease to bowl alone. Now is the time to make a new beginning, to give modeling some formality, structure, and direction. We should form a community of modelers as we move into the 21st century.

CORRESPONDENCE

Please address all correspondence to:
Ray Boston
Biomathematics Unit
New Bolton Center, School of Veterinary Medicine
University of Pennsylvania
382 West Street Road
Kennett Square, PA 19348
boston@cahp2.nbc.upenn.edu

REFERENCES

Anderson, R.M., Medley, G.F., May, R.M., and Johnson, A.M., 1986, A preliminary study of the transmission dynamics of the human immunodeficiency virus (HIV), the causative agent of AIDS, *IMA J. Math. Appl. Med. Biol.* 3:229-263.

Avioli, L.V., and Berman, M., 1966, Mg^{28} kinetics in man, *J. Appl. Physiol.* 21:1688-1694.

Avioli, L.V., and Berman, M., 1968, Role of magnesium metabolism and the effects of fluoride therapy in Paget's disease of bone, *J. Clin. Endocr. Metab.* 28:700-710.

Berman, M., Hoff, E., Barandes, M., Becker, D.V., Sonenberg, M., Benua, R., and Kourtras, D.A., 1968, Iodine kinetics in man–a model, *J. Clin. Endocr. Metab.* 28:1-14.

Birge, S.J., Peck, W.A., Berman, M., and Whedon, G.D., 1969, Study of calcium absorption in man. A kinetic analysis and physiologic model, *J. Clin. Invest.* 48:1705-1713.

Boston, R.C., Greif, P.C., and Berman, M., 1981, Conversational SAAM-an interactive program for the kinetic analysis of biological systems, *Comput. Prog. Biomed.* 13:111-119.

Boston, R.C., Greif, P.C., and Berman, M., 1982, CONSAM (Conversational version of the SAAM modeling program), in: *Lipoprotein Kinetics and Modeling,* M. Berman, S.M. Grundy, and B.V. Howard, eds., Academic Press, New York.

Boston, R.C., and McNabb, T., 1990, *The Analysis of SAAM's Optimizer: A Critical Appraisal with Reference to Optimizers of Other Well-known Data Analysis Systems*, Report to the Resource Facility for Kinetic Analysis, Seattle.

Boston, R.C., McNabb, T., Greif, P.C., and Zech, L.A., 1996, Essential numerical supports for kinetic modeling software: linear integrators, in: *Mathematical Modeling in Experimental Nutrition,* S.P. Coburn and D.W. Townsend, Academic Press, New York.

Boston R.C., and Wastney M.E., 2000, New features for population studies analysis in the WinSAAM modeling software, FASEB J. 14: A441.

Brauer, E.W., Kopf, A.W., Witten, V.H., Berman, M., and Cave, V.G., 1960, Radioactive phosporous in the in vivo diagnosis of melanoma of the skin, *J. Am. Med. Assoc.* 172: 139/1753-144/1758.

Foster, D.M., Aamodt, R.L., Henkin, R.I., and Berman, M., 1979, Zinc metabolism in humans: a kinetic model, *Am. J. Physiol.* 237:R340-R349.

France, J., and Thornley, J.H.M., 1984, *Mathematical Models in Agriculture* (chapter 10), Butterworth Press, London.

Goebel, R., Berman, M., and Foster, D., 1982, Mathematical model for the distribution of isotopic carbon atoms through the tricarboxylic acid cycle, *Fed. Proc,* 41:96-103.

Greif, P.C., Wastney, M.E., Linares, O., and Boston, R.C., 1998, Balancing needs, efficiency, and functionality in the provision of modeling software: a perspective of the NIH WinSAAM project, in: *Mathematical Modeling in Experimental Nutrition,* A.J. Clifford and H.G. Müller, eds., Plenum Press, New York.

Hall, S., Goebel, R., Barnes, I., Hetenyi, G., and Berman, M., 1977, The turnover and conversion to glucose of alanine in new born and grown dogs, *Fed. Proc.* 36:239-244.

Insel, P.A., Kramer, J.J., Sherwin, R.S., Liljenquist, J.E., Tobin, J.D., Andres, R., and Berman, M., 1974, Modeling insulin-glucose system in man, *Fed. Proc.* 34:311-314.

Leme, P.R., Creven, P.J., Allen, L.M., and Berman, M., 1975, Kinetic model for the disposition and metabolism of moderate and high dose methotrexate in man, *Canc. Chemo. Rep.* 59:811-817.

Lewallen, C.G., Berman, M., and Rall, J.E., 1959, Studies of iodoalbumin metabolism. I. A mathematical approach to the kinetics, *J. Clin. Invest.* 38:66-87.

Lyne, A., Boston, R.C., Pettigrew, K., and Zech L., 1992, EMSA-A SAAM service for the estimation of population parameters based on model fits to identically replicated experiments, *Comp. Meth. Prog. Biomed.* 38:117-151.

Neer, R., Berman, M., Fisher, L., and Rosenberg, L., 1967, Multicompartmental analysis of calcium kinetics in normal adult males, *J. Clin. Invest.* 46:1364-1379.Pacini, G., and Bergman, R.N., 1986, MINMOD: a computer program to calculate insulin sensitivity and pancreatic responsivity from the frequently sampled intravenous glucose tolerance test, *Comp. Meth. Prog. Biomed.* 23:113-122.

Phillips, D.R., Moate, P. J., and Boston, R.C., 1994, A modelling procedure for the analysis of dynamic drug-DNA interactions probed during active transcription of the DNA, *Anti-Canc. Drug Design* 9:209-219.

Putnam, R.D., 2000, *Bowling Alone: The Collapse and Revival of American Community*, Simon and Schuster, New York.

Schwartz, C.C., Berman, M., Halloran, L.G., Swell, L., and Vlahcevic, Z.R., 1982, Cholesterol disposal in man: special role of HDL free cholesterol, in: *Lipoprotein Kinetics and Modeling,* M. Berman, S.M. Grundy, and B.V. Howard, eds., Academic Press, New York.

Sherwin, R.S., Kramer, K.J., Tobin, J.D., Insel, P.A., Liljenquist, J.E., Berman, M., and Andres, R.A., 1974, A model of the kinetics of insulin in man, *J. Clin. Invest.* 53:1481-1492.

Temple, R., Berman, M., Robbins, J., and Wolff, J., 1972, The use of lithium in the treatment of thyrotoxicosis, *J. Clin. Invest.* 51:2746-2756.

Wastney, M.E., Young, D.C., Andretta, D.F., Blumenthal, J., Hylton, J., Canolty, N., Collins, J.C., and Boston, R.C., 1998, Distributing working versions of published mathematical models for biological fields via the Internet, in: *Mathematical Modeling in Experimental Nutrition,* A.J. Clifford and H.G. Müller, eds., Plenum Press, New York.

PART II
DEVELOPMENT AND USE
OF MATHEMATICAL MODELS

ASPECTS OF EFFECTIVE MODELING

Judah Rosenblatt[*]

INTRODUCTION

Too many people, including a substantial number of scientists and teachers, view science as the search for truth - in the form of *the immutable laws of nature*. For proof of this assertion, you need only notice some of the more prominent controversies: creationism *vs.* evolution, arguments over the big bang theory, and Dr. Atkins *vs.* Dr. Dean Ornish. One of the older such controversies, which in many ways is as relevant now as ever, is the controversy between Galileo and the Catholic Church. It is sometimes summarized as follows: "In the middle ages, people thought the sun revolved around the earth. Now *we know* that the earth *really* revolves around the sun." If this assertion is what most people currently believe, it would indicate that little has been learned about the real nature of science in the last 500 years. *Science is concerned with describing variables* (which we will consider to be the same as the measurements used to observe them) and especially with describing the relations between measurements for the purposes of prediction and (under appropriate circumstances) control[†]. This may sound somewhat less dramatic than searching for truth, but if you examine just about any scientific investigation, you'll see that the assertion above captures the essence of science.

For instance, in searching for a healthy, safe, weight-loss diet, we care about the amounts of various foods ingested at various times, health indicators (such as blood pressure, electrolyte values, etc.), and body weight. All of these can be classified as measurements, and our interest is to use some of them which we can specify (those listed first) to indirectly control the remaining ones. We will call the quantities that we can control directly the *independent variables;* the quantities to be indirectly controlled by the independent variables will be called the *dependent variables.*

Controlling or curing any disease can be put in the same framework as in the example on diets. Here the independent variables would include medications, diet, surgical and other procedures as well as exercise, etc. The dependent variables would include disease

---

[*] Judah Rosenblatt, University of Texas Medical Branch, Galveston, TX 77555.

[†] Here we will restrict our discussion to the case of control, noting that prediction is important in situations where we want to be prepared to act on our predictions - for example, where hospital supplies must be ordered in response to expected demand.

signs and symptoms. Problems in chemistry, physics, and engineering only differ from those just discussed in the overall choice of independent and dependent variables.

In the framework of description of measurements, Galileo's controversy with the church could be meaningfully phrased as a question of which of the following descriptions of relative planetary motion is preferable: one in which the center of the coordinate system is based on earth or one where the sun is the central reference point. Either description can be used, generating essentially the same answers. But computations are orders of magnitude harder if earth is the chosen center. So, from a practical viewpoint the choice is obvious: nothing is to be gained, save needless complexity, by making earth the center. This conclusion has been abstracted in *Occam's Razor*[*], which may be stated in the form

> It's unreasonable to use a complicated model if a simpler one yields essentially the same results.

Most scientific and engineering questions are concerned with choosing adequate models. But here one has to be very careful, because the adequacy of the model we choose for some system may very well depend on the use planned for the model. So, for example, if we're interested mainly in how long some drug remains in the bloodstream after a single intravenous bolus, a one-compartment model yielding simple exponential decay may suffice. But if this drug was radioactive and we wanted to get an idea of the total dose of radioactivity the body is subjected to, such a model is probably grossly inadequate. In that case, a multicompartment model, with one of the compartments including fatty tissue into which the drug may enter much faster than it washes out, seems more suitable.

By their very nature, just about all measurements are influenced by a variety of factors. This means that not only is any given model likely to be imperfect, but that any such model only reasonably applies to some limited range of the measurements it purports to describe. Some of the greatest misuses of models consist of extrapolating far beyond their ranges of adequacy. Using a single compartment model, as in the example just mentioned, can result in underestimating a body's total radiation exposure by a factor of many hundreds, even though the one-compartment model's absolute error in estimating the amount of drug in the body at any time may tend to be quite small. This type of extrapolation abounds - from drawing unwarranted conclusions based on the results of animal experimentation (recall the Delaney clause) to committing our resources in the far future based on current budget surplus projections which change drastically every year. It occurs as well in trusting ultrasmall p-values produced by most statistical packages based on the acceptance of an exact Gaussian distribution of error. So we should always try to recognize the limitations of any model we use in order to keep our conclusions trustworthy.

---

[*] Occam's Razor is sometimes misstated in the form: if two models give essentially the same results, the true one is the simpler of the two. Convenience and adequacy, not truth, are the relevant issues.

COMMONLY ENCOUNTERED ASSUMPTIONS AND THEIR JUSTIFICATIONS

Given some questions concerning a system of interest, how do we go about building a useful model? One approach we shouldn't use is the kind I've often faced from clients who ask me to help prove the result they want. The other side of this coin, being asked to *analyze data,* with no clue as to the problem being addressed, is equally unreasonable. Any sensible approach should include

- a reasonably specific system, with associated measurements
- some questions about these measurements or a stated objective such as determining the accuracy with which dependent variables can be controlled
- some plausible assumptions about the measurements

In a nutrition context, the system is frequently the human body. The associated measurements could range from

- caloric intake (independent variable) and weight (dependent variable) at various times
- protein, carbohydrate, and fat intake (independent variables) with weight at various times (the dependent variable)
- measurements of various blood lipid levels (dependent variables) under strictly controlled diets or controlled infusion of specified nutrients (independent variables)

With such measurements, we might want to determine respectively what daily caloric intake would produce some desired steady weight; or what mixtures of protein, carbohydrate and fat intake would yield some particular weight without unhealthy side effects; or the relation between diets or infusions of nutrients and various lipid levels.

Next comes the critical issue of the assumptions to be made. First, we should be aware that, in the initial stages of investigating a system, the assumptions we make almost never specify a single model; rather, they usually specify a collection of candidates for a model (i.e., the form of the model or, equivalently, a *class* of models).

MODELS BASED ON FORMULAS

The kind of model involving one independent and one dependent variable most frequently encountered in elementary investigations is a straight line. It is a *class* of models (i.e., the form of a model) because initially we don't know which line or lines we'll narrow down to from all those being considered.

Probably the main reason for the interest in straight line models is that most relations we encounter between a single independent and a single dependent variable look pretty straight over a short enough stretch, as seen in Figure 1. However, *actual data* might not really look very straight even over short stretches, as seen in Figure 2, but the curve you

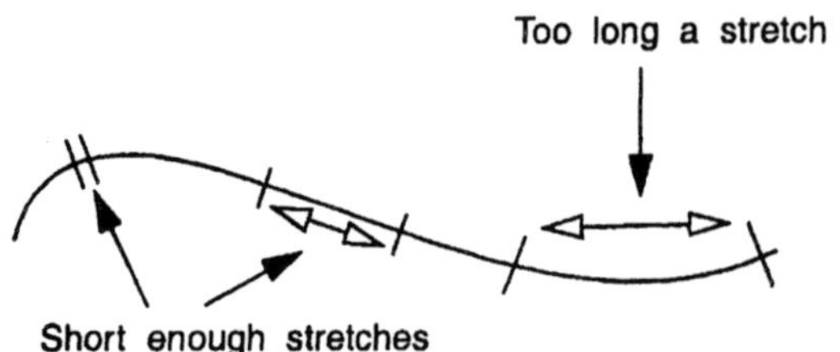

Figure 1. Relations between variables look straight over a short stretch.

would *eyeball* to try to fit such data probably would. Reasons why raw data don't usually tend to lie exactly on slowly varying smooth looking curves will be discussed shortly.

If the data don't look straight enough over the range of interest, one might try for a better-fitting second degree polynomial. An even better fit might be had if we try an n-th degree polynomial for some $n > 2$. This can be justified on theoretical grounds, because it can be shown that, given any function f, continuous at each point of an interval which includes its end points, there is a polynomial whose graph is within any specified nonzero vertical distance from f, as illustrated in Figure 3. That is, there is some polynomial whose graph lies within the shaded area enclosing the graph of f. This sounds appealing, but actually, high degree polynomials usually do a poor job of approximating data; in fact, so long as no two data values have the same first coordinate, one can find a polynomial which fits the data perfectly but oscillates wildly between data values, as illustrated in Figure 4.

Another class of curves used to fit data consists of *splines,* which can be *patched-together* straight line segments or smoothly *patched-together* second or third-degree polynomials, as illustrated in Figure 5. Splines are especially useful when there is a change in mechanism in the system being examined. (Polynomials have a hard time approximating such changes). Whatever the class of curves (or *surfaces,* if you have two independent variables, or *hypersurfaces* with more than two), unless you're just eyeballing a fit to the data, you need some sort of mathematical criterion to choose the *best* fit from the class of candidates you're using. Here Occam's Razor would indicate use of the most convenient criterion yielding sensible answers. The simplest mathematical criterion which often does this is *least squares.* If your data points were

Figure 2. Data points do not look straight.

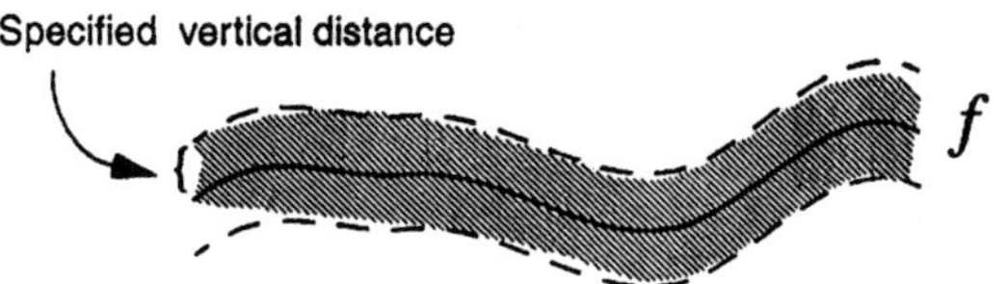

Figure 3. The polynomial f.

$$(x_1, y_1, z_1), \ (x_2, y_2, z_2), \ \ldots \ (x_n, y_n, z_n)$$

and you are trying to fit a plane whose formula is

$$z = m + ax + by \tag{1}$$

(i.e., you're trying to choose values m, a, and b so that, in some sense for *all i*, $m + ax_i + by_i$ is simultaneously close to all z_i), the *least squares choice* for m, a and b is the triple *(m, a, b)* minimizing the expression

$$\sum_{i=1}^{n} (z_i - [m + ax_i + by_i])^2 \tag{2}$$

Least squares is easy to implement mathematically when the unknowns occur as simple weights[*], and there are many computer packages to do this (MLAB, WinSAAM, S-PLUS, Mathematica etc.). However, it can be seriously deficient if even a few data points are *far out in left field (outliers)*. This is shown in Figure 6. There is always the temptation to toss out such outliers if they have too great an effect on the fit (and many computer packages tag these outliers), but such action should not be taken unless there is real

Figure 4. n-th degree polynomial going through $n+1$ points.

---

[*] Least squares can also be applied to cases where the unknowns do not occur as simple weights, such as they do in Eq. (2).

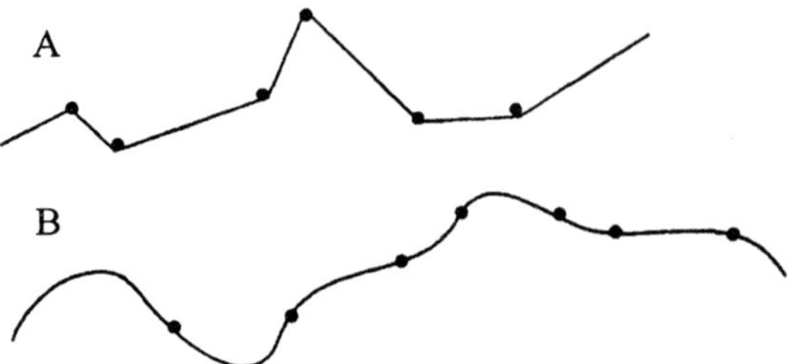

Figure 5. Example of splines. (A) Patched-together straight-line segments. (B) Patched-together 2nd or 3rd degree polynomials.

justification (e.g., proven laboratory contamination, or a value that is physiologically impossible, or a proved misplaced decimal point, etc.).

The curve/hypersurface fitting we've been discussing so far can be thought of as *generic empirical*. But in a number of important cases, the class of functions used to fit certain types of data is based on reliable theory. A most frequently encountered case is that of data (t_1, x_1) $(t_2, x_2), ..., (t_n, x_n)$ which arise in enzyme kinetics. Here the t_i are time points, and the x_i may be the corresponding concentration in the bloodstream of some chemical not produced in the body which was infused or ingested earlier (a tracer). In the simplest cases, the theory of compartmental modeling (see Jacquez, 1985) based on the assumption of certain diffusion properties of such chemicals leads to the conclusion that the concentration x_t at time t is of the form

$$x_t = \sum_{j=1}^{k} a_j e^{-\lambda_j t} \tag{3}$$

where the $\lambda_j \geq 0$, a_j are constants to be estimated from the observed data (t_i, x_i), and k is the number of compartments that the chemical can access. In most experimental situations, there does not exist any choice for the λ_j, a_j for which the observed data

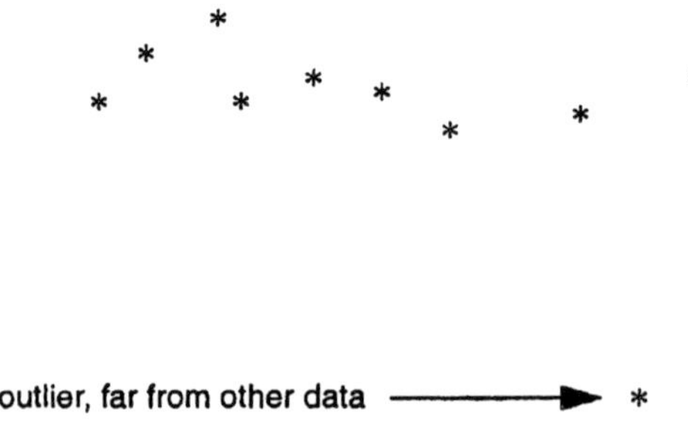

Figure 6. An example of an outlying data point.

perfectly satisfies the equation

$$x_i = \sum_{j=1}^{k} a_j e^{-\lambda_j t_i} \qquad (4)$$

for all $i = 1,2,...,n$. The usual explanation for this discrepancy is *experimental error* - measuring devices aren't perfect. Although this may account for a good deal of the discrepancy, it seems to me that much of it should be attributed to the influence of other factors which are not taken into account by the model but which may well affect the concentrations significantly. In this model, only the assumed constant fractional transfer coefficients and the initial concentrations in the various compartments determine the later concentrations. But it's conceivable that such factors as changes in temperature could alter the rate of flow of this chemical between compartments and that some fluctuating chemical reactions not considered could alter the concentration of the given chemical. Both of these could contribute to the discrepancies between the simple fitted compartmental model and the observed data. In such situations, the most common approach is the addition of a statistical term to the deterministic formula model, writing the expanded model as

$$x_t = \sum_{j=1}^{k} a_j e^{-\lambda_j t} + e_t \qquad (5)$$

where the indicated summation is the deterministic part and the statistical term e_t is assumed unknown but possesses some statistical (long run average) properties. Frequently, it is assumed that the factors not being explicitly taken into account tend to balance out, leading to the conclusion that averages

$$\frac{1}{m} \sum_{s=1}^{m} e_{t_s}$$

become small when m becomes large, and that the mean squared deviation

$$\frac{1}{m} \sum_{s=1}^{m} e_{t_s}^2$$

approaches some constant σ^2 for large m.

Going even further, it is common to assume that the $e_{t_1},...,e_{t_n}$ are statistically independent (knowing the values of any of them doesn't alter your description of the remaining ones) with a Gaussian distribution[*]. This assumption has some empirical and

---

[*] This means that the probabilities are computed by finding appropriate areas under the Gaussian density whose formula is

some theoretical justification but is far from being universally applicable. More sophisticated and flexible assumptions, going under the labels of *generalized linear modeling* and *nonparametric methods,* are available in the search for more suitable descriptions. See Venables and Ripley (1999).

While least squares can be used for fitting the deterministic portion of the above statistical models, an even more suitable approach, custom fitted to the statistical interpretation, is the method of *maximum likelihood* (which, however, usually chews up considerably more computer resources). See Venables and Ripley (1999, page 261) for examples of how maximum likelihood can be implemented in the S-PLUS statistical package using their *nls* (nonlinear least squares) program.

MODELS BASED ON FIRST PRINCIPLES

When we don't have a very basic understanding of a system, we refer to it as a *black box,* meaning that all we can do to characterize it is to measure its inputs (independent variables) and corresponding outputs (dependent variables). The fitting procedures referred to in the previous section were mainly applied to black boxes. Once a system has been studied for some time, it can often be better described by a *local description* (*i.e.,* one which describes what goes on in a typical small portion of the system). The reason for restricting to a small portion (of space, time, or any region of small variation of the independent variables) is that, in small portions, the corresponding changes in the dependent variable are likely to be close to linear (recall the discussion of short stretches of most curves) and anyway, it's probably easier to get a handle on what's happening in such small regions.

For example, in the simplest compartmental modeling, it is not unreasonable to assume that the rate of *flow*[*] $f_{ij}(t)$ from compartment j to compartment i near time t is well approximated by a fixed multiple m_{ij} of the amount $a_j(t)$ of chemical in compartment j at time t. Thus, if the amount of chemical in compartment j is doubled, the flow rate out of compartment j is doubled. This is reasonable so long as the molecules of this chemical don't get in each other's way, clogging up the works. In more refined compartmental modeling, to characterize a clogging effect, m_{ij} might be chosen to decrease as $a_j(t)$ increases.

A simple two compartment model of the bloodstream/fatty tissue aspects of the body is graphically presented in Figure 7. Here the model is characterized by the three transfer coefficients m_{21}, m_{12}, and m_{01}. If we know these transfer coefficients and the amount of the chemical being described by this model at some given time, we can calculate from this local description how much of the chemical will be in each compartment at succeeding times. We do this as follows. Suppose we start off knowing the chemical amounts $a_1(0)$,

---

$$\frac{1}{\sqrt{2\pi\sigma^2}}\,e^{-x^2/(2\sigma^2)}$$

[*] The rate of flow out of one compartment into another at time t is approximately the ratio

$$\frac{\text{the amount of chemical exiting from this compartment in a short time period which includes time t}}{\text{the duration of this time period}}$$

$a_2(0)$ at time 0. From the definition of flow rate, the change $a_1(\Delta t) - a_1(0)$ may be written as

$$a_1(\Delta t) - a_1(0) \cong -m_{21}\Delta t a_1(0) + m_{12}\Delta t a_2(0) - m_{01}\Delta t a_1(0) \qquad (6)$$

amount flowing amount flowing amount lost
from cmpt 1 from cmpt2
to cmpt 2 to cmpt 1

A similar approximation can be written for $a_2(\Delta t) - a_2(0)$. From these, we can compute approximately $a_1(\Delta t), a_2(\Delta t)$. This process may be repeated to then find $a_1(2\Delta t)$, $a_2(2\Delta t)$, $a_1(3\Delta t)$, $a_2(3\Delta t)$, etc. Generally, we write the following approximations for small Δt.

$$\frac{a_1(t + \Delta t) - a_1(t)}{\Delta t} \cong -m_{21}a_1(t) + m_{12}a_2(t) - m_{01}a_1(t) \qquad (7)$$

$$\frac{a_2(t + \Delta t) - a_2(t)}{\Delta t} \cong m_{21}a_1(t) + m_{12}a_2(t) \qquad (8)$$

which we abbreviate as the differential equations

$$a_1'(t) = -m_{21}a_1(t) + m_{12}a_2(t) - m_{01}a_1(t) \qquad (9)$$

$$a_2'(t) = m_{21}a_1(t) + m_{12}a_2(t) \qquad (10)$$

The basic local compartmental model descriptions allow us to compute a numerical solution for the amount of chemical in each compartment at later times, if we know the transfer coefficients m_{ij} and values of the amounts of chemical in each compartment at some specified initial time (called time 0 for convenience). If we don't know the transfer coefficients but have measured data $(t_k, a_j(t_k))$ for all j and some set of time values t_k, we can search for the transfer coefficients which yield the least squares (or some other) best

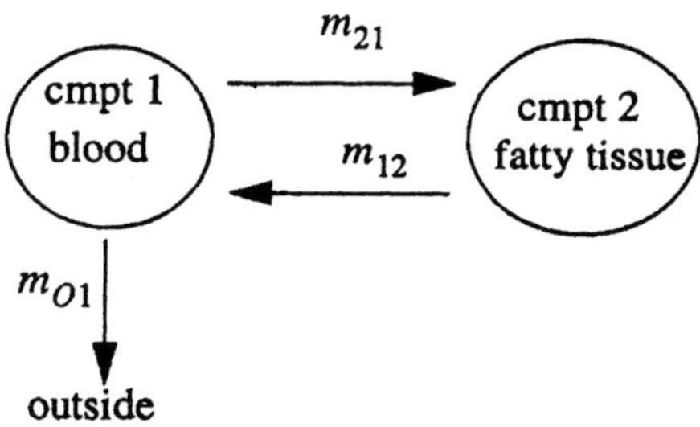

Figure 7. A simple two-compartment model of blood and fatty tissue.

fit, by starting off with an initial guess for the transfer coefficients and systematically trying out changes from the initial guess to obtain an improvement. We continue until no further improvement seems possible. What's worth realizing is that fit criteria such as least squares or maximum likelihood are not restricted to functions provided by formulas. They can be applied to choose the best fit from a class of functions computable from an algorithm of the type just presented. So if the class of functions from which you want to choose a best fit is supplied by some sort of local description such as differential Eqs. (9) and (10), the *best fitting* coefficients can be hunted for as long as a good approximation to the function corresponding to any guess can be calculated. This reasoning applies even if the transfer coefficients are allowed to come from some not-too-complicated class of functions of time or chemical amounts, such as

$$m_{12}(a_2) = \frac{b_0}{1 + b_1 a_2} \tag{11}$$

where b_0 and b_i are unknown constants and $a_2(t)$ is the amount of chemical in compartment 2 at time t. The constants b_0 and b_1 would need to be estimated from the data in most cases. The case $b_1 = 0$ leads to the case of constant m_{12}. Equation (11) can be used to represent clogging at high concentrations of the chemical and is referred to as *Michaelis-Menten kinetics*. Although it doesn't usually yield simple formulas, it is easy to handle numerically and doesn't present any conceptual difficulties.

An overview of the types of mathematics used to carry out the fitting process can be found in Rosenblatt (1998) and was presented at the 6th Conference on Mathematical Modeling in Nutrition in 1997. A much more thorough development is available in Rosenblatt and Bell (1999). The MLAB program (Civilized Software, http:// www. civilized.com) is designed to do such fitting when the class of functions is specified by differential equations; MLAB supplies estimates of the statistical characteristics of the unknown parameters. More information about the functionality of MLAB can be found in Knott (1979). The WinSAAM program (http://www.winsaam.com) and the SAAMII program (http://www.saam.com) are specially designed to do fitting of compartmental models and also supply such statistical estimates.

First principles models based on local descriptions (ordinary and partial differential equations) can try to take account of a great variety of factors beyond the compartmental kinetic models we have looked at. Such factors include

- the effects of different energy sources (e.g., heat and vibration)
- statistical responses to inputs to a system (e.g., mutations due to ultra violet radiation)
- responses to statistically characterized inputs (e.g., responses to statistically described arrivals to some processing system - such as an airport's runways, a computer system's central processor, or stimuli to the brain)

We are only beginning to scratch the surface of the applications of such models to medical problems. It remains to be seen how much progress we will be able to make, due to the formidable complexity of the models which have been built so far and the difficulty of obtaining accurate measures of the independent variables which are needed.

PITFALLS AND GUIDELINES

Applicability of Research Studies

There are two natural tendencies in modeling which need to be balanced against each other. In the desire to adequately model any real situation, we want to take account of all important factors. This can easily lead to a model so complex as to be intractable. To avoid this difficulty, it is all too easy to fail to pay attention to some very significant factors. The careful scientist will try to hold constant those factors not being explicitly taken into account in the model, to keep down the so-called *unexplained* variability. But with this care, it should be recognized that nothing is learned about the effect of the independent variables held constant. This means than any results determined from such data may apply to only a very restricted population.

It used to be common practice to make observations (such as those on treatment of high blood pressure) on some middle-aged group of males and then to extrapolate to the entire population. Experience has shown that these extrapolations (to females, to the elderly, etc.) are not generally justified. But a trial on a group that is representative of the entire population gets more expensive the more representative it is.

There is no fully satisfactory solution to this dilemma. The best that can be hoped is that the most important independent variables have been explicitly taken into account by the model, so that the unaccounted for variables have relatively little effect. If this is the case, the results should hold reasonably well for the population of interest.

Basic Internal Standard Adjustment

Sometimes we can use measurements on an artificially introduced quantity to adjust for unwanted, uncontrollable variability in other quantities. This is referred to as *compensation* or *adjustment* or *correction via an internal standard*.

For example, in studying the effect of ultraviolet radiation on genetic material, one of my colleagues attempted to fill each of a number of wells to be exposed to this radiation with the same amount of a given homogeneous mixture containing some genetic material. Pipetting not being amenable to great accuracy, the amount of loaded mixture tended to vary so much from well to well that using only the measured amounts of undamaged DNA from the various wells yielded very poor estimates of the proportion of DNA undamaged by various doses of radiation. He recognized the problem and solved it by himself, obtaining much better results.

What was needed was an accurate measure of how much mixture was actually loaded into each well[*]. One way to obtain this information is to add to the mixture a chemical

---

[*] If, additionally, we wanted to compensate for variability in the radiation level to which the wells are exposed, we could also add a chemical whose known response per unit amount is proportional to the radiation level

(the standard) which emits an accurately measurable signal whose level S_u per unit volume of standard is precisely known. By adding a known amount of this standard to the entire mixture before it was subdivided into wells, the relative error in the amount added can be kept much smaller than the relative error in loading the wells. The observed signal measurement on this standard associated with a given well can be used to correct its raw measured amount of undamaged DNA to what it would have been had the desired (nominal) amount of mixture been loaded. Here are a few details. Let V_0 be the volume of the original mixture prior to adding the standard; let V_S be the volume of the standard to be added; let V_{S+0} be the volume of the combined mixture plus standard (possibly different from $V_0 + V_S$); let V_N be the nominal volume of mixture plus standard per well; let v_a be the actual volume of mixture plus standard for a typical well; let s_a be the actual signal measured for this well; and let s_n be the nominal (desired) signal for this well. Recall that S_u is the (assumed known) signal level per unit volume of pure standard. Note that s_u, the signal level per unit volume of standard plus mixture, is given by

$$s_u = \frac{V_s}{V_{s+0}} S_u \tag{12}$$

Let G_u be the number of DNA strands per unit volume of the original mixture (without the standard) which may all be assumed undamaged (because any original damaged ones are not of interest and would not be amplified by the polymerase chain reaction that is used). Then g_u, the number of DNA strands per unit volume of mixture with added standard, is given by

$$g_u = \frac{V_0}{V_{s+0}} G_u \tag{13}$$

(This equation is only given so that one can determine the nominal number of DNA strands to be loaded per well.)

Let U_a be the raw (uncorrected) measurement of the actual amount of undamaged DNA measured on a typical well after exposure to some amount of ultraviolet radiation, and let U_c be the corresponding corrected amount of DNA. We see that

$$\frac{U_c}{U_a} = \frac{v_n}{v_a} \tag{14}$$

so that

$$U_c = \frac{v_n}{v_a} U_a \tag{15}$$

We also have

$$\frac{v_n}{v_a} = \frac{s_n}{s_a} \tag{16}$$

---

and to the volume. We eliminate the dependence on volume by correcting for volume variability with the first standard. Then use the second standard to correct for exposure variability to the radiation.

But

$$s_n = v_n s_u \tag{17}$$

Substituting Eq. (17) into Eq. (16) and substituting this latter into Eq. (15) yields

$$U_c = \frac{v_n s_u}{s_a} U_a \tag{18}$$

or, with the aid of Eq. (12),

$$U_c = \frac{v_n}{s_a} \frac{V_s}{V_{s+0}} S_u U_a \tag{19}$$

which shows how to obtain the corrected amount of undamaged DNA after exposure Uc in any well. Here v_n is the nominal volume of mixture + standard to be put in the well, V_S is the volume of the standard to be added to the original mixture, V_{S+0} is the combined volume of original mixture + standard, s_a is the actual measured signal from the standard in this well, S_u is the accurately known signal level per unit volume of the original standard (if this is radioactive, it could also be per unit time, so that longer observation increases accuracy), and U_a is the raw (uncorrected) measure of the amount of undamaged DNA in the well.

COPING WITH UNWELCOME DISCREPANCIES: TWO EXAMPLES

Determining the DNA Damage Resulting from Ultraviolet Radiation

A mixture of DNA strands is to be uniformly loaded (to the extent this is possible) into $n*k$ wells. All wells are to be exposed to a uniform ultraviolet radiation field: the i-th set of k wells for a time period of duration t_i minutes, $0 = t_0 < t_1 < .... < t_{n-1}$. The object is to describe the relationship between duration of exposure and relative amount of DNA damage. But under the given experimental conditions, the undamaged DNA can be amplified by a known factor (using polymerase chain reactions) and then the final product quantified reasonably accurately while damaged DNA will not be amplified by this process. Knowing the amplification factor and the amount of the final product then yields the amount of *undamaged* DNA that was initially present. From this and the known total amount of initial DNA, we can find the relative amount of damaged DNA. So it is assumed that we can obtain k independent measurements of the amount of undamaged DNA remaining in wells which have been exposed to ultraviolet radiation for the times t_i, $i = 0, 1, ..., n-1$. These measurements would be reasonably accurate if an internal standard is used (see previous section).

If we denote the j-th measurement made at time t_i by M_{ij}, the commonly used model to describe the behavior of these measurements is to represent M_{ij} as cX_{ij}, where the X_{ij} are statistically independent Poisson random variables having mathematical expectations e^{-rt_i}, with r and c to be estimated from the data. We shall shortly see the origin of this model, noting its theoretical and practical shortcomings. In fact, we'll see how unsuitable it is for the problem at hand and will develop a preferable model.

To start, examine one strand of DNA, and assume (due to the nature of the ultraviolet radiation) that, in each short period of time of duration Δt, the probability that an initially undamaged strand suffers radiation damage is (with small relative error) $r\Delta t$. We also assume that, once a strand is damaged, it remains damaged under the conditions of the experiment (i.e., no repair mechanism is operating). Then, using the rules governing conditional probability, we find the probability

$$P\{\text{no damage up to time } j\,\Delta t\} \cong (1 - r\Delta t)^j \tag{20}$$

Now pick some fixed time $t = J\,\Delta t$ and note that

$$(1 - r\Delta t)^J = (1 - r\Delta t)^{t/\Delta t} \tag{21}$$

or, letting n = $1/(\Delta t)$,

$$(1 - r\Delta t)^J = \left(1 - \frac{r}{n}\right)^{nt} \tag{22}$$

Letting Δt become small, n becomes large. We next determine how the above expression behaves. Using the properties of the exponential and logarithm functions and the mean value theorem [see Rosenblatt and Bell (1999)], we find

$$\left(1 - \frac{r}{n}\right) = \exp\left(\ln\left(\left(1 - \frac{r}{n}\right)^{nt}\right)\right) = \exp\left(nt \ln\left(1 - \frac{r}{n}\right)\right) = \exp\left(-rt\,\frac{1}{1 - \vartheta\frac{r}{n}}\right) \tag{23}$$

where $0 \le \vartheta \le 1$, so that $\left(1 - \dfrac{r}{n}\right)^{nt}$ gets close to e^{-rt} as n grows large. Thus, we expect that

$$P\{\text{no damage up to time J } \Delta t\} = P\{\text{no damage up to time } t\} = \exp(-rt) \tag{24}$$

If we assume that the DNA strands are indistinguishable (i.e., that they all behave alike), then the number of undamaged DNA strands would have a binomial distribution $B(n,p)$ with n being the initial number of DNA strands and $p = e^{-rt}$. See Blum and Rosenblatt (1972). This distribution has mathematical expectation ne^{-rt} and standard deviation

$$\sqrt{ne^{-rt}(1 - e^{-rt})}$$

The commonly assumed Poisson distribution has mathematical expectation ne^{-rt} and standard deviation

$$\sqrt{ne^{-rt}}$$

When t is small ($<< 1/r$), the binomial distribution's standard deviation is considerably smaller than that of the Poisson. At $t = 0$, the binomial distribution has standard deviation 0 (which is reasonable if we knew the initial number of DNA strands perfectly) while the Poisson distribution has standard deviation $\sqrt{n}$. For large t (e.g., $t > 3/r$), the standard deviations are, for all practical purposes, equal.

All of this is largely academic for the data we were considering, because the observed sample standard deviations at the t_i of the data were considerably larger than is predicted by either model; see McCarthy, Rosenblatt and Lloyd (1996). (The fact that, rather than the number of undamaged DNA strands, an unknown multiple of this number is measured, does not affect the validity of this claim.) The explanation is that the errors in our measuring methods swamped those induced by the randomness of the combination of bombarding ultraviolet radiation and placement of the DNA. These errors seemed to have the same statistical characteristics at all of the times at which data were observed.

The next difficulty was that a single exponential didn't seem to fit the data that well. The best fitted single exponential had a sample standard deviation over twenty percent bigger than a curve which went through the mean values at each of the t_i. So what type of curve should be used to fit the data? Adding a constant to the single exponential was suggested, but it had no apparent physical meaning (because such a function would not approach 0 for large times, while the number of undamaged DNA strands must approach 0 as time grows large). Next, a weighted sum of exponentials

$$a_1 e^{-r_1 t} + a_2 e^{-r_2 t}$$

was considered, where a_1, a_2, r_1, and r_2 are to be chosen to provide the best least squares fit. The S-PLUS nls (nonlinear least squares) program was used for this purpose. It showed that the best fit possible could be just about achieved by a suitable weighted sum of two exponentials. But what physiological meaning might this have? We concluded that, far from consisting of only one type of DNA strand, there could be two types (maybe with the same genetic structure, but, possibly due to geometrical differences, sensitive at two different locations or with different sensitivities at the same location). It should be observed that, for this two-exponential model, for $r_1 > r_2$, the r_1 strands get almost completely wiped out before the r_2 strands. Note how this would imply violation of the original model assumption that the probability of damage to an undamaged strand in the interval from t to $t + \Delta t$ is always $r\Delta t$ independent of t. Here, if for large t a strand is undamaged, chances are it's an r_2 type of strand. Presumably the microbiologists will investigate these results further.

Note how similar this problem is to determining whether a two compartment model is a better fit than a one compartment model for kinetic data.

Separating a Mixture of Normal and Cancerous Cells

One approach to treating certain types of blood cancer is to examine, one by one, each blood cell of the type of interest and toss out those deemed to be cancerous. The task is clearly hopeless using people examining blood cells under microscopes. But machines can now be built to make a sizeable number (1 to 15) of laser measurements on a sequence of cells each enclosed in its own droplet, speeding by in a stream of many thousands of droplets per second. (We're hoping to get this stream up to speeds near a million droplets per second.)

If we had a reasonable scheme for rapidly deciding whether or not each cell in the stream is cancerous, a droplet containing a cell classified as cancerous could be given an electric charge on the fly, and then, if all the droplets entered a magnetic field, only the cancerous ones would be deflected. This would purge the blood of the cancerous cells leaving the normal cells to be re-infused back to the individual from whom they were taken. The question for the modeler is how to use the measurements for classification of cells as cancerous or normal. One technique, called *linear discriminant analysis,* uses a weighted sum *WS* of the measurements, normalized by an estimate V of the variability (standard deviation of this sum) for this classification. The theory assumes multivariate normal distributions with common covariance matrix for both normal and cancerous cell measurements. Based on these assumptions, the theory is able to come up with a value c such that a cell is classified as cancerous if and only if $WS/V > c$. Under the assumptions above, c can be chosen to obtain the most acceptable compromise for cell classification: one that, for a fixed probability p of misclassifying a normal cell, maximizes the probability of correctly classifying a cancerous cell. (Presumably p is chosen so that the individual will have sufficiently many normal cells to stay alive.)

This procedure runs into trouble because the assumed distributions of the two cell types bears little relation to reality. Nonetheless, in many cases, it turns out that, although it may not be optimal, the classifier *WS/W* seems a reasonable one for the given purpose. However, the critical value c that is calculated from the given assumptions is way off the mark. Unfortunately, even if we had a good idea of the statistical behavior (joint distributions) of the measurements, both for normal and cancerous cells, theoretically or numerically using standard deterministic methods, calculating the two possible distributions of the quantity *WS/V* is usually a hopeless task. (Only under the unworkable assumptions of the theory does a theoretical computation seem feasible.)

However, not all is lost, provided we have reasonable size samples of data from cancerous cells and separately from normal ones (or if we had the measurement vectors probability distribution for each cell type. In this latter case we could easily generate samples behaving as if they were described by these distributions, using computer generated pseudo-random numbers). From these samples, we can compute

$$X_{ij} = \left(\frac{WS}{V}\right)_{ij} \tag{25}$$

i = normal, cancerous j = $1,...,n_{normal}$ when i = normal, j = $1,...,$ $n_{cancerous}$ when i = cancerous.

We draw histograms of $X_{normal,j}$, j = $1,...,n_{normal}$ and $X_{cancerous,j}$, j = $1,...,n_{cancerous}$ normalized to have unit area, for two simulated cases, as shown in Figures 8 and 9. Since these two histograms estimate the probability density of

$$\left(\frac{WS}{V}\right)_{normal} \text{ and } \left(\frac{WS}{V}\right)_{cancerous}$$

intuitively, we could use them to choose approximately a critical value C fixing at, say, a small value a, the probability of discarding a cell which is normal (noncancerous). To do this, we choose C so that the area under the normal cell histogram to the right of C is equal to α. The probability of discarding a cancerous cell is then approximately the area under the cancer cell histogram to the right of C. We can use these approximate probabilities to judge whether or not we have an acceptable criterion for purging the blood of cancer cells (i.e., one which purges a suitable proportion of cancer cells without losing too many normal ones).

With only minor modifications, this procedure can be converted to one which is justified, as follows. Instead of choosing C so that the area to its right under the normal cell histogram is α, we find a high level $(1 - \gamma)$ upper confidence limit $C_{1-\gamma,\,\alpha}$ for C_α (the desired critical value). Provided the normal cell histogram is based on enough data, it can be shown that the value $C_{1-\gamma,\,\alpha}$ is obtained by first finding the smallest integer k satisfying the inequality

$$\sum_{r=0}^{n-k}\binom{n}{k}\alpha^r(1-\alpha)^{n-r} \leq \gamma$$

where n is the number of data values used to construct the normal cell histogram. The value $X_{normal,\,[k]}$ is used as $C_{1-\gamma,\,\alpha}$, where $X_{normal,\,[k]}$ is the k-th element of the sequence

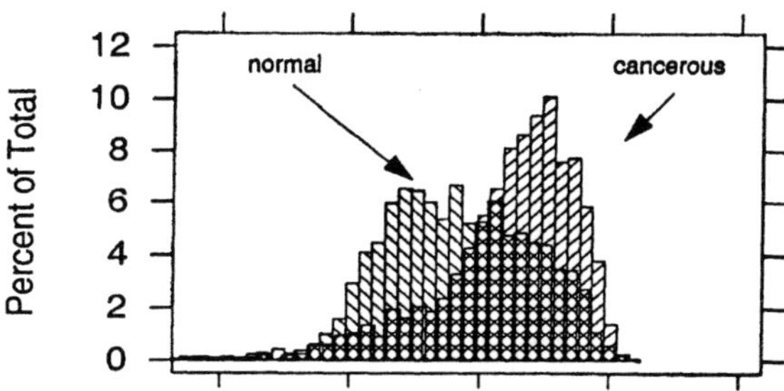

Figure 8. Histogram for simulated case 1, showing normal and cancerous cells.

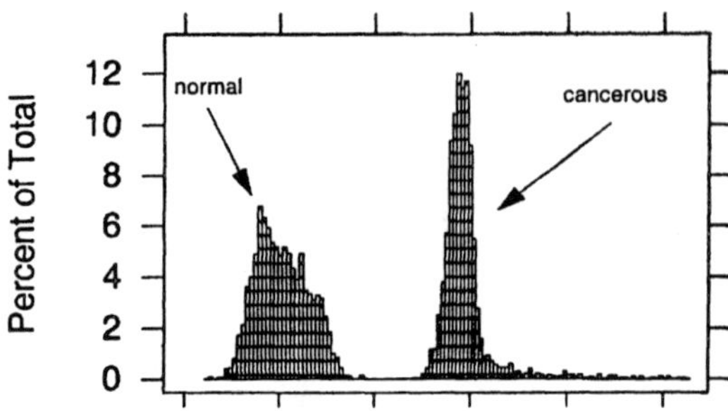

Figure 9. Histogram for simulated case 2, showing normal and cancerous cells.

$X_{normal,j}$ arranged in numerically non-decreasing order. Then with probability at least 1 - γ, the probability of purging a normal cell doesn't exceed α (J. Rosenblatt, J. Leary and J. Hokanson, manuscript in preparation).

Provided that the cancer cell histogram is based on enough data, from the number of cancer cells whose $X_{cancer,\,j}$ values are to the right of $C_{1-\gamma,\,\alpha}$, a lower confidence limit of specified confidence can be found for the probability that a cell will be purged if it is cancerous, using the standard binomial confidence limits; see Blum and Rosenblatt (1972).

FINAL COMMENTS

Good mathematical modeling is a demanding field, because it requires both a substantial knowledge of the science of the area being investigated and an understanding of how to develop applicable mathematical and statistical theory and methods. It requires the judgement to decide which factors are important enough to be included in the model for the problem being tackled and which ones should be overlooked so that the model will be tractable as well as useful.

In addition, familiarity with *current* computing methods is essential for investigation and implementation of solutions (essential because older computer packages that one may have gotten used to are not likely to be supported). Particularly important are the pseudo-random number-based Monte Carlo simulation techniques, because of their ability to provide easily obtainable approximate answers to otherwise theoretically intractable problems.

The payoff for the extra effort needed to carry out effective modeling is the ability to formulate and solve problems which don't fit into the standard mold (models applied in a routine fashion that are available in the widely used computer and statistical packages). The resulting improvements are similar to those given by custom-made clothing compared with outfits directly off the rack. The payoff allows more appropriate and more powerful solutions, which can be the difference between mediocre and excellent solutions to the research problems being addressed.

CORRESPONDENCE

Please address all correspondence to:
Judah Rosenblatt
University of Texas Medical Branch
301 University Boulevard
Galveston, TX 77555
jirosenb@utmb.edu

REFERENCES

Rosenblatt, J.I., and Blum, J.R., 1972, *Probability and Statistics,* W.B. Saunders, Philadelphia.

Knott, G., 1979, MLAB-- A mathematical modeling tool, *Prog. Biomed.,* 10: 271-280.

Jacquez, J. A., 1985, *Compartmental Analysis in Biology and Medicine,* University of Michigan Press, Ann Arbor.

McCarthy, M.J., Rosenblatt, J.I., and Lloyd, R.S., 1996, A modified quantitative polymerase chain reaction assay for measuring gene-specific repair of UV photoproducts in human cells, *Mutation Res.,* 363: 57-66.

Rosenblatt, J.I., 1998, The mathematics behind modeling, in: *Mathematical Modeling in Experimental Nutrition,* A. J. Clifford and H. G. Müller, eds., Plenum Press, New York.

Rosenblatt, J.I., and Bell, S., 1999, *Mathematical Analysis for Modeling,* CRC Press, Boca Raton.

Venables, W.N., and Ripley, B.D., 1999, *Modern Applied Statistics with S-PLUS,* 3rd ed., Springer-Verlag, New York.

FITTING A MATHEMATICAL MODEL TO BIOLOGICAL DATA: INTRACELLULAR TRAFFICKING IN NIEMANN-PICK C DISEASE

Meryl E. Wastney, Peter G. Pentchev, and Edward B. Neufeld[*]

INTRODUCTION

Mathematical models are powerful and heuristic aids for investigating complex systems. Once a model has been developed, it encompasses a large body of knowledge, and it can be used to make predictions about how a system would react under various conditions. This approach applies whether the system being modeled is physical (e.g., an automobile) or biological.

Unlike physical systems for which the model is often a prototype for the system, biological models need to be developed based on observed data from the system. To develop a biological model, it is necessary to understand the underlying biology, the modeling software tools, kinetics, and statistics. Learning to *use* modeling software *vs.* learning to *apply* the software to represent a system are two different tasks. As an analogy, knowing how to saw, nail, and sand wood does not mean one knows how to make fine furniture!

The purpose of this paper is to show how a model was developed for a system based on observed data. We will describe how an initial model was selected, how the model was modified to fit the data, and some results from the modeling. The software used was WinSAAM (Boston et al., 1998), although the same principles apply regardless of the package used.

The example that will be used to illustrate how to develop a model from data is intracellular trafficking in Niemann-Pick C (NPC) disease. This is a rare neurodegenerative disorder of children (Patterson et al., 2001) that is characterized biologically by accumulation of cholesterol within cells (Pentchev et al., 1987; Blanchette-Mackie et al., 1988; Sokol et al., 1988). The defective gene in NPC codes for the protein *NPC* (Carstea et al., 1997) that has a sterol-sensing domain and a motif for targeting late endocytic

---

[*] Meryl E. Wastney, Metabolic Modeling Services, Dalesford, Hamilton, New Zealand. Peter G. Pentchev and Edward B. Neufeld, National Institute of Neurological Disorders and Stroke, National Institutes of Health, Bethesda, MD 20891.

compartments. Possible roles for the protein in cholesterol trafficking have been discussed (Neufeld, 1998). In addition to cholesterol, a chemically heterogeneous group of other compounds, including phospholipids, glycoproteins, and gangliosides, accumulate in NPC cells (Yano et al., 1996; Neufeld et al., 1999). This raises the question of whether there is a generalized defect in intracellular transport in NPC cells. To test this theory, a kinetic study was performed using sucrose as a marker for intracellular transport (Neufeld et al., 1999).

Cells can take up compounds such as cholesterol by endocytosis. An invagination in the cell membrane forms a small vesicle that moves through the cell, off-loading compounds to specific destinations and eventually picking up compounds from lysosomes. The vesicles then return material from the lysosomes back to the cell surface (called retrograde transport). Sucrose taken up by cells via endocytosis remains within the vesicular system, as it cannot cross the membrane and it is not degraded. It is exported from cells by retrograde vesicular transport. Because it remains within vesicles, following the cellular uptake and release of radiolabeled sucrose provides a marker for intracellular transport. The aim of our studies was to determine whether intracellular transport of sucrose, a generic marker of intracellular vesicular trafficking, is defective in NPC cells.

METHODS

Methods have been described in detail (Neufeld et al., 1999). Briefly, fibroblasts from normal (n=5) and NPC (n=3) subjects were cultured in lipoprotein–free medium so that results were not confounded by the storage of cholesterol in NPC cells. Cells were loaded with ^{14}C-sucrose for 3 hours. Initial studies were longer, up to 18 hr, but once the defect was shown to be associated with a pool that turned over in about an hour, model simulations showed that shorter loading studies would maximize the amount of tracer in this pool and therefore highlight differences between normal and NPC cells (see later). Cells were washed to remove excess tracer, and samples of cells and medium were taken over a 24-hour washout period. The samples were analyzed for ^{14}C-sucrose. Data were expressed as uptake (dpm/mg protein) and as dose remaining in cells (expressed as % of tracer present at the start of the washout period). Eight experiments were performed in pairs of one normal cell type and one NPC cell type.

KINETIC DATA ANALYSIS

Data were analyzed by compartmental analysis using the WinSAAM software (Boston et al., 1998), available from http://winsaam.nci.nih.gov/. With this approach, it was assumed that the marker moved between intracellular compartments, or pools, that could be distinguished by their turnover rates.

Figure 1 shows the uptake of label at the end of the 180 min loading period, and Figure 2 presents data on release of tracer from cells over the next 24-hour period. The steps in developing the model were to 1) develop the simplest model to fit data from the normal cells, 2) fit data from NPC cells, and 3) compare the kinetics. Note that, during

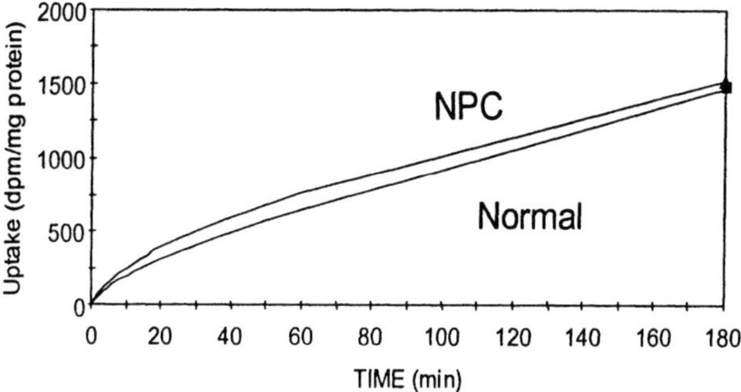

Figure 1. Uptake of ^{14}C-sucrose by fibroblasts of normal subjects and NPC patients after a 3 hr incubation. The lines are the model-predicted time course for the 3-hr loading period.

model development, we consulted the model of Blomhoff et al. (1989) for sucrose transport in rabbit hepatocytes. As described later, we chose to set our model up slightly differently.

Fitting Data from Normal Cells

The initial model was a one-compartment model (i.e., one pool inside the cells). All models had one pool representing extracellular medium. The medium compartment was required to represent uptake during loading and to account for the possible re-entry of tracer that had left the cell during the 24-hr washout. A one-compartment model was not able to fit the whole curve (Figure 3). It overestimated the early data and underestimated the later points. Slowing the rate of loss increased the misfit at the start of the curve, and increasing the loss improved the fit to the first part of the curve but increased the misfit at the end. We concluded that this model was too simplistic.

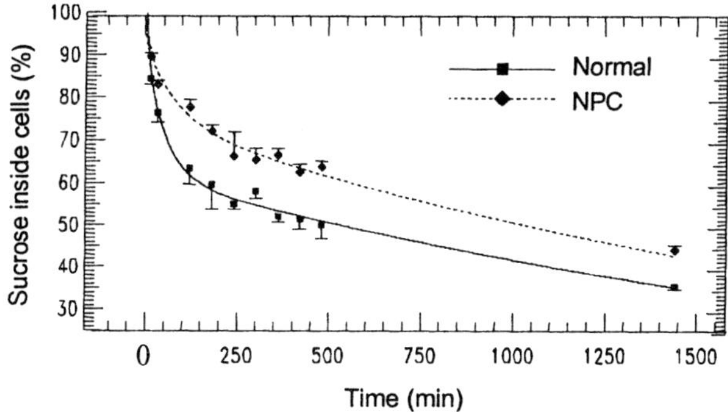

Figure 2. Release of ^{14}C-sucrose from fibroblasts of a normal subject and an NPC patient (mean ± SD of triplicate samples) over a 24-hour period. Symbols are observed data and lines are calculated by the model in Figure 9. Reprinted with permission from Neufeld at al., 1999.

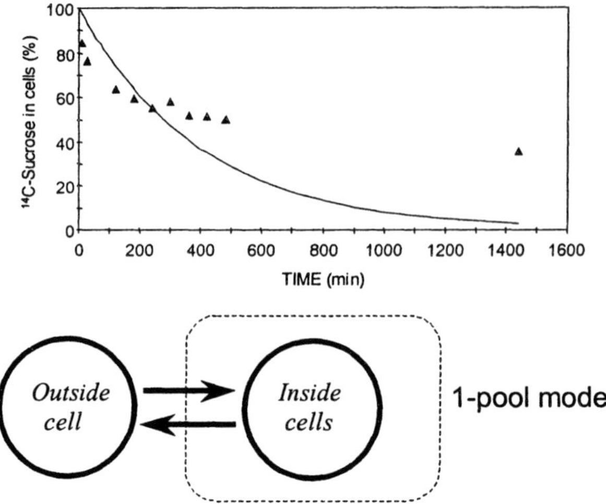

Figure 3. Calculated fit of a one-pool model (line) to observed data (symbols) from normal fibroblasts. The model is unable to fit the data.

The next model consisted of two intracellular compartments (Figure 4). Tracer in the two intracellular compartments was added together and the sum compared with the observed data. With this model, we were able to fit the early and late data, but this model did not fit data between 100 and 400 min. Again, the model was deemed too simplistic.

Adding a third compartment provided a good fit to all the data (Figure 5). The parameter values and a biological interpretation of the model are shown in Figure 6. The turnover times were calculated as the reciprocal of all pathways out of a compartment and were 6 min for compartment 1, 1 hr for compartment 2, and 20 hr for compartment 3.

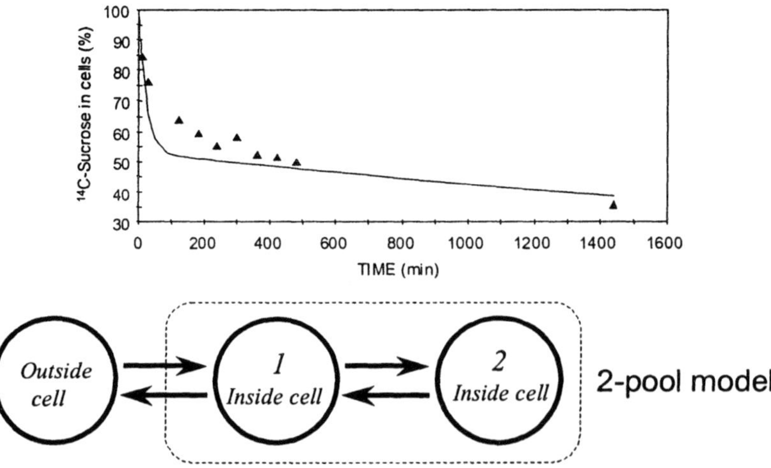

Figure 4. Calculated fit of a two-pool model (line) to observed data (symbols) from normal fibroblasts. The model is unable to fit the data between 100 and 400 min.

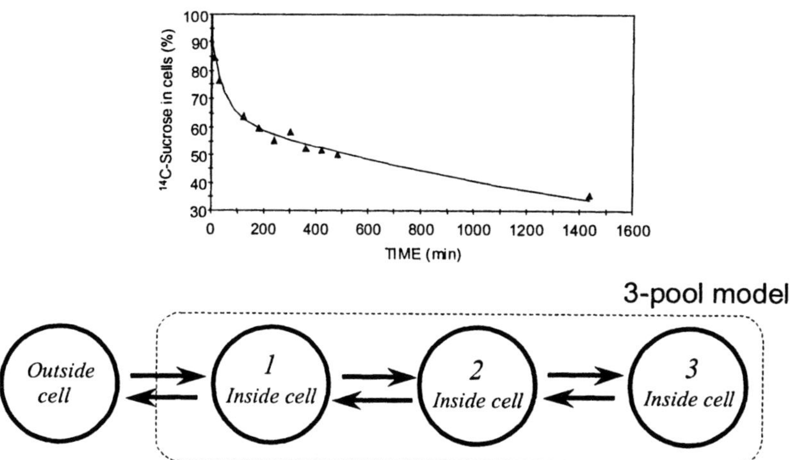

Figure 5. Calculated fit of a three-pool model (line) to observed data (symbols) from normal fibroblasts. The model is able to fit the data.

These times correspond to literature values for endosomes (compartment 1), late endosomes (compartment 2), and lysosomes (compartment 3). Because material from late endosomes is considered to be lost from cells by a direct pathway (i.e., not through endosomes) (Mukherje et al., 1997), the pathway from compartment 2 into compartment 1 was changed to loss directly to the medium. This change did not affect the fit of the model to the data.

Fitting Data from NPC Cells

The NPC data are compared in Figure 7 to the fitted data for normal cells. To fit the model shown in Figure 6 to these data, we set about identifying the minimum number of changes in parameter values that would fit the NPC data. Altering the turnover of the fastest compartment, compartment 1, had a very small effect on the fit (Figure 8, upper panel). This is because that compartment turns over in about 6 min, a time too rapid to affect the later part of the curve. The middle panel shows that altering loss from the lysosomes (compartment 3) changed the shape of the tail of the curve and did not fit the NPC data. Only by changing the turnover of compartment 2 were we able to fit the NPC data (Figure 8, lower panel).

Comparing ^{14}C-Sucrose Kinetics in NPC and Normal Cells

The final parameter values are shown in Figure 9. The rate of loss from compartment 2 in NPC cells is half that of normal cells. Data could be fitted equally well by increasing transfer into compartment 3 from compartment 2 in NPC cells, but the loss to the outside was more consistent with other aspects of the trafficking lesion defined in the mutant cells.

We used the model to calculate the size of the intracellular compartments (Figure 10). These values are relative because there is no way to measure the mass of any of

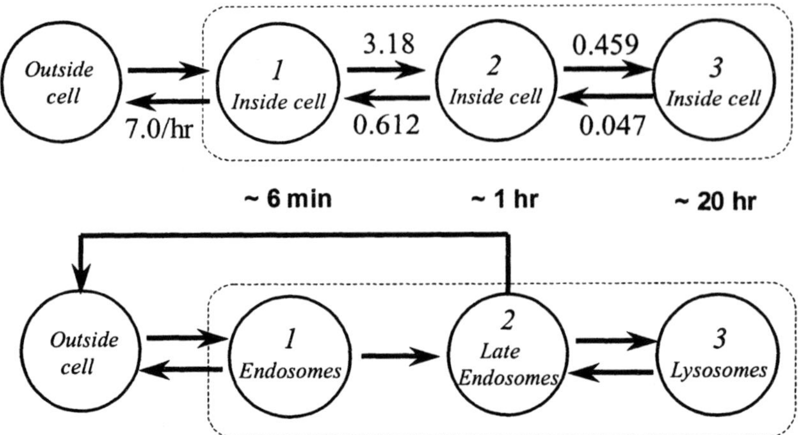

Figure 6. Biological interpretation of the model used to fit the data (Figure 5). Upper part of the figure shows the 3-pool model with the calculated parameter values (fraction/hr). Underneath are the turnover times of each pool (calculated as the reciprocal of all the pathways leading out of that compartment). The lower part of the figure assigns a biological entity to each compartment, based on information about its turnover time. The pathway from compartment 2 to the outside was rerouted to be direct, based on current biological understanding of this process.

these compartments. We arbitrarily assigned a mass of 1 to compartment 1 and calculated the relative mass of the other compartments. For normal cells, compartment 2 was predicted to be five times the size of compartment 1 and compartment 3 was 57 times larger. Compartments 2 and 3 were twice as large in NPC cells as in normal cells. Because most of the mass was associated with the very slowly turning-over compartment 3, or lysosomes, this compartment had appeared to be the one most perturbed when viewed microscopically. The kinetics revealed that, although the perturbation is most obvious microscopically in the lysosomes of NPC cells, the defect is associated with the more rapidly turning-over late endosome compartment.

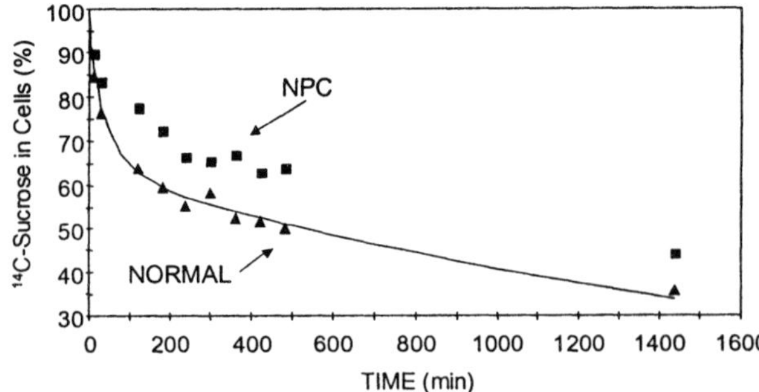

Figure 7. Release of ^{14}C-sucrose from fibroblasts of a normal subject and an NPC patient. The line is the calculated value determined from fitting the model (Figures 5 and 6) to the data from normal cells.

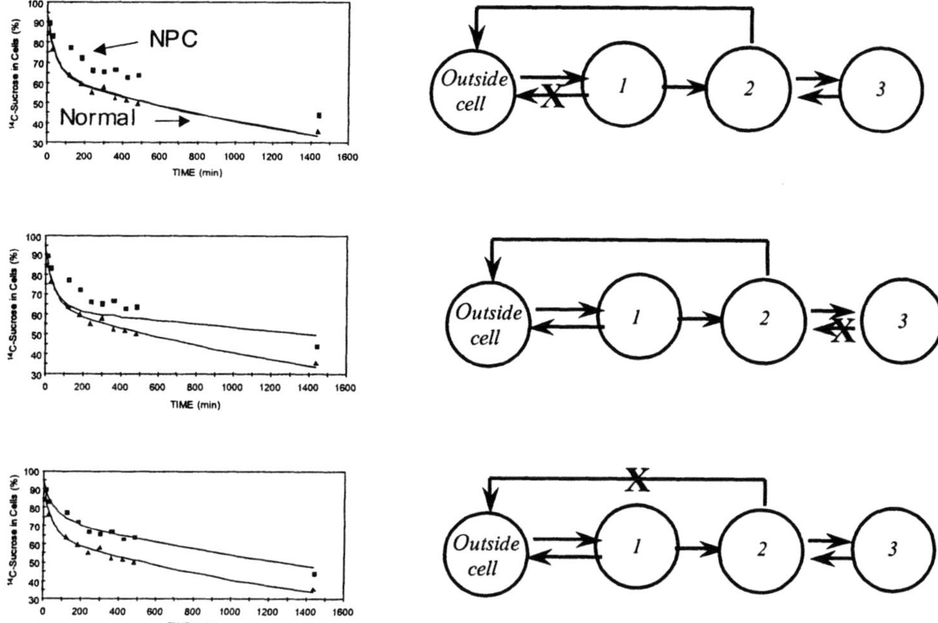

Figure 8. Fit of the model for normal cells (Figure 6) to data from NPC cells. Upper panel: if release of sucrose from compartment 1 was changed, the curve did not fit the NPC data. Middle panel: if release from compartment 3 was changed, the model did not fit the NPC data. Lower panel: if release from compartment 2 was changed, the model was able to fit the NPC data.

Other Aspects of a Modeling Project

This description of fitting a model to data has only covered some aspects of the development of a model. Other important aspects of a modeling project, including a description of the modeling tools required, assessing the model fit to the data statistically, and publishing a model, are described further in Wastney et al. (1999).

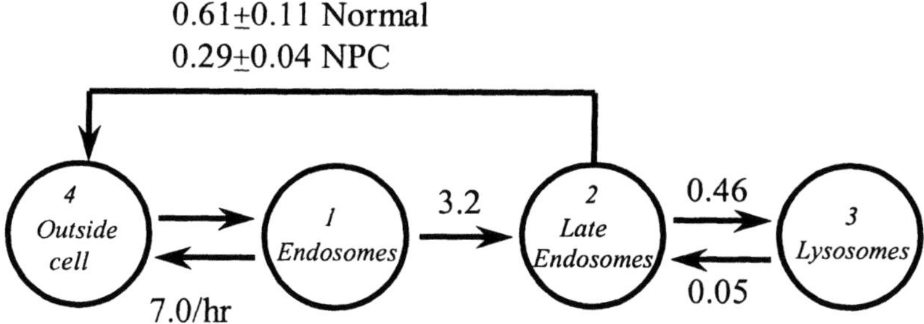

Figure 9. Parameter values determined by fitting the model to normal and NPC data. Differences in the data (Figure 2) are explained by a difference in release of material from compartment 2 (late endosomes) to the medium. Reprinted with permission from Neufeld at al., 1999.

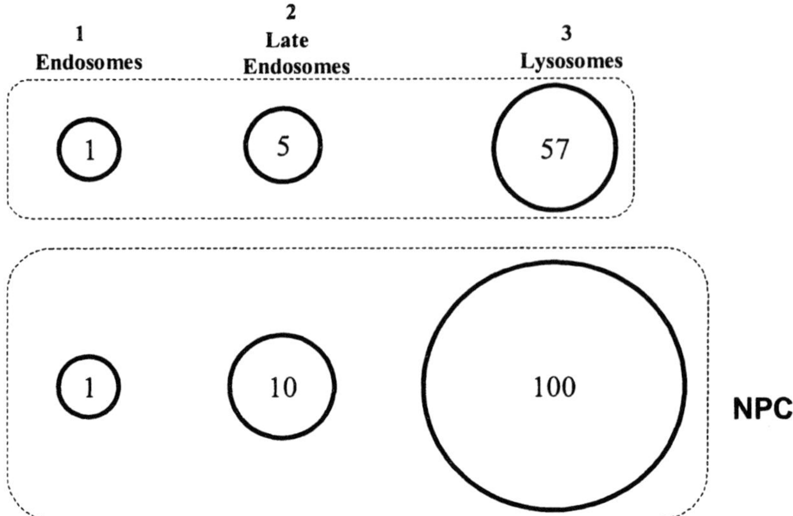

Figure 10. Calculated mass of the compartments in normal and NPC cells. Values are relative units, based on a mass of unity for compartment 1. Values for compartments 2 and 3 are twice as large in NPC cells as in normal cells. Because of the slow turnover of compartment 3, it is ten times the size of compartment 2. The major effect of altered kinetics in NPC cells is predicted to be a large increase in lysosome size.

PROGRAMMING NOTES

The WinSAAM code used to fit data from one experiment is shown in Appendix 1 to illustrate how aspects of the experiment were simulated. Some points are discussed below.

Modeling Tracer Uptake

We began the data analysis by first determining whether models existed for this type of study. We found a relevant one (Blomhoff et al., 1989), although a different cell type was studied in those experiments. Blomhoff et al. let the initial conditions, or distribution of tracer among the intracellular pools, be represented by variable parameters whose values were determined by the program. In contrast, we used the model to simulate both uptake of tracer during the loading phase *and* release during the washout phase (Figures 1 and 8). In this way, the distribution of tracer between the intracellular compartments at the start of the washout was constrained by the uptake data. The initial condition in the compartment outside the cells (called compartment 4) was set equal to the dpm added.

End of Loading

After the loading period ended (i.e., after 180 min), the contents of compartment 4 were set to zero to simulate washing the cells to remove the external medium. This was achieved by the function QO in Appendix 1.

Tracer Distribution in Cells at the End of Loading/Start of Washout

The dpm in compartments 1, 2, and 3 were summed by WinSAAM to reflect the total dpm measured inside cells. Then the fraction of the counts in each compartment was calculated and expressed as a percent. As seen in Figure 11, the percent predicted to be in each compartment varied over time. At 180 min, the relative distribution was 20% in compartment 1 and about 40% each in compartments 2 and 3. The code to calculate the uptake of tracer, convert units from dpm in cells to percent of dose at the start of the washout, and fitting of the normal and NPC data are shown in Appendix 1.

Fitting Washout Data

Because we wanted the washout data to be plotted starting at time zero (although the experiment had been running for 180 min of loading), the *time-change* (TC) feature of WinSAAM was used to 'reset' time during the simulation to zero at the start of the washout. Because we did not explicitly reset any values for compartment contents, the program continued solving the washout phase with the tracer values calculated for the end of washout.

DISCUSSION

We have described one approach to fitting a model to biological data. The data (Figure 2) could have been fitted alternatively by a mathematical function, such as a sum of exponentials. With that approach, we would have learned that NPC cells lost material at a slower rate than normal cells and that three exponentials were required to fit the data. We were able to obtain *additional* information by fitting a model to the data as summarized here. Specifically, with modeling, we were able to determine or identify

- number of intracellular compartments,
- connectivity (or pathways) between compartments,
- turnover rates of the compartments,
- rates of transfer between compartments,
- relative compartment sizes,

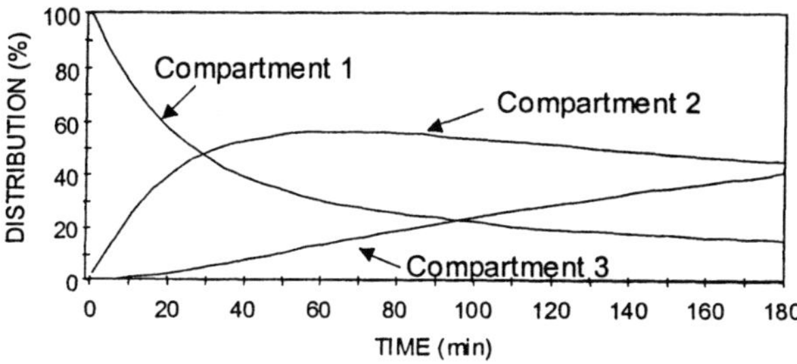

Figure 11. Model-predicted distribution of ^{14}C-sucrose between intracellular compartments at the end of the 3-hr loading in normal cells.

- pathway that is defective in NPC cells,
- difference in rate of trafficking between normal and NPC cells, and
- rationalization of how the difference in transfer rates accounted for cell phenotype (cholesterol storage in lysosomes).

We concluded from these modeling studies that there is a generalized defect in vesicular trafficking in NPC cells. A defect has also been detected in *in vivo* studies (Shamburek et al., 1997). The cellular defect is at the site of a compartment that turns over in about 1 hr, putatively considered to be the late endosomes. Our studies identified the site of the defect and helped to focus experimental and biological attention on this compartment for further study for both unraveling the disease and targeting therapeutic strategies.

CORRESPONDENCE

Please address all correspondence to:
Meryl E. Wastney
Metabolic Modeling Services, Ltd
PO Box 23008
Dalesford, Hamilton
New Zealand
wastneym@drc.co.nz

REFERENCES

Blanchette-Mackie, E.J., Dwyer, N.K., Amende, L.M., Kruth, H.S., Butler, J.D., Sokol, J., Comley, M.E., Vanier, M.T., August, J.T., Brady, R.O., and Pentchev, P.G., 1988, Type-C Niemann-Pick disease: low-density lipoprotein uptake is associated with premature cholesterol accumulation in the Golgi complex and excessive cholesterol storage in lysosomes, *Proc. Natl. Acad. Sci. USA* 85:8022-8026.

Blomhoff, R., Nenseter, M.S., Green, M.H., and Berg, T.A., 1989, Multicompartmental model of fluid-phase endocytosis in rabbit liver parenchymal cells, *Biochem. J.* 262:605-610.

Boston, R.C., Greif, P., Wastney, M., and Linares, O., 1998, Balancing needs, efficiency and functionality in the provision of modeling software: a perspective of the NIH WinSAAM project, *Adv. Exp. Med. Biol.* 445:3-20.

Carstea, E.D., Morris, J.A, Coleman, K.G., Loftus, S.K., Zhang, D., Cummings, C., Gu, J.Z., Rosenfeld, M.A., Pavan, W.J., Krizman, D.B., Nagle, J., Polymeropoulos, M.H., Sturley, S.L., Ioannou, Y.A., Higgins, M.E., Comly, M., Cooney, A., Brown, A., Kaneski, C.R., Blanchette-Mackie, E.J., Dwyer, N.K., Neufeld, E.B., Chang, T.-Y., Lisum, L., Strauss III, J.F., Ohno, K., Zeigler, M., Carmi, R., Sokol, J., Markie, D., O'Neill, R.R., van Diggelen, O.P., Elleder, M., Patterson, M.C., Brady, R.O., Vanier, M.T., Pentchev, P.G., and Tagle, D.A., 1997, Neimann-Pick C1 disease gene: homology to mediators of cholesterol homeostasis, *Science* 277:228-231.

Mukherjee, S., Ghosh, R.N., and Maxfield, F.R., 1997, Endocytosis, *Physiol. Rev.* 77:759-803.

Neufeld, E.B., 1998, What the Niemann-Pick Type C gene has taught us about cholesterol transport, in: *Intracellular Cholesterol Trafficking*, T.Y. Chan and D.A. Freeman, eds., Kluwer Acad. Pub., Norwell.

Neufeld, E.B., Wastney, M., Patel, S., Suresh, S., Cooney, A.M., Dwyer, N.K., Roff, C.F., Ohno, K., Morris, J.A., Carstea, E.D., Incardona, J.P., Strauss III, J.F., Vanier, M.T., Patterson, M.C., Brady, R.O., Pentchev, P.G., and Blanchette-Mackie, E.J., 1999, The Niemann-Pick C1 protein resides in a vesicular compartment linked to retrograde transport of multiple lysosomal cargo, *J. Biol. Chem.* 274:9627-9635.

Patterson, M.C., Vanier, M.T., Suzuki, K., Morris, J.A., Carstea, E., Neufeld, E.B., Blanchette - Mackie, E.J., and Pentchev, P.G., 2001, Niemann-Pick disease, type C: a lipid trafficking disorder, in: *The Metabolic and Molecular Bases of Inherited Disease, 8th ed.,* C.R. Scriver, K.W. Kinzler, D.Valle, B. Childs, B. Vogelstein, A.L. Beaudet, and W.S. Sly, eds., McGraw-Hill, New York.

Pentchev, P.G., Comly, M.E., Tokoro, T., Butler, J., Sokol, J., Filling-Katz, M., Quirk, J.M., Marshall, D.C., Patel, S., and Vanier, M.T., 1987, Group C Niemann-Pick disease: faulty regulation of low-density lipoprotein uptake and cholesterol storage in cultured fibroblasts, *FASEB J.* 1:40-45.

Shamburek, R.D., Pentchev, P.G., Zech, L.A., Blanchette-Mackie, E.J., Carstea, E.D., VandenBroek, J.M., Cooper, P.S., Neufeld, E.B., Phair, R.D., Brewer Jr., H.B., Brady, R.O., and Schwartz, C.C., 1997, Intracellular trafficking of the free cholesterol derived from LDL cholesteryl ester is defective in vivo in Niemann-Pick C disease: insights on normal metabolism of HDL and LDL gained from the NP-C mutation, *J. Lipid Res.* 38:2422-2435.

Sokol, J., Blanchette-Mackie, E.J., Kruth, H.S., Dwyer, N.K., Amende, L.M., Butler, J.D., Robinson, E., Patel, S., Brady, R.O., Comley, M.E., Vanier, M.T., and Pentchev, P.G., 1988, Type C Niemann-Pick disease. Lysosomal accumulation and defective intracellular mobilization of low-density lipoprotein cholesterol, *J. Biol. Chem.* 263:3411-3417.

Wastney, M.E., Patterson, B.H., Linares, O.A., Greif, P.C., and Boston, R.C., 1999, *Investigating Biological-Systems using Modeling: Strategies and Software,* Academic Press, San Diego.

Yano, T., Taniguchi, M., Akaboshi, S., Vanier, M.T., Tai, T., Sakuraba, H., and Ohno,K., 1996, Accumulation of GM2 ganglioside in Niemann-Pick disease type C fibroblasts, *Proc. Japan Acad.* 72:214-219.

APPENDIX 1. WinSAAM Code From One Study


```
A SAAM31                    NPC-E35.saam            4/1/99
2         2
C EXPT 35
C COMPARTMENTS 1-3 NORMAL CELLS (4 is media)
C COMPARTMENTS 21-23 NPC CELLS (24 is media)
C DPM AND PROTEIN - ALL IN 0.8 ML
C CELLS PRELOADED WITH 14C-SUCROSE FOR 180 MIN AND
C THEN WASHED OUT FOR 1440 MIN
C
C  ++++++++++++ NORMAL ++++++++++++
H PAR
C P(1) IS MG OF PROTEIN/0.8 ML
   P(1)        1.589000E-01
C IC UNITS ARE DPM in 1 ml converted to DPM/0.8ml; L(I,J) are /min
   IC(4) = 10E6*0.8
   L(1,4)      7.652366E-07   3.666667E-07   3.300000E-06
   L(2,1)      5.309897E-02
   L(3,2)      6.856965E-03   3.333333E-03   3.000000E-02
   L(4,1)      1.176950E-01
   L(4,2)      8.960547E-03   1.429678E-03   2.000000E-02
   L(2,3)      7.552562E-04   1.523333E-04   1.371000E-03
H DAT
C TO END PRELOAD, SET VALUE OF COMPARTMENT 4 TO ZERO AT 180 min
104QO
            180.          0.
C G(6) IS ACCUMULATED DOSE IN CELL
XG(6)=F(1)+F(2)+F(3)
C G(7) IS ACCUMULATION NORMALIZED/(MG PROTEIN IN 0.8 ML)
XG(7)=G(6)/P(1)
C G(9) IS % INSIDE CELL
XG(9)=100*G(6)/(F(4)+G(6))
C TO SIMULATE POINTS FOR UPTAKE OF DOSE DURING LOADING
107G(7)
            0.
2           2.                      10.
2           10                      4.
H DAT                       TC(2)
C TC(2) IS WASHOUT PHASE. TIME IS RESET TO ZERO
   T       0.
```

```
C THESE LINES GENERATE POINTS FOR A SMOOTH CALCULATED CURVE
109G(9)
                0.
2               1.                              10.
2               10.                             9.
2               100.                            11.
C ++++++++++++++ NPC
H PAR
C P(21) IS MG OF PROTEIN/0.8 ML
   P(21)      1.500000E-01
C IC UNITS ARE DPM
   IC(24) = 10E6*0.8
   L(21,24)  5.942401E-07  3.666667E-07   4.999999E-06
   L(22,21)=L(2,1)
   L(23,22)=L(3,2)
   L(24,21)=L(4,1)
   L(24,22)  4.914901E-03  1.429678E-03   5.000000E-02
   L(22,23)=L(2,3)
H DAT
C TO END PRELOAD
124QO
                180.            0.
C G(26) IS ACCUMULATED DOSE IN CELL
XG(26)=F(21)+F(22)+F(23)
C G(27) IS ACCUMULATION NORMALIZED/(MG PROTEIN IN 0.8 ML)
XG(27)=G(26)/P(21)
C G(29) IS % INSIDE CELL
XG(29)=100*G(26)/(F(24)+G(26))
H DAT                            TC(2)
   T         0.
129G(29)
                0.
2               1.                              10.
2               10.                             9.
2               100.                            11.
C ******************** END OF MODEL
C EXPT 35
C NORMAL (5565) 3 hr preload
H DAT
C Time, min; ACCUMULATION (DPM/MG PROTEIN)
107G(7)                                         FSD=.05
                0.
                180.            1523.
C TIme in min, data are % IN CELL
H DAT                            TC(2)
   T         0.
109G(9)                                         FSD=.1
                0.
                10              84.6
                30              76.4
                120             64.3
                180             59.7
                240             55.3
                300             58.2
                360             52.3
                420             51.5
                480             50.1
                1440            35.5
C
C +++++++++++++ NPC
H DAT
C Uptake (DPM/MG PROTEIN)
127G(27)                                        FSD=.05
                0.
                180.            1488.
C UNITS,  TIME, MIN; % IN CELL
H DAT                            TC(2)
   T         0.
129G(29)                                        FSD=.1
```

```
       0.
       10              89.7
       30              83.2
       120             77.7
       180             72.3
       240             65.5
       300             65.5
       360             66.6
       420             62.7
       480             63.9
       1440            44.2
C ************ END OF DATA
```

A SATURATION KINETIC MODEL TO TEACH BALANCE OF ESSENTIAL FATTY ACIDS IN NUTRITION EDUCATION

James L. Hargrove, Jinah Hwang, and Diane K. Hartle[*]

INTRODUCTION

New computer software provides educators and students with dramatic capabilities for teaching theory in the classroom (Hargrove, 1998; Wastney et al., 1999). Computers confer the ability to create scenarios that facilitate hypothesis testing, and they can be used to explore change over time when the balance of a system is altered. Most hypotheses lead to quantitative predictions that can be stated as one or more equations, and computers allow these to be solved as part of classroom presentations even when the underlying equations appear difficult. This paper shows how to use software for modeling to make predictions about the tissue effects of dietary essential fatty acids. Using STELLA® Research software,[†] we converted a set of equations that predicts the effect of dietary n-3 and n-6 fatty acids on membrane lipids into a dynamic model. The outcomes can be used to discuss the consequences of consuming balanced or imbalanced patterns of essential fatty acids.

TISSUE EFFECTS OF NUTRIENTS AND MINIMUM RISK FOR DISEASE

One of the principles of nutrition is that effects on health take place at the level of the tissues. In many cases, changes in diet must produce corresponding changes in tissue

---

[*] James L. Hargrove and Jinah Hwang, Department of Foods and Nutrition, The University of Georgia, Athens, GA 30602. Diane K. Hartle, Department of Pharmaceutical and Biomedical Sciences, College of Pharmacy, The University of Georgia, Athens, GA 30602.

[†] STELLA is a copyright of High Performance Systems, Inc., 45 Lyme Road, Suite 300, Hanover, NH 03755. The Internet address is http://www.hps-inc.com/. STELLA is an acronym for Systems Thinking, Experiential Learning Laboratory with Animation.

composition before any improvements can be sustained. For example, in countries with a low incidence of ischemic heart disease, the ratio of consumption of n-6 to n-3 polyunsaturated fatty acids (PUFA) is less than 4. This is true for traditional diets in Japan, Iceland, Greenland, and Greece. In the U.S. and northern Europe, there is a higher incidence of heart disease, and the ratio ranges from 8-12 (Lands, 1991; Skuladottir et al., 1995).

A SATURATION MODEL OF MEMBRANE LIPID COMPOSITION

Lands et al. (1990, 1992) derived equations for tissue saturation that predict the abundance of highly unsaturated, 20-22-carbon fatty acids (HUFA) in membrane phospholipids as a function of dietary polyunsaturated fatty acids. The assumption underlying the equations is that different 18-carbon precursors compete with one another for elongation, desaturation, and insertion into membranes as follows:

$$\text{n-3 as \% HUFA} = \frac{100}{1 + C_3/en\%3[1 + en\%6/C_6 + en\%0/C_0 + en\%3/K_S]} \tag{1}$$

$$\text{n-6 as \% HUFA} = \frac{100}{1 + C_6/en\%6[1 + en\%3/C_3 + en\%0/C_0 + en\%6/K_S]} \tag{2}$$

$$\text{n-9 as \% HUFA} = \frac{100}{1 + K(1 + en\%3/C_3 + en\%6/C_6]} \tag{3}$$

where En%3 = dietary n-3 fatty acids expressed as kcal per 100 kcal of diet, En%6 = dietary n-6 fatty acids expressed as kcal per 100 kcal of diet, and En%0 = dietary saturated fatty acids plus monounsaturated fatty acids expressed as kcal per 100 kcal of diet. These parameters are values that fit the above equations to a hyperbolic saturation curve in which PUFA interact competitively for elongation, desaturation, and insertion into membranes. Further, C_6 is a parameter that fits a saturation curve for n-6 PUFA; it ranges from 0.0215 to 0.0590 for data derived from rats and takes similar values for humans (Lands et al., 1992). C_3 is a parameter that fits a saturation curve for n-3 PUFA; it ranges from 0.0310 to 0.0850 for rats (slightly greater than C_6). C_O and K_S are parameters that fit a saturation curve for other fatty acids (saturated, n-7, and n-9 unsaturated); C_O ranges from 5.0 to 9.0 for rats, and K_S equals about 0.15.

These equations represent a static or steady state model of membrane composition. The model was solved using STELLA® Research software. This package contains several tools that are suitable for education: 1) a high level map that enables a tutorial to be created with flow charts, text, imported images, video clips, and controls for changing input and viewing results; 2) a middle level designed to allow models to be built using icons chosen from a display panel plus a list of built-in mathematical, logical, and probabilistic functions; and 3) a lower level that lists equations.

In order to convert the static model to a kinetic one, information concerning rates of change in human adipose tissue was required. The rate at which membranes can achieve a new steady state will be limited primarily by exchange of the abundant 18-carbon PUFA into the large reservoir of adipose tissue.

REFERENCE PATTERNS FOR RATES OF CHANGE IN HUMAN ADIPOSE TISSUE

The effect of dietary fat on composition of adipose tissue in humans fits a simple mono-exponential model. Dayton et al. (1966) monitored the composition of lipids in human adipose during a five-year feeding period with a diet in which linoleic acid (LA) comprised 37% of total fat. The content of LA increased from about 11% to an asymptote of 35.6% with a half-time of 680 days. This value is equivalent to a daily fractional turnover of 0.1%. Both the initial and final contents of LA were very similar to dietary intake and could be predicted from the equation

$$y = a + be^{-kt} \tag{4}$$

where y equals % LA in adipose, a is the final percentage of LA in adipose, b is the difference between initial and final values, k is fractional turnover, and t is time in days. Schäfer and Overvad (1990) observed that human intake of PUFA (x) was correlated with the concentration in adipose (y) by the relationship

$$y = 0.38x + 9.1 \tag{5}$$

At low levels of intake, Eq. (5) predicts that the concentration of LA in adipose will be maintained above levels in the diet, but the converse is true when dietary levels are above 14.6%. The good agreement between composition of 18-carbon fatty acids in the diet and in adipose suggests that there is very little net synthesis of lipid in humans and little preferential oxidation of 18-carbon PUFA *vs.* saturated or monounsaturated fatty acids (SFA and MUFA, respectively). This is not true of longer-chain PUFA, which tend to be retained in lipoprotein particles and preferentially oxidized in the liver (Havel, 1985).

It is reasonable to assume that uptake of α-linolenic acid (18:3 n-3) into adipose tissue also is a linear function of the amount or proportion in the diet. However, HUFA such as 20:4 n-6 and 20:5 n-3 are principally transported in phospholipids rather than triacylglycerols. Although Lands (1991) reports that α-linolenic acid in adipose typically equals 1-2% in human subjects, Calder et al. (1992) were unable to detect it in post-mortem tissue from the United Kingdom. In contrast, Handelman et al. (1988) reported that the linoleic acid (18:2 n-6) content of adipose tissue has steadily increased from about 8% in 1960 to as much as 20% in the 1980s. Thus, the ratio between 18-carbon n-6 and n-3 precursors in the adipose reservoir can exceed 20:1!

The good fit of data for human adipose composition to a simple exponential function greatly simplifies a minimal model of human fat metabolism for the present purposes. There is no need to attempt to model lipoprotein kinetics, the pool of non-esterified fatty acids generated through lipolysis, or uptake and oxidation of fatty acids within the tissues. What permits the model to be simplified is the very long time frame of exchange relative to the rapid turnover of membrane lipids and the small pools of fatty acids in the blood.

A SATURATION KINETIC MODEL OF LIPID EXCHANGE

Define the predicted values of the three steady state equations at time $t = 0$ with specified initial dietary fatty acids to be $n3_0$, $n6_0$, and $n9_0$, respectively. Let the corresponding steady state values after changing to a diet with new fatty acid composition equal $n3_\infty$, $n6_\infty$, and $n9_\infty$, respectively, and denote any intermediate values with the subscript t.

If one assumes that intermediate values can be characterized by a fractional exchange rate denoted by k (with units of $time^{-1}$), then the predicted concentration of each type of HUFA in cell membrane phospholipids can be written as an equation of the following form:

$$n3_t = n3_0 + (n3_\infty - n3_0)(1 - e^{-kt}) \tag{6}$$

This equation predicts that the en% of n-3 fatty acids in HUFA at any time equals the initial concentration plus the difference between the final steady state value and the initial steady state value. The term $(1 - e^{-kt})$ begins with a value of zero and increases to a value of 1.0 as time increases. The time course of change is determined by the rate parameter k which can be assigned a value equal to ln 2 (0.693) divided by 4 days in rats or about $0.17325\ d^{-1}$ (Klein et al., 1980). In humans, the fractional exchange rate for adipose is about $0.001\ d^{-1}$. However, the exchange into membrane phospholipids is much faster. For the present purposes, we will assume that the half-life in phospholipids is 14 days or $0.05\ d^{-1}$. Therefore, the rate of change should not fit a simple exponential function but should contain at least one rapid and one slow component.

RESULTS

How Much Dietary Essential Fatty Acid Is Needed to Saturate Phospholipids?

Shown in Figure 1 is the predicted effect of increasing the daily intake of linoleic acid (18:2 n-6) at different levels of intake of n-3 fatty acids (0.3 to 4.0 en%). The intake of saturated and monounsaturated fatty acids was kept constant at 20% of total calories. The model predicts that phospholipids are half-saturated when linoleic acid equals about 0.1% of calories and saturated at about 1% of calories when the intake of n-3 fatty acids is low. As the intake of n-3 fatty acids increases, higher amounts of n-6 fatty acids are required to achieve a similar level of membrane saturation. This ratio is a very sensitive control point.

Effect of Changing Total Fat Intake with no Change in Composition

The model predicts that the ratio between n-3 and n-6 fatty acids in membrane phospholipids will be maintained at a constant value as total fat consumption decreases from 30% of calories to 10% of calories. In contrast to the effect of the n-3:n-6 ratio, total fat intake is not a sensitive control point for modifying membrane composition.

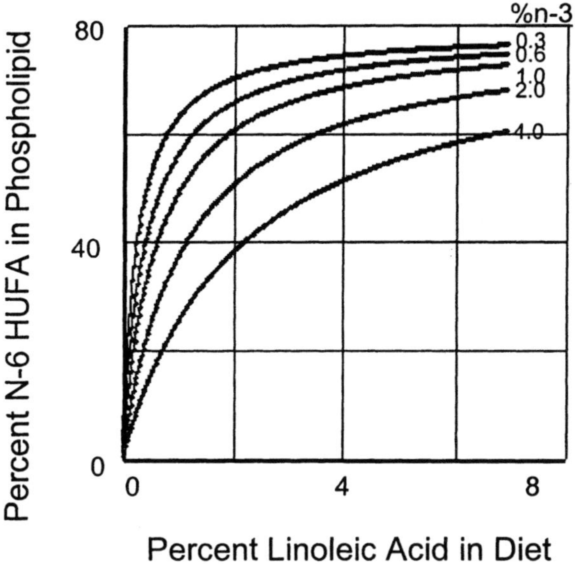

Figure 1. Predicted effects of dietary (n-6) linoleic acid on the saturation of membrane phospholipids with n-6 HUFA. The five curves represent outcomes when the amount of dietary n-3 α-linolenic acid equals 0.3, 0.6, 1.0, 2.0, or 4.0 percent of dietary energy, respectively.

Effect of Substituting MUFA for PUFA

Oleic acid competes poorly for esterification in phospholipids when compared to LA (Lands et al., 1990). However, if an oil with a high content of MUFA is substituted for one that is high in n-6 PUFA, the abundance of n-6 HUFA in membranes decreases owing to the decreased abundance of n-6 PUFA relative to competing fatty acids (Hwang, 2000). The saturation equations indicate the poor competition by oleic acid as a high value for the constant C_O as compared to C_6 and C_3. For example, C_O ranges from 2-5 en%, whereas C_3 and C_6 range from 0.02-0.10 en% in different tissues (Lands et al., 1990). Under these circumstances, n-3 fatty acids may increase because of reduced competition by n-6 PUFA (Navarro et al., 1994).

Kinetics of Exchange

Eighteen-carbon PUFA enter a large, slowly-exchanging reservoir in adipose that has a half-time of about 680 days in humans (Dayton et al., 1966) and about 5-7 days in rats (Klein et al., 1980). Eighteen-carbon PUFA also enter the rapidly-exchanging phospholipid pool. In contrast, a larger proportion of 20- and 22-carbon HUFA enter the phospholipid pool, with a half-time of exchange that ranges from a few hours to a few days, depending on tissue type (Zuijdgeest-van Leeuwen et al., 1999). Thus, dietary eicosapentaenoic acid and docosahexaenoic acid equilibrate with phospholipids within days to weeks.   In contrast, 18:2 n-6 and 18:3 n-3 require years to equilibrate fully because of the slow kinetics of the adipose pool (Dayton et al., 1966).

The effects of diet on membrane phospholipid composition therefore depend on the percentage of calories derived from 18- *vs.* 20-22-carbon n-3 fatty acids and on the rates of exchange within phospholipids and adipose. Figure 2 shows an example with kinetics based on human subjects. The rapidly-exchanging pool was assumed to have a half-time of 14 days, and the slowly exchanging pool equilibrated with a 680 day half-time. Consumption of linoleic acid was decreased from 5 to 2% of calories, and consumption of n-3 fatty acids was increased to 1%, with equal contributions from 18-carbon and 20-22-carbon sources.

STORAGE OF DIETARY PUFA IN ADIPOSE TRIACYLGLYCEROL

Although 20- and 22-carbon HUFA accumulate primarily in membrane phospholipids, 18-carbon precursors of n-3 and n-6 HUFA are stored in adipose in proportion to their abundance in the diet. Figure 3 shows this relationship based on data obtained in

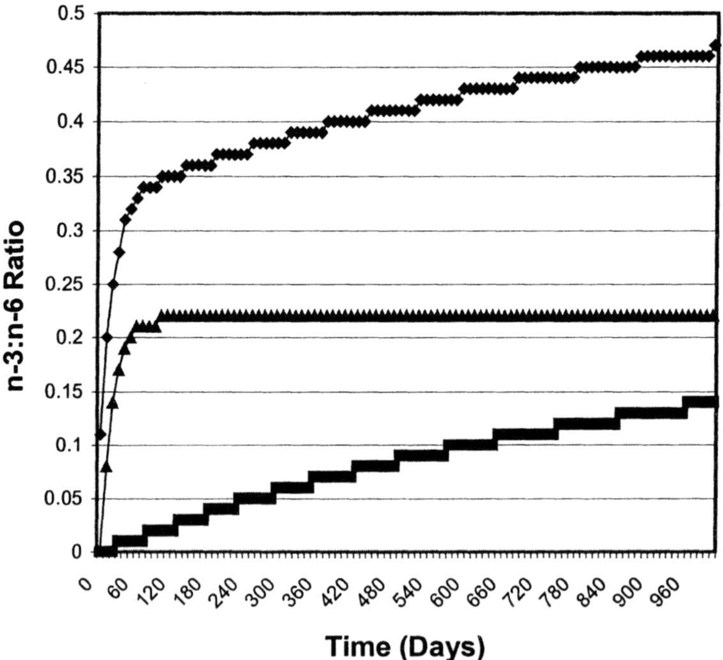

Figure 2. Predicted rate of equilibration of membrane phospholipids after a change in dietary n-3: n-6, assuming a slow component (adipose) and a rapid component (phospholipid). Top curve is a composite for change in phospholipid; the bottom two curves depict the rapid and slow components, respectively.

rats (Lands et al., 1990). Dams were fed several defined diets that varied in content of 18:2 n-6 and 18:3 n-3 fatty acids, and they were then mated with an unrelated male. After the pups were born, they were weaned and then maintained on the same diets that had been fed to the dams. Samples of adipose were taken when the animals were 20 weeks of age, ensuring that the fat depots were fully equilibrated with dietary fatty acids. The results suggest that adipose stores all 18-carbon fatty acids approximately in proportion to dietary abundance. This is not true for the corresponding HUFA (Lands et al., 1990).

During a fast, lipolysis is activated and free fatty acids are released from triacylglycerol. The free fatty acids are taken up by the liver and used for synthesis of phospholipids in lipoprotein particles. For this reason, the fatty acid composition of plasma phospholipids in blood plasma of fasting animals and humans represents a biomarker for prior dietary patterns (Lands, 1995).

CONCLUSIONS

Nutrition scientists are keenly interested in the education of graduate students who are training to become the next generation of scientists and who should be taught the art of hypothesis testing. As collegiate instructors, we also train undergraduates who may become nutrition educators. We may be asked to serve on boards that set policy or to discuss recent findings with representatives of broadcast or print media. Many members of the public, however, believe that dietary advice changes too often to be of value. In each case, improved grounding in theory can help avoid making overstatements that tend to confuse audiences.

Sound grounding in theory requires a quantitative approach. Empirical science is based upon measurement and comparison, and it produces numerical data. The proper framework for scientific thinking involves mathematics, which is probably the weakest element in most biological training. At a minimum, our students should be able to solve the equations they encounter in textbooks and journals. Ideally, they should also be able

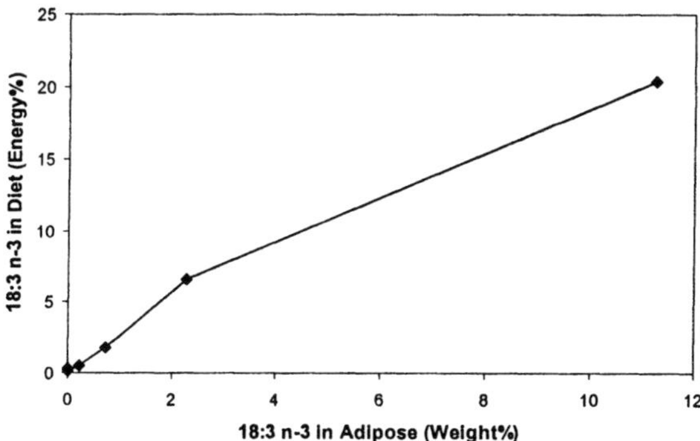

Figure 3. Relationship between 18:3 n-3 fatty acids in rodent diets and accumulation in adipose triacylglycerol. Values were plotted based on tabular data reported by Lands et al. (1990) for female rats fed chemically defined diets with different ratios of 18-carbon n-3 and n-6 unsaturated fatty acids.

to phrase hypotheses in a numerical form so that outcomes can be predicted quantitatively. Without this ability, it is very difficult to prove an hypothesis to be wrong or to say how widely it misses the mark. This leads to the question of whether computer software can assist with understanding of theory and with testing hypotheses.

This paper asks whether biochemical theory may suggest answers to the age-old question of what constitutes a healthy diet. The equations discussed here are fundamentally simple because they are based on Michaelis-Menten kinetics with competitive inhibition. However, their form is so complex that their predictions would be meaningless to anyone who lacked a means to solve them. One function of software for mathematical modeling is that it can be used to solve equations, and the results can be saved and reused at any time without needing to rewrite the equations. New parameter values can be tested at any future time when new questions may arise. In the present paper, the original equations for a steady-state model (Lands et al., 1990; Lands, 1991) were modified slightly to incorporate the kinetics of change. Evidence suggests that fatty acids equilibrate with a very slowly exchanging pool in adipose tissue and with another very rapidly exchanging pool associated with phospholipids. The real situation is far more complex, but the present solution is simple enough to answer what the outcome will be after consuming a particular diet for a stated length of time. It may be argued that the exact answer can never be provided without accurate assessment using biochemical measures. This is true, but most people make decisions concerning food without any monitoring for most of their lives. Therefore, it is important to use every available tool to help motivate students to make better choices in the light of theory.

The model agrees with world-wide epidemiological evidence that total fat intake is not an important control point for phospholipid metabolism. Japanese maintain excellent heart health when consuming diets in which fat intake is only 10-12% of calories, and residents of Crete and Iceland manage very well at fat intakes of more than 30%. People in countries with the highest life expectancy traditionally consume diets in which the ratio of n-6 to n-3 fatty acids is less than 4-5 (de Lorgeril et al., 1994; Renaud et al., 1995). Simopoulos (1999) has reviewed extensive literature suggesting that human diets were historically richer in n-3 fatty acids and that the food supply would be improved if a better balance with n-6 fatty acids were restored. The present kinetic model suggests that the ratio between these eicosanoid precursors is crucial in restoring balance within membrane phospholipids. However, 18-carbon precursors are distributed and metabolized differently than 20-22-carbon preformed HUFA, and any strategy aimed at improving health should take this difference into account. There is merit to including sources of precursors such as canola oil and walnuts, as well as consuming cold-water fish with high contents of n-3 HUFA.

The half-life of fatty acids in human adipose is about 680 days (Dayton et al., 1966), although this value depends on the amount of fat present and daily energy usage. This consideration is very important for people who wish to improve their heart health. α-Linolenic acid, the precursor to 20:5 n-3 and 22:6 n-3 fatty acids, is stored in adipose. For that reason, it is predicted that tissues would not truly equilibrate with an increased intake of this fatty acid for several years. In contrast, HUFA are not readily stored in adipose but are delivered to tissues as components of plasma phospholipids. The half-life of phospholipids in humans is a matter of hours to days in most tissues (Gibney and Hunter, 1993; Emken et al., 1998). The most effective outcome would result from taking in less 18:2 n-6 (linoleic acid) and more 18:3 n-3 (20:5 and 22:6) fatty acids. The practice of eating two or more servings of fish per week, or of taking in EPA and DHA supplements, is effective.

The saturation model of dietary fat and phospholipid composition makes a second very important point. Namely, a ratio of n-6 to n-3 fatty acids of about 3-4 can be achieved at low levels of intake such as occurs in typical Japanese diets, because saturation occurs when the diet contains about 1% energy from these fatty acids. The model suggests that current U.S. intake of 5-7% n-6 linoleic acid is superfluous. Because n-3 fatty acids are not abundant in most foods, an improved balance in phospholipids can only be achieved by simultaneously increasing intake of good sources of n-3 fatty acids and decreasing intake of n-6 fatty acids. Individuals who prefer to consume higher levels of fat should note that the stearic acid and oleic acid in olive oil do not compete favorably for phospholipids. It is interesting that traditional diets in Iceland, Greenland, Crete, and Japan conform well to the saturation model and are associated with excellent longevity.

The personal computer has created a revolution in instruction, but its ability to serve as a mathematical assistant and to facilitate quantitative thinking is just beginning to be appreciated. Most scientists already use spreadsheets for performing statistical analysis and as a means to graph outcomes prior to publication. Spreadsheets are also capable of setting up mathematical models. However, there are also several high quality software packages designed for computer-assisted mathematical modeling. The ability to conduct numerical studies adds an exciting dimension to experimentation because it allows investigators to make numerical predictions that are falsifiable. Models can be used in classrooms to add the dimensions of dynamics and interactivity to computer-assisted presentations. It is also feasible to post developed models on the World Wide Web and exchange them by file transfer (Wastney et al., 1998). More than this, modeling software is fun to use.

ACKNOWLEDGMENTS

Work concerning this model was sponsored in part by a grant from the S.E. Regional Affiliate of the American Heart Association and by funds from the Georgia Agriculture Experiment Station, Hatch number H633.

CORRESPONDENCE

Please address all correspondence to:
James L. Hargrove
Department of Foods and Nutrition
Dawson Hall
The University of Georgia
Athens, GA 30602
jhargrov@fcs.uga.edu

REFERENCES

Calder, P.C., Harvey, D.J., Pond, C.M., and Newsholme, E.A., 1992, Site-specific differences in the fatty acid composition of human adipose tissue, *Lipids* 27:716-720.

Dayton, S., Hashimoto, S., Dixon, W., and Pearce, M.L., 1966, Composition of lipids in human serum and adipose tissue during prolonged feeding of a diet high in unsaturated fat, *J. Lipid Res.* 7: 103-111.

de Lorgeril, M., Renaud, S., Mamelle, N., Salen, P., Martin, J.-L., Monjaud, I., Guidollet, J., Touboul, P., and Delaye, J., 1994, Mediterranean alpha-linolenic acid-rich diet in secondary prevention of coronary heart disease, *Lancet* 343:1454-1459.

Emken, E.A., Adlof, R.O., Duval, S.M., and Nelson, G.J., 1998, Effect of dietary arachidonic acid on metabolism of deuterated linoleic acid by adult male subjects, *Lipids* 33:471-480.

Gibney, M.J., and Hunter, B., 1993, The effects of short- and long-term supplementation with fish oil on the incorporation of n-3 polyunsaturated fatty acids into cells of the immune system in healthy volunteers, *Europ. J. Clin. Nutr.* 47:255-259.

Handelman, G.J., Epstein, W.L., Machlin, L.J., van Kuijk, F.J.G.M., and Dratz, E.A., 1988, Biopsy method for human adipose with vitamin E and lipid measurements, *Lipids* 23:598-604.

Hargrove, J.L., 1998, *Dynamic Modeling in the Health Sciences,* Springer, New York.

Havel, R.J., 1985, The role of the liver in atherosclerosis, *Arteriosclerosis* 5:569-90.

Hwang, J., 2000, *Effects of Dietary Fats on Fatty Acid Composition of Membrane Phospholipids and Drug Toxicity: the Saturation Kinetic Model,* Ph.D. dissertation, The University of Georgia, Athens.

Klein, R.A., Halliday, D., and Pittet, P.G., 1980, The use of 13-methyltetradecanoic acid as an indicator of adipose tissue turnover, *Lipids* 15: 572-578.

Lands, W.E.M., 1991, Biosynthesis of prostaglandins, *Ann. Rev. Nutr.* 11: 41-60.

Lands, W.E.M., 1995, Long-term fat intake and biomarkers, *Am. J. Clin. Nutr.* 61:721S-725S.

Lands, W.E.M., Morris, A., and Libelt, B., 1990, Quantitative effects of dietary polyunsaturated fats on the composition of fatty acids in rat tissues, *Lipids* 25:505-516.

Lands, W.E.M., Libelt, B., Morris, A., Kramer, N., Prewitt, T.E., Bowen, P., Schmeisser, D., Davidson, M.H., and Burns, J.H., 1992, Maintenance of lower proportions of (n-6) eicosanoid precursors in phospholipids of human plasma in response to added dietary (n-3) fatty acids, *Biochim. Biophys. Acta* 1180:147-162.

Navarro, M.D., Periago, J.L., Pita, M.L., and Hortelano, P., 1994, The n-3 polyunsaturated fatty acid levels in rat tissue lipids increase in response to dietary olive oil relative to sunflower oil, *Lipids* 29:845-849.

Renaud, S., de Lorgeril, M., Delaye, J., Guidollet, J., Jacquard, F., Mamelle, N., Martin, J.-L., Monjaud, I., Salen, P., and Toubol, P., 1995, Cretan Mediterranean diet for prevention of coronary heart disease, *Am. J. Clin. Nutr.* 61 (Suppl):1360S-1367S.

Schäfer, L., and Overvad, K., 1990, Subcutaneous adipose-tissue fatty acids and vitamin E in humans: relation to diet and sampling site, *Am. J. Clin. Nutr.* 52: 486-489.

Simopoulos, A.P., 1999, New products from the agri-food industry: the return of n-3 fatty acids into the food supply, *Lipids* 34 (Suppl):S297-S301.

Skuladottir, G.V., Gudmundsdottir, S., Olafsson, G.B., Sigurdsson, S.B., Sigfusson, N., and Axelsson, J., 1995, Plasma fatty acids and lipids in two separate, but genetically comparable, Icelandic populations, *Lipids* 30:649-655.

Wastney, M.E., Yang, D.C., Andretta, D.F., Blumenthal, J., Hylton, J., Canolty, N., Collins, J.C., and Boston, R.C., 1998, Distributing working versions of published mathematical models for biological systems via the Internet, *Adv. Exp. Med. Biol.* 445:131-135.

Wastney, M.E., Patterson, B.H., Linares, O.A., Greif, P.C., and Boston, R., 1999, *Investigating Biological Systems Using Modeling,* Academic Press, New York.

Zuijdgeest-van Leeuwen, S.D., Dagnelie, P.C., Rietveld, T., van den Berg, J.W., and Wilson, J.H., 1999, Incorporation and washout of orally administered n-3 fatty acid ethyl esters in different plasma lipid fractions, *Brit. J. Nutr.* 82:481-488.

PART III
THEORETICAL MODELING ISSUES

MODELING PROCESSES FROM PROBABILITIES

James H. Matis and Thomas R. Kiffe[*]

INTRODUCTION

John A. Jacquez passed away in October 1999 after a very distinguished career in biomedical modeling. Among his most influential accomplishments is the book *Compartmental Analysis in Biology and Medicine*, first published in 1972 and as the third edition in 1996. Others will no doubt give comprehensive reviews of his many lifetime achievements. The objective of this paper is to show how his work lays a strong foundation for the present topic, namely modeling processes from probabilities. We recognize that this topic is only a small part of his book, and we leave it to others to point out the richness of his work in additional areas of great importance in biomedical modeling.

This paper is expository. The principle objectives are to develop the theory and application of the stochastic compartmental model. To facilitate the exposition, we will refer to a data set on calcium clearance which we have previously considered in the literature (Matis and Wehrly, 1998). Sections 1 and 2 review the deterministic compartmental model, with Sections 3 and 4 outlining an analogous stochastic formulation. Sections 5 and 6 present an equivalent stochastic formulation based on exponential transit times, and Sections 7 and 8 generalize the stochastic model to non-exponential transit times. Numerous references will be given to corresponding results in Jacquez (1996).

SECTION ONE: ONE-COMPARTMENT DETERMINISTIC MODEL

Consider first the one-compartment deterministic model. Let $X(t)$ denote the amount of tracer (e.g., labeled calcium) in the compartment at time t, let $\dot{X}(t)$ denote the derivative of $X(t)$, and let k denote the constant fractional (or proportional) flow rate. The one-compartment model is defined as

$$\dot{X}(t) = -kX(t) \tag{1.1}$$

---

[*] James H. Matis, Department of Statistics and Thomas R. Kiffe, Department of Mathematics, Texas A&M University, College Station, TX 77843.

This differential equation has the familiar solution

$$X(t) = X(0)\exp(-kt) \tag{1.2}$$

where $X(0)$ is the initial amount (i.e., the dose).

Typically one measures a concentration $C(t)$. The model for $C(t)$ may be obtained by dividing $X(t)$ by the apparent volume of distribution V. That is,

$$C(t) = (X(0)/V)\exp(-kt) \tag{1.3}$$
$$= C(0)\exp(-kt)$$

The parameters k and V, or k and $C(0)$, may be estimated from data using least squares procedures.

SECTION 2: MULTICOMPARTMENT DETERMINISTIC MODEL

The subsequent generalization holds for any n-compartment model. Consider now the following expanded set of notation. Let $X_i(t)$ be the amount of tracer in compartment i at time t; let $\mathbf{X}(t) = [X_1(t),\ldots,X_n(t)]'$ be the column-vector of amounts at time t; let k_{ij} for $i = 0,1,\ldots,n$, $j = 1,\ldots,n$, $i \neq j$ denote the fractional flow rate to i from j, where 0 represents the system exterior; let $k_{jj} = \Sigma_{i \neq j} k_{ij}$ denote the total outflow rate from j; and let $\mathbf{K} = (k_{ij})$ be the $n \times n$ matrix of k_{ij} coefficients.

One could now assume a linear compartmental model, namely one where each derivative $\dot{X}_i(t)$ is a linear function of the $X_i(t)$ s. Such linear compartmental models may be expressed in matrix form as follows:

$$\dot{\mathbf{X}}(t) = \mathbf{K}\mathbf{X}(t) \tag{2.1}$$

This vector of differential equations has the following solution:

$$\mathbf{X}(t) = \exp(\mathbf{K}t)\mathbf{X}(0) \tag{2.2}$$

with the matrix exponential defined as $\exp(\mathbf{K}t) = \mathbf{I} + \sum_{i=1}^{\infty}\mathbf{K}^i t^i / i!$ Comparable expressions are given in Jacquez' 1996 book (pages 48 and 53).

In practice, one may find a closed form solution by finding first the eigenvalues, say $\lambda_1,\ldots,\lambda_n$ of the matrix $\mathbf{K}$. One then has the following corollary: assuming that the eigenvalues of $\mathbf{K}$ are distinct and real, Eq. (2.2) implies that

$$X_i(t) = \sum_{j=1}^{n} A_{ij}\exp(\lambda_j t), \text{ for } i=1,\ldots,n \tag{2.3}$$

This defines the "sums-of-exponentials" model which generalizes Eq. (1.2) and which is a direct consequence of the assumed linear (or homogeneous) compartmental model. The

A_{ij} and λ_j are often called the "macro-parameters," and they are usually involved functions of the k_{ij} or "micro-parameters."

SECTION 3: ONE-COMPARTMENT STOCHASTIC MODEL BASED ON TRANSFER PROBABILITIES

Consider the following rationale. On page 235, Jacquez (1996) discusses the previous deterministic theory.

> ....the material in a compartment is treated as a continuum, i.e. it is infinitely divisible. But matter is atomic, not continuous, and cells, animals and people come in discrete units. Thus....a compartment has in it an integral number of units and in any transfer only an integral number of units can be transferred. Consequently it is important to develop the theory for such systems in which transfers occur in discrete numbers of units and that is done in terms of the probabilities of transfers of one unit from a compartment to another or to the outside. After doing that, we want to compare the results of the stochastic theory with those of the deterministic theory of compartmental systems.

This leads to a so-called "particle" model. Let $P(t)$ denote the probability that a particular particle introduced at time 0 is still in the compartment at time t, with $\dot{P}(t)$ as its derivative, and let $X(t)$ denote the number of particles introduced at $t=0$ which are still present at time t. Assume

$$\text{Prob [any random particle present at } t \text{ leaves by } t + \Delta t \text{ , where } \Delta t \text{ is small]} = k\Delta t \qquad (3.1)$$

and that all $X(0)$ particles are "equivalent and independent." The constant k in Eq. (3.1) is a probability intensity coefficient, also called a hazard rate, which defines a conditional probability of particle transfer, rather than the proportional transfer rate as before in Eq. (1.1). This corresponds to Jacquez' stochastic formulation on page 239. The differential equation model for $P(t)$ is

$$\dot{P}(t) = -kP(t) \qquad (3.2)$$

This is a stochastic analog to the deterministic model; see Eq. (1.1). The solution to the model with initial condition $P(0)=1$ is

$$P(t) = \exp(-kt) \qquad (3.3)$$

$P(t)$ may be viewed as a common "survival" probability of the particles, and Eq. (3.3) implies that the individual particle "lifetimes" follow an exponential distribution. Moreover, because all of the particles are independent by assumption, the number of particles which survive to time t follows a binomial distribution. This fact leads to the following corollary: the distribution [Eq. (3.4)] and mean [Eq. (3.5)] of the particle count $X(t)$ are

$$X(t) \sim \text{binomial } [n = X(0), p = P(t)] \qquad (3.4)$$

$$E[X(t)] = X(0)P(t) = X(0)\exp(-kt) \tag{3.5}$$

Jacquez (page 242) derives Eq. (3.5) from a set of so-called Kolmogorov equations. This formulation is helpful for subsequent generalization, but it is not presented here for simplicity. The fundamental difference between the stochastic model for $X(t)$ in Eq. (3.4) and the deterministic model in Eq. (1.2) arises from the assumed chance mechanism in Eq. (3.1), which generates so-called "process" uncertainty or process error. The observed data from a system assumed to follow this stochastic model may be fitted to the mean value function in Eq. (3.5), which is identical to the model in Eq. (1.2). Several applications of the stochastic one-compartment model are given in Matis and Kiffe (2000).

SECTION 4: MULTICOMPARTMENT STOCHASTIC MODEL BASED ON TRANSFER PROBABILITIES

Jacquez (1996) proceeds to develop a stochastic n-compartment model in his Sections 12.4-12.8. Consider the following definitions. Let $P_{ij}(t)$ be the probability that a particle starting in j at time 0 will be in i at time t, $\mathbf{P}(t) = [P_{ij}(t)]$ be an n x n matrix of probabilities, and $X_{ij}(t)$ be the number of the $X_{jj}(0)$ particles starting in j at time 0 which are in i at time t. For subsequent convenience, the present formulation tracks separately the distribution of particles for each compartment of origin. The assumptions defining the chance (or probability) mechanism for this model are

Prob [a random particle in j at time t will be in i at time $t + \Delta t$ for small Δt]$= k_{ij}\,\Delta t$ $\qquad$ (4.1)

and that all particles are "equivalent and independent." These assumptions, which generalize Eq. (3.1), define a particular stochastic model called a Markov process. As in Section 3, the k_{ij} constants may be regarded as rate coefficients which define the instantaneous transfer probabilities. A set of differential equations generalizing Eq. (3.2) may be constructed for any n. Using the matrix definition of $\mathbf{K}$ given in Section 2, these equations give the following model:

$$\dot{\mathbf{P}}(t) = \mathbf{K}\mathbf{P}(t) \tag{4.2}$$

This matrix model for probabilities has similar form to that of Eq. (2.1). The initial condition is $\mathbf{P}(0) = \mathbf{I}$, which leads to the following solution:

$$\mathbf{P}(t) = \exp(\mathbf{K}t) \tag{4.3}$$

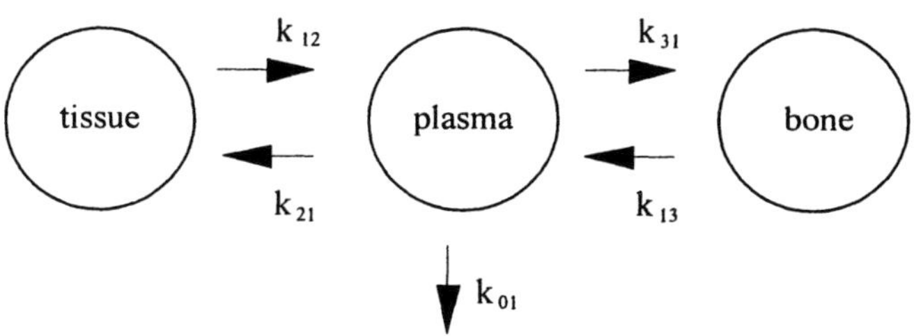

Figure 1. Schematic for calcium clearance model.

A corollary which generalizes the previous result follows: if the λ s are distinct and real, then the probabilities have form

$$P_{ij}(t) = \sum_{\ell} A_{ij\ell} \exp(\lambda_i t) \text{ for } i, j = 1\ldots, n \tag{4.4}$$

For any λ_i, the mean (i.e., expected) values for the counts are

$$E[X_{ij}(t)] = X_{jj}(0)P_{ij}(t) \tag{4.5}$$

The mean value function, defined in Eqs. (4.4) and (4.5) and derived from stochastic assumptions, has the same form as Eq. (2.4) which was derived from deterministic assumptions. Comparable results for the means are given by Jacquez (page 249), who also proceeds to find the variance and covariance of the random counts. In practical applications, one could fit the identical sum-of-exponentials functions in Eqs. (2.4) or (4.5) to experimental data, the difference being only in the interpretation of model foundations.

The application of this stochastic Markov process model may be illustrated with a data set on calcium clearance (Matis and Wehrly, 1998). The underlying physiological model is represented by the compartmental schematic in Figure 1. In this postulated compartmental system, compartment 1 represents calcium in the plasma. The calcium is assumed to exchange with two other compartments, which we call soft tissue and bone. The regression function in Eqs. (4.4) and (4.5) for this model would have six macroparameters, consisting of As and λ s. The regression function could also be stated in terms of six microparameters, five from the $P_{11}(t)$ occupancy probability in Eq. (4.4) (i.e., $k_{21}, k_{12}, k_{31}, k_{13}$ and k_{01}) and one from a scaling factor in Eq. (4.5) (in this case, $C_{11}(0) = X_{11}(0)/V_1$). The model was fit using KINETICA (Allen and Matis, 1990). The estimates of the microparameters, with estimated standard errors in parentheses, are $\hat{C}_{11}(0) = 693.9 \ (17.0)$, $\hat{k}_{21} = 2.611 \ (0.251)$, $\hat{k}_{12} = 2.998 \ (0.289)$, $\hat{k}_{31} = 0.389 \ (0.036)$, $\hat{k}_{13} = 0.162 \ (0.014)$, and $\hat{k}_{01} = 0.0652 \ (0.0022)$. The fitted curve using macroparameters is given by the program as

$$\hat{C}_{11}(t) = 351.39e^{-5.835t} + 197.20e^{-0.375t} + 145.31e^{-0.0144t} \qquad (4.6)$$

which is plotted with the data in Figure 2.

It seems remarkable that one can estimate, with acceptable precision, all six model parameters from such data on a single compartment. The estimated curve fits the data well, with a mean squared error of 27.6. However, as indicated in Weiss et al. (1994), there are systematic, albeit small, deviations from the fitted curve.

Clearly, the stochastic solution has provided an alternative, rich theoretical framework for compartmental modeling. Yet, as Jacquez observes (page 236), "the solutions for the mean values for such stochastic systems are the same as the solutions for the corresponding (linear) deterministic systems." It is our contention that much of the practical contribution of stochastic modeling lies in its subsequent flexibility in utilizing various transit time distributions of interest, as indicated subsequently.

SECTION 5: ONE-COMPARTMENT STOCHASTIC MODEL BASED ON EXPONENTIAL TRANSIT TIME DISTRIBUTIONS

In his Chapter 7 entitled "Residence, Exit and Transit Times," Jacquez (1996) develops these concepts from the deterministic model, using underlying principles of moments from basic kinetic systems in physics. In retrospect, it is surprising that neither he nor we recognized the close parallels between our work until 1999 when he reviewed our text (Matis and Kiffe, 2000). Commenting on our mean residence time results from a scholastic model, Jacquez made the following observations (personal communication to J.H.M.).

Deterministic compartmental models also give residence time distributions and for linear compartmental models with constant coefficients, one obtains [our stochastic results], ... see Chapter 7 of my book, especially pages 144-145. Thus the deterministic and stochastic formulations for linear systems with constant coefficients give the same mean residence times. ...I suspect the two also have the same pdfs [i.e., probability density functions] of residence times. I have not thought about

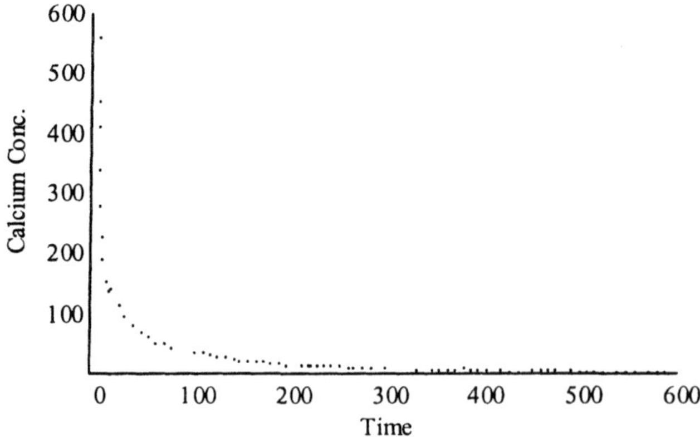

Figure 2. Observed data with fitted curve from Figure 1 model.

this problem before, but here is a simple demonstration that the deterministic and stochastic one compartment, linear models have the same pdfs for residence times, counterintuitive as that might seem to be.

Jacquez then derives Eq. (5.2) that follows. Once he pointed out these facts, correspondences between the two formulations were clear. We present this section from our perspective, as statisticians, and point out similarities in Jacquez' Chapter 7 throughout the following two sections.

In addition to the definition based on probabilities, as in Eq. (3.1), a stochastic model may also be defined on the basis of its transit time (also called retention time or sojourn time) distributions. In some ways, this latter conceptualization of the inherent chance mechanism is more natural because it relies on a continuous time probability distribution, rather than on a conditional flow probability in discretized time units.

Let R be the transit time of a random particle in the system, let $F(a) = $ Prob $[R \leq a]$ be the distribution function of R (i.e., the probability that the particle will leave prior to attaining "age" a in the compartment), let $f(a) = dF(a)/da$ be the density function, and let $S(a) = $ Prob $[R > a] = 1 - F(a)$ be the survivorship function (i.e., the probability that the particle "survives" to age a in the compartment). Then, suppose one assumes an exponential distribution for R. Assuming $R \sim$ exponential distribution (k), this implies

$$F(a) = 1 - \exp(-ka) \tag{5.1}$$

$$f(a) = k \exp(-ka) \tag{5.2}$$

$$S(a) = \exp(-ka) \tag{5.3}$$

Jacquez obtained Eq. (5.2) from the deterministic model, noting on page 135 that $\phi(t) = ke^{-kt}$ is a proper density function for the time a random particle leaves the system. Yet his perspective is different, and he does not explicitly identify this as an exponential distribution.

Our approach is to assume the exponential distribution and then show its equivalence to the previous stochastic formulation. Clearly, survivorship function $S(a)$ in Eq. (5.3) is equivalent to the previous function $P(t)$ solved in Eq. (3.3). The argument of this function is changed for subsequent convenience from t to a.

The relationship to assumption (3.1) may be established using the conditional flow probability, or hazard rate, of the assumed exponential distribution. The exponential distribution leads to the following model: the conditional probability that a particle present at age a leaves by $a + \Delta t$ is

$$1 - \exp(-k\Delta t) = k\Delta t + o(\Delta t) \tag{5.4}$$

where $o(\Delta t)$ denotes higher order terms of Δt.

In the model derivation, as Δt approaches 0, assumption (5.4) is equivalent to assumption (3.1); see, for example, Bailey (1964, page 85). Moreover, a well-known characterization of the exponential distribution is that it is the *only* continuous distribution with the constant conditional transfer probability, as given in assumption (5.4). Therefore, it is the *only* continuous distribution with the so-called "lack of memory" property which is characteristic of the transit time distributions in Markov processes. One practical consequence of this model is that the transfer mechanism must not discriminate on the basis of the accrued age of a particle in the compartment. The common conceptualization of "well-mixed" compartments satisfies this constraint. In our experience, particularly as it relates to kinetic modeling in ruminants, experimenters relate easily to the concept of a possible "lack of memory" in particle transfer, and hence they are able to conceptualize whether or not the assumption of an exponential distribution for the transit time variable R is plausible in any given application.

In summary, it is clear that the formulations in Sections 3 and 5 are equivalent. It follows that all the results in Section 3 hold for the present formulation which we call the "exponential" compartment formulation. The focus on transit times may be exploited to determine various statistical moments of R, such as its mean and variance. This leads to the following well-known corollary that gives the mean [Eq. (5.5)] and variance [Eq. (5.6)] for the assumed exponential distribution R.

$$E[R] = \int af(a)da = k^{-1} \tag{5.5}$$

$$V[R] = \int a^2 f(a)da - (E[R])^2 = k^{-2} \tag{5.6}$$

The mean and variance are denoted as $\bar{t}$ and V, respectively, by Jacquez (page 136).

SECTION 6: MULTICOMPARTMENT STOCHASTIC MODEL BASED ON EXPONENTIAL TRANSIT TIME DISTRIBUTIONS

Many new random variables, such as residence times, may be defined for a stochastic multicompartmental model to describe the dynamics of particle transfer among the various compartments. We will again define these in statistical terms. Let R_{ij} denote the transit time during a single visit in j of a particle transferring next to I, and let R_j denote the transit time during a single visit of a random particle in j prior to its next transfer out of j (either to another compartment or to the exterior). Also, let N_{ij} denote the random number of visits that a particle starting in j will make to i, let S_{ij} be the total residence time that a particle originating in j will accumulate in i during all of its N_{ij} visits ($S_{ij} = \sum_1^{N_{ij}} R_{ij}$), and let $S_j = \sum_i S_{ij}$ be the residence time (also called the exit time) of a particle originating in j.

Consider now the following assumption: let $R_{ij} \sim$ exponential distribution (k_{ij}) for all i and j. The assumption implies that each hazard rate is a constant; hence there are constant (i.e., time-invariant) transfer probabilities between compartments. Consequently, a Markov-chain model could be used to describe sequential particle transfers among the compartments. A useful result on the expected values of the S_{ij} residence times from Matis et al. (1983) follows. Let $\mathbf{E}(S)$ be an $n \times n$ matrix of mean residence times with elements $E[S_{ij}]$ and let $\mathbf{E}.(S)$ be an n-vector of expected residence times with elements $E[S_i]$. This results in the following theorem. The mean residence times (MRT) are

$$E(S) = -\mathbf{K}^{-1} \tag{6.1}$$

$$\mathbf{E}.(S) = \mathbf{E}(S)\mathbf{1} \tag{6.2}$$

where $\mathbf{1}$ is a vector of ones. Clearly, Eq. (6.1) is a matrix analog to Eq. (5.5). The negative inverse in Eq. (6.1) results from the fact that the diagonal elements of $\mathbf{K}$ are negative by construction. Jacquez gives comparable results on pages 137 and 139. Many other results, for example the variances of the residence times, and the means and variances of the N_{ij} number of transfers, are easy to derive from classical theory of Markov chains and Markov processes.

Equation (6.1) involves only simple matrix inversion and therefore is easy to apply. These mean residence times have been very useful in practical applications where the compartmental structure is assumed known. Some advantages of using mean residence times to describe system kinetic behavior are discussed in Matis et al. (1983).

Consider the following illustration. The estimated rates from the sample data set and model may be substituted into the coefficient matrix $\mathbf{K}$ in Eq. (4.2) which defines a linear system. The estimated matrix is

$$\hat{\mathbf{K}} = \begin{bmatrix} -3.0652 & 2.999 & 0.162 \\ 2.611 & -2.999 & 0 \\ 0.389 & 0 & -0.162 \end{bmatrix}$$

Substituting $\hat{\mathbf{K}}$ into Eq. (6.1), the matrix of mean residence times (MRT) is

$$\hat{\mathbf{E}}(S) = \begin{bmatrix} 15.344 & 15.344 & 15.344 \\ 13.362 & 13.359 & 13.692 \\ 36.860 & 43.038 & 36.858 \end{bmatrix}$$

As a summary, under this stochastic model, a calcium molecule introduced into the plasma compartment (compartment 1) has estimated mean residence times of 15.34, 13.36, and 36.86 hours in the three compartments, for a mean exit time of 50.22 hours.

SECTION 7: ONE-COMPARTMENT MODEL BASED ON NON-EXPONENTIAL TRANSIT TIME DISTRIBUTIONS

It was observed in Section 5 that the assumption of an exponential transit time is equivalent to the assumption of an age-independent hazard rate. The generalization to age-dependent hazard rates would obviously lead to non-exponential transit time distributions. Let $k(a)$ denote the hazard rate function at time a. By definition, the conditional probability that a particle that has remained in the compartment for age a

leaves by $a + \Delta t$ is then $k(a)\Delta t$. As given in Section 5, the hazard function for any nonnegative distribution may be defined as $k(a) = f(a)/S(a)$. Using this hazard rate concept, a new generalized one-compartment model may be specified by generalizing Eq. (3.2) as follows:

$$\dot{P}(a) = -k(a)P(a) \tag{7.1}$$

Alternatively, one could specify a generalized compartmental model directly through an assumed non-exponential transit time distribution, as is commonly done in statistical survival analysis. The Weibull, gamma, and Erlang distributions are three common non-exponential transit time distributions used as survival models.

Consider first the Erlang distribution: let $R \sim \text{Erlang } (n, \lambda);\quad n = 1, 2, \ldots;\quad \lambda > 0$. This model, a special case of the gamma with an integer-valued shape parameter n, has the following density [Eq. (7.2)], survivorship function [Eq. (7.3)], hazard rate [Eq. (7.4)], mean [Eq. (7.5)], and variance [Eq. (7.6)]:

$$f(a) = \lambda^n a^{n-1} \exp(-\lambda a)/(n-1)! \tag{7.2}$$

$$S(a) = \exp(-\lambda a)\sum_{i=1}^{n-1}(\lambda a)^i / i! \tag{7.3}$$

$$k(a) = [\lambda^n a^{n-1}/(n-1)!]/\left[\sum_{i=1}^{n-1}(\lambda a)^i / i!\right] \tag{7.4}$$

$$E[R] = n/\lambda \tag{7.5}$$

$$V[R] = n/\lambda^2 \tag{7.6}$$

The rate function in Eq. (7.4) has two qualitative properties which make the Erlang distribution a useful transit time model for practical applications. First, for $n>1$, the rate function at age 0 is $k(0) = 0$, after which the rate increases. This provides an initial dampening of the passage probability of newly introduced particles. Secondly, the rate asymptotes to λ as the age a increases. This implies that the age discrimination within the compartment diminishes, either relatively rapidly or slowly depending on n, as the transit time increases. Both of these qualitative features are characteristic of data from non-homogeneous compartments and/or compartments with non-instantaneous initial mixing. The model also has a survival function which is easy to fit to data. In the special case of $n = 1$, one has the exponential distribution, and the formulas reduce to Eqs. (5.1) through (5.6).

In addition, the Erlang distribution has a mathematical property which makes it relatively easy to apply; namely, an Erlang transit time variable may be obtained as the sum of identical and independent exponential transit time variables. The theorem is as follows. Let U_i be independent exponential (λ) transit time variables for $i = 1,2,\ldots,n$. Then $R = \sum_{i=1}^{n} U_i$ is distributed as an Erlang (n,λ) transit time variable.

The Erlang transit time concept is mentioned in Jacquez also (pages 254 - 255). He writes "[consider] the one-way catenary [model] with inflow only into the first compartment and outflow only from the last compartment...If all the k_{ij} [rates] are the same, the mean outflow from compartment n has the Erlang distribution in times of outflow."

The tractability and desirable share properties of the Erlang transit time models have led to their widespread practical application, including in ruminant nutrition modeling (see Matis, 1972; Ellis et al., 1994; Ellis et al., 1999). When these Erlang variables are represented as a sequence of exponential compartments, our experience is that experimenters have a natural tendency to give a mechanistic interpretation to each exponential stage. Consequently, we call the individual components "pseudo-compartments," and we explain that the pseudo-compartments in all likelihood are not anatomical (or mechanistic) subdivisions but instead are merely a mathematical artifice to generate the desired overall Erlang transit time variable.

Although the Erlang variable may be a flexible first approximation to many non-exponential distributions, a natural question is how to generate other transit time variables. In particular, there is a practical need to generate distributions with longer tails than those provided by the Erlang models. A very general approach within a compartmental modeling context is the concept of phase-type (PH) distributions, whose first application is credited to Neuts (1981). A phase-type distribution is defined as the distribution of exit times from a linear, stochastic n-compartment model (of exponential compartments). Note that the Erlang distribution is a PH distribution because it is the exit time distribution of a uni-directional catenary model with equal rates. Other immediate generalizations for PH distributions would include the exit time distributions from a uni-directional catenary model with unequal rates, as well as those from reversible catenary models. When one considers also all possible so-called mammillary models and mixtures of mammillary and catenary models, it is clear conceptually that the PH distributions must be a very rich family with a great diversity of shape characteristics. Indeed, Neuts (1981) shows that any positive, non-degenerate distribution may be represented as a PH distribution. This implies that, in theory, one may use a compartmental model to generate any desired transit time distribution. In practice, the result is used to yield easy approximations of any desired transit time distributions using simple compartmental models. However, the practical utility of this modeling approach is not in single compartment modeling but rather in the multicompartment modeling which is discussed in the following section.

SECTION 8: MULTICOMPARTMENT MODELS BASED ON NON-EXPONENTIAL TRANSIT TIME DISTRIBUTIONS

Consider now a physiological system which lends itself to compartmental modeling but where one or more of the compartments is known to have non-exponential transit times. For example, in the calcium clearance problem, the bone compartment would not be "well-mixed;" hence its transit time would have a non-exponential distribution. In general, consider the following assumption: each R_j transit time has some arbitrary nondegenerate distribution. The specification of a nondegenerate distribution rules out R_j being only a fixed time delay. Therefore, one could restate the assumption as each R_j transit time follows some phase-type distribution.

One problem using PH distributions in general is that the PH representation of a given distribution may not be unique. Consequently, the selection of some particular PH distribution to describe an arbitrary, completely specified distribution remains a challenging problem. However, in most compartmental problems, the exact forms of the non-exponential distributions are not known but must instead be estimated from data. In such cases, one may utilize compartmental submodels to estimate the underlying PH retention time distributions. Conceptually, this modeling approach enables one to represent any generalized compartmental model as an expanded linear system, though obviously the dimensionality of the expanded system may increase substantially with the creation of many pseudo-compartments. Consider the following notation. Let $P_{uv}^*(t)$ denote the probability that a particle starting in pseudo-compartment v at time 0 will be in pseudo-compartment u at time t, for u, $v=1,2,\ldots$ $n=\sum n_i$ and let $\mathbf{P}^*(t) = [P_{uv}^*(t)]$ be the matrix of occupancy probabilities for the pseudo-compartments. Also, let $\mathbf{E}^*(S)$ be the matrix of mean residence times for the pseudo-compartments, and let $\mathbf{K}^*$ denote the expanded matrix rate coefficients among the pseudo-compartments. It follows from Eq. (4.2) that the model for the expanded (linear) pseudo-compartment representation of a multicompartment model with non-exponential transit times is

$$\dot{\mathbf{P}}^*(t) = \mathbf{K}^*\mathbf{P}^*(t) \tag{8.1}$$

It also follows from Eq. (4.3) that the solution is

$$\mathbf{P}^*(t) = \exp(\mathbf{K}^*t) \tag{8.2}$$

In applying the above solution, one must consider the fact that the coefficient matrix $\mathbf{K}^*$ in all likelihood has a special pattern form; hence, in practical applications, $\mathbf{K}^*$ might yield equal and/or complex eigenvalues. Consequently the $P_{uv}^*(t)$ solutions will probably not be the sums-of-exponential models given in Eq. (4.6). Instead, the solutions will tend to have other algebraic forms involving powers of time and/or periodic functions. This greatly increases the model flexibility without adding additional parameters. Although $\mathbf{K}^*$ is a pattern matrix, possibly with equal and/or complex eigenvalues, the structure does not complicate the derivation of mean residence times. $\mathbf{K}^*$ may be substituted directly into formulas (6.1) and (6.2) to obtain these means for the pseudo-compartments. Appropriate linear combinations will then give the solutions of the generalized compartments.

Consider applying this modeling approach to the calcium clearance data presented in Section 6. We assume the same physiological model as in Figure 1 (i.e., a system with three compartments), with connections as given. However, we assume that the bone compartment is not well mixed but rather has a transit time distribution with a long tail. This long-tailed distribution is modeled empirically using a mixture of two Erlangs. The expanded linear representation of the overall model is given in Figure 3. Although the

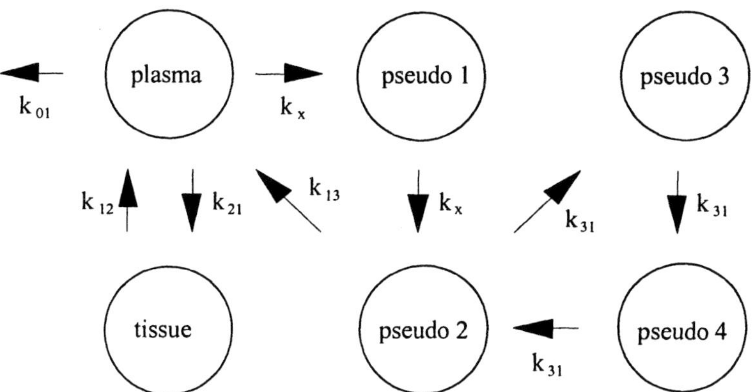

Figure 3. Schematic with expanded calcium clearance model, with $k_x = k_{31} + k_{13}$.

number of pseudo-compartments has increased, the model still has only six parameters once the structure, which relates to the shape of the distribution, is assumed.

The model was fit to data using KINETICA (Allen and Matis, 1990). The mean squared error is 4.94, which is a substantial improvement over the exponential compartmental model in Section 4. The parameter estimates and their estimated standard errors are $\hat{C}_{11}(0) = 722.8$ (7.1), $\hat{k}_{21} = 3.131$ (0.233), $\hat{k}_{12} = 3.735$ (0.141), $\hat{k}_{31} = 0.0302$ (0.0019), $\hat{k}_{13} = 0.414$ (0.016), and $\hat{k}_{01} = 0.0607$ (0.0010). The estimate of the linear combination $k_x = k_{31} + k_{13}$ is $\hat{k}_x = 0.444$ (0.016). The estimated concentration-time curve in plasma is

$$\hat{C}_{11}(t) = 356.16e^{-7.105t} + 68.50e^{-0.0078t}$$

$$+ [44.98\sin(0.2852t) + 187.85\cos(0.2852t)]e^{-0.5659t}$$

$$+ [32.85\sin(0.0162t) + 110.26\cos(0.0162t)]e^{-0.0380t} \tag{9.4}$$

which is plotted with the data in Figure 4. The expanded coefficient matrix $\mathbf{K}^*$ has complex eigenvalues, which give the damped oscillations in Eq. (9.4). In Figure 4, the residuals (i.e., the deviations between the observed data and the fitted curve) have no apparent systematic lack of fit. The estimated expanded coefficient matrix is

$$\hat{\mathbf{K}}^* = \begin{bmatrix} -3.6357 & 3.131 & 0 & 0.414 & 0 & 0 \\ 3.131 & -3.735 & 0 & 0 & 0 & 0 \\ 0.444 & 0 & -0.444 & 0 & 0 & 0 \\ 0 & 0 & 0.444 & -0.444 & 0 & 0.030 \\ 0 & 0 & 0 & 0.030 & -0.030 & 0 \\ 0 & 0 & 0 & 0 & 0.030 & -0.030 \end{bmatrix} \tag{9.5}$$

The estimated mean residence times (MRT) obtained by substituting Eq. (9.5) into Eq. (6.1) are

$$
\mathbf{E}^*[S] = \begin{bmatrix}
16.469 & 16.469 & 16.469 & 16.469 & 16.469 & 16.469 \\
13.805 & 14.073 & 13.805 & 13.805 & 13.805 & 13.805 \\
16.469 & 16.469 & 18.718 & 16.469 & 16.469 & 16.469 \\
17.668 & 17.668 & 20.081 & 20.081 & 20.081 & 20.081 \\
17.668 & 17.668 & 20.081 & 20.081 & 53.230 & 20.081 \\
17.668 & 17.668 & 20.081 & 20.081 & 53.230 & 53.230
\end{bmatrix} \tag{9.6}
$$

The MRTs for a calcium particle introduced into plasma are found as before in the first column. The MRT in the plasma and soft tissue compartments of this better-fitting stochastic model are 16.5 and 13.8, respectively, which are close to the corresponding estimates of 15.3 and 13.4 from the exponential model in Figure 1. However, the MRT for bone is the sum of the MRT in the four pseudo-states in the rest of the first column [i.e., MRT=16.5 + 3 (17.7) = 69.5]. This estimate is nearly a 100% increase from the estimate of 36.9 hours from the alternative model.

It seems remarkable that the estimated MRT in bone would differ so drastically in light of the fact that both curves fit the data quite well. Conceptually, the substantial increase in the MRT for the bone is attributable to the improved fit in the far right tail of the clearance data. An assumed exponential bone compartment, even with a small parameter k, is not sufficiently flexible to generate the desired shape characteristic of a long tail. However, the fitted PH retention time distribution has the desired long tail property. This feature is generated conceptually by its second Erlang distribution, which adds a long retention time to the small proportion of particles.

The analysis in this section assumes a particular PH distribution based on a specific configuration of four pseudo-compartments. The shape and analytical expression for the

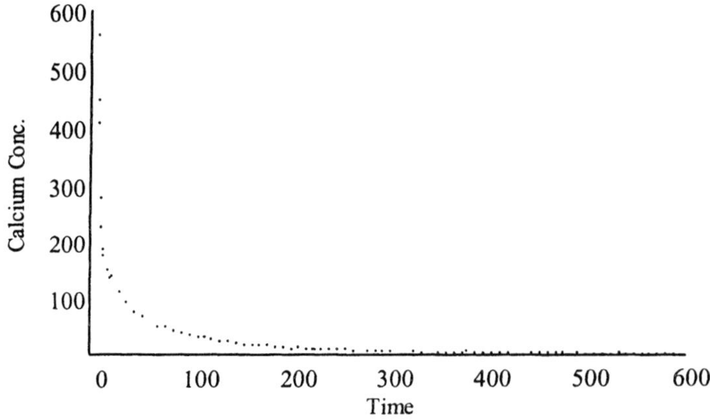

Figure 4. Observed data with fitted curve from Figure 3 model.

estimated PH distribution are given in Matis and Wehrly (1998). Although this particular assumed PH model obviously provides an excellent fitting of the data, there is no assurance that it is either unique or the best. Indeed, recent results indicate that a similar PH model with *five* pseudo-compartments improves the goodness-of-fit slightly, but the use of *six* pseudo-compartments apparently does not. The results are the same qualitatively when weighted nonlinear least squares is used (Matis and Kiffe, 2000).

DISCUSSION

This paper shows that Jacquez (1996) lays a strong foundation for modeling processes with probabilities and specifically for the stochastic compartmental model. He makes three contributions of present interest, namely establishing the theoretical case for using a probabilistic model, deriving the mean residence times, and recognizing the Erlang distribution as an alternative transit time distribution. He also outlines other properties of stochastic models which are not of present interest, including analyzing the models using generating functions, saying "the beginner needs to know their definitions and the more advanced student should learn how to use them," as well as deriving so-called population kinetic models in which the rates are random variables. We pay tribute to him for his great pioneering work in stochastic modeling, a role which has not been widely recognized.

This paper also shows how one can extend the stochastic foundation to include arbitrary transit time distributions in the compartments and outlines methodology whereby this generalized stochastic model may be implemented in practice. To appreciate the model utility, consider first common deterministic modeling practice.

A common approach in obtaining a compartmental model from a single concentration-time curve is to determine first the number of exponential terms required to fit the data adequately with a sum-of-exponentials model such as Eq. (2.3). Typically, one cannot estimate more than three exponentials with adequate precision because of statistical multicollinearly problems (Bates and Watts, 1988). In general, the resulting exponential compartments produced by the mathematical conceptualization are then matched with some mechanistic conceptualization of the underlying physiological system. The process is similar when data are available from multiple compartments over overlapping time periods. The assumed mathematical structure will again yield exponential compartments which are then associated with some conceptual physiological model.

In criticizing this approach, Cohen (1980) makes these observations.

> First, how does one decide what are the compartments? To carve a chicken nicely, one must cut at the joints. How does one find the joints of [a physiological system]? Second, is there any evidence to support, or any procedure to test, the memoryless or ahistorical assumption built into every compartmental model?

We interpret these rhetorical questions to imply that a rigid mathematical Eq. (i.e., sum-of-exponentials model) may be a poor device to find chicken joints (i.e., natural compartments).

The stochastic model provides the flexibility to approach the problem differently. In our analogy, one would feel the chicken first to find the joints, and then use a versatile mathematical approach to describe where to cut the chicken. One might accomplish this by first specifying the underlying mechanistic model of the desired physiological system with a set of phenomenological compartments and their interrelationships. If the phenomenological compartments have apparent non-exponential transit times, one can use mathematical pseudo-compartments to create PH distributions. In principle, this could satisfy Cohen, at least in part, by checking the hypothesis and, if need be, using this flexible mathematical tool to cut at the desired point. The data are then fit using an expanded linear system and the results transformed back to the basic phenomenological compartments. As in the illustration in this paper, the improvement in the mean squared error might not be large; however, the differences in estimating mean residence times may be substantial.

One might ask whether this approach requires a stochastic formulation. That is, would a deterministic formulation do just as well? We suggest that the success of the present approach is in adding additional pseudo-compartments without increasing the number of parameters in order to create a flexible mathematical representation of the transit time distributions within the compartments. This seems contrary to the spirit of the deterministic model foundation, and indeed we are not aware of the deterministic model being used to mimic the present approach. We are optimistic, therefore, that the new approach has great future potential, and we are developing additional tools to implement it.

ACKNOWLEDGMENTS

This paper is presented in memory of John A. Jacquez. The research described here was supported in part by grants 010366-0346 and 000517-0371 from the Texas Advanced Research and Advanced Technology Programs. Passages from Jacquez (1996) and Cohen (1980) are quoted with permission from the publishers.

CORRESPONDENCE

Please address all correspondence to:
James H. Matis
Department of Statistics
Texas A&M University
College Station, TX 77843
matis@stat.tamu.edu

REFERENCES

Allen, D.M., and Matis, J.H., 1990, *KINETICA, A Program for Kinetic Modeling in Biological Sciences*, Department of Statistics, University of Kentucky, Lexington.

Bailey, N.T.J., 1964, *The Elements of Stochastic Processes with Applications to the Natural Sciences*, Wiley, New York.

Bates, D.M. , and Watts, D.G., 1988, *Nonlinear Regression Analysis and Its Applications*, Wiley, New York.

Cohen, J.E., 1980, Review of: Compartmental Analysis of Ecosystem Models, *Quart. Rev. Biol.* 55:453-454.

Ellis, W.C., Matis, J.H., Hill, T.M., and Murphy, M.R., 1994, Methodology for estimating digestion and passage kinetics of forages, in: *Forage Quality, Evaluation and Utilization*, G.C. Fahey, Jr. ed., American Society of Agronomy, Madison.

Ellis, W.C., Poppi, D.P., Matis, J.H., Lippke, H., Hill, T.M., and Rouquette, F.M. Jr., 1999, Dietary-digestive-metabolic interactions determining potential of ruminant diets, in: *Nutritional Ecology of Herbivores,* H.J.G. Jung and G.C. Fahey, Jr., eds., American Society of Animal Science, Savoy.

Jacquez, J.A., 1996, *Compartmental Analysis in Biology and Medicine*, 3rd ed., BioMedware, Ann Arbor.

Matis, J.H., 1972, Gamma time-dependency in Blaxter's compartmental model, *Biometrics* 28: 597-602.

Matis, J.H., and Kiffe, T.R., 2000, *Stochastic Population Models - A Compartmental Approach*, Springer, New York.

Matis, J.H., and Wehrly, T.E., 1998, A general approach to non-Markovian compartmental models, *J. Pharmacokin. Biopharm.* 26:437-456.

Matis, J.H., Wehrly, T.E., and Metzler, C.M., On some stochastic formulations and related statistical moments in pharmacokinetic models, *J. Pharmacokin. Biopharm.* 11:77-92.

Neuts, M. F., 1981, *Matrix-Geometric Solutions in Stochastic Models: an Algorithmic Approach*, John Hopkins University Press, Baltimore.

Weiss, G.H., Goans, R.E., Gitterman, M., Abrams, S.A., Vieira, N.E., and Yergey, A.L., 1994, A non-Markovian model for calcium kinetics, *J. Pharmacokin. Biopharm.* 22:367-379.

UNDERSTANDING THE RELATIONSHIP BETWEEN CARCINOGEN-INDUCED DNA ADDUCT LEVELS IN DISTAL AND PROXIMAL REGIONS OF THE COLON

Jeffrey S. Morris, Naisyin Wang, Joanne R. Lupton, Robert S. Chapkin, Nancy D. Turner, Mee Young Hong, and Raymond J. Carroll[*]

INTRODUCTION

A number of studies document strong epidemiological and clinical links between diet and colon cancer incidence. In particular, a corn oil-supplemented diet (rich in n-6 polyunsaturated fatty acids) has been shown to be less protective against colon tumorigenesis (Haenszel and Kurihara, 1968; Blot et al., 1975; Bang et al., 1976; Boyle et al. 1985; Chang et al., 1997; Hong et al. 2000) than a fish oil-supplemented diet (rich in n-3 polyunsaturated fatty acids). The mechanisms contributing to this apparent diet effect are still largely unknown, however, and they are currently being investigated (see, for example, Hong et al., 2000).

In studying the etiology of colon cancer, it is important to understand the mechanism of cellular DNA damage in the distal and proximal regions of the colon in rats exposed to a carcinogen and fed certain types of diet. After being injected with the carcinogen azoxymethane (AOM), cellular DNA damage is known to be the first event in carcinogenesis, taking place in the first few hours after exposure. If this damage becomes fixed, a mutation may occur; this can potentially lead to the formation of polyps. *DNA adduct level* is a measure of the amount of DNA cell damage that has occurred (Rogers and Pegg, 1977; Swenberg et al., 1979). Important insights into the effects of diet on colon carcinogenesis can be gained by understanding the relationships between DNA adduct levels in the distal and proximal regions of the colon in rats fed particular diets.

Colonic cells replicate and grow in discrete units called *crypts*. In Figure 1, we show representative crypts from cross-sections of the distal and proximal regions of the rat colon. In each crypt, there are stationary, permanent cells called *stem cells* that generate

* Jeffrey S. Morris, Department of Biostatistics, The University of Texas M.D. Anderson Cancer Center, Houston, TX 77030. Naisyin Wang and Raymond Carroll, Department of Statistics, The University of Texas M.D. Anderson Cancer Center, Houston, TX 77030. Joanne R. Lupton, Robert S. Chapkin, Nancy Turner and Mee Young Hong, Faculty of Nutrition, Texas A&M University, College Station, TX 77843.

all of the cells in that crypt. Daughter cells are formed at the crypt depths where stem cells are located. Then, as more cells are produced, cells move up the crypt until they are finally exfoliated into the lumen. Thus, a cell's relative depth in the crypt, or *relative cell position,* is closely related to its age. Cells near the bottom of the crypt are younger, while cells near the top are older. Once they reach the uppermost third of the crypt, cells will normally not divide (Lipkin et al., 1963; Lipkin, 1974).

The special cell-life sequence along the colon crypt indicates that cells at the same relative cell positions in different crypts share common characteristics. As a result, knowledge about the relationship of DNA adduct levels in the distal and proximal regions of the colon as a function of relative cell position helps to answer an important question. That is, do rats that have elevated DNA adduct levels at particular depths in the distal crypts also have elevated DNA adduct levels at the same depth in the proximal crypts? Further, does this relationship depend on diet?

To study these questions, we estimated the correlation function for DNA adduct levels in the distal and proximal regions of the colon as a function of relative cell position in AOM-treated rats fed fish oil- or corn oil-supplemented diets (Hong et al., 2000). Fifteen rats each were fed two diets: fish oil-supplemented or corn oil-supplemented. After two weeks, three rats from each group were euthanized to serve as controls. Remaining rats were injected with AOM, a carcinogen known to induce colon cancer. Three rats from each group were euthanized at 3, 6, 9, and 12 hours after exposure to the carcinogen. Approximately 20 crypts were selected from both the proximal and distal regions of the colon of each rat, and the DNA adduct levels were measured in all cells lining the left side of a cross-section of each crypt (~40 cells/crypt). For each cell, there was a corresponding level of the covariate, relative cell position, which is a proxy for the true cell position. Values of the covariate were equally spaced within each crypt, ranging

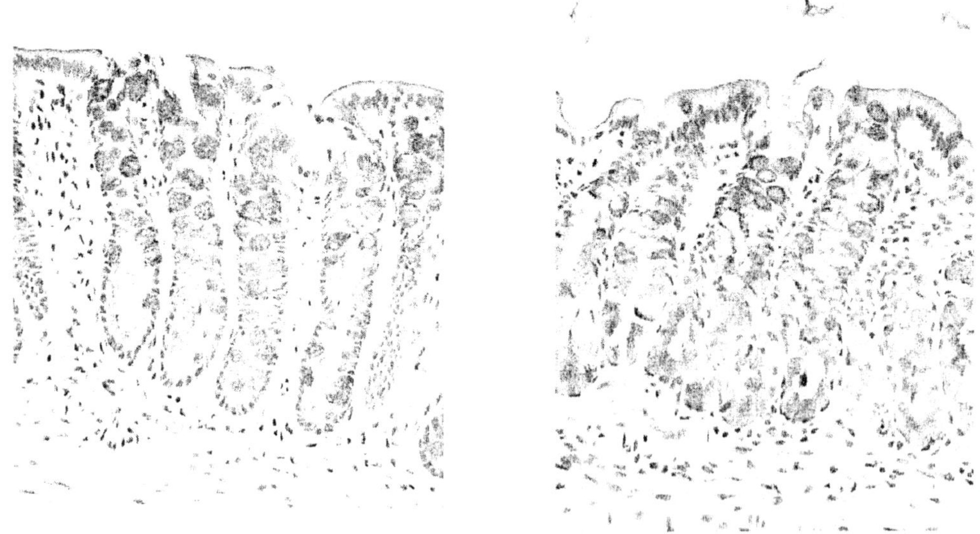

Figure 1. Images of distal and proximal crypts, respectively, in rat colon.

from 0 at the very bottom of the crypt to 1 at the very top. One difficulty that prevented us from estimating the correlation of interest using conventional methods is that the two responses were not measured simultaneously on the same basic unit of measurement (cells within crypts). They were, however, both measured on the same experimental unit (rats), so we can obtain an estimate of the correlation at the experimental unit level.

As discussed later, surprising results were obtained using both parametric and nonparametric methods. For rats in the fish oil-supplemented group, the correlation was positive near the bottom of the crypts but close to zero in the middle and upper depths. For corn oil-fed rats, the correlation was strongly negative throughout all depths of the crypts. Both the negative correlation in corn oil-fed rats and the diet differences were unexpected. These results may have potentially important implications for our understanding of colon carcinogenesis.

GENERAL MODEL AND CORRELATION

Each of the two responses is assumed to follow a nested mixed model with a fixed mean structure and random, mean-zero processes at the levels of experimental unit, subsampling, and residual error. In our example, the two responses are the DNA adduct levels in the distal and proximal regions of the colon, the experimental units are rats, and the subsampling units are crypts. The mean structure and processes are functions of the covariate X, the relative cell position. The fixed mean structure may differ for different diets as well as different times after carcinogen exposure. The response (DNA adduct level) is denoted by Y. Consider time group $j=1,...,5$, rat $r=1,...,R$, and cell number i. The models for the DNA adduct levels at relative cell position X for distal crypt c_d and proximal crypt c_p of rat r in time group j are:

$$Y_{jrc(p)} = m_j(X,p) + m_{jr}(X,p) + m_{jrc(p)}(X,p) + \epsilon_{jrc(p)}(X,p) \tag{1}$$

$$Y_{jrc(d)} = m_j(X,d) + m_{jr}(X,d) + m_{jrc(d)}(X,d) + \epsilon_{jrc(d)}(X,d) \tag{2}$$

where $m_j(X,p)$ and $m_j(X,d)$ denote fixed mean functions for time group j in the proximal and distal regions, respectively. The functions $m_{jr}(X,p)$ and $m_{jr}(X,d)$ are the realizations of rat-level random effects, the functions $m_{jrc(p)}(X,p)$ and $m_{jrc(d)}(X,d)$ are the realizations of crypt-level random effects, and the ϵ functions are residual errors. Conceptually, these models give us curves for the two responses as functions of relative cell position for each crypt. The different subscripts for the distal and proximal crypts emphasize the fact that, while both measurements are taken from the same rat, they are from different crypts. We assume that, except for a possible non-zero correlation $\rho(X)$ depending on cell position, between $m_{jr}(X,p)$ and $m_{jr}(X,d)$, the random variables $m_{jr}(X,p)$, $m_{jr}(X,d)$, $m_{jrc(p)}(X,p)$, $m_{jrc(d)}(X,d)$ and the ϵ functions are mutually independent. For simplicity, we will assume that the ϵ functions are also independent within crypts. This effectively holds in our example, although the assumption is not necessary, and the methods can be modified to weaken the assumption. All random processes have mean zero. Note that the independence assumptions imply that measurements from the same rat but different crypts are dependent on each other only through the rat-level random effect.

While the fixed effects $m_j(X,p)$ and $m_j(X,d)$ are of interest, here we are primarily interested in estimating the correlation $\rho(X)$ between the distal adduct level and the proximal adduct level as a function of relative cell position X. The reason, while perhaps obvious, is worth emphasizing. Our interest is in understanding whether abnormally high DNA adduct levels for a given rat in the proximal region of the colon are associated with increased adduct levels in the distal region. If adduct levels could be measured on the same cells from the same crypts, then we could calculate the estimated correlation function using a simple generalization of the usual variance and covariance formulas (Ramsay and Silverman, 1997). In our case, adduct levels cannot be measured for the same cells or crypt. Making use of the fact that we have both distal and proximal measurements for each rat, however, we can estimate the correlation at the rat level.

The quantity of interest, namely the correlation at the rat level having accounted for overall trends, as a function of relative cell position X is

$$\rho(X) = \mathrm{corr}\{m_{jr}(X,p), m_{jr}(X,d)\} \tag{3}$$

Because of the unequal number of cells per crypt, we cannot obtain an estimate of this function directly using a naïve approach but instead must obtain it through specifying and fitting the model presented earlier. Our first approach is to represent this model assuming parametric structure on the various cell position functions.

PARAMETRIC APPROACH

In our Gaussian mixed model, parametric structure is assumed for the fixed mean function as well as for the rat- and crypt-level random effects. For the proximal response, we let β_p, β_{pr}, and β_{prc} denote the fixed effect and rat- and crypt-level random effects, respectively. For the vector of responses $Y_{jrc(p)}$, with a particular proximal crypt c at time j and for rat r, we can in general write the parametric version of the model as

$$Y_{jrc(p)} = Z_{1,jrc(p)}\,\beta_p + Z_{2,jrc(p)}\,\beta_{pr} + Z_{3,jrc(p)}\,\beta_{prc} + \epsilon_{jrc(p)} \tag{4}$$

where $Z_{1,jrc(p)}$, $Z_{2,jrc(p)}$ and $Z_{3,jrc(p)}$ are the corresponding design matrices. The random variables β_{pr} and β_{prc} are assumed to be independent, mean-zero Gaussian random variables with covariance matrices $\Sigma_{1,p}$ and $\Sigma_{2,p}$, respectively. Equivalent model and structures are assumed for the distal region. While, as described earlier, the crypt-level random effects β_{prc} and β_{drc} are assumed to be mutually independent, the rat-level random effects β_{pr} and β_{dr} have covariance matrix $\Sigma_{p,d}$. We assume that $\epsilon_{jrc(p)}$ and $\epsilon_{jrc(d)}$ are mean-zero Gaussian random variables with diagonal covariance matrices.

If, at a given value of X, the component of the design matrix $Z_{2,jrc(p)}$ is denoted by the vector $Z_p(X)$, and $Z_d(X)$ is defined similarly, then the correlation function $\rho(X)$ defined above becomes

$$\rho(X) = \{Z_p^T(X)\Sigma_{pd}Z_d(X)\}\{Z_p^T(X)\Sigma_{1p}Z_p(X)\}^{-1/2}\{Z_d^T(X)\Sigma_{1d}Z_d(X)\}^{-1/2} \tag{5}$$

In order to estimate this correlation function, we need to estimate the rat-level co-variance matrices Σ_{1p}, Σ_{1d}, and Σ_{pd}. For this presentation, restricted maximum likelihood (REML) estimates of these matrices were obtained using a ridge-stabilized Newton-Raphson algorithm in PROC MIXED in SAS. Minimum variance quadratic unbiased (MIVQUE) starting values, which are the default for PROC MIXED, were used.

Point-wise confidence bounds for the correlation function estimate were found using a parametric bootstrap procedure (Efron and Tibshirani, 1993). In this procedure, a large number of bootstrap samples were simulated from the assumed parametric distribution of the data based on the ML estimates of the fixed effect parameters and REML estimates of the covariance parameters from the original data set. Then, a correlation function estimate was obtained for each bootstrap sample based on rat-level covariance estimates from PROC MIXED. In order to make this procedure computationally feasible, MIVQUE rather than REML estimates of the rat-level covariance matrices were used in the parametric bootstrap. This gave us error bounds that were conservative for REML estimates, but spot checks indicated that the two estimates were nearly identical.

This parametric method was applied separately to two data sets: one for data from the fish oil group, the other from the corn oil group. Initial analyses suggested that the covariate effect was quadratic in shape, so the initial full model included quadratics for the fixed mean, random rat, and random crypt portions of the model. It was determined via likelihood ratio testing that, for both diets, the rat-level quadratic term was not necessary in the proximal region, so these effects were modeled as linear. The parabolic coefficients in the fixed mean were allowed to differ for the five time levels. To minimize colinearity in the coefficients, orthogonal polynomials were used in the design matrices. The rat- and crypt-level covariance matrices were allowed to be unstructured in form.

Figure 2 shows the parametric estimates of the fish- and corn oil correlation functions between DNA adduct levels in the distal and proximal regions of the colon as well as the 90% point-wise confidence bounds from the parametric bootstrap. For rats fed fish oil diets, the correlation estimate was positive near the bottom of the crypt but close to zero in the middle and upper portions. Based on a likelihood ratio test, there was weak evidence of the existence of a non-zero correlation ($p = 0.056$). These results suggest that, in the bottom of the crypts, fish oil-fed rats with higher than average DNA adduct levels in the distal region of the colon also tended to have higher than average DNA adduct levels in the proximal region of the colon. Such effects diminished as the cells further divided.

The correlation function for corn oil-fed rats was negative throughout the crypts. From a likelihood ratio test, there was evidence of a non-zero correlation ($p = 0.018$), and point-wise bootstrap confidence intervals were similarly statistically significant for the middle 60% of the crypts. These results suggest that, uniformly down the crypts, corn oil-fed rats with higher than average DNA adduct levels in the distal region of the colon tended to have lower than average DNA adduct levels in the proximal region of the colon and *vice versa*. Based on the parametric bootstrap results, the difference between the fish oil- and corn oil-fed groups' correlations was significant ($p < 0.05$) for cells in the bottom 10% of the crypts.

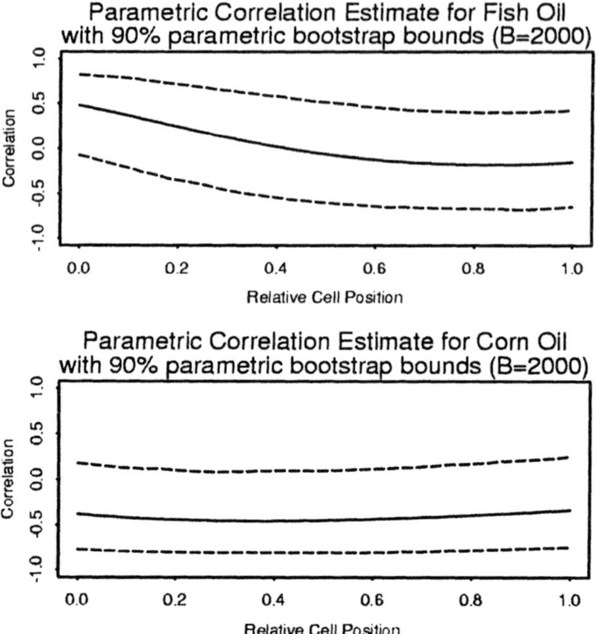

Figure 2. Parametric estimates of the correlation function between distal and proximal DNA adduct level for rats fed fish oil- (top panel) and corn oil-supplemented diets (bottom panel), with 90% bounds from parametric bootstrap of size 2000.

NONPARAMETRIC APPROACH

The correlation results from the parametric method, specifically the negative correlations encountered for corn oil-fed rats, were unexpected. In order to check the model robustness in the parametric approach, we developed a nonparametric method for estimating the correlation function. There are a variety of ways to fit the correlation function nonparametrically. We adopted a functional data analysis approach (FDA; Ramsay and Silverman, 1997), modified for our particular problem. In scalar form, for a cell i with relative cell position $X_{jrc(p)i}$, our models are

$$Y_{jrcpi} = m_j(X_{jrcpi}, p) + m_{jr}(X_{jrcpi}, p) + m_{jrcp}(X_{jrcpi}, p) + \epsilon_{jrcpi} \tag{6}$$

$$Y_{jrcdi} = m_j(X_{jrcdi}, p) + m_{jr}(X_{jrcdi}, p) + m_{jrcd}(X_{jrcdi}, p) + \epsilon_{jrcdi} \tag{7}$$

where ε_{jrcdi} and $\varepsilon_{jrc(d)i}$ are independent and identically distributed mean-zero random variables with variances $\sigma^2_{e,p}$ and $\sigma^2_{e,d}$, respectively.

Our approach was to estimate the crypt-level functions on a fixed grid of cell positions and then separately fit a mixed model at each point on the grid. Specifically, within each crypt, we used nonparametric regression techniques to estimate the crypt-level function, based on a sample size of ~40 cells, and we then evaluated the fit on the grid.

While the choice of the method of nonparametric regression is not crucial, the numerical results displayed in Figure 3 used a Gasser-Müller estimator using an Epanechinov kernel to estimate the crypt-level curves.

Theoretical results suggest that, at each grid point, the nonparametric estimates follow a linear mixed model with solely rat-level random effects and crypt-level random errors. We thus obtained REML estimates of $\rho(X)$ by fitting a mixed model with the suggested structure to the nonparametrically estimated crypt-level functions. Numerical and theoretical analyses suggest that the variability in the nonparametric regression at the crypt level is sufficiently small compared to the rat-to-rat variability so that the estimation error can be ignored for practical purposes. Thus, for each value on the grid, each model is simply a linear mixed model with rat-level random effects and independent, identically distributed crypt-level random errors. The estimated correlation function was then obtained using the REML estimates of the rat-level covariance parameters. We developed a nonparametric bootstrap procedure to obtain confidence bounds on the nonparametric estimate of the correlation function, and we tested various hypotheses of interest.

Just as was done for the parametric method, the nonparametric method was applied separately to the fish oil and corn oil data sets. The bandwidth for the nonparametric regression was selected for each crypt in such a way as to prevent oversmoothed estimation of the objective correlation function. This was done by multiplying the determined optimal bandwidth at the crypt level (0.23) by a factor equal to the number of crypts for a given rat. Sensitivity analyses indicated that a wide range of possible bandwidths resulted in nearly identical results.

Figure 3 shows the nonparametric estimates of the correlation functions between DNA adduct levels in the distal and proximal regions of the colon for both fish oil and corn oil diets, with 90% confidence bounds from the nonparametric bootstrap. For rats fed the fish oil diet, the nonparametric correlation function estimate behaves similarly to that from the parametric method. There was a positive estimated correlation at the bottom of the crypts that decreased to near zero for the middle and upper portions of the crypts.

For rats fed the corn oil diet, the nonparametric method yielded correlation results that were negative at all depths in the crypts. The negative correlation was significant ($p < 0.05$) for all cell positions in the middle 60% of the crypts. The fish oil and corn oil correlation functions were marginally significantly different ($p < 0.10$) over the bottom 35% of the crypts. These results suggest that the surprising conclusions obtained using the parametric method were not due to the parametric modeling assumptions.

BAYESIAN APPROACH BASED ON TRUE CELL POSITION MEASURE

The relative cell position measure used for the aforementioned analyses is only a proxy for the true relative cell depths or true cell positions. It is possible, at least in principle, that the surprising negative correlation in corn oil-fed rats resulted from using the proxy measure for relative (nominal) cell position. That is, if the true cell positions were used, the correlation function might differ substantially. To investigate this possibility, an internal validation data set was collected which contained true cell

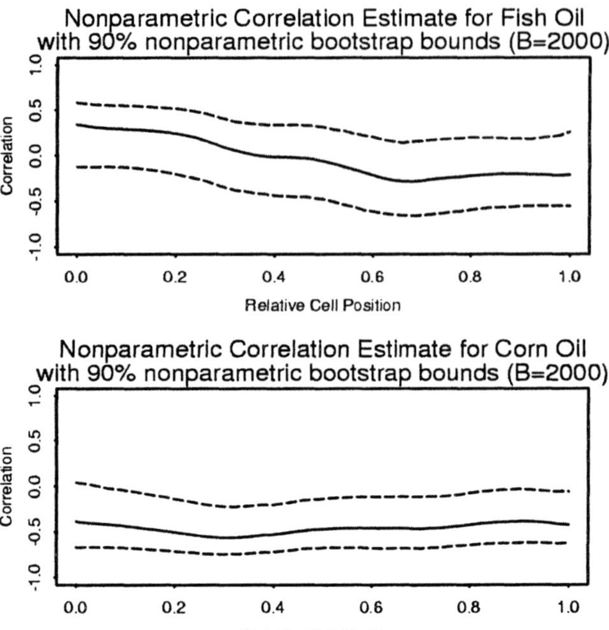

Figure 3. Nonparametric estimates of the correlation function between distal and proximal DNA adduct levels for rats fed fish oil- (top panel) and corn oil-supplemented diets (bottom panel), with 90% bounds from a parametric bootstrap of size 2000.

positions, and a correlation function estimate based on these data was determined for the corn oil-fed rats. The estimation was done using a Bayesian approach and required dealing with missing data issues.

The true cell position is defined as the ratio between the actual distance from the bottom of the crypt to the middle of the cell and the length of the crypt. The proxy was computed to be roughly the proportion of cells below the current cell in the crypt, as is customary in cellular studies of the colon. For example, given n cells lining the left wall of a cross-section of a particular crypt, the proxy relative cell position for the ith cell from the bottom was taken to be $i/(n+1)$. This proxy measure may give a general indication of the true cell position but will differ in practice for a number of reasons: cells may differ in size, shape, spacing, and orientation at various depths in the crypts. The true cell position is typically not measured because it is tedious and time-consuming to do so, whereas the proxy can be obtained without further measurements.

The internal validation data set contained the actual cell depths for all cells in three randomly selected distal and proximal crypts for each rat in the AOM study. From these actual depths, the true cell positions were constructed for the selected crypts, representing ~30% of the data, with the true cell positions for the remaining unselected crypts considered missing covariate data.

Because of the missing data, it was necessary to specify a model for the true cell positions in order to estimate the correlation function based on them rather than the proxy measure. Because of the ordering inherent in the positioning of cells within crypts, care must be taken in specifying a model for the joint distribution of the true cell positions for a given crypt. Rather than specifying the marginal distributions for each cell, which would require that various ordering constraints be placed, we instead modeled all cells within the crypt jointly as the order statistics of a Beta (a,b) distribution. This modeling choice accounts for the inherent ordering constraints and naturally accommodates a flexible class of possible structures. For the sake of parsimony, we assumed the same Beta distribution for all crypts within each of the proximal and distal regions but allowed the distributions in the two regions to differ from each other.

Note that, in the case where a and b are both one, the true cell positions are uniformly distributed [i.e., distributed as the order statistics of a Uniform (0,1) distribution]. In that case, the proxy measurements would be unbiased for the true measurements. A Bayesian analysis performed on the internal validation data suggested that, in the proximal region of the colon, the cells were essentially uniformly distributed within the crypts whereas in the distal region, they were not. There was a systematic departure from uniformity observed in cells within the distal crypts. Specifically, these cells were more concentrated at the bottom of the crypts. Our results confirm the possibility that the correlation function based on true cell positions may differ from that for the proxy, since the different cell distributions in the two regions would result in a different effective pairing of the distal and proximal measurements that contribute to the correlation. Thus, we set out to re-estimate the corn oil-fed rats' correlation function, this time based on the true cell position rather than the proxy, to check whether the negative correlation would still occur.

Estimation of this correlation function required estimation of the true cell position values for the missing crypts as well. We used a hierarchical model for the DNA adducts, analogous to the model used earlier ("General Model and Correlation") but containing the additional assumptions needed to model the unknown true cell positions. A Bayesian estimate of the correlation function was obtained, with the model fitting done using a Gibbs sampler with two Metropolis-Hastings steps (see Metropolis et al., 1953; Hastings, 1970; Geman and Geman, 1984; Gilks et al., 1996; Tanner, 1996). Vague priors were used, and sensitivity analyses indicated that the results were robust to the priors chosen.

Figure 4 shows the parametric Bayesian estimates of the correlation between DNA adduct levels in the distal and proximal regions of the colon as a function of both the proxy and the true cell position measure, with 90% posterior bounds. We see that, for these data, the correlation function based on the true cell positions matched very closely that based on the nominal cell positions (i.e., proxy measure). The estimated correlation was still negative at all depths within the crypts. In spite of the fact that the distal and proximal cell distributions differ, we still obtained virtually identical results. Thus, we conclude that the observed negative correlation was not due to using the proxy measure.

DISCUSSION OF RESULTS

Our analysis of the rat AOM data suggests that, for rats fed fish oil-supplemented diets, there may be a positive relationship between the DNA adduct level measurements

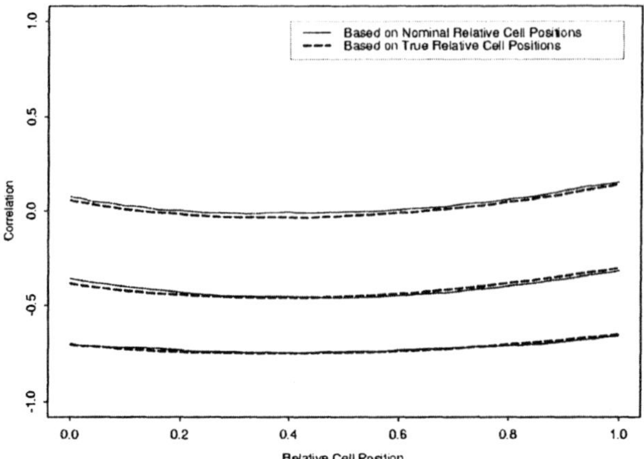

Figure 4. Bayesian estimates of the correlation between distal and proximal DNA adduct levels as a function of true and nominal relative cell position for rats fed corn oil-supplemented diets, along with 90% posterior confidence bands.

in the distal and proximal regions of the colon near the bottom of the crypts. For rats fed corn oil-supplemented diets, however, the DNA adduct level measurements in the distal and proximal regions of the colon were negatively related. This implies that rats that have higher than average DNA adduct levels in the distal crypts tend to have lower than average DNA adduct levels in the proximal crypts and *vice versa*. This result is counter-intuitive, since one would naturally expect the relationship, if any existed, to be positive.

A variety of statistical tools, including parametric mixed model and Bayesian methods as well as nonparametric FDA-based methods, based on both the proxy and the true cell position measures, were used in reaching these conclusions. All methods confirmed the surprising results for rats fed the corn oil-supplemented diet.

Care must be taken in interpretation of the results presented here, given the marginal level of statistical significance in many cases. At this point, we shouldn't read too much into the conclusions. However, the marginal statistical significance does not invalidate our conclusions, considering the relatively large magnitude of the estimated correlations and the small effective sample sizes and subsequently low statistical power for detecting significant non-zero correlations (e.g., power for detecting negative corn oil correlation at $X=0$ like that encountered is approximately 38%).

One possible implication of our results is that the differing effects of fish oil- and corn oil-supplemented diets might have caused different activating patterns for the carcinogen. The negative relationship between the distal and proximal DNA adduct level measurements implies that the activation of the carcinogen/bacterial interaction may occur more locally for rats fed corn oil-supplemented diets and may possibly intensify its effect on the colonic cells in that region. Carefully designed experiments are needed before such a conjecture can be verified.

ACKNOWLEDGMENTS

This research was supported by the Texas A&M Center for Environmental and Rural Health via a grant from the National Institute of Environmental Health Sciences (P30-E509106) and by NIH grants CA57030, CA74552, CA61750, and CA74552. R.J.C. also gratefully acknowledges the support of an ARC grant to the Centre for Mathematics and its Applications, Australian National University, Canberra, where part of this research took place.

CORRESPONDENCE

Please address all correspondence to:
Jeffrey S. Morris
Department of Biostatistics
University of Texas M.D. Anderson Cancer Center
1515 Holcombe Blvd., Box 447
Houston, TX 77030
Jeffmo@mdanderson.org

REFERENCES

Bang, H.O., Dyerberg, J., and Hijorne, N., 1976, The consumption of food consumed by Greenland eskimos, *Acta Med. Scand.* 200:69-73.

Blot, W.J., Lanier, A., Fraumeni, J.F., and Bender, T.R., 1975, Cancer mortality among Alaskan natives, 1960-69, *J. Natl. Canc. Inst.* 55:547-554.

Boyle, P., Zaridze, D.G., and Smans, M., 1985, Descriptive epidemiology of colorectal cancer, *Int. J. Canc.* 36:9-18.

Chang, W.C.L., Chapkin, R., and Lupton, J.R., 1997, Predictive values of proliferation, differentiation and apoptosis as intermediate markers for colon tumorigenesis, *Carcinogenesis* 18:721-730.

Efron, B., and Tibshirani, R.J., 1993, *An Introduction to the Bootstrap,* Chapman and Hall, New York.

Geman, S., and Geman, D., 1984, Stochastic relaxation, Gibbs distributions and the Bayesian restoration of images, *IEEE Trans. Pattern Anal. Machine Intell.* 6:721-741.

Gilks, W.R., Richardson, S., and Spiegelhalter, D.J., 1996, *Markov Chain Monte Carlo in Practice,* Chapman and Hall, New York.

Haenszel, W., and Kurihara, M., 1968, Studies of Japanese migrants. I. Mortality from cancer and other diseases among Japanese in the United States, *J. Natl. Canc. Inst.* 40:43-68.

Hastings, W.K., 1970, Monte Carlo sampling methods using Markov chains and their applications, *Biometrika* 57:97-109.

Hong, M.Y., Lupton, J.R., Morris, J.S., Wang, N., Carroll, R.J., Davidson, L.A., Elder, R.H., and Chapkin, R.S., 2000, Dietary fish oil reduces O^6-methylguanine DNA adduct levels in the rat colon in part by increasing apoptosis during tumor initiation, *Canc. Epidemiol. Biomarkers Preven.* 9:819-826.

Lipkin, M., 1974, Proliferative changes in the colon, *Dig. Dis.* 19:1029-1032.

Lipkin, M., Bell, B., and Sherlock, P., 1963, Cell proliferation kinetics in the gastrointestinal tract of man. I. Cell renewal in colon and rectum, *J. Clin. Invest.* 42:767-776.

Metropolis, N., Rosenbluth, A.W., Rosenbluth, M.N., Teller, A.H., and Teller, E., 1953, Equations of state calculations by fast computing machine, *J. Chem. Phys.* 21:1087-1091.

Ramsay, J.O., and Silverman, B.W., 1997, *Functional Data Analysis,* Springer-Verlag, New York.

Rogers, K.J., and Pegg, A.E., 1977, Formation of O^6-methylguanine by alkylation of rat liver, colon and kidney DNA following administration of 1,2-dimethylhydrazine, *Canc. Res.* 37:4082-4087.

Swenberg, J.A., Cooper, H.K., Bucheler, J., and Kleihues, P., 1979, 1,2-Dimethylhydrazine-induced methylation of DNA bases in various rat organs and the effect of pretreatment with disufiram, *Canc. Res.* 39:465-467.

Tanner, M.A., 1996, *Tools for Statistical Inference: Methods for the Exploration of Posterior Distributions and Likelihood Functions,* Springer-Verlag, New York.

PART IV
EXPERIMENTAL ASPECTS OF KINETIC DATA

METHODOLOGICAL ISSUES IN STABLE ISOTOPE-BASED KINETIC STUDIES IN CHILDREN

Ian J. Griffin and Steven A. Abrams[*]

INTRODUCTION

Childhood and adolescence are times of rapid growth and development. This growth includes both an increase in body mass and changes in body composition, with a rapid increase in lean tissues. The nutritional requirements needed to meet these challenges are extremely high, particularly for those minerals found predominantly in bone and muscle, and in those subjects whose dietary intake is marginal.

Nutritional inadequacies during this critical period can have obvious and immediate effects. However, there are other, more insidious, long-term effects that need to be considered. Even marginal nutritional deficiencies during critical periods can have profound effects on later health and well being. Adolescence, for example, is a period of unparalleled calcium accretion. Small differences in calcium balance during early puberty can have a significant effect on peak bone mineral mass and can greatly increase the risk of later mortality and morbidity from osteoporosis. Other deficiencies, in minerals such as zinc, appear to be particularly common during adolescence. Presumably, this relates to the extremely high zinc requirement that results from the rapid expansion of lean tissue mass during this period, as 60% of body zinc is present in muscle and 30% in bones.

Zinc is an area of intense research activity. For many years, zinc deficiency was felt to be extremely unlikely in free-living humans. This belief changed in the 1960s when Prasad (1993a) described zinc deficiency in adolescents in Iran and Egypt. The deficiency was characterized by growth retardation ("dwarfism") and delayed puberty. Subsequently, the effects of more moderate zinc deficiency have become increasingly clear (Prasad, 1993b). Studies in developing countries have confirmed the importance of occult zinc deficiency in mortality from infectious diseases, especially respiratory illnesses and diarrhea.

---

* Ian J. Griffin and Steven A. Abrams, USDA/ARS Children's Nutrition Research Center and Section of Neonatology, Department of Pediatrics, Baylor College of Medicine, Houston, TX 77030.

Zinc remains a poorly understood mineral and the dietary requirement for zinc needs clarification. This is, at least in part, due to the lack of any reliable measure of zinc "status." The most widely used measure continues to be plasma zinc, but no measures are currently adequate (Aggett, 1991; Prasad, 1993a, b).

Compartmental models are a powerful tool with which to study mineral metabolism. They allow a more complete, dynamic description of metabolism than can be achieved by a single "static" measure of status. They also allow estimation of mineral absorption, and urinary and fecal excretion. Finally, they permit the measurement of *kinetically-distinct* compartments and the fluxes between them. Such models have produced important insights into calcium metabolism in children (Yergey et al., 1990; Abrams and Ellis, 1998; Abrams, 1999; Smith et al., 1999; Abrams et al., 2000) and promise similar advances in our understanding of zinc metabolism. For that promise to be met, the many difficulties of carrying out multicompartmental stable isotope studies in children need to be addressed.

Stable isotopes are well suited to the study of metabolism in children, as they overcome the ethical difficulties associated with the use of radioisotopes and the resulting exposure to ionizing radiation. Although the limited use of radioisotopes to study pathological conditions in children may be ethically permissible, their use in healthy children has been consigned to the past.

As pediatric caregivers, we seek not only to treat disease but to optimize the health and wellbeing of children now and as they age. To do this, we need to understand the metabolic adaptations that occur in healthy children and in those with disease. These insights cannot be achieved by extrapolation from adult data, nor from experiments in other species. To understand the unique physiology of children, children themselves must be studied. Despite the safety of mineral stable isotopes, the design and conduct of stable isotope studies in children are fraught with many technical and practical difficulties.

OBJECTIVE

The objective of this manuscript is to discuss the specific difficulties involved in carrying out stable isotope-based compartmental modeling studies in children, with particular reference to recent studies we have conducted examining the applicability of such methods to the study of zinc metabolism in children (Griffin et al., 2000).

MODEL SELECTION

Radioisotope-Based Models

The most widely used zinc compartmental models are those that were initially described by Foster et al. (1979), and gradually expanded and refined by Wastney et al. (1986). These complex *radioisotope-based* models describe zinc kinetics in terms of at least a dozen kinetically-distinct compartments. These models make use of the advantages offered by radioisotopes (such as whole-body counting and external scanning) to estimate sizes of "liver," "bone," and "muscle" compartments. Obviously, radioisotopes cannot be used in healthy children, but attempts have been made to use such models in, for example, preterm infants (Wastney et al., 1986). However, this involves considerable extrapolation from the literature to estimate the mass of zinc in some compartments that cannot be directly sampled.

Stable Isotope-Based Models

We therefore decided to evaluate a *stable isotope model* as described by Lowe et al. (1997), which is in turn based on radioisotope models developed by Foster et al. (1979) and Wastney et al. (1986, 1992). The stable isotope-based model has been shown to have *a priori* identifiablity (Lowe et al., 1997; Saccomani et al., 1994) such that, for any given data set, all of the unknown parameters have one unique solution.

This model is shown in Figure 1, in which circles represent kinetically-distinct compartments and arrows represent the transfers between them. The model consists of 6 compartments – 3 gastrointestinal compartments, a central "plasma" compartment, and 2 kinetically-distinct tissue compartments. Dietary zinc enters the first of the three gastrointestinal compartments (#4, #5 and #6) and is finally excreted in the feces from compartment #6. Zinc is absorbed from the second gastrointestinal compartment (#5) to a central "plasma" compartment (#1). From there zinc undergoes bidirectional transfers to and from two kinetically-distinct compartments (#2 and #3), denoted "fast" and "slow" to reflect their different turnover rates. Zinc from the central compartment is lost in urine (urinary excretion) and into the gastrointestinal tract (endogenous fecal zinc excretion). There is a further loss of zinc from the central compartment, which represents transfer of tracer to a large, slowly turning over compartment (#7) (Lowe et al., 1997). Zinc in this compartment has been estimated to have a half-life of approximately 50 days and studies

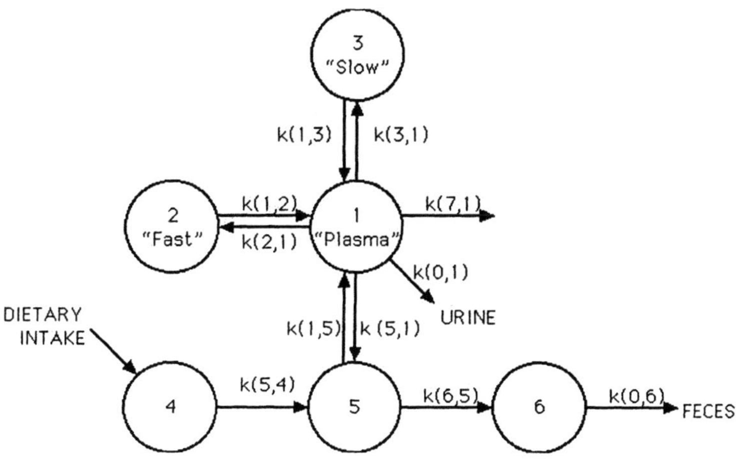

Figure 1. Diagrammatic representation of the basic model. Circles represent kinetically-distinct compartments and arrows represent fluxes between compartments. Modified from Griffin et al. (2000).

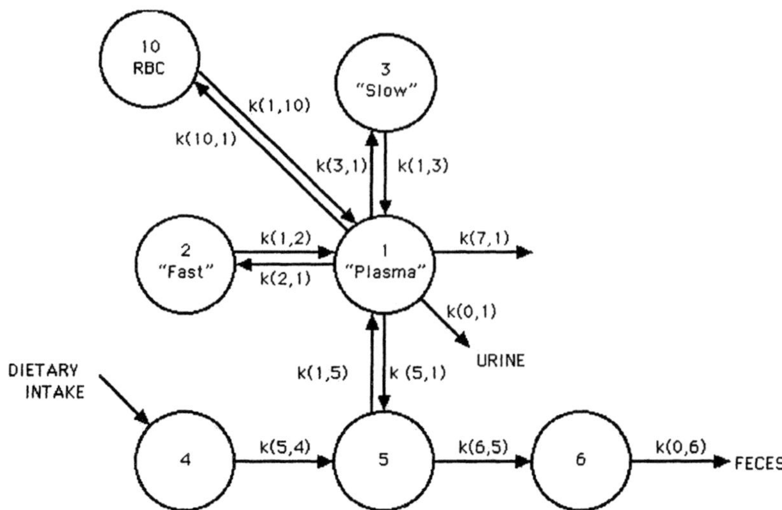

Figure 2. Modification of the basic model to include a single red blood cell compartment (#10).

of many months' duration are needed to identify it (Lowe et al., 1997). In the current study, transfer of zinc to this compartment was modeled as an irreversible loss from the system.

One limitation of this model is that it does not include any compartments representing red blood cell zinc. We therefore modified the model to include one red blood cell compartment (#10, Figure 2). This was also based on data from earlier radioisotope studies (Foster et al., 1979; Wastney et al., 1986, 1992).

These models can be expressed as a set of multiple simultaneous exponential equations. Tracer is given orally (into compartment #4) and intravenously (into compartment #1), and serial samples of plasma, red blood cells, urine, and feces are taken. The resulting data can be used to solve the multiple exponential equations using the SAAM II software package (SAAM Institute, 1997). This allows estimation of compartmental masses (where M_i is the mass of compartment i) and intercompartmental fluxes [where $k(i,j)$ is the fraction of compartment j entering compartment i per unit time].

In selecting a model, we wanted to use one in which all parameters could be estimated from the data collected during the study, rather than using extrapolations from other human populations or animal models. In this way, we hoped to develop a useful and reliable tool for the study of zinc metabolism in children.

TECHNICAL ISSUES

We have for many years carried out studies of calcium kinetics in children. These require the simultaneous administration of oral and intravenous tracers. Blood samples are then drawn at 6, 15, 30, 45, 60, 120, 180, 240, and 480 minutes from an indwelling intravenous catheter, and a complete 6-day urine and fecal collection is carried out (Abrams et al., 1994; Abrams and Ellis, 1998; Abrams, 1999). However, the current studies presented several new problems, including the number of blood samples required and the amount of blood that can safely be drawn from small children.

Limitations in Blood Volume

Children, due to their smaller body mass, have lower blood volumes than adults. Most research ethics committees therefore limit the total amount of blood that can be drawn from children in a single study to approximately 5 mL/kg. In the original protocol of Lowe et al. (1997), 21 blood samples were drawn over 7 days (5, 10, 15, 30, 45, 60, 75, 90, 120, 180, 240, 300, 360, and 420 minutes from an indwelling intravenous catheter, then daily for seven days). When we first considered studying zinc kinetics in children, up to 10 mL of blood was required to measure zinc isotope enrichment by thermal ionization magnetic sector mass spectrometry. Such a large requirement for blood precluded such studies from being carried out in children. However, through gradual improvements in separation and purification of plasma samples, we were finally able to reduce the required volume to 3 mL per sample or less. This reduced the amount of blood required to a manageable and ethically-permissible volume. Modifications of the sampling protocol further reduced blood requirements by decreasing the number of blood samples that needed to be drawn.

Effect of Changes in Turnover Rates

At the outset of our study, we hypothesized that turnover between compartments was likely to be higher in children than in adults. Because of this, we modified the times at which blood samples were drawn to include more samples early in the study and fewer in the later stages. Specifically, we added a blood sample at 20 minutes, omitted the samples at 45, 75, 90, 300, and 420 minutes, and added a sample at 480 minutes. In this manner, we hoped to model the turnover of the rapidly-exchanging compartments with greater precision.

Sampling Times

Based on the work of Lowe et al. (1997), the need to reduce the total number of blood samples, and our concerns that turnover would be higher in children than in adults, we decided to collect blood from an intravenous catheter at 5, 10, 15, 20, 30, and 60 minutes, then at 2, 3, 4, and 8 hours, and then daily to 6 days (Griffin et al., 2000). In addition, complete 6-day urine and fecal collections were carried out (Griffin et al., 2000).

Clearly, these sampling times overlap with those required for calcium kinetic assessments (Abrams and Wen, 1999), and it would be theoretically possible to carry out calcium and zinc kinetic studies simultaneously, just as we have previously combined calcium and magnesium kinetic studies (Abrams and Ellis, 1998). In this respect, calcium kinetics are somewhat easier, as there is isotopic equilibrium between plasma and urine within 24 hours. However, even after 6 days, there is no such equilibrium for zinc. In other words, the isotope enrichment in plasma at, for example, 120 hours is still significantly higher than in a pooled urine sample collected between 116 and 124 hours. Because of this, we needed to continue blood draws for the full six days of the study. A similar situation has previously been described for magnesium (Abrams and Ellis, 1998).

Isotope Doses

When calculating isotope doses in children, it is tempting to give the same dose *per kilogram* as has been used in adults. This, however, is fraught with difficulties. In the case of calcium, for example, high-turnover calcium body pool masses during adolescence are so large that higher calcium isotope doses are required in adolescents than in adults (Abrams, 1999).

We estimated the ideal dose of oral and intravenous tracer based on previous data from our laboratory on the urinary, fecal, and plasma enrichment of zinc tracers in this population. Our objective was that enrichment at the end of the study would be at least threefold greater than the limit of precision of thermal ionization magnetic sector mass spectrometry. We chose doses of 0.4 – 0.5 mg of an intravenous tracer highly enriched in zinc-70 and 1.0 – 1.1 mg of an oral tracer highly enriched in zinc-67. These produced adequate enrichment in our subjects (Griffin et al., 2000).

Steady-State Considerations

As noted above, childhood is a time of rapid growth and by definition, therefore, is not a steady-state condition. However, the software used in compartmental modeling assumes steady-state conditions (SAAM Institute, 1997). Even during rapid growth, however, the gain in body mass during a 1-week period is less than 1%, so that the deviation from steady-state conditions is so small as to be insignificant. Thus, it can, for practical purposes, be ignored.

PRACTICAL ISSUES

The preceding section has examined some of the more technical or "scientific" issues associated with kinetic modeling in children. However, more difficult issues can be encountered with regard to the practicalities of doing these studies.

Ethical Considerations

Children clearly have a limited ability to give consent for potentially uncomfortable procedures. Prior to the initiation of the study, children are interviewed and a screening blood sample is taken. This provides an opportunity to assess whether the individual subjects are sufficiently comfortable with the process of venipuncture to allow kinetic studies to be carried out. This is especially important in the case of zinc kinetics, where daily blood samples are required. We may choose not to enroll a subject who wishes to participate in a zinc study because of possible difficulties with repeated phlebotomy. However, such subjects will often tolerate calcium kinetic studies which can be carried out using a single venipuncture to place an intravenous catheter and from which all subsequent blood samples can be taken.

If subjects express interest in the study, they and their parents are given full and complete details, and they independently sign consent forms. It is emphasized that enrollment in the study is optional, and that subjects can either refuse to participate or leave the study after enrollment.

In practice, feedback from subjects has been overwhelmingly positive. We strive to make their "research" experience an enjoyable one. In many cases, children have used the research experience as an informative way of learning more about nutrition and science. Indeed, most tears during the study period occur at discharge!

Blood Draws

Although we have been able to reduce the *number* of blood draws, successfully collecting the desired samples can be difficult. For the first 8 hours, samples are collected from an intravenous catheter. We have found that a 22G Jelco™ or Angiocath™ works well, but given the narrow diameter of such a catheter (necessitated by the small size of the vessel), blood sampling is relatively slow. It is not uncommon to take 30 seconds to draw the required blood volume from such a catheter (in that case, the sample time is taken as the midpoint of the blood draw). Given the additional time to apply a tourniquet to the arm prior to sampling and flush the catheter after sampling, it is unlikely that samples could be taken more frequently than every 4 to 5 minutes. This apparently minor difficulty was one of the reasons that an alternative stable isotope model (Miller et al., 1998) was not used, as it was based on adult data and blood samples taken every 2 minutes. Although such sampling intervals are achievable in adults, they cannot be expected in studies of children. One advantage of a smaller intravenous catheter is that it is shorter. This allows more mobility and movement of the elbow. This is important after the first couple of hours of the study, as it lets the subjects feed themselves and play.

Clearly, young children tolerate blood draws less well than adults. We always use a local anaesthetic cream (EMLA™) on the skin. This has been very effective in reducing the discomfort associated with blood draws or intravenous catheter placement and has made studies much more acceptable to subjects, as well as easier for investigators. To date, we have carried out over 50 zinc kinetic studies in children without any subjects withdrawing from the study, and we have experienced minimal difficulties with the blood draws.

Dietary Modification

In previous studies, we have examined the effect of different calcium intakes on model parameters. The dietary requirements for the current study were less involved: we wished all children to remain on a zinc intake of about 12 mg/d (the current recommended dietary allowance for zinc in this age group) for 2 weeks before the study (adaptation period) and for the 6 day study period (Griffin et al., 2000). In adults, this manipulation would be relatively easy to carry out; patients could be admitted for the adaptation period and their entire diet supervised. Alternatively, they could be seen daily and all food provided as "packed meals." In children, however, such intensive dietary manipulations cannot be carried out except under exceptional circumstances. Instead, a more flexible approach is needed. We have found that a careful dietary history can give a good idea of the current dietary intake. A list of high zinc foods can then be provided to the parents and a target zinc intake given. Overwhelmingly, parents are good at increasing their child's zinc intake to the target level. Compliance can be improved by regular phone calls to assist parents and help them through difficulties with the diet. We check compliance with a 3-day food record (2 weekdays and one weekend day) carried out by the parents at home. The dietary intake for these 3 days and the 6 days of study admission are averaged to estimate the child's zinc intake.

We have found such methods to be highly successful in our studies of calcium metabolism as well as in the current study. We have also successfully reduced zinc intake in children to 3 – 4 mg/d by using an expanded list of zinc-containing foods, giving more detailed dietary instructions, and keeping in closer contact with parents by phone. These changes have improved the accuracy of our dietary modifications during both increases and decreases in dietary zinc intake.

Schooling

One of the usual requirements of our institutional ethics committee is that children should miss little, or ideally, no school time as a result of participation in studies. Zinc kinetic studies take 6 days, and we have been reluctant to attempt complete urine and fecal collections at home because of doubts about possible compliance. The need for daily blood draws also makes this approach difficult. Therefore, all our studies are carried out on in-patients. We can, of course, conduct them during school vacations, but scheduling can be very difficult during these periods due to the limited number of in-patient beds, competition from other research protocols, and family issues (wanting to go on vacation during these periods). We have, in part, overcome these difficulties by recruiting from the increasingly large population of home-schooled children. These subjects can be admitted at any time, and the parent/teacher can continue the child's schooling as though he or she was at home. To do this, however, accommodations must be made for other siblings. Such families are often highly motivated and follow study protocols very closely. They frequently use the period of dietary adaptation as an opportunity to teach the child about nutrition and health, the interpretation of food labels, etc. The admission period can also produce many educational benefits. The children meet many dieticians, physicians, and scientists, and they learn about the way scientific studies are carried out. For example, they can have their body composition measured (as part of other approved protocols) and learn about the importance of nutrition in childhood and its impact on long-term health (such as later osteoporosis). Such experiences help demystify the scientific process and can be entertaining for the subjects, as well as educational.

Other Issues

Children can become bored if restricted to the Metabolic Research Unit for six days. To prevent this, we try to arrange trips most days to the cinema, park, science museum, etc. Urine and fecal collections need to continue during these periods and a significant number of "helpers" are required to accompany the children. Helpers take containers to continue the collections, should the subjects need them, and bags to transport samples back to the Metabolic Research Unit. These activities take a considerable amount of time, effort, and money but they are particularly important for subjects who may be returning several times for studies. By helping make the research experience positive and enjoyable, such activities encourage subsequent recruitment. Many study subjects hear about the center and its research by word of mouth from subjects who have enjoyed their experience in prior studies.

STUDIES OF ZINC KINETICS IN GIRLS

In order to demonstrate how these complex issues can be addressed and overcome, we will now discuss a recent study we completed on zinc kinetics in girls (Griffin et al., 2000). The main objectives were to address the issues of

- suitability of the selected model to studies in children
- optimum isotope dosing
- optimum timing of blood samples and the type of practical and technical issues discussed above.

Secondary objectives were to

- compare kinetic data in children with data previously reported for adults
- examine the effect of inclusion of red blood cell data into the model.

Study Population

Seven healthy girls aged 9.9 ± 0.8 years were studied (Griffin et al., 2000). They started a 21-day adaptation period at home, designed to provide about 12 mg/d zinc. They were then admitted for 6 days to the Metabolic Research Unit of the Children's Nutrition Research Center in Houston, TX. During that 6-day period, zinc kinetic studies were carried out.

Isotope Administration

Zinc-67 (74% enrichment by mass) and zinc–70 (90% enrichment by mass) were produced in the former Soviet Union and obtained as the oxide from Tracer Sciences Inc. (Toronto, Canada). Aqueous solutions of the tracers, prepared by the Investigational Drug Service of Texas Children's Hospital, Houston, TX, were tested for pyrogenicity and sterility prior to use.

EMLA™ cream was applied to both antecubital fossae, and, an hour later, a 22G intravenous catheter was placed for blood sampling. Subjects then ate a breakfast which included 1.0 mg of the zinc-67 enriched tracer that had been added to 40 mL of orange juice 18 hours earlier. After breakfast, 0.5 mg of the zinc-70 enriched tracer was infused by venipuncture of the other arm. Immediately after this, a 6-day urine and fecal collection was started.

Sample Collection, Preparation, and Analysis

Blood samples were collected in zinc-free tubes (Monovette AH, Sarstedt Inc., Newton, NC). Samples were drawn from the indwelling intravenous catheter at 5, 10, 15, 20, 30, 60, 120, 180, 240, and 480 minutes following intravenous isotope administration. Further blood samples were drawn by venipuncture of the arm or hand daily for 6 days. Once again, EMLA™ cream was applied to the venipuncture site at least an hour earlier. Venipuncture was successful in about 90% of attempts. Blood samples were centrifuged to separate plasma from red blood cells. The red blood cells were then washed 3 times in an equal volume of normal (0.9%) saline.

A complete 6-day urine collection was carried out starting immediately after isotope administration; urine was analyzed in 8-hour pools. A complete 6-day fecal collection was also carried out. Each stool was stored and analyzed individually.

Samples were prepared by acid digestion and anion exchange columns, as described in detail elsewhere (Griffin et al., 2000). Isotope enrichments were measured by thermal ionization magnetic sector mass spectrometry (Finnigan MAT 261, Bremen, Germany) (Griffin et al., 2000) and converted to tracer:tracee ratios (Cobelli et al., 1992; Lowe et al., 1997; Griffin et al., 2000). Zinc concentrations in plasma, red blood cells, urine, and feces were measured by atomic absorption spectroscopy.

Modeling

Data were modeled using SAAM II (SAAM Institute, 1997) using the two models described above and shown in Figures 1 and 2 (Griffin et al., 2000). Data were fit to two separate models – one excluding red blood cell data ["simplified model" identical to that of Lowe et al. (1997), Figure 1] and one including a single red blood cell compartment ("complete model," Figure 2). Plasma data were entered as tracer:tracee ratios, urine and fecal data as cumulative losses of tracer and tracee, and red blood cell data as total tracee. For further details, see Griffin et al. (2000).

Table 1. Precision of estimates of the mass of all 7 compartments, shown as mean ± standard deviation and range

Compartment	Coefficient of Variation (%) Mean ± SD (range)
#1, "plasma"	5.9 ± 1.2% (3.8 – 7.6)
#2, "fast"	16.0 ± 8.1 (8.6 – 30.8)
#3, "slow"	4.3 ± 1.5 (2.5 – 7.2)
#4	10.0 ± 5.6 (0.4 – 19.5)
#5	19.8 ± 6.1 (2.7 – 21.4)
#6	8.1 ± 2.6 (3.3 –0.6)
#10, "red blood cell"	5.5 ± 1.5 (3.9 – 7.5)

Applicability of the Model to Children

The models were adequate to fully describe the observed data in all cases (Griffin et al., 2000). The observed plasma enrichment and modeled fit for a single subject are shown in Figure 3. Precision of the measurements of compartment masses in each individual was estimated by the modeling software and expressed as a coefficient of variation. Table 1 summarizes the mean coefficients of variation for the entire study population. The modified model was able to measure the mass of all compartments with good precision. Of the non-gastrointestinal compartments (#1, 2, 3 and 10), the poorest precision was for compartment #2. This may be due to issues related to sample timing and compartment turnover (see later).

Effect of Inclusion of Red Blood Cell Data

The effect of inclusion of red blood cell data was examined by comparing results of the six-compartment model (Figure 1), which excluded red blood cell data, and the seven-compartment model (Figure 2), which included those data.

Inclusion of red blood cell data led to a significant decrease in the masses of compartments #1 and #2. Therefore, the total mass of the tissue compartments (#1, 2 and 3) was also significantly lower. However, once the mass of the red blood cell compartment was added, the mass of non-gastrointestinal compartments (#1, 2, 3, and 10) actually increased. The magnitude of the differences was, however, small (Table 2).

Comparison with Adult Data

Finally, results from the simplified model were compared to data in adults, using an identical model (Lowe et al., 1997) (Tables 3 and 4). The mass of all of the non-gastrointestinal compartments was significantly greater in children (Table 3) compared to

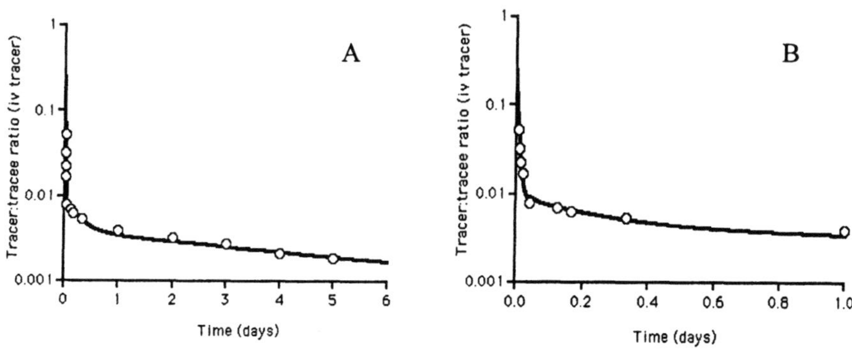

Figure 3. Plasma tracer:tracee ratio for intravenous tracer in a single representative subject. Observed data (circles) and modeled data (line) are shown. Panel A shows data for the entire study; panel B shows the data for day one on an expanded scale.

adults. There was, however, little difference in turnover between the tissue compartments (Table 4).

Effect of Increased Turnover Rates

The only significant difference in turnover (Table 4) was from the plasma to the fast compartment (#2); this was about 50% higher in children than in adults. The mass of compartment #2 was the least precise estimate of any of the tissue compartments (Table 1). We speculate that improvement in the precision of measurement of the mass of compartment #2 will require additional blood samples at time periods when the flux of tracer into compartment #2 is high (i.e., within the first 15 to 20 minutes after iv isotope administration). To do this, the technical difficulties associated with frequent blood draws in children will need to be overcome.

CONCLUDING COMMENTS

Clearly, the issues involved with studies in children are complex and variable. We hope that this chapter has introduced readers to these problems. Many of the problems

Table 2. Differences in compartment masses (mg/kg, mean ± standard deviation) between the simplified model (excluding red cell data) and the complete model (7 compartments including red cells)

Compartment	Simplified model	Complete model	p-value
M1	2.5 ± 0.9	2.4 ± 0.9	0.047
M2	17 ± 8.8	15.6 ± 8.9	0.007
M3	80 ± 22	78 ± 20	0.12
M10	•	5.5 ± 3.1	•
M4	2.5 ± 0.8	2.5 ± 0.8	0.49
M5	1.3 ± 1.2	1.5 ± 1.7	0.55
M6	36 ± 19	36 ± 16	0.53
Sum (M1+2+3)	99 ± 30	96 ± 28	0.020
Sum (M1+2+3+10)	99 ± 30	102 ± 30	0.050

Table 3. Compartment masses (mg/kg) from the current study compared to previous work in adults (Lowe et al., 1997)

Compartment	Current Study	Adult data	p-value
M1	0.06 ± 0.02	0.04 ± 0.01	0.006
M2	0.43 ± 0.22	0.14 ± 0.05	0.010
M3	2.0 ± 0.6	1.4 ± 0.2	0.032
Sum (M1+2+3)	2.5 ± 0.8	1.6 ± 0.1	0.027

Table 4. Turnover (day^{-1}) between the plasma (#1), fast (#2), slow (#3) and very slow (#7) compartments in children (current study) and previous data for adults (Lowe et al., 1997)

Turnover (day^{-1})	Current study	Adult data	p-value
k(1,2)	16.7 ± 3.9	22.3 ± 17.4	0.42
k(1,3)	2.2 ± 1.0	1.5 ± 0.4	0.16
k(2,1)	105 ± 39	65 ± 21	0.044
k(3,1)	64 ± 17	58 ± 10	0.58
k(7,1)	7.2 ± 1.8	7.3 ± 1.5	0.91

are technical in nature, but some of the hardest ones include the practicalities of carrying out such intensive and involved studies in young children.

The ethical issues surrounding such studies in children are clearly complex. It is obviously important that studies involving healthy volunteers, who gain little direct benefit from their participation, be rewarding, educational, and fun, as well as safe. This can only be achieved by a flexible approach that allows compromises in study design to improve tolerance in young subjects.

A team approach, including nursing, dietary, laboratory, medical, scientific, and ancillary staff (those who, for example, supervise meals, chaperone children during trips) is an essential prerequisite. In this way, studies can be carried out that not only provide important scientific information, but also are viewed by the participants as a positive and enjoyable experience, the kind that they would recommend to their families and friends.

ACKNOWLEDGMENTS

This work was supported in part with federal funds from the USDA/ARS under Cooperative Agreement No. 58-6250-6-001. It is a publication of the U.S. Department of Agriculture (USDA)/Agricultural Research Service (ARS) Children's Nutrition Research Center, Department of Pediatrics, Baylor College of Medicine and Texas Children's Hospital, Houston, Texas. Contents of this publication do not necessarily reflect the views or policies of the USDA, nor does mention of trade names, commercial products, or organizations imply endorsement by the U.S. Government.

CORRESPONDENCE

Please address all correspondence to:
Ian J. Griffin
USDA/ARS Children's Nutrition Research Center
1100 Bates Street
Houston, TX 77030
igriffin@neo.bcm.tmc.edu

REFERENCES

Abrams, S.A., 1999, Using stable isotopes to assess mineral absorption and utilization by children, *Am. J. Clin. Nutr.* 70:955-964.

Abrams, S.A., and Ellis, K.J., 1998, Multicompartmental analysis of magnesium and calcium kinetics during growth: relationships with body composition, *Magnes. Res.* 11:307-313.

Abrams, S.A., and Wen, J.P., 1999, Methodologies for using stable isotopes to assess magnesium absorption and secretion in children, *J. Am. Coll. Nutr.* 18:30-35.

Abrams, S.A., Copeland, K.C., Gunn, S.K., Gundberg, C.M., Klein, K.O., and Ellis, K.J., 2000, Calcium absorption, bone mass accumulation, and kinetics increase during early pubertal development in girls, *J. Clin. Endocrinol. Metab.* 85:1805-1809.

Abrams, S.A., Schanler, R.J., Yergey, A.L., Vieira, N.E., and Bronner, F., 1994, Compartmental analysis of calcium metabolism in very-low-birth-weight infants, *Pediatr. Res.* 36:424-428.

Aggett, P.J., 1991, The assessment of zinc status: a personal view, *Proc. Nutr. Soc.* 50:9-17.

Cobelli, C., Toffolo, G., and Foster, D.M., 1992, Tracer-to-tracee ratio for analysis of stable isotope tracer data: link with radioactive kinetic formalism, *Am. J. Physiol.* 262:E968-E975.

Foster, D.M., Aamodt, R.L., Henkin, R.I., and Berman, M., 1979, Zinc metabolism in humans: a kinetic model, *Am. J. Physiol.* 237:R340-R349.

Griffin, I.J., King, J.C., and Abrams, S.A., 2000, Body weight specific zinc compartmental masses in girls significantly exceed those reported in adults: a stable isotope study using a kinetic model, *J. Nutr.* 130:2607-2612.

Lowe, N.M., Shames, D.M., Woodhouse, L.R., Matel, J.S., Roehl, R., Saccomani, M.P., Toffolo, G., Cobelli, C., and King, J.C., 1997, A compartmental model of zinc metabolism in healthy women using oral and intravenous stable isotope tracers, *Am. J. Clin. Nutr.* 65:1810-1819.

Miller, L.V., Krebs, N.F., and Hambidge, K.M., 1998, Human zinc metabolism: advances in the modeling of stable isotope data, *Adv. Exp. Med. Biol.* 445:253-269.

Prasad, A.S., 1993a, Historical aspects of zinc deficiency, in: *Biochemistry of Zinc*, A.S. Prasad, ed., Plenum Press, New York.

Prasad, A.S., 1993b, Clinical spectrum of human zinc deficiency, in: *Biochemistry of Zinc*, A.S. Prasad, ed., Plenum Press, New York.

SAAM II User's Guide, 1997, SAAM Institute, Seattle, WA.

Saccomani, M.P., Audolym, S., Angio, L.D., Sattier, R., and Cobelli, C., 1994, PRIDE: a program to test a priori global identifiability of linear compartmental models, in: *Proceedings of the 10th IFAC Symposium on System Identifiability*, Danish Automation Society, Copenhagen.

Smith, S.M., Wastney, M.E., Morukov, B.V., Larina, I.M., Nyquist, L.E., Abrams, S.A., Taran, E.N., Shih, C.-Y., Nillen, J.L., Davis-Street, J.E., Rice, B.L. and Lane, H.W., 1999, Calcium metabolism before, during and after a 3-month spaceflight: kinetic and biochemical changes, *Am. J. Physiol.* 277:R1-R10.

Wastney, M.E., Aamodt, R.L., Rumble, W.F., and Henkin, R.I., 1986, Kinetic analysis of zinc metabolism and its regulation in normal humans, *Am. J. Physiol.* 251:R398-R408.

Wastney, M.E., Ahmed, S., and Henkin, R.I., 1992, Changes in regulation of human zinc metabolism with age, *Am. J. Physiol.* 263:R1162-R1168.

Yergey, A.L., Abrams, S.A., Vieira, N.E., Eastell, R., Hillman, L.S., and Covell, D.G., 1990, Recent studies of human calcium metabolism using stable isotopic tracers, *Canad. J. Physiol. Pharmacol.* 68:973-976.

INTRINSIC LABELING OF PLANTS FOR BIOAVAILABILITY STUDIES

Janet A. Novotny, Steven Britz, Frances Caulfield, Gary Beecher, and Beverly Clevidence[*]

INTRODUCTION

Fruits and vegetables are dietary components which convey numerous health benefits. Not only do fruits and vegetables contain many essential nutrients, but they also contribute to disease prevention. In the last decade, nutrition research has made great strides in identifying specific nutrients which have beneficial health effects and the mechanisms by which those health benefits are achieved. However, the ability of a specific nutrient to impart health benefits depends on the gastrointestinal tract's ability to extract the nutrient from the plant material. The presence of a specific nutrient in a plant food is insufficient for providing health benefits if the bioavailability of that nutrient is very low. For example, spinach contains a high calcium content, but the presence of phytates and oxalates in spinach prevents the calcium from being absorbed in the gastrointestinal tract (Weaver et al., 1987; Heaney and Weaver, 1989; Peterson et al., 1992). Thus, calcium has a very low bioavailability from spinach.

Values for nutrient bioavailability have widespread impact on the nutrition community. The most profound impact of bioavailability studies is on the nutrient intake recommendations of the National Academy of Sciences' (NAS) expert panels. When the Dietary Reference Intakes (which update the Recommended Dietary Allowances) are determined, a factor for bioavailability is included. The National Academy of Sciences' nutrient recommendations have broad impact in nutrition. First, government-approved food programs are required to meet the appropriate level of many nutrients. Second, national surveys of nutrient intake are evaluated against the NAS intake recommendations. Third, discrepancies between nutrient intakes and recommended intakes drive nutrition policy and education programs in the United States. Therefore, reliable values for the bioavailability of nutrients are crucial.

* Janet A. Novotny, Steven Britz, Francis Caulfield, Gary Beecher, and Beverly Clevidence, USDA, Beltsville Human Nutrition Research Center, Beltsville, MD 20705.

TECHNIQUES FOR STUDYING BIOAVAILABILITY

There are several ways to investigate nutrient bioavailability. The approaches fall into two categories: either what goes into the body (i.e., penetrates the GI tract) is compared to a known dose, or what comes out of the body (i.e., is unabsorbed) is compared to the dose. The first is determined by sampling blood and/or body tissues and the second by sampling feces.

Balance studies compare unabsorbed material to a known dose. This is a mathematically simple technique in which the amount of a nutrient collected in the feces is subtracted from the amount of that nutrient ingested. The amount of nutrient which is not recovered is assumed to have been absorbed. Two problems arise from this assumption. First, it is possible for nutrients to pass through the small intestine unabsorbed and become metabolized by the intestinal flora, thus leading to overestimation of nutrient bioavailability. Second, it is possible for the intestinal bacteria to synthesize some nutrients in the colon (such as biotin, pantothenic acid, and vitamin K), leading to underestimation of nutrient bioavailability. Further, balance studies are labor-intensive and fecal collection is burdensome to human subjects.

Some pharmacokinetic applications evaluate plasma response to determine bioavailability. In this case, a compound is ingested, and plasma is sampled to follow the rise and fall of that compound in the blood. The plasma decay curve is extrapolated back to the ordinate, and the intercept is used to calculate the mass of the compound absorbed. One problem with this approach is that absorption of nutrients or other compounds which are transported to the liver via the portal vein may be underestimated. Further, endogenous levels of the compound in the blood will interfere with the calculation. Finally, if the compound is converted to one or more metabolites with an unknown stoichiometry, it is impossible to translate the appearance of the metabolite in blood to bioavailability.

Stable isotope methodology in combination with compartmental modeling solves some of the problems encountered by the other methods. Administering an isotopically-labeled nutrient to an individual allows discernment of that nutrient from the endogenous nutrient in biological samples. And compartmental modeling provides a wealth of information about rates of absorption and metabolism of ingested compounds.

ISOTOPIC LABELING OF NUTRIENTS IN PLANTS

To use stable isotopes to study bioavailability of nutrients from plants, the nutrients must be labeled with an isotope. There are five main options by which this may be achieved. Each technique involves delivery of the isotope to the plant tissue by a different mechanism.

Extrinsic labeling of plants entails mixing a stable isotope-labeled nutrient with food under the assumption that the labeled nutrient will exchange with the unlabeled nutrient in the food. This is a relatively simple and inexpensive means of labeling food nutrients, but the assumption of nutrient exchange is not always valid. For example, calcium is readily exchangeable in some food sources but not others (Schwartz et al., 1982). In contrast, ^{51}Cr is present in alfalfa in the same complexed form whether produced intrinsically or extrinsically (Starich and Blincoe, 1982). Because the labeled nutrient is often more loosely associated with the food matrix, studies by extrinsic labeling usually overestimate bioavailability (Bjorn-Rasmussen et al., 1972, 1973; Monsen, 1974; Weaver et al., 1984).

Another option is to introduce the label to the plant *through the nutrient solution.* In this case, the plant may be grown hydroponically or in a solid particulate growth medium, and the nutrient solution is enriched with a rare isotope which is a component of the nutrient of interest. The plant then incorporates the label into its tissues during growth. Hydroponics with an enriched nutrient solution has been used for labeling of peas with zinc (Fox et al., 1991), bean pods with calcium (Grusak et al., 1996), soybean seeds with nitrogen (Grusak and Pezeshgi, 1994), leafy vegetables with magnesium (Schwartz et al., 1980), and wheat with zinc (Starks and Johnson, 1985b). An enriched nutrient solution has also been used to label wheat grown in sand with zinc (Fox et al., 1991). For hydroponic labeling of organic compounds, one can use deuterium oxide in the nutrient solution. The deuterium is taken up by the roots and incorporated into molecules throughout the plant. Since hydrogen enters a plant almost exclusively from water and is subsequently distributed throughout the plant, in theory one can control the hydrogen composition of the plant through the composition of water to which the plant is exposed. However, the isotope effect of deuterium limits the level of labeling one may achieve. Because the mass of deuterium is much larger than the mass of hydrogen, and because of the importance of hydrogen bonding in biological molecules, replacement of hydrogen with deuterium can affect biological function. In animals, deuterium oxide administration representing 20% or more of total water intake can cause severe organ malformation and sterility (Katz, 1960). In plants, high levels of deuterium enrichment can also be deleterious. Deuterium enrichment above 50 atom percent excess results in stunted growth rates (Gruzak, 1997). For this reason, only partial random labeling can be achieved with deuterium. Partial random labeling means that each molecule containing hydrogen would be only partially labeled with deuterium, and the positions and numbers of deuterium atoms on a given molecule would be random. This complicates the laboratory analysis significantly, since the nutrient of interest cannot be assessed through monitoring a single isotopomer (labeled form of a molecule) but rather a random distribution of molecules must be monitored. The random labeling of the molecules also prevents MS-MS analysis of the compound of interest to ensure that only the compound of interest is being measured rather than a mix of interfering compounds with the same mass.

Foliar application, a method which has been used in agricultural systems to correct mineral deficiencies, has been successfully used to label plants. This method involves application of a nutrient solution to viable plant leaves. Wheat has been labeled with both copper and zinc by this method (Starks and Johnson, 1985b, 1986). This technique is limited to nutrients which may enter the plant through the leaves.

Stem injection is a labeling approach which allows specific targeting of the isotope to the plant tissues under study. For this approach, the labeled nutrient or its precursor is injected into the stem using a hypodermic needle and syringe. This method targets the label to the edible portion of the plant. Since isotopic labels are expensive, it is advantageous to minimize the loss of label on non-edible portions of the plant. The reliability of this method for bioavailability studies requires that the form of the nutrient accumulated in the plant be identical to that which occurs naturally. Starks and Johnson (1985a, b, 1986) showed that incorporation of stem-injected zinc and stem-injected copper into wheat and incorporation of stem-injected copper into peanuts reflects that which occurs naturally.

The final alternative is to present the label to the plant *atmospherically*. By exposing the plant to labeled carbon dioxide, organic molecules can be labeled as the carbon is fixed photosynthetically and distributed to tissues through biosynthetic pathways. This method is fairly difficult and complicated, due to the requirement for specialized equipment including a gas-tight chamber and special instruments for gas handling and analysis. Further, this method entails great expense if the plant is uniformly labeled with carbon-13, a more reliable labeling scheme than random partial labeling. However, the lack of an isotope effect of carbon-13 and the methodological advantage of a uniformly labeled molecule is a distinct advantage of fully labeling a plant with carbon-13. Atmospheric labeling has been used to assess carbon assimilation in photosynthesis (Yoneyama et al., 1980; Gordon et al., 1980, 1982; Kouchi, 1982; Kouchi and Yone-yama, 1984). Atmosheric labeling has also been investigated as a means of producing carbon-13 labeled starch, glucose, fructose, and sucrose in kilogram quantities from tobacco leaves (Kollman et al., 1973).

In this study, we used atmospheric labeling to produce carbon-13 labeled nutrients *in situ* in plant tissues for human consumption. We grew kale in a sealed growth chamber so that it would be entirely labeled with carbon-13. Kale was chosen because of its high nutrient content (including beta-carotene, lutein, folate, ascorbate, and others), its fairly short growth cycle (several weeks), and its high edible to inedible ratio (to minimize loss of isotopic label in inedible portions of the plant). Our initial nutrient focus was beta-carotene because of its role in providing vitamin A. To our knowledge, this is the first study to use atmospheric labeling to produce carbon-13 labeled plants for bioavailability studies.

METHODS: PLANT GROWTH CHAMBER AND CONDITIONS

The plants were grown in an airtight plexiglass box inside a stainless steel chamber of approximately 10 feet x 10 feet x 12 feet (see Figure 1). The inner plant growth chamber was constructed of 1/8 inch plexiglass and consists of two sections. The top section is an open-bottom box of 46 inches x 44 inches x 24 inches. The bottom section of the box is a plate of plexiglass with a 1 ½ inch by 1 ½ inch trough around the perimeter. The top section fits into the trough, which is filled with water, providing an air-tight connection between the top and bottom portions of the box. A Dayton Model 4C550 Axal fan was mounted at the center of the box, 2 inches below the top surface, directed for upflow. Four ports were drilled just outside the fan's perimeter for placement of gas lines. The top surface also contains a Teel Model 1P995 two-way check valve placed over a 1 ½ inch PVC pipe for pressure release. The bottom surface of the box contains four tubes for supplying nutrient solutions to four 17 inch x 20 ½ inch x 6 inch Nalgene bins containing the plants.

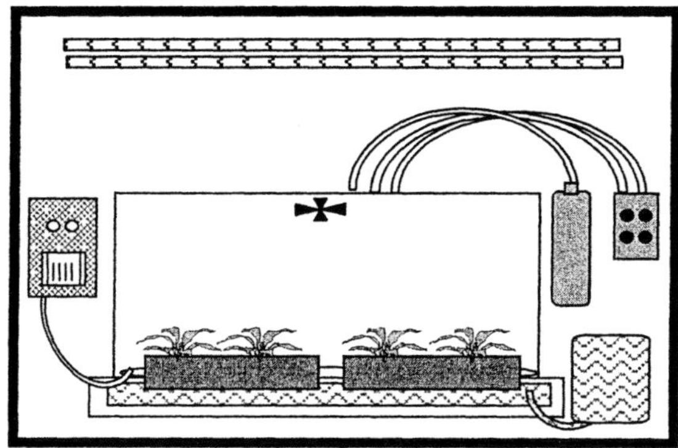

Figure 1. Schematic of the kale growth chamber. See text for a full description.

The only carbon dioxide available to the plants was carbon-13 dioxide (>99% carbon-13; Isotec Inc., Miamisburg, OH). Carbon dioxide partial pressure was controlled inside the kale growth chamber by a WMA-3 PP System and a WMA-3 PID controller (PP systems, Haverhill, MA), which injected $^{13}CO_2$ if levels dropped below 36 Pa. Calibration of the CO_2 analyzer for carbon-13 dioxide was performed prior to plant introduction. Carbon dioxide was sampled at a rate of 0.5 L/min through a tube placed in one of the ports at the top of the chamber. The sampled air was returned through another air line placed opposite the sample line. On the top inner surface of the outer chamber were mounted high pressure sodium and metal halide lamps. Plants experienced a 24 h photoperiod of 900 μmol m^{-2} s^{-1} of photosynthetically active radiation (PAR, 400-700 nm). Temperature in the chamber was controlled to be 24°C by a shaded thermocouple mounted at the center of the plexiglass box.

Kale seedlings were transplanted into vermiculite in Nalgene bins 8 days after germination. The bins were placed in the plexiglass box inside the outer growth chamber. Plants were harvested once the leaf canopy closed (approximately 2 weeks after transfer to the growth chamber). Leaf material was separated from roots, and leaf material was cleaned, weighed, and blanched. The blanched kale was chopped into 1 inch x 1 inch pieces, mixed for batch uniformity, and frozen at −80°C until consumption.

METHODS: KALE ANALYSIS

Extraction of carotenoids from kale was performed according to a previously published method (Khachik et al., 1986). A sample of kale was weighed and then transferred to a glass mixer jar. To the kale was added 0.5 g sodium carbonate, 0.75 g celite, and 25 mL tetrahydrofuran (THF) containing 0.05% butylated hydroxytoluene (BHT). The mixer jar was surrounded by ice and mixed for 20 min. The mixture was filtered and rinsed with THF. The remaining cake and filter paper were transferred to the mixer jar, and the process was repeated until the cake was colorless. The filtrate was concentrated and then partitioned into CH_2Cl_2 and saturated salt water in a separatory funnel.

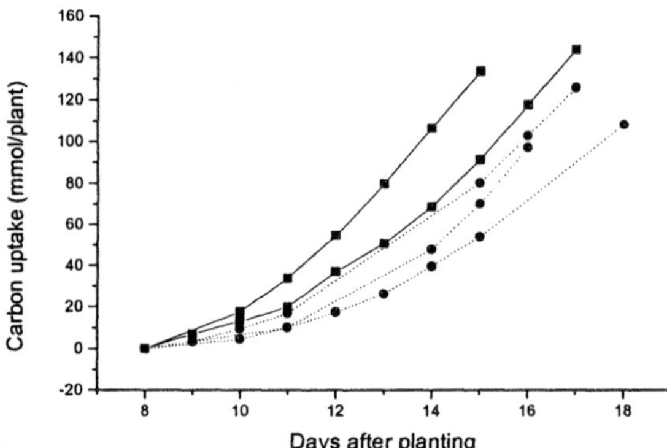

Figure 2. Comparison of carbon uptake by kale plants exposed to either carbon-13 dioxide or carbon-12 dioxide. Squares with solid lines represent carbon-13 dioxide uptake; circles with dotted lines represent carbon-12 uptake. Each line corresponds to a different kale growth trial.

Total beta-carotene content of the kale sample was determined by HPLC with an isocratic mobile phase (CH_3CN:CH_2Cl_2:CH_3OH at 65:25:10 plus 0.1% diisopropylethylamine) and a Microsorb C-18 column (Rainin Instrument Co., Woburn, MA) on a Hewlett-Packard Model 1100 HPLC system. The beta-carotene isotopomer distribution was determined by HPLC-MS (Pawlosky et al., 2000). Briefly, HPLC was performed using the mobile and stationary phases stated above on a Beckman model 110 HPLC system. The effluent of the beta-carotene fraction (appearing around 30 min) was delivered directly into the particle beam interface of a Hewlett-Packard 5989B mass spectrometer. A stream of helium at a pressure of 60 psi nebulized the incoming solvent stream. The carotenoids were desolvated and ionized using negative chemical ionization with methane. A full scan was performed for m/z (ion mass to charge ratio) less than 600 amu. Selected ion monitoring was used to evaluate mass to charge ratios of 536 and 537 for carbon-12 beta-carotene and mass to charge ratios of 576, 575, and 574 for carbon-13 beta-carotene.

RESULTS: CARBON UPTAKE AND INCORPORATION BY KALE

The replacement of $^{12}CO_2$ with $^{13}CO_2$ did not appear to impact growth of the kale plants. The kale plants grew as well in the carbon-13 environment as kale plants tested in trial runs in a carbon-12 environment. Carbon incorporation into kale plants was very similar for plants grown in $^{13}CO_2$ compared to plants grown in $^{12}CO_2$ environments, as shown in Figure 2.

Isotopic labeling of beta-carotene was nearly complete. A full mass scan showed that all carbon-13 labeled beta-carotene existed with mass to charge ratios of 576 amu (all 40 carbons labeled), 575 amu (39 carbons labeled, 1 unlabeled), or 574 amu (38 carbons labeled, 2 unlabeled). These three isotopomers accounted for 99.9% of the beta-carotene present in the kale. The fully labeled molecule (m/z of 576) comprised approximately

60% of the isotopomer distribution. Expressed as carbon sites labeled, the labeling was 99% complete. The reader may wonder how 99% of carbon sites may be labeled while only 60% of the labeled beta-carotene molecules existed with all 40 carbons as carbon-13. This can occur because of the large number of carbons comprising a beta-carotene molecule. A beta-carotene molecule contains 40 carbons. Therefore, 1000 beta-carotene molecules contain 40,000 carbon sites. If 1% of those sites are not labeled, then there are 400 carbon sites unlabeled. If each one of those unlabeled sites occurs on a different molecule, then 400 molecules of the original 1000 (that is, 40%) will contain 39 carbon-13 atoms and 1 carbon-12 atom. Thus, 99% of carbon sites may be labeled, while simultaneously only 60% of molecules have just 39 carbon sites labeled.

This isotopically labeled kale will be administered to subjects in a clinical feeding study to determine the bioavailability of carotenoids. Once the clinical study is complete and the plasma analysis of carotenoids is finalized, compartmental modeling will be conducted using the WinSAAM software program (Berman and Weiss, 1978; Wastney et al., 1999). The initial model will be that published by Novotny et al. (1995, 1996). The first simulations will be performed using the rate constants generated from the previous beta-carotene model (Novotny et al., 1995) after the model is modified to include the dose and subject characteristics for this experiment. The modeling process will continue with comparison of model predictions and experimental observations, and adjustments will be made to the fractional transfer coefficients to improve the model fit. Once the model prediction fits the observed data fairly well by eye, the fractional standard deviations of the transfer coefficients will be examined. If any parameters are not identified by this model and data set, those parameters will be removed, and the model structure will be simplified as necessary. On the other hand, if the data suggest that additional compartments are required, those will be added in physiologically reasonable ways, and the fitting process will recommence. The resulting model will yield two very important pieces of information: the level of absorption of beta-carotene from kale and the rate of conversion of beta-carotene to vitamin A in human subjects.

In conclusion, we have shown that a plant can be successfully grown to be very highly enriched in carbon-13. Such intrinsic labeling of plants offers unique opportunities for the study of nutrient bioavailability. We feel that this is a very powerful approach to studying bioavailability, and this study and future studies of its kind will shed new light on the body's utilization of nutrients from plants.

ACKNOWLEDGEMENTS

The authors acknowledge Mr. G. Wiley and Ms. D. Harrison for extraction and carotenoid HPLC analysis of kale; Mr. W. Harris for key assistance with plant growth; and Dr. R. Pawlosky and Mr. V. Flanagan for mass spectral analysis of the carotenoid isotopomer distribution in the kale extract.

CORRESPONDENCE

Please address all correspondence to:
Janet A. Novotny
USDA, BHNRC, DHPL
Beltsville Human Nutrition Research Center
Building 308, Room 201 BARC-East
Beltsville, MD 20705
novotny@bhnrc.arsusda.gov

REFERENCES

Berman, M., and Weiss, M.F., 1978, *SAAM 27 Manual,* DHEW Publ. (NIH) 78-180, U.S. Government Printing Office, Washington, DC.

Bjorn-Rasmussen, E., Hallberg, L., and Walker, R.B., 1972, Food iron absorption in man. I. Isotopic exchange between food iron and inorganic iron salt added to food: studies on maize, wheat, and eggs, *Am. J. Clin. Nutr.* 25:317-323.

Bjorn-Rasmussen, E., Hallberg, L., and Walker, R.B., 1973, Food iron absorption in man. II. Isotopic exchange of iron between labeled foods and between a food and an iron salt, *Am. J. Clin. Nutr.* 26:1311-1319.

Fox, T.E., Fairweather-Tait, S.J., Eagles, J., and Wharf, G., 1991, Instrinsic labelling of different foods with stable isotope of zinc (^{67}Zn) for use in bioavailability studies, *Br. J. Nutr.* 66:57-63.

Gordon, A.J., Ryle, G.J.A., Powell, C.E., and Mitchell, D.F., 1980, Export, mobilization, and respiration of assimilates in uniculm barley during light and darkness, *J. Exp. Botany* 31:461-473.

Gordon, A.J., Ryle, G.J.A., Mitchell, D.F., and Powell, C.E., 1982, The dynamics of carbon supply from leaves of barley plants grown in long and short days, *J. Exp. Botany* 33:241-250.

Grusak, M.A., 1997, Intrinsic stable isotope labeling of plants for nutritional investigations in humans, *J. Nutr. Biochem.* 8:164-171.

Grusak, M.A., and Pezeshgi, S., 1994, Uniformly ^{15}N-labeled soybean seeds produced for use in human and animal nutrition studies: description of a recirculating hydroponic growth system and whole plant nutrient and environmental requirements, *J. Sci. Food Agric.* 64:223-230.

Grusak, M.A., Pezeshgi, S., O'Brien, K.O., and Abrams, S.A., 1996, Intrinsic ^{42}Ca-labelling of green bean pods for use in human bioavailability studies, *J. Sci. Food Agric.* 70:11-15.

Heaney, R.P., and Weaver, C.M., 1989, Oxalate: effect on calcium absorbability, *Am. J. Clin. Nutr.* 50:830-832.

Khackik, F., Beecher, G.R., and Whittaker, N.F., 1986, Separation, identification, and quantification of the major carotenoid and chlorophyll constituents in extracts of several green vegetables by liquid chromatography, *J. Agric. Food Chem.* 34:603-616.

Katz, J.J., 1960, Chemical and biological studies with deuterium, *Am. Scientist* 48:544-580.

Kollman, V.H., Hanners, J.L., Hutson, J.Y., Whaley, T.W., Ott, D.G., and Gregg, C.T., 1973, Large-scale photosynthetic production of carbon-13 labeled sugars: the tobacco leaf system, *Biochem. Biophys. Res. Comm.* 50:826-831.

Kouchi, H., 1982, Direct analysis of ^{13}C abundance in plant carbohydrates by gas chromatography-mass spectrometry, *J. Chromatog.* 241:305-323.

Kouchi, H., and Yoneyama, T., 1984, Dynamics of carbon photosynthetically assimilated in nodulated soya bean plants under steady-state conditions. 1. Development and application of ^{13}CO$_2$ assimilation system at a constant ^{13}C abundance, *Ann. Botany* 53:875-882.

Monsen, E.R., 1974, Validation of an extrinsic iron label in monitoring absorption of nonheme food iron in normal and iron deficient rats, *J. Nutr.* 104:1490-1495.

Novotny, J.A., Dueker, S.R., Zech, L.A., and Clifford, A.J., 1995, Compartmental analysis of the dynamics of beta-carotene metabolism in an adult volunteer, *J. Lipid Res.* 36:1825-1838.

Novotny, J.A., Zech, L.A., Furr, H.C., Dueker, S.R., and Clifford, A.J., 1996, Mathematical modeling in nutrition: constructing a physiologic compartmental model of the dynamics of beta-carotene metabolism, *Adv. Food Nutr. Res.* 40:25-54.

Pawlosky, R.J., Flanagan, V.P., and Novotny, J.A., 2000, A sensitive procedure for the study of beta-carotene-d8 metabolism in humans using high performance liquid chromatography-mass spectrometry, *J. Lipid Res.* 41:1027-1031.

Peterson, C.A., Eurell, J.A., and Erdman, J.W. Jr., 1992, Bone composition and histology of young growing rats fed diets of varied calcium bioavailability: spinach, nonfat dry milk, or calcium carbonate added to casein, *J. Nutr.* 122:137-144.

Schwartz, R., Belko, A.Z., and Wein, E.M., 1982, An in vitro system for measuring instrinsic dietary mineral exchangeability: alternative to intrinsic isotopic labeling, *J. Nutr.* 112:497-505.

Schwartz, R., Grunes, D.L., Wentworth, R.A., and Wien, E.M., 1980, Magnesium absorption from leafy vegetables intrinsically labeled with the stable isotope ^{26}Mg, *J. Nutr.* 110:1365-1371.

Starich, G.H., and Blincoe, C.J., 1982, Properties of a chromium complex from higher plants, *J. Agric. Food Chem.* 30:458-468.

Starks, T.L., and Johnson, P.E., 1985a, Growth and intrinsic labeling of peanuts with ^{65}Cu for use in human bioavailability studies, *J. Agric. Food Chem.* 33:774-775.

Starks, T.L., and Johnson, P.E., 1985b, Techniques for intrinsically labeling wheat with ^{65}Zn, *J. Agric. Food Chem.* 33:691-698.

Starks, T.L., and Johnson, P.E., 1986, Evaluation of foliar application and stem injection as techniques for intrinsically labeling wheat with copper-65, *J. Agric. Food Chem.* 34:23-26.

Wastney, M.E., Patterson, B.H., Linares, O.A., Grief, P.C., and Boston, R.C., 1999, *Investigating Biological Systems Using Modeling: Strategies and Software,* Academic Press, New York.

Weaver, C.M., Nelson, W., and Elliott, J.G., 1984, Bioavailability of iron to rats from processed soybean fraction determined by intrinsic labeling techniques, *J. Nutr.* 114:1042-1048.

Weaver, C.M., Martin, B.R., Ebner, J.S., and Krueger, C.A., 1987, Oxalic acid decreases calcium absorption in rats, *J. Nutr.* 117:1903-1906.

Yoneyama, T., Arai, K., and Totsuka, T., 1980, Transfer of nitrogen and carbon from a mature sunflower leaf-^{15}NO$_2$ and ^{13}CO$_2$ feeding studies, *Plant Cell Physiol.* 21:367-1381.

ELEMENTAL MASS SPECTROMETRY FOR COMPARTMENTAL BIOLOGICAL MODELING

Assad Al-Ammar and Ramon M. Barnes[*]

INTRODUCTION

Elemental/inorganic mass spectrometry is a valuable and appealing technique to obtain stable isotope data for compartmental modeling of biological processes. "Biological processes" encompass biochemical reactions, enzyme kinetics, nutrient metabolism, drug metabolism, and mineral and trace element metabolism. The complexity of these biological processes requires special techniques for their characterization. Mathematical simulation by compartmental modeling is the computational concept that is most widely used to systemize biological processes, and stable isotope analysis by plasma source mass spectrometry (PSMS) has become a reliable and versatile means to provide data for these models (Barnes, 1996, 1998a, b; Becker and Dietz, 1998; Vanhaecke et al., 1999a; Becker and Dietz, 2000). Effective PSMS systems shorten analysis time, extend element isotope coverage, and enhance detection sensitivity, precision, and accuracy. Some noteworthy applications combining compartmental modeling with stable isotope/inorganic mass spectrometry to study trace metal metabolism, drug and nutrient metabolism, and biokinetics are described here.

ANALYTICAL TECHNIQUES USED WITH COMPARTMENTAL MODELING

As with all mathematical modeling techniques, experimental data needed for fitting must be generated. Several analytical chemistry techniques are used to provide these data for biological modeling. Injection of a study compound to the biological system followed by chromatographic, radiochemical, or spectroscopic determination of the compound or its metabolic products are widely used approaches, and stable (^{2}H, ^{13}C, ^{18}O) or radioisotope tags are commonplace. The diversity of analytical techniques supporting biological compartmental modeling is growing, especially through the use of enriched stable isotope tracers measured by elemental mass spectrometry with a plasma ion source.

---

[*] Assad Al-Ammar, Department of Chemistry, University of Massachusetts, Amherst, MA 01003. Ramon M. Barnes, University Research Institute for Analytical Chemistry, Amherst, MA 01002.

Applications of isotope analysis are represented in Figure 1 as three major groups: isotope ratio measurements, tracer experiments, and isotope dilution calibration. The latter two depend on enriched stable isotope tracers. Stable isotope tracers permit, for example, estimation of absorption and turnover rates of nutritional and toxic elements. Typically, a small quantity of the enriched stable isotope (a single tracer) is injected into the blood or ingested in labeled food into a larger natural pool of the element, resulting in relatively small changes (<10%) in the isotope ratios. In the dual tracer approach, one tracer is administered orally and another intravenously. Consequently, high accuracy and precision in sampling, sample preparation, and analysis are needed to obtain reliable data. In this paper, the application of plasma source mass spectrometry, especially the inductively coupled plasma (ICP) ion source, is discussed.

Advances made in biological compartmental modeling systems have been aided significantly by commercial development of quadrupole and sector-based ICP – mass spectrometry (ICP-MS) for the determination of enriched stable isotope tracers in biological and agricultural/botanical materials. This instrumentation is very sensitive, rapid and easy to operate, and cost effective relative to alternatives. Sample pretreatment is less complicated and time consuming than required for other mass spectrometry ionization sources like thermal ionization MS (TIMS). Typically, for TIMS isotope ratio determinations, the analyte must be isolated from the matrix by laborious and lengthy sample preparation steps. Although the cost of ICP-MS instrumentation has fallen steadily since its introduction in 1983, as with most mass spectrometers, an ICP-MS system is still relatively expensive compared to equipment used for other elemental spectroscopic techniques (e.g., flame spectrophotometry). Supporting facilities including clean rooms also can be costly.

Application of ICP-MS allows detailed kinetics to be described in at least three ways (Barnes, 1998b). First, the high sensitivity for stable isotope quantification permits small tracer doses to be administered, reducing both the experimental cost and risk of the dose perturbing the biological system. Second, methods are being developed to couple ICP-

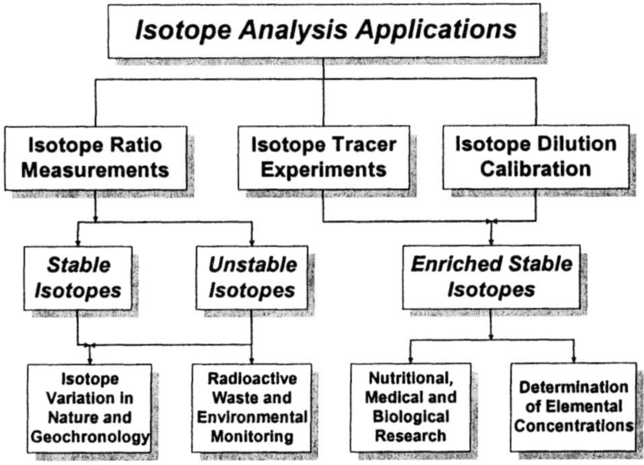

Figure 1. Applications of isotope analysis. After Becker and Dietze (1998).

MS with other analytical techniques (e.g., chromatography, electrophoresis, field-flow fractionation) to measure isotope enrichment in separated metal-binding species (e.g., metalloproteins, metalloenzymes, selenomethionine, seleno amino acids) in blood, urine, and specific tissues (Lu et al., 1995; Lu and Barnes, 1996; Vanhaecke et al., 1999a). Knowing the kinetics of metal bound to specific molecular species may provide insight into the forms of metal-binding species that are most physiologically active. Third, the simultaneous multielement measurement capability of ICP-MS makes possible studies in which the kinetics of several element tracers are determined simultaneously in the same subject. Furthermore, elements with multiple stable isotopes (e.g., Mo, Zn, Cd, Sn) can be applied simultaneously in isotope ratio studies for modeling and/or quantification. Two isotopes might be used as tracers, whereas a third could be used for isotope dilution MS (IDMS) (Vanhaecke et al., 1999a) to enhance the accuracy of the quantitative determination.

The measurement accuracy and precision of an ICP-quadrupole MS (ICP-QMS) system are limited by the sequential measurement mode, mass discrimination, and spectral overlap interferences. The ICP-QMS instrument illustrated in Figure 2 includes the ICP source (ICP torch) fed by a solution aerosol produced by a pneumatic or ultrasonic nebulizer (spray chamber); a system of sampling orifices between the ICP ion source and the QMS that stepwise reduce the pressure from atmospheric at the source to $<10^{-6}$ Torr at the MS; an arrangement of ion optics (lenses) to steer ions to the MS aperture and remove unwanted neutrals and photons; the quadrupole analyzer that selects a single ion according to its mass to charge ratio (m/z); and an ion detector (CEM channel electron multiplier) which converts the mass analyzed ions to electron pulses.

An ICP is diagrammed in Figure 3. The hot, atmospheric pressure argon discharge is generated in a quartz tube assembly (18 mm id torch) surrounded by an induction coil. Acting like a transformer, the coil supplied by a radio frequency current at ~40 MHz generates a magnetic field, which in turn induces an electron flow in the ionized argon in the quartz tube to create the plasma. An annular-shaped argon discharge is produced and

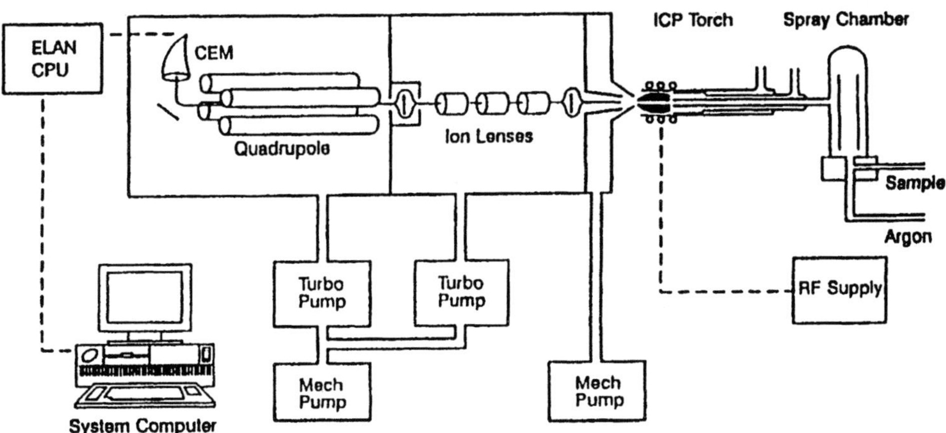

Figure 2. ICP quadrupole mass spectrometer arrangement (Perkin-Elmer Sciex Elan 5000).

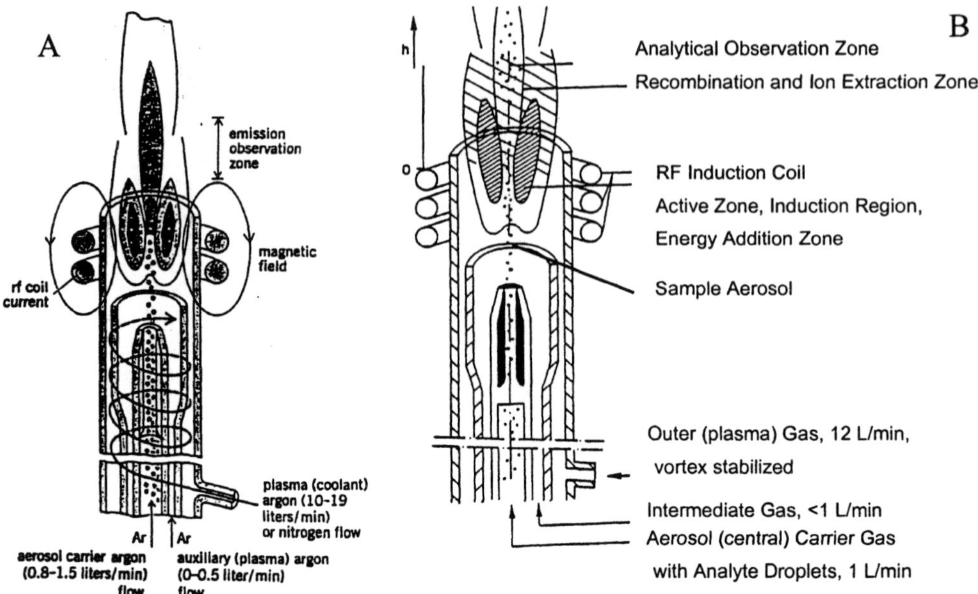

Figure 3. Inductively coupled plasma torch. (A) Induction coil and magnetic field lines, argon gas supplies, and gas flow circulation, sample aerosol pathway. (B) Plasma zone details and ion extraction region. Ions are extracted from the centerline of the discharge about 5 mm above the induction coil.

maintained by electron collisions. Sample aerosol interacts with the discharge as it flows into the central volume in the annular plasma. It is heated and desolvated; the resulting salt is rapidly dissociated into atoms, and the atoms are excited and ionized in the >5000 K plasma environment. The ICP is an excellent photon source for optical emission spectrometry as well as a superb ion source for MS. The ions (typically singly charged) are extracted from the plasma centerline through the sampling cone orifice of the mass spectrometer. Detailed descriptions and applications of ICP spectrochemistry are included in several recent books (Montaser, 1998; Hill, 1999).

Treatment of blood, feces, tissues, and other biological samples may require complete decomposition with nitric acid or only simple dilution. Alternative sample introduction techniques for solids, powders, and gases including laser ablation, slurry nebulization, or gas chromatography extend the applications of the ICP source. Compared to TI or electron impact ionization MS sources, the ICP is relatively noisy [≥ 0.05% relative standard deviation (RSD)] owing to variations in sample introduction, nebulization, and atomization; plasma gas dynamics; and temporal energy transfer from the field to the discharge to the sample (Vanhaecke et al., 1999a). Source noise can be minimized by appropriate selection of data acquisition operating parameters for QMS or scanning single collector high resolution (HRMS) systems. Dual or multiple simultaneous detectors also reduce source noise contributions if output signals are ratioed when using an internal reference technique or isotope ratios.

All mass spectrometers separate ions on the basis of their ratio of mass to charge (i.e., m/z or m/e), often called "ion mass." Since QMS (and single collector high resolution MS) instruments measure one ion at a time with a single detector, the mass/charge range (spectrum) must be scanned during a relatively short period. The uncertainty of isotope ratios obtained therefore include, among other things, temporal imprecision in the source during the two separate isotope determinations. This imprecision can be minimized with simultaneous ion detection using multiple detector/collector (MC) systems.

Mass discrimination bias is the preferential transmission of one ion over another from the ion source to the detector (Montaser, 1998). An abundance of sample matrix ions, for example, introduces a defocusing of the ion beam in the ion optics stage ("space charge effect") that can lead to an underestimation of the low-mass ion and an error in the isotope ratio measurement. Typically, the mass discrimination is about 1% per mass unit at m/z 100. Compared to TIMS, ICP-MS mass discrimination often is larger, especially for light elements, but systematic time-dependent mass fractionation observed with TIMS does not occur in ICP-MS (Vanhaecke et al., 1999a).

Spectral overlap interferences from background (argon, water), solvent (water, acid ions), and matrix (the sample) molecular ions restrict ICP-QMS measurement of some (e.g., S, Si, K, Ca, Cr, V, Fe, and Se) isotopic ratios. The objective of a mass analyzer is to separate ions of different mass. Two signal peaks are resolved if the height of the valley between them is less than the peak height expressed as some percentage (e.g., 50% or 10% valley). The resolution (R, resolving power with 10% valley) is defined as $R = m/\Delta m$, where m is the m/z of a peak, and Δm is the peak width (in m/z) at some fraction of peak height. Different mass analyzers are capable of different resolution. The quadrupole mass analyzer is commonly operated to separate ions that differ by one m/z (i.e., $\Delta m = 1$, unit resolution). The absolute resolution changes with mass (i.e., $R_{1/2} \approx 48$ for ^{24}Mg and ^{25}Mg; $R_{1/2} \approx 414$ for ^{207}Pb and ^{208}Pb). Consequently, with unit resolution, important elements in agricultural, biological, and botanical studies are affected by spectral interferences at the same nominal mass. Some interferences are listed in Table 1. For the spectral interference-free measurement of some isotope ratios by ICP-QMS (e.g., Zn in plasma and feces), analyte separation from the matrix is required.

Abundance sensitivity is a measure of how much of one mass peak overlaps with an adjacent peak and is especially important when a small peak is measured next to a major peak (e.g., when measuring a trace element or minor isotope). For quadrupole mass analyzers, the peaks tend to tail toward the low mass for the most common operating parameters, and abundance sensitivity typically is $>10^7$ in a well designed commercial system. A mass analyzer with high resolution does not necessarily have high abundance sensitivity.

To improve the precision, eliminate spectral overlap interferences, and extend the isotopic ratio measurement range and their applications, plasma source high-resolution (double focusing, magnetic sector field) mass spectrometers (HR ICP-MS or SF-MS) with single and multiple detectors (collectors) (MC-ICP-MS) have been developed and marketed (Vanhaecke et al., 1997; Becker and Dietze, 1998; Turner et al., 1998; Vanhaecke et al., 1999a; Becker and Dietze, 2000). These ICP MC mass analyzers are similar to TIMS systems with $R_{1/2} \approx 400$ to 10,000, and their measurement precision and accuracy match. The main advantage of high-resolution mass analyzers, in general, is the unequivocal identification of measured components and the mass separation of interfering ions. However, if the HR ICP-MS is operated in the high-resolution mode to

Table 1. Some spectral interferences on important biological elements at unit mass analyzer resolution

Interfering Ion Signals	Analyte Ion
$^{12}C_2^+$, $^{23}NaH^+$, $^{48}Ca^{2+}$	^{24}Mg
$^{12}C^{14}N$	^{26}Mg
$^{13}C^{14}N$	^{27}Al
$^{12}C^{16}O$	^{28}Si
$^{14}N^{15}N$	^{29}Si
$^{16}O_2$	^{32}S
^{40}Ar	^{40}Ca
$^{38}Ar^1H$	^{39}K
$^{35}Cl^{16}O$	^{51}V
$^{40}Ar^{12}Cl$ or $^{16}O^{35}ClH$	^{52}Cr
$^{40}Ar^{35}ClH$	^{53}Cr
$^{40}Ar^{16}O$ or $^{40}Ca^{16}O$	^{56}Fe
$^{44}Ca^{16}O$	^{60}Ni
$^{40}Ar^{35}Cl$	^{75}As
$^{40}Ar^{37}Cl$ and $^{36}Ar^{40}ArH$	^{77}Se
$^{38}Ar^{40}Ar$	^{78}Se
$^{40}Ar^{40}Ar$	^{80}Se

minimize interferences, the isotope ratio precision may be limited by both ion-counting statistics and ion beam instability (Turner et al., 1998). The ion transmission of HRMS is inversely proportional to the mass resolution, since the ion beam is restricted at the source and collector slits. Following the introduction of these commercial ICP-MS systems, numerous recent applications have reported isotope ratio measurements (Halliday et al., 1995; Vanhaecke et al., 1997; Becker and Dietz, 1998; Marechal et al., 1999; Vanhaecke et al., 1999a; Becker and Dietze, 2000). However, MC-ICP-MS instrumentation is expensive, and it loses sensitivity and analysis speed when operated in high-resolution mode.

One alternative approach for eliminating spectral interferences, called the dynamic reaction cell (DRC) ICP-MS, was developed recently based on a conventional ICP-QMS. A second enclosed, rf-only quadrupole pressurized with gas (10-20 mTorr) is installed as an isolated reaction cell between the ion optics and the mass analyzer. This unique approach was introduced recently (Perkin-Elmer SCIEX ELAN 6100 DRC ICP-MS) (Denoyer et al., 1999; Tanner and Baranov, 1999; Becker and Dietze, 2000) and is especially valuable when common sector field instruments have insufficient mass resolution. Reaction cells with ICP-MS utilize electron-transfer reactions of argon-containing plasma ions with lower-ionization potential neutrals (e.g., NH_3) to remove isobaric interferences (e.g., $^{40}Ar^{16}O$ on ^{56}Fe). When the ionization potential of the interference is lower than the analyte ion, alternative ion-molecule chemistry by reactions with N_2O, O_2, $N_2O + CH_4$, CH_3F, or CH_3Cl can be substituted. These reactive collisions also produce new molecular ions resulting from the reactive gas (NH_3) or its impurities and elemental ions that may be added to the mass spectrum. These unwanted molecular ions are removed by adjusting the reaction cell operating (Mathieu) parameters to limit the mass transmission band pass, which generally reduces molecular ions transferred to the mass analyzer. The concept of a band pass-limited DRC permits extremely efficient removal of many argon-based interferences (up to 9 orders of magnitude) and has the potential of eliminating other polyatomic interferences. Instrument background is dramatically reduced to less than 1 count/second (cps), while normal quadrupole ICP-MS sensitivity and scanning speed are maintained. The collisions broaden the ion residence time, and short-term fluctuations of the ion signal are decreased leading to improved isotope ratio measurement precision. Owing to removal of isobaric spectral interferences, the biologically important elements like K, Ca, Cr, Fe, As, and Se can be detected at their most abundant masses. This leads to detection limits that are typically an order of magnitude better than conventional ICP-QMS. Furthermore, the $^{87}Sr^+$ and $^{87}Rb^+$ signals cannot be resolved by sector field ICP-MS instrumentation, but the DRC ICP-MS with appropriate reaction gas can provide reasonable signals owing to the selective reaction of one but not both ions. Thus, the DRC approach can successfully replace HR-ICP-MS for some applications.

Another recent approach (Micromass IsoProbe multi-collector ICP-MS) employs only a magnetic analyzer preceded by a hexapole collision cell to reduce the ion energy (to < 1 eV) of plasma-generated ions and Ar and Ar molecular ion interferences (Turner et al., 1998; Becker and Dietze, 2000). The hexapole acts both as reaction cell and ion guide but is unable to provide band pass-limited transmission of the DRC. An electro-static energy analyzer is not required. The argon collision gas provides the best thermalization conditions for most masses, and hydrogen (or helium) can be introduced to produce electrons to neutralize argide ions. Isotope ratios can be measured with 0.01% RSD precision with only a few hundred nanograms of sample. Thus, a number of novel instrumental avenues exist with ICP-MS to generate data for inorganic nutrient metabolism modeling studies.

COMPARTMENTAL MODELING

Application of compartmental modeling of biological systems can be positively combined with stable isotope tracer-mass spectrometry. Models are simplified representations of systems. The processes occurring in the biological system and their interactions can be represented mathematically by a set of equations (i.e., the model). By solving the equations simultaneously, the model solution will mimic the behavior of the biological system. Mathematical models therefore predict the behavior of a system before actual experiments on the system. Models are often developed because probing a real system may not be possible or may be too costly.

To establish a model for any system, the target material (chemical compound, inorganic trace element, enzyme, etc.) is specified. The modeler is interested in studying the distribution, transfer, and reaction of this material in the biological system. The system is then divided into its basic components, each component of which represents a "compartment." Compartmental models assume that the target material is distributed throughout the system in discrete entities (compartments). A compartment is considered to contain the target material homogeneously distributed and undergoing a specific kinetic process. The compartment may be defined physically (e.g., specific body organs like liver, intestine, etc. or specific body pools like blood, plasma) or conceptually (e.g., all materials that turn over at a particular rate). As an example, modeling zinc metabolism in the human body may involve blood, intestine, bone, liver, kidney, urine, and feces as separate compartments, and a labeled zinc compound as the tracer material.

The initial system subdivision to its basic components is based on available information on the system properties and the mode of its interactions with the target material. This initial compartmental subdivision helps to start modeling a system. The results of this initial modeling process, when fit with the experimental data, are used to reshape the system by eliminating some of the proposed compartments when the modeling process reveals their unimportant role and by introducing new compartments to improve the fit. Modeling might even reveal the necessity of subdividing a specific compartment into separate compartments. This type of subdivision should be based on the existence of two different mechanistic processes (or transformations) that occur consecutively in the subdivided compartment.

Following the initial compartmental subdivision of the system, the model then describes mathematically the connections between the system compartments. This mathematical description is expressed as one or several ordinary differential equations that describe the rate, direction, and manner (i.e., linear or nonlinear) of exchange of target material among compartments, inputs (i.e., input of target material from outside the study system), and excretion (i.e., irreversible losses) out of the system. Compartmental modeling exclusively uses ordinary differential equations compared to non-compartmental modeling, which involves other types of equations (e.g., partial differential equations, algebraic, stochastic). This makes compartmental models very useful for describing biological systems, because these systems are often visualized as pools (or compartments), and some of their chemical or biological processes can be described by ordinary differential equations. A two-compartment model is shown in Figure 4. The mathematical description of this model as ordinary differential equations takes the form

$$dF_1/dt = -(L_{21} + L_{01})\, F_1 + L_{12}\, F_2 + U_1 \tag{1}$$

$$dF_2/dt = L_{21}\, F_1 - (L_{12} + L_{02})\, F_2 + U_2 \tag{2}$$

In the steady state, both dF_1/dt and dF_2/dt are equal to zero.

Assuming that the compartmental model in Figure 4 is the modeler's initial representation of the system (i.e., the starting model), then the result expected from running the model are the values of L_{ij}, F_1, and F_2. Initial values of L_{ij} must be assumed to run the starting model. The magnitudes of the assumed values are based on prior knowledge of the properties of the system under study. However, since this knowledge is not always available, estimating the initial values of L_{ij} is usual from experimental data obtained from measurement of F_1 or F_2 values as they change with time when the system is not under steady state. The slopes of this curve at different values of t can be used to calculate an estimate of L_{ij}. These estimated values could be used to solve the differential equations that describe the starting model by employing available modeling software.

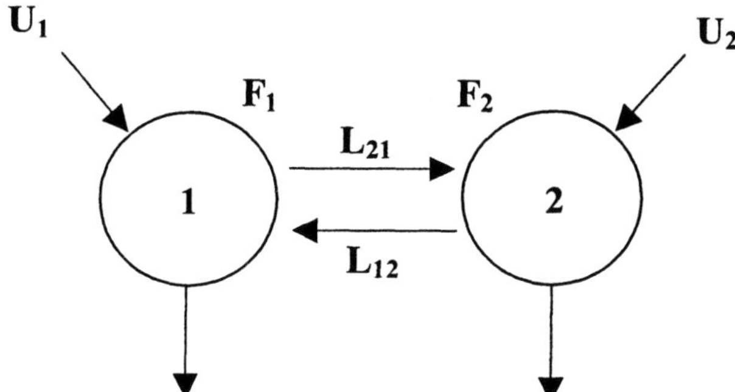

Figure 4. A two-compartment model. F_1, F_2, dF_1/dt and dF_2/dt represent the concentration or the total amount of the target material and their derivatives in compartments 1 and 2, respectively. L_{ij} is the fractional rate of material transfer from compartment j to compartment i. L_{0j} is the fractional rate of material transfer from compartment j to outside (i.e., material excretion). U_i is the rate of continuous infusion (introduction of the target material from outside into compartment i).

Several commercial software packages are available for this purpose. The calculated F_1 vs. t values are compared with the experimental data. Disagreement between the two sets is applied to adjust the values of L_{ij}, which then are used to run the model to obtain a second fit between the calculated and measured data. This fitting process is repeated, sometimes with elimination or addition of new compartments, until a good fit is obtained.

This model fitting process is difficult and dependent on the insight and mathematical skill and experience of the modeler. Some useful guidelines are available elsewhere (Wastney et al., 1999). The F_1 vs. t data obtained by introducing the target material such as an isotopically labeled species to the system vastly improves the estimation of the correct model.

ISOTOPICALLY LABELED TRACERS

Modeling biological systems often involves measuring the rate of transfer of isotopically enriched tracer among the system compartments (Crews et al., 1994; Fairweather-Tait et al., 1997; Coudray and Fairweather-Tait, 1998; Crews et al., 1998; Lund et al., 1998; Minihane and Fairweather-Tait, 1998; Vanhaecke et al., 1999a). This methodology is used commonly during modeling, because biological systems mostly exist in steady state. At steady state, the rate of input and output of the target material into the system are equal and constant. To generate the experimental data to model this system, the system must be perturbed by introducing a noticeable quantity of the target material. This step is followed by experimentally measuring the rate of transfer of the target material throughout different system compartments. However, introducing a large quantity of test material leads to a change in the magnitude of the fractional rate of transfer (L_{ij}) in the biological system. This happens because biological systems, in most cases, are nonlinear systems. Nonlinear systems are characterized by having L_{ij} parameters that are variable with a difference in concentration or quantity of the target material in the compartments.   Often this functional dependence is complex, and therefore extensive perturbation of the system by the introduced material makes modeling very difficult.   To circumvent the nonlinearity of the biological system, a very small amount of isotopically enriched tracer is introduced to the system.   This causes no significant system perturbation.   Thus, the system can be treated as linear and can be modeled easily by regarding L_{ij} as constant parameters.   The experimental data needed for model fitting can be considered as representing an undisturbed system.

The tracer data can then be used in the model fitting process to determine the number of compartments, the connection between these compartments, and the fractional transfer constants or L_{ij}.   Once these values have been determined, the steady state values for the system, such as the mass of the target material in some compartments, the input of the target material (e.g., intake in the diet), or the loss of this material (e.g., rate of urinary excretion) can be determined.

Often studies are performed under several conditions (for example, with two or more tracers).   All data should be modeled simultaneously, because each study provides information about different system aspects.   For instance, an intravenous tracer will provide information on the distribution of a compound from blood, while an oral tracer provides information on absorption.   As an example, modeling calcium metabolism involves the administration of an oral (^{44}Ca) and an intravenous tracer (^{42}Ca).   The parameter values obtained from fitting the intravenous data are used to fit the oral data by equating all the parameters in the oral model to the values in the intravenous model. Frequently, data are obtained at two conditions to identify the differences between the conditions.   This ability to determine the differences between sets of kinetic data is a powerful feature of modeling.   Some examples compare normal and abnormal conditions, metabolism before and after treatment, or two physiological conditions.   Models can be set up as duplicate models or using time-interrupts.   With both methods, data from one condition are fit, and then data from the second condition are fit by systematically introducing differences in parameter values between the first and second condition.   This is called the "minimal principle."   It assumes that a minimal number of differences in parameter values between two conditions exist that will explain the kinetics.   The aim is to identify the changes that are necessary and sufficient to explain the differences.

The logic behind this approach is that regulation of a system normally occurs at specific points.   The problem is to identify the key intersections or pathways.   This approach was used to identify five zinc metabolism regulation sites *in vivo* (Wastney et al., 1986).   Radioisotope tracer data were obtained from human volunteers over 9 months while they consumed their regular diets with 10 mg ^{65}Zn/d and for another 9 months while they consumed an additional 100 mg Zn/d.   The tracer was measured in seven sites: plasma, red blood cells, urine, feces, and over liver, thigh, and whole body.   By systematically changing parameter values, differences between the normal zinc and high intake periods were explained by changes in five parameters representing five zinc regulation sites (Wastney et al., 1986).

Modeling based on stable isotopes is theoretically similar to modeling when applying radioisotopes.   In addition, obtaining the tracer-to-tracee ratio directly is an advantage, since it is not as easy with radioisotopes.   Administering radioisotopes (e.g., ^{59}Fe) is not advised in experiments involving human subjects owing to radiation hazards, but it is done often. Consequently, measuring stable isotopes with plasma source MS has been especially applicable for human subjects.

APPLICATIONS OF STABLE ISOTOPE- MASS SPECTROMETRY COMPARTMENTAL MODELING

Most applications involve reaction kinetics and metabolism of some important biological compounds, drugs, enzymes, or inorganic trace elements (e.g., bromine, calcium, copper, iron, magnesium, molybdenum, nickel, selenium, zinc) in humans and animals (Mellon and Sandström, 1996; Platzner, 1997; Vanhaecke et al., 1999a). Stable isotopes in human nutrition have been measured with different mass spectrometry approaches including electron impact MS (EIMS) and gas chromatography (EI-GC-MS), fast atom bombardment MS (FABMS), TIMS, and PSMS (Mellon and Sandström, 1996; Platzner, 1997). In the classical EIMS determination of light element isotope enrichment (δ^2H, $\delta^{13}C$, $\delta^{15}N$, $\delta^{18}O$) of organic molecules, high precision, dedicated gas inlet MS instruments are used. Alternatively, GC-MS is applied with reduced measurement precision, typically about 5% RSD (Mellon and Sandström, 1996). A comparison of mass spectroscopic isotope measurement precision is summarized in Table 2 (Heumann et al., 1998). For stable isotope measurement of heavier elements (e.g., inorganic nutrients), FABMS, TIMS, or neutron activation analyses are employed (Mellon and Sandström, 1996; Heumann et al., 1998). The introduction of ICP-MS represents an attractive alternative for most stable isotopes (Mellon and Sandström, 1996; Becker and Dietze, 2000), although few light elements ($Z < 10$) like B and S (Menegário et al., 1998) have been studied by ICP-MS.

Stable isotope mass spectrometry and compartmental modeling techniques have been used to study the metabolism of some trace elements and inorganic nutrients (Mellon and Sandström, 1996). Some applications of compartmental modeling with experimental data from stable isotopes measured by elemental mass spectrometry are summarized in the next paragraphs.

Lowe et al. (1997) developed a zinc metabolism model based on SAAM-CONSAM in six healthy women using stable zinc isotopes. After equilibration on a constant diet containing 7 mg Zn/d, an oral tracer highly enriched in ^{67}Zn and an intravenous tracer

Table 2. Typical ranges of good precision reported for isotope ratio measurements by different plasma source MS. Adapted from Heumann et al. (1998)

Mass Spectrometer Type	RSD (%)
Glow Discharge MS	0.1-1
Quadrupole ICP-MS	0.1-0.5
HR ICP-MS	0.05-0.2
ICP-TOFMS	0.05-0.1
MS/MC ICP-MS	0.005-0.02
TIMS	0.006-0.01
Electron Impact Gas MS	<0.005

highly enriched in ^{70}Zn were administrated simultaneously. Tracer-tracee ratios in plasma, urine, and feces were calculated from ^{67}Zn/^{66}Zn and ^{70}Zn/^{66}Zn isotope ratios measured by ICP-QMS. Wastney and colleagues (Serfass et al., 1996; Wastney et al., 1996, 2000) also studied kinetics of zinc metabolism and its variation with diet, genetics, and disease by adding a zinc stable isotope to the system and measuring its movement in the system over time by ICP-QMS. The compartmental model was used to determine pool sizes and transport rates. Other recent applications of stable isotope-mass spectrometry-compartmental modeling for zinc metabolism have been described (Miller et al., 1996; Hambridge et al., 1998; Miller et al., 1998; Bonnel et al., 1999; Griffin et al., 1999a, b; Kopp-Hoolihan et al., 1999; Krebs et al., 1999; Lowe et al., 2000).

Applications of stable isotope-mass spectrometry-compartmental modeling for trace element metabolism with other isotopes have developed recently based on ICP-MS determinations (Barnes, 1996; Mellon and Sandström, 1996). Smith et al. (2000a, b) studied lead. The effects of outpatient succimer treatment of lead-poisoned children on the absorption and retention of lead from exposures suffered concurrently with treatment were not well known. A juvenile non-human primate model of childhood lead toxicity and a sensitive double stable isotope tracer methodology was used to investigate the effects of oral succimer on GI lead absorption and retention. Seventeen rhesus monkeys (Macaca mulatta) aged 15 months were fasted for 18 h and then dosed in rapid succession with ^{204}Pb (5 μg IV) and ^{206}Pb (72.6 μg orally), followed by treatment with vehicle (n = 8) or succimer (n = 9; 30 mg/kg/d) for 5 days. Blood and daily total fecal and urinary outputs (24 h) were collected over treatment days -1 to 5 and analyzed for ^{204}Pb, ^{207}Pb, and ^{208}Pb using magnetic-sector ICP-MS. Results indicate (i) GI Pb absorption was not measurably altered in the succimer group compared to the vehicle; (ii) succimer treatment significantly increased the urinary excretion of IV-administered ^{204}Pb but not ^{206}Pb that had been orally administered; (iii) whole body retention of the orally administered Pb dose was not measurably different between treatments; and (iv) succimer treatment reduced fecal excretion of IV-administered Pb in favor of increased urinary excretion. Only 11 and 16% (mean vehicle and control, respectively) of the IV ^{204}Pb dose was recovered in urine and feces over the 5 days of treatment, while only 41 and 49% of the orally administered ^{206}Pb dose was recovered.

Patriarca et al. (1997) examined nickel metabolism. Four healthy adult subjects (2 male, 2 female) fasted overnight before ingesting oral doses of 10 mg/kg of the stable Ni tracer ^{62}Ni. Blood samples were withdrawn at fixed intervals, and the total daily output of urine and feces were collected for five days following dose ingestion. The determination of ^{62}Ni in plasma, urine, and feces was performed by ICP-MS after isotope dilution with ^{61}Ni. On the assumption that there was negligible excretion of absorbed Ni by the gut, the percentage absorption was calculated from the ^{62}Ni in the feces and ranged from 28% to 40%. The percentage of the absorbed dose excreted in the urine over 5 days ranged from 51% to 82%. Plasma ^{62}Ni peaked between 1.5 and 2.5 h after dose ingestion and reduced by a factor of ten or more over the next few days. The tracer data provided the input to a three-compartment model developed using SAAM II (Resource Facility for Kinetic Analysis, Seattle). The model consisted of a plasma compartment with renal output and two other compartments bi-directionally linked to the plasma compartment. An excellent fit to the tracer data was obtained. Initial analysis suggests that it can take up to 1 month to excrete 90% of the absorbed tracer dose.

Cantone et al. (Cantone et al., 1997; Giussani et al., 2000) studied molybdenum biokinetics. An experiment was performed with three volunteer subjects to compare the biokinetic behavior of an intrinsic and an extrinsic tag. For the intrinsic tagging, cress was grown on an isotopically enriched ^{95}Mo solution. For the evaluation of intestinal Mo absorption, an aliquot of 10.5 g of cress, corresponding to about 500 µg ^{95}Mo, was mixed with 150 µL (i.e., about 280 µg ^{96}Mo) of an isotopically enriched ^{96}Mo solution used as extrinsic tag and then administered to each of the volunteers. A series of blood plasma samples was collected for 500 min post-injection and total urine was obtained for 2 days. The analysis was carried out by proton nuclear activation (PNA) for plasma and by ICP-MS for urine samples. The results show slightly different values for intestinal absorption (77 ± 5% for the intrinsic tag *vs.* 90 ± 4% for the extrinsic tag) but similar biokinetic behavior once the tracers were absorbed in the intestine. Excretion in the urine was very fast: between 85% and 90% of the total amount excreted within 2 d was lost in the first 12 h. The results obtained by mass spectrometry and activation analysis agreed quite well.

Koch et al. (1996a, b) investigated bismuth pharmacokinetics. Utilization of a sensitive ICP-MS analysis and appropriate clinical study designs increased assurance that therapeutic use of new bismuth compounds would produce systemic exposure below levels that can be associated with development of toxicity to this heavy metal. Bismuth compounds have been used to treat a variety of gastrointestinal disorders, and bismuth acts locally in the gut to protect the mucosa and inhibit the growth of intestinal pathogens. Systemic exposure was therefore important for toxicological rather than pharmacological reasons. Clinical pharamcokinetic data on bismuth are limited. Oral absorption was quite variable and did not appear to be extensive (<1% of the dose), with the vast majority of administered bismuth recovered in the feces. Plasma concentrations of bismuth following oral dosage were quite low, and care was taken to avoid bioanalytical interference from environmental sources (e.g., cosmetics). Bismuth was tightly bound to serum albumin and was widely distributed to various tissues, where it bound to inducible metal-binding proteins. Bismuth persisted in the body, displaying terminal distribution half-lives of several months, although elimination half-lives were on the order of a few weeks. Bismuth has not been observed to undergo human metabolic transformation. Elimination from the systemic circulation occurred primarily by renal excretion.

Buckley et al. (1996) studied the kinetics of direct-reacting (readily exchanged) copper in blood plasma. The approach involved addition of enriched ^{65}Cu to plasma, extraction with diethyldithiocarbamate in mineral oil, and determination by ICP-MS. After intravenous infusion of enriched ^{65}Cu into two healthy men, biological half-lives of 8.7 and 12.3 min were determined for direct-reacting Cu.

In these studies, the isotope ratios were determined using ICP-MS because of its high sensitivity, ease of operation, and the cost effectiveness of this technique. However, the relatively low precision isotope ratio and sample matrix effects on measurement accuracy led to examination of other mass analyzers. For example, Vanhaecke et al. (1999b) evaluated the isotope ratio performance of a commercial axial time-of-flight ICP mass spectrometer (TOF-ICPMS) for compartmental modeling. This approach provides almost simultaneous detection of the entire mass spectrum, and isotope ratios correlate well temporally leading to acceptable isotope ratio precision of ≤0.05% RSD (n = 10) for

30 s per replicate acquisition time. This isotope ratio precision can be obtained for many ratios simultaneously. Mass discrimination was ~1% at midmass, but TOF-ICPMS have been reported to be somewhat less sensitive than QMS or HRMS systems.

High-resolution ICP-MS systems have begun to play an important role in stable isotope tracer studies (Platzner, 1997; Becker and Dietze, 1998; Vanhaecke et al. 1999; Becker and Dietze, 2000; Smith et al., 2000a, b; Stürup, 2000), although few compartmental modeling applications for trace element metabolism have yet been published. Quadrupole-based MS instruments provide precisions of 0.1 to 0.5% RSD for short term and 0.05% RSD for long-term measurements (Table 2). Double-focusing sector-field ICP-MS yields 0.04% RSD at low mass resolution and 0.1% RSD at medium resolution. When the HR ICP-MS is operated in the high-resolution mode to minimize spectral overlap, the isotope ratio precision may be limited by both ion-counting statistics and ion beam instability (Turner et al., 1998). Precisions of 0.4 to 0.8% RSD were reported, for example, for ^{42}Ca/^{44}Ca, ^{25}Mg/^{24}Mg (Dombrovári et al., 2000), ^{41}K/^{39}K, ^{57}Fe/^{56}Fe, ^{65}Cu/^{63}Cu, and ^{67}Zn/^{64}Zn in biological samples (Becker and Dietze, 2000).

Stürup (2000) reported the simultaneous measurement of zinc isotope ratios (^{64}Zn/^{66}Zn, ^{67}Zn/^{66}Zn, ^{68}Zn/^{66}Zn, ^{70}Zn/^{66}Zn) in human feces, urine, and serum by HR-ICP-MS and found that all isotopes could be measured (0.7% RSD) without isobaric interferences, except for ^{64}Zn which required a mathematical correction for ^{64}Ni. This correction degraded the isotope ratio precision to 1.2% RSD. This high-resolution measurement eliminated the need for extended zinc purification and extraction steps required for ICP-QMS measurements. Mass discrimination of ions resulting from space-charge effect (i.e., mass bias) degrades isotope ratio precision and accuracy, but mass discrimination decreases with increasing isotope mass (Becker and Dietze, 2000). This affects light elements most strongly. Collision and dynamic reaction cell ICP-MS systems appear to provide isotope ratio precisions of 0.03% to 0.06% RSD. Only with multi-ion collector detectors has ICP-MS precision matched TIMS values. Precisions of 0.002% for ^{82}Se/^{80}Se, 0.009% for ^{42}Ca/^{40}Ca, and 0.008% for ^{11}B/^{10}B have been reported (Becker and Dietze, 2000). Eagles et al. (2000) determined that the accuracy and precision of MC-ICP-MS calcium isotope ratio (^{42}Ca/^{40}Ca, ^{44}Ca/^{40}Ca) measurements were sufficient for nutrient metabolism studies, sample consumption was at least 4 times lower than required for MC-TIMS, and measurement was at least 6 times faster than TIMS. However, careful sample preparation techniques were needed.

CONCLUSION

The difficulty in modeling nonlinear biological systems can be circumvented by infusion of very small amounts of the target material as an isotopically labeled and enriched compound coupled with sensitive detection technologies. This causes no significant perturbation of the system. Accordingly, the system can be treated as a linear system and can be modeled easily by regarding L_{ij} as constant parameters. The experimental data needed for model fitting can be considered as representing an undisturbed system.

Recent commercial introduction of multiple collector, double focusing, magnetic sector field, high resolution mass spectrometers and dynamic reaction cell or collision cell mass spectrometry with ICP sources shall extend the application of compartmental modeling to elements that are otherwise very difficult or impossible to measure by ICP-MS owing to spectral overlap interferences (e.g., ^{24}Mg, ^{26}Mg; ^{28}Si, ^{29}Si, ^{30}Si; ^{32}S, ^{34}S; ^{40}Ca; ^{50}Cr, ^{52}Cr, ^{53}Cr; ^{77}Se, ^{78}Se, ^{80}Se). These and other "difficult" or overlooked elements like B, Ca, K, Ti, V, Fe, and Ni are important in biological, environmental, and nutritional studies. The precision and accuracy of enriched stable isotopic tracer analyses with these advanced ICP-MS systems can be comparable to TIMS results, and their applications in geochronology and biological modeling with stable isotope measures is expected to grow significantly during the coming decade. Extension of stable isotope ICP-MS analyses and biological modeling to toxic and heavy metals (Cd, Hg, Pb, Pt, Pd, Sn, Sb, Te, Tl) also can be expected to develop.

Stable isotope intrinsically labeled materials or compounds (e.g., ^{118}Sn-enriched dibutyltin [Crews, 1998]) will become more widely applied in studies of metal metabolism, because of the growing development of sensitive and accurate compound-specific (speciation) ICP-MS analyses (Crews, 1998) following chromatographic, electrophoretic, or other separations especially with MC-ICP-MS and TOF-ICP-MS. These developments can be expected to fuel the continued improved specificity of biological system modeling.

ACKNOWLEDGEMENT

ICP Information Newsletter, Inc. (Hadley, MA) supported preparation of this paper.

CORRESPONDENCE

Please address all correspondence to:
Ramon Barnes
University Research Institute for Analytical Chemistry
P.O. Box 666
Hadley, MA 01035
rmbarnes@chem.umass.edu

REFERENCES

Barnes, R.M., 1996, Analytical plasma source mass spectrometry in biomedical research, *Fresenius J. Anal. Chem.* 355:433-441.

Barnes, R.M., 1998a, Capillary electrophoresis and inductively coupled plasma mass spectrometry: status report, *Fresenius J. Anal. Chem.* 361:246-251.

Barnes, R.M., 1998b, Plasma source mass spectrometry in experimental nutrition, in: *Mathematical Modeling in Experimental Nutrition,* A.J. Clifford and H.-J. Müller, eds., Plenum Press, New York.

Becker, J.S., and Dietze, H.-J., 1998, Ultratrace and precise isotope analysis by double-focusing sector field inductively coupled plasma mass spectrometry, *J. Anal. Atomic Spectrom.* 13:1057-1063.

Becker, J.S., and Dietze, H.-J., 2000, Precise and accurate isotope ratio measurements by ICP-MS, *Fresenius J. Anal. Chem.* 368:23-30.

Bonnel, E., Woodhouse, L., Shames, D., Lowe, N., Jackson, M.J., Turnlund, J., and King, J.C., 1999, Effect of acute zinc depletion on a compartmental analysis of zinc kinetics, *FASEB J.* 13:A569.

Buckley, W., Vanderpool, R., Godfrey, D., and Johnson, P.E., 1996, Determination, stable isotope enrichment and kinetics of direct-reacting copper in blood plasma, *J. Nutr. Biochem.* 7:488-494.

Cantone, M.C., deBartolo, D., Giussani, A., Ottolenghi, A., Pirola, L., Hansen, Ch., Roth, P., and Werner, E., 1997, A methodology for biokinetic studies using stable isotopes: results of repeated molybdenum investigation on a healthy volunteer, *Appl. Radiat. Isotopes* 48:333-338.

Coudray, C., and Fairweather-Tait, S.J., 1998, Do oligosaccharides affect the intestinal absorption of calcium in humans? *Am. J. Clin. Nutr.* 68:921-922.

Crews, H.M., Ducros, V., Eagles, J., Mellon, F.A., Kastenmayer, P., Luten, J.B., and Macgaw, B.A., 1994, Mass-spectrometric methods for studying nutrient mineral and trace-elements absorption and metabolism in humans using stable isotopes-a review, *Analyst* 119:2491-2514.

Crews, H.M., 1998, Speciation of trace elements in foods, with special reference to cadmium and selenium: is it necessary? *Spectrochim. Acta Part B* 53:213-219.

Denoyer, E., Tanner, S.D., and Voellkopf, V., 1999, A new dynamic reaction cell for reducing ICP-MS interferences using chemical resolution, *Spectroscopy* 14:43-45.

Dombrovári, J., Becker, J.S., and Dietze, H.-J., 2000, Isotope ratio measurements of magnesium and determination of magnesium concentration by reverse isotope dilution technique on small amounts of ^{26}Mg-spiked nutrient solutions with inductively coupled plasma mass spectrometry, *Int. J. Mass Spec.* 202:231-240.

Eagles, J., Fairweather-Tait, S., Mellon, F., Teucher, B., Meffan-Main, S., and Thirlwall, M., 2000, Accuracy and precision of multi-collector ICP-MS calcium isotope ratio measurement in human metabolic studies, *10th Biennial National Atomic Spectroscopy Symposium*, Sheffield.

Encinar, J.R., Garcia Alonso, J.I., and Sanz-Medel, A., 2000, Synthesis and application of isotopically labeled dibutyltin for isotope dilution analysis using gas chromatography-ICP-MS, *J. Anal. Atomic Spectrom.* 15:1233-1239.

Fairweather-Tait, S.J., Minihane, A.M., Eagles, J., Owen, L., and Crews, H.M., 1997, Rare earth elements as nonabsorbable fecal markers in studies of iron absorption, *Am. J. Clin. Nutr.* 65:970-976.

Giussani, A., Cantone, M.C., deBartolo, D., Roth, P., and Werner, E., 2000, Internal dose for ingestion of molybdenum radionuclides based on a revised biokinetic model, *Health Phys.* 78:46-52.

Griffin, I., Shames, D., King, J.C., and Abrams, S., 1999a, A compartmental model of zinc kinetics in children, *J. Invest. Med.* 47:134A.

Griffin, I., Shames, D.M., King, J.C., and Abrams, S., 1999b, A multi-compartmental zinc kinetic model in children, *FASEB J.* 13:A214.

Halliday, A.N., Lee, D., Christensen, J., Walder, A., Freedman, P., Jones, C., Hall, C., Yi, W., and Teagle, D., 1995, Recent developments in inductively coupled plasma magnetic sector multiple collector mass spectrometry, *Int. J. Mass Spec. Ion Proc.* 146/147:21-33.

Hambidge, K., Krebs, N.F., and Miller, L.V., 1998, Development of compartmental model to fit zinc stable isotope tracer data, *FASEB J.* 12:A345.

Heumann, K.G., Gallus, S.M., Rädlinger, G., and Vogl, J., 1998, Precision and accuracy in isotope ratio measurements by plasma source mass spectrometry, *J. Anal. Atomic Spectrom.* 13:1001-1008.

Hill, S.J., ed., 1999, *Inductively Coupled Plasma Spectrometry and Its Application*, Sheffield Academic Press, Sheffield.

Koch, G.M., Kerr, B.M., and Gooding, A., 1996a, Pharmacokinetics of bismuth and ranitidine following multiple doses of ranitidine bismuth citrate, *Brit. J. Clin. Pharamcol.* 42:207-211.

Koch, K.M., Kerr, B.M., and Gooding, A., 1996b, Pharmacokinetics of bismuth and ranitidine following single doses of ranitidine bismuth citrate, *Brit. J. Clin. Pharmacol.* 42:201-205.

Kopp-Hoolihan, L., Van Loan, M., Wong, W.W., and King, J.C., 1999, Fat mass deposition during pregnancy using a four-component model, *J. Appl. Physiol.* 87:196-202.

Krebs, N.F., Westcott, J., and Miller, L.V., 1999, Localization of secretion and reabsorption of endogenous zinc by compartmental modeling of intestinal data, *FASEB J.* 13:A214.

Lowe, N.M., Shames, D., Woodhouse, L., Matel, J., Roehl, R., Saccomani, M., Toffolo, G., Cobelli, C., and King, J.C., 1997, A compartmental model of zinc in healthy women using oral and intravenous stable isotope tracers, *Am. J. Clin. Nutr.* 65:1810-1819.

Lowe, N.M., Woodhouse, L.R., Matel, J.S., and King, J.C., 2000, Comparison of estimates of zinc absorption in humans by using 4 stable isotopic tracer methods and compartmental analysis, *Am. J. Clin. Nutr.* 71:523-529.

Lu, Q., Bird, S.M., and Barnes, R.M., 1995, Interface for capillary electrophoresis and inductively coupled plasma mass spectrometry, *Anal. Chem.* 67:2949-2956.

Lu, Q., and Barnes, R.M., 1996, Evaluation of an ultrasonic nebulizer interface for capillary electrophoresis and inductively coupled plasma mass spectrometry, *Microchem. J.* 54:129-143.

Lund, E.K., Wharf, S.G., Fairweather-Tait, S.J., and Johnson, I.T., 1998, Increases in the concentrations of available iron in response to dietary iron supplementation are associated with changes in crypt cell proliferation in rat large intestine, *J. Nutr.* 128:175-179.

Marechal, C.N., Telouk, P., and Albarede, F., 1999, Precise analysis of copper and zinc isotopic compositions by plasma-source mass spectrometry, *Chem. Geol.* 156:251-273.

Mellon, F.A., and Sandström, B., eds., 1996, *Stable Isotopes in Human Nutrition: Inorganic Nutrient Metabolism*, Academic Press, London.

Menegário, A.A., Giné, M.F., Bendassolli, J.A., Bellato, A.C.S., and Trivelin, P.C.O., 1998, Sulfur isotope ratio (^{34}S:^{32}S) measurements in plant materials by inductively coupled plasma mass spectrometry, *J. Anal. Atomic Spectrom.* 13:1065-1067.

Miller, L.V., Krebs, N.F., Jefferson, M., Easley, D., and Hambidge, K., 1996, Modeling of human zinc metabolism: model development using multiple stable isotopes, *FASEB J.* 10:1117.

Miller, L.V., Krebs, N.F., and Hambidge, K., 1998, Human zinc metabolism: advances in the modeling of stable isotope data, in: *Mathematical Modeling in Experimental Nutrition*, A.J. Clifford and H.-J. Müller, eds., Plenum Press, New York.

Minihane, A.M., and Fairweather-Tait, S.J., 1998, Effect of calcium supplementation on daily nonheme-iron absorption and long-term iron status, *Am. J. Clin. Nutr.* 68:96-102.

Montaser, A., ed., 1998, *Inductively Coupled Plasma Mass Spectrometry*, Wiley-VCH, New York.

Patriarca, M., Lyon, T.D., and Fell, G.S., 1997, Nickel metabolism in humans investigated with oral stable isotope, *Am. J. Clin. Nutr.* 66:616-621.

Platzner, I.T., 1997, *Modern Isotope Ratio Mass Spectrometry*, Wiley, New York.

Serfass, R., Fang, Y.Y., and Wastney, M.E., 1996, Zinc kinetics in weaned piglets fed marginal zinc intake: compartmental analysis of stable isotope data, *J. Trace Elements Exp. Med.* 9:73-86.

Smith, D.R., Woolard, D., and Luck, M.L., 2000a, Succimer and the reduction of tissue lead in juvenile monkeys, *Toxicol. Appl. Pharmacol.* 166:230-240.

Smith, D.R., Calacsan, C., Luck, M., Cremin, J., and Laughlin, N.K., 2000b, Succimer and the urinary excretion of essential elements in a primate model of childhood lead exposure, *Toxicol. Sci.* 54:473-480.

Stürup, S., 2000, Application of HR-ICP-MS for the simultaneous measurement of zinc isotope ratios and total zinc content in human samples, *J. Anal. Atomic Spectrom.* 15:315-321.

Tanner, S.D., and Baranov, V.I., 1999, Theory, design and operation of a dynamic reaction cell for ICP-MS, *At. Spectrosc.* 20:45-52.

Turner, P.J., Mills, D.J., Schroder, E., Lapitaja, G., Jung, G., Iacone, L.A., Haydar, D.A., and Montaser, A., 1998, Instrumentation for low- and high-resolution ICPMS, in: *Inductively Coupled Plasma Mass Spectrometry*, A. Montaser, ed., Wiley-VCH, New York.

Vanhaecke, F., Moens, L., Dames, R., Papadakis, I., and Taylor, P., 1997, Applicability of high-resolution ICP-mass spectrometry for isotope ratio measurements, *Anal. Chem.* 69:268-273.

Vanhaecke, F., Moens, L., and Taylor, P., 1999a, Use of ICP-MS for isotope ratio measurements, in: *Inductively Coupled Plasma Spectrometry and Its Application*, S.J. Hill, ed., Sheffield Academic Press, Sheffield.

Vanhaecke, F., Moens, L., Dames, R., Allen, L., and Georgitis, S., 1999b, Evaluation of the isotope ratio performance of an axial time-of-flight ICP mass spectrometer, *Anal. Chem.* 71:3297-3303.

Wastney, M.E., Aamodt, R.L., Rumble, W., and Henkin, R., 1986, Kinetic analysis of zinc metabolism and its regulation in normal humans, *Am. J. Physiol.* 251:R398-R408.

Wastney, M.E., Angelus, P., Barnes, R.M., and Subramanian, K.N., Zinc kinetics in preterm infants: a compartmental model based on stable isotope data, 1996, *Am. J. Physiol.* 40:R1452-R1459.

Wastney, M., Patterson, B., Linares, O., Greif, P., and Boston, R., 1999, *Investigating Biological Systems Using Modeling: Strategies and Software*, Academic Press, New York.

Wastney, M.E., House, W.A., Barnes, R.M., and Subramanian, K.N., 2000, Kinetics of zinc metabolism: variation with diet, genetics and disease, *J. Nutr.* 130:1355S-1359S.

PART V

MODELING OF VITAMINS, MINERALS, AND CHOLESTEROL

THE USE OF MODEL-BASED COMPARTMENTAL ANALYSIS TO STUDY VITAMIN A METABOLISM IN A NON-STEADY STATE

Michael H. Green and Joanne Balmer Green[*]

INTRODUCTION

Over the past 20 years, we have collaborated with several laboratories in using mathematical modeling to describe and quantitate whole-body vitamin A metabolism in the rat. Steady state models have been developed for animals at different levels of vitamin A status (Green et al., 1985; Lewis et al., 1990; Green and Green, 1994a; Kelley and Green, 1998) and in response to other variables (Kelley et al., 1998; Jang et al., 2000; Kelley et al., 2000). Limited experimental and mathematical evidence (Green and Green, 1994b; Novotny et al., 1995; v Reinersdorff et al., 1998) suggests that there are many similarities in vitamin A metabolism between rats and humans. As discussed at the 5th Conference on Mathematical Modeling in Experimental Nutrition in 1994 (Green and Green, 1996), modeling studies have uniquely contributed to current understanding of whole-body vitamin A metabolism. In particular, interpretation of kinetic data has revealed previously unrecognized complexities in vitamin A dynamics that facilitate homeostatic control of plasma (and probably tissue) vitamin A levels.

In most of our vitamin A kinetic experiments, we have used slowly growing adult male Sprague-Dawley rats and manipulated dietary vitamin A intake so that animals would be near a steady state with respect to vitamin A balance during the turnover study. The assumption of system steady state underlies many applications of the Simulation, Analysis and Modeling computer programs (SAAM; Berman and Weiss, 1978), allowing for the computation of a number of useful parameters (e.g., input and disposal rates, transfer rates, and compartment sizes). Also, the assumption of time-invariant fractional transfer coefficients to define a model allows computation of other parameters of interest (e.g., transit times, residence times, recycling number, and recycling time).

---

[*] Michael H. Green and Joanne Balmer Green, Nutrition Department, The Pennsylvania State University, University Park, PA 16802.

However, under many experimental conditions, vitamin A is not in a steady state, and some of the model-dependent fractional transfer coefficients may be time variant. We recently studied two such cases. In one, we investigated the influence of iron deficiency on vitamin A metabolism (Jang et al., 2000); in the other, we studied the effects of the environmental contaminant 2,3,7,8-tetrachlorodibenzo-*p*-dioxin (TCDD) on vitamin A biodynamics (Kelley et al., 1998). Iron deficiency can be associated with a lowered plasma retinol concentration and increases in liver vitamin A levels, while TCDD exposure produces depletion in liver vitamin A stores and sometimes causes an increase in plasma retinol concentration. By making certain assumptions and corrections, we were able to use a steady state approach to model both of these data sets and provide useful information and predictions. However, we were left with the nagging knowledge that a more complex solution would be required to describe the time course of changes in liver vitamin A and to accurately calculate input and disposal rates.

In this manuscript, we describe the use of a non-steady state approach to modeling data for representative rats from these two experiments using WinSAAM (Wastney et al., 1999), the Windows version of the SAAM software. This analysis allows us to predict likely sites of alteration in vitamin A metabolism due to iron deficiency or TCDD exposure. Such predictions may be useful in designing experiments to look for responsible enzymes or transport proteins.

OVERVIEW OF VITAMIN A METABOLISM

As background, we first summarize current understanding of whole-body vitamin A metabolism. See reviews by Blomhoff et al. (1991) and Blaner and Olson (1994). As indicated in Figure 1, ingested vitamin A is absorbed as retinol and packaged as retinyl esters into triglyceride-rich absorptive lipoproteins (chylomicrons) in the enterocytes of the small intestine. Chylomicrons are secreted into lymph and then transported into plasma where most of the triglycerides are cleared into peripheral tissues as a result of the action of the enzyme lipoprotein lipase. Chylomicron remnants containing most of the dietary vitamin A are then primarily taken up into hepatocytes by receptor-mediated endocytosis. There, resulting retinol is either secreted into plasma bound to retinol-binding protein (RBP) for transport to tissues, stored temporarily in hepatocytes as retinyl esters, or transferred to liver perisinusoidal stellate cells for re-esterification and storage. Thus, in the liver, there are two main cellular sites of vitamin A (hepatocytes and stellate cells) and two primary chemical forms (retinol and retinyl esters). Conventional wisdom is that more than 80% of whole-body vitamin A is found in the liver. Modeling has shown that RBP-bound retinol recycles many times among plasma and tissues before uptake into vitamin A-requiring tissues for irreversible utilization or into the liver for oxidation and excretion. In many vitamin A-requiring tissues, retinol may be oxidized to retinoic acid, which interacts with nuclear receptors to regulate gene transcription. In addition, several lines of evidence point to vitamin A oxidation and excretion that may not be tied to vitamin A function (i.e., nonfunctional utilization).

OVERVIEW OF *IN VIVO* KINETIC STUDIES AND STEADY STATE MODEL

In experiments designed to model whole-body vitamin A metabolism in rats [see Green and Green (1990a) for details], we prepare a dose of tritiated vitamin A in one of its physiological transporters (either RBP or chylomicrons). Doses are prepared *in vivo* using donor rats. Non-perturbing aliquots of the labeled doses (plasma or lymph) are administered to recipient rats, and serial blood samples are collected, typically at geometrically increasing intervals beginning 10 minutes after dose administration. Plasma samples are extracted and analyzed for radioactivity by liquid scintillation spectrometry; selected samples are also analyzed for retinol concentration. At the end of the kinetic study (35-115 days, depending on liver vitamin A levels), rats are killed. Liver vitamin A tracer and tracee levels, and carcass radioactivity, are determined. In some studies, tracer is measured in urine and feces, and/or tracer and tracee are measured in organs other than liver.

To begin modeling with the SAAM program, we calculate fraction of the injected dose remaining in plasma *vs.* time, fraction of the dose in liver and carcass, and fraction of the dose that was irreversibly lost [1- (fraction of dose left in plasma + liver + carcass at the end of the study)]. Data are plotted semilogarithmically; for an example, look ahead to Figure 3. In developing a steady state model for whole-body vitamin A metabolism as viewed from the plasma, processes with similar kinetic behavior are lumped into the same compartment; see Green and Green (1990a, b) and Adams et al. (1995) for more details. This is the approach that will be used here. We fit plasma tracer data to a 3 or 4-component exponential equation and use the sum of the exponential constants (i.e., intercepts) to normalize individual plasma data points to an initial fraction of dose in plasma at time zero of 1 and thereby correct the initial estimate of plasma volume. An F-statistic (Landaw and DiStefano, 1984) is used to determine whether 3 or 4 components are required to fit the data. The number of exponential components then determines the number of compartments in the compartmental model.

Figure 2 shows the four-compartment model which has provided a good fit to several data sets on vitamin A kinetics. In this model, compartment 1 represents RBP-bound retinol in plasma, compartment 2 represents rapidly turning-over pools of extravascular vitamin A,

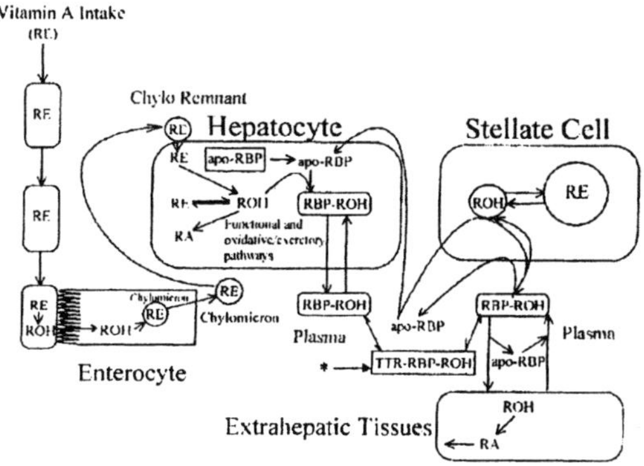

Figure 1. Current understanding of whole-body vitamin A metabolism. RE, retinyl esters; ROH, retinol; chylo, chylomicron; RBP, retinol-binding protein; TTR, transthyretin; RA, retinoic acid.

and compartments 3 and 4 are more slowly turning-over extravascular pools. Based on kinetic behavior and current understanding of whole-body vitamin A metabolism, we have postulated that vitamin A in compartment 2 corresponds to retinol in interstitial fluid, retinol filtered by the kidneys, and more rapidly turning-over intracellular retinol pools. Further, we have assumed that compartments 3 and 4 correspond to vitamin A stored mainly in liver but also in extrahepatic tissues, with the most slowly turning-over pools being retinyl esters in liver perisinusoidal stellate cells. The fractional transfer coefficients [L(I,J)s], or the fraction of tracer or tracee in compartment J transferred to compartment I per day, are adjusted systematically using WinSAAM until a visually-good fit is obtained between the observed data and the model simulation. Finally, weighted nonlinear regression analysis in SAAM is used to obtain a final fit of the observed data to model-predicted values, to determine model parameters (fractional transfer coefficients), to generate statistical uncertainties for the model parameters, and to provide evidence (or lack of evidence) for multicolinearity and autocorrelation. Then, the estimated plasma retinol pool size and the site of tracee entry into the system are added to the problem deck which contains the model-predicted L(I,J)s, and this deck is used to calculate other kinetic parameters in a steady state solution to the model. In a steady state, all of the fractional transfer coefficients are time invariant, and the vitamin A input rate [U(1), Figure 2] equals the vitamin A disposal (utilization) rate [R(0,3), Figure 2].

VITAMIN A DYNAMICS IN IRON-DEFICIENT RATS: NON-STEADY STATE SOLUTION

It has been previously observed that iron deficiency is associated with decreases in plasma retinol concentration in rats and humans; for references, see Jang et al. (2000). To investigate whether the decreased plasma retinol concentration is due to a decreased movement of vitamin A between liver and plasma or to an increase in vitamin A disposal rate, we did a vitamin A kinetic study in control *vs.* iron-deficient rats (Jang et al., 2000). To eliminate body size as a variable, food intake by control rats was limited to match their body weights with the iron-deficient animals. By limiting iron intake (4 µg/g diet *vs.* 45 for controls), hematocrit, blood hemoglobin, and liver iron were decreased by 61, 80, and 75% in

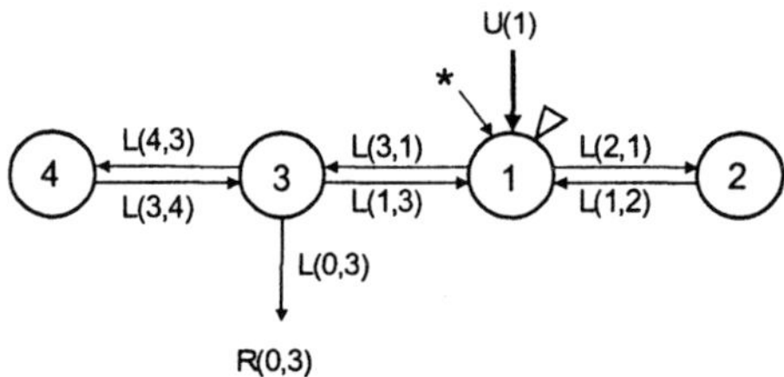

Figure 2. Four-compartment model describing whole-body vitamin A metabolism in rats. Compartment 1 represents plasma RBP-bound retinol, compartment 2 corresponds to rapidly turning-over extravascular vitamin A pools, and compartments 3 and 4 are more slowly turning-over extravascular pools, primarily in the liver. The asterisk represents the site of tracer input; the open triangle indicates the site of sampling. Fractional transfer coefficients [L(I,J)s] are the fraction of tracer or tracee is compartment J transferred to compartment I per day; U(I) is the system input rate (nmol/day), and R(0,3) is the vitamin A disposal rate (nmol/day).

the iron-deficient group. For the kinetic study, rats were injected with [³H]retinol-labeled plasma, and serial blood samples were collected until the rats were killed 48 days later; for general experimental details, see Green and Green (1990a). Data were treated as summarized above. In spite of a vitamin A intake (24 nmol/day in controls and 29 nmol/day in iron-deficient rats) that was designed to maintain balance in control rats, liver vitamin A levels in iron-deficient rats were ~7 times higher and plasma retinol concentrations were ~40% of control values by the end of the turnover study. On average, the iron-deficient rats were in a positive liver vitamin A balance of 6.8 nmol/day during the kinetic study.

Plasma tracer response curves for a representative control and iron-deficient rat (ID #4) are shown in Figure 3. The initial disappearance of tracer from plasma was the same in both groups. The dramatically lower fraction of injected dose remaining in plasma of the iron-deficient rat after 1 day indicates either that iron deficiency decreased recycling of retinol from the slow turning-over pools to plasma or that the pool it recycled from was larger in the iron-deficient rat.

As described by Jang et al. (2000), we used WinSAAM to develop a four-compartment model to fit the plasma and loss data for the iron-deficient rat. Once the model-predicted L(I,J)s were identified with acceptable precision and there was a good fit between the model-predicted and observed data (Figure 3), the estimated plasma pool size [M(1) = 5.936 nmol] and the site of new vitamin A input into the system [U(1)] were added to the SAAM deck, and a steady state solution was calculated (Figure 4). The model predicted that the slower turning-over pools of vitamin A [M(3) + M(4)] contained 607 nmol of retinol (89% of the measured amount of vitamin A in the liver of ID rat #4 at the end of the study). The model-predicted input rate for vitamin A [U(1) = 13.34 nmol/day] implied an absorption efficiency (45%) that was low compared with previous estimates for control rats (75%). Based on our intuition about the system, previous work on the relationship between plasma retinol level and disposal rate (Kelley and Green, 1998), and the positive vitamin A balance in this rat, we assumed that the model-predicted disposal rate (13.34 nmol/day) was higher than the actual value. Thus, we decided to develop a non-steady state model that would better predict the actual input and disposal rates and also would account for both the observed liver vitamin A level at the end of the experiment and the estimated positive vitamin A balance (~6 nmol/day) during the kinetic study.

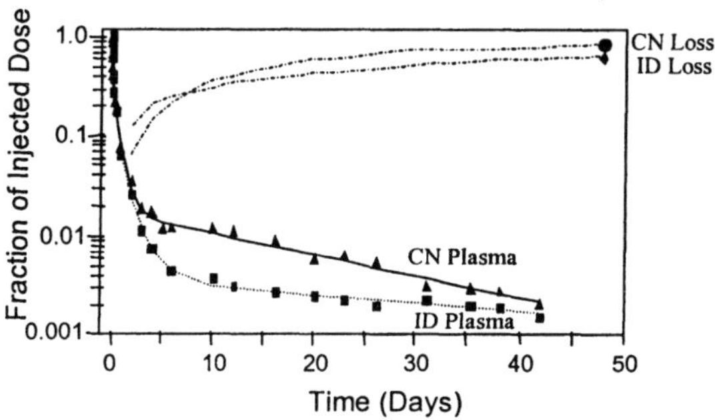

Figure 3. Plasma disappearance curves and tracer loss for a representative control and iron-deficient rat (ID # 4) from the study by Jang et al. (2000). Symbols are observed data and lines are model-predicted values for the four-compartment steady state model.

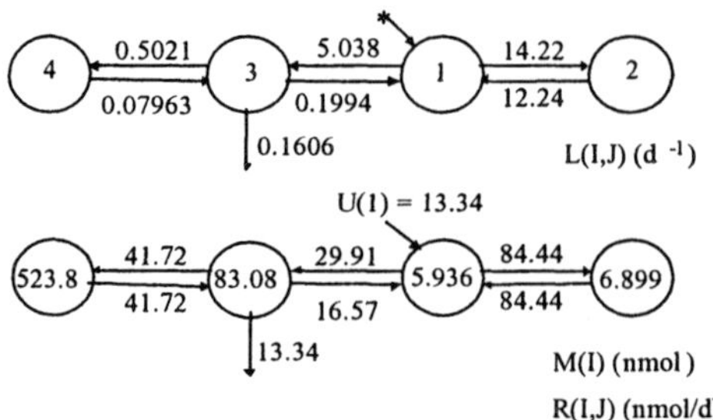

Figure 4. Steady state solution to the four-compartment model for iron-deficient rat #4. The top panel shows the model-predicted fractional transfer coefficients [L(I,J)s]; the asterisk is the site of tracer input. The lower panel shows the model-predicted transfer rates [R(I,J)s] and compartment masses [M(I)]s; U(1) represents input of vitamin A.

To develop a model that would accurately describe the vitamin A non-steady state in this iron-deficient rat, we started with the steady state solution and developed a parallel model for tracee in the system. The model parameters [L(I,J)s] were set equal in the tracer and tracee models. We then looked for minimal changes in the steady state model that would accurately describe the change in liver vitamin A while still fitting the tracer data for plasma and loss. We hypothesized that the part of the system that was most likely not in a steady state was the large, slowly turning-over retinyl ester storage pool (compartment 4) located primarily in the liver. Thus, in the four-compartment model shown in Figure 2, the fractional transfer coefficient most likely to be affected would be L(3,4), representing the mobilization of vitamin A from the slowest turning-over extravascular pool (presumably retinyl esters in liver stellate cells). We also hypothesized that the rest of the system remained in a steady state. This assumption is based on the fact that observed values for plasma retinol concentration were statistically time invariant during the kinetic study.

The time variance for L(3,4) (Figure 5) was developed by setting L(3,4) equal to R(3,4), where R(3,4) = L(3,4) * M(4) and then making L(3,4) function-dependent by dividing it by M(4). R(4,3) and R(3,4) were set to create a positive balance in compartment 4. For example, in the final model (see below), L(3,4) was initially 30.59; when divided by the time-zero value for M(4) (336.8 nmol), L(3,4) / M(4) = 0.09085. As M(4) increased during the

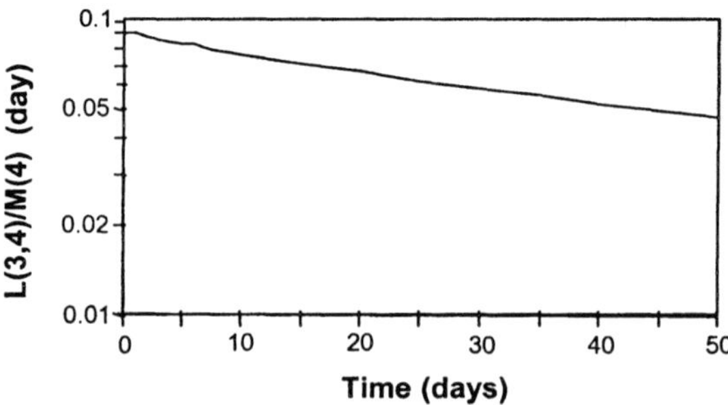

Figure 5. Model-predicted time variance in L(3,4) / M(4) in the non-steady state model for iron-deficient rat #4.

kinetic study, L(3,4) / M(4) decreased and R(3,4) remained constant. By 48 days, M(4) = 634 nmol and L(3,4) / M(4) = 0.04822 (Figure 5). The initial models were unstable [i.e., not only M(4) varied with time, but other M(I)s and R(I,J)s were time variant until they came into a new steady state]. By allowing the compartment masses and input rate [U(1)] to vary within a likely range and then adjusting the L(I,J)s, we were able to develop a physiologically meaningful non-steady state model that fit the tracer data for plasma and loss for iron-deficient rat #4 (Figure 6) and predicted observed changes in liver tracee (Figure 7). Note in Figure 7 that M(1), M(2), and M(3) remain in a steady state during the turnover study, while M(4) and liver vitamin A increase with time.

The non-steady state model is presented in Figure 8. When the steady state and non-steady state models are compared, there is little difference in exchange between compartments 1 and 2 or in the transfer to compartment 3 from 1. The size of compartment 3 was reduced by 40%, and compartment 4 expanded from 337 nmol at the beginning of the turnover study to 634 nmol on day 48. The size of compartments 3 + 4 was 687 nmol, very close to the observed value of 685 nmol for liver vitamin A. The input rate [U(1)] increased from 13.3 in the steady state model to 16.8 nmol/day in the non-steady state model. The latter value equates to an efficiency of vitamin A absorption of 57%. The model thus predicts that vitamin A absorption efficiency is decreased in iron deficiency. The disposal rate [R(0,3)] decreased from 13.3 in the steady state model to 10.6 nmol/day in the non-steady state model, resulting in a positive vitamin A balance of 6.2 nmol/day. For the non-steady state model, the value for [U(1) + R(0,3)] / 2 is 13.7 nmol/day, a figure close to that predicted by the steady state model for those two parameters (13.34 nmol/day). That is, it appears that the steady state model predicted the average of the input and output rates, while the non-steady state model better predicts the dynamics of the system in a non-steady state. The fact that it was necessary and sufficient to have L(3,4) / M(4) decreasing with time to fit the tracee and tracer data implies that something involved in the mobilization of liver retinyl ester stores (Figure 1) is inhibited by iron deficiency. Thus, our modeling tells us that a good place to look biochemically would be the retinyl ester hydrolases in perisinusoidal stellate cells in the liver.

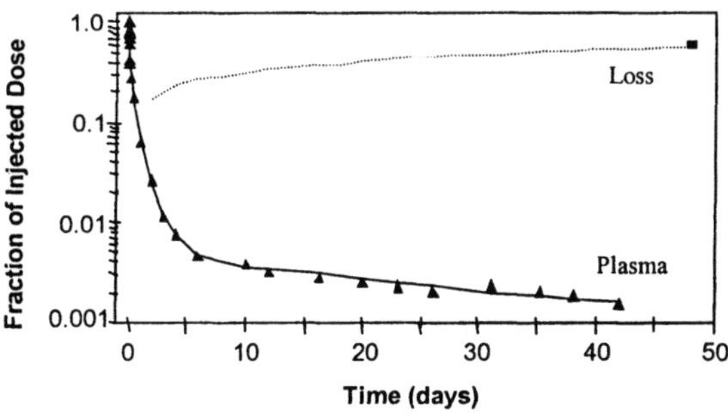

Figure 6. Fit of the non-steady state model to the working hypothesis/four-compartment model for iron-deficient rat #4. Symbols are observed data and lines are model-predicted values.

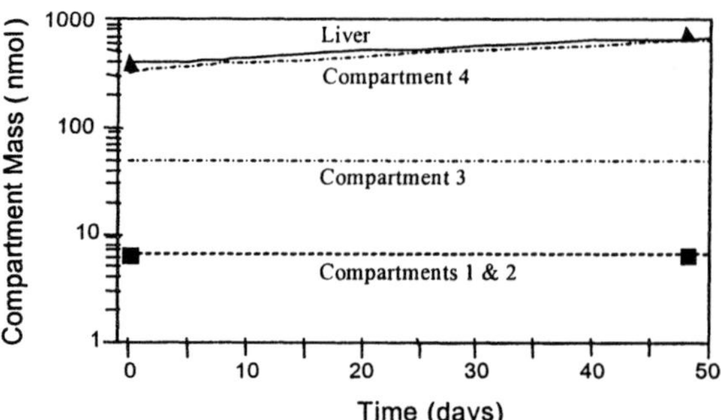

Figure 7. Predictions of the non-steady state model for compartment masses *vs.* time in iron-deficient rat #4.

VITAMIN A DYNAMICS IN TCDD-TREATED RATS: NON-STEADY STATE SOLUTION

Exposure to the persistent environmental contaminant 2,3,7,8-tetrachlorodibenzo-*p*-dioxin (TCDD) has dramatic effects on vitamin A homeostasis; for references, see Kelley et al. (1998). Rat liver vitamin A levels fall precipitously after TCDD exposure, and kidney vitamin A concentrations increase; plasma retinol levels remain at control values or are increased. To pinpoint the site(s) of action of TCDD on vitamin A dynamics, we did vitamin A kinetic studies in control rats *vs.* those treated with repeated oral doses of TCDD (Kelley et al., 1998). At the end of a 42-day turnover study, plasma retinol concentrations were 32% higher in TCDD-treated rats, and liver vitamin A levels were decreased by 93% (from 964 ± 216 nmol in 3 rats killed at the beginning of the kinetic study to 95 nmol 42 days later in the TCDD-treated rat to be discussed here).

Plasma tracer response curves for a representative control and TCDD-treated rat (TCDD #2) are shown in Figure 9. The plasma data between 2 and 10 days indicate that the control rat had more vitamin A in the liver to exchange with the tracer since the fraction of the dose in plasma was lower. The steeper terminal slope in the TCDD-treated rat indicates a higher

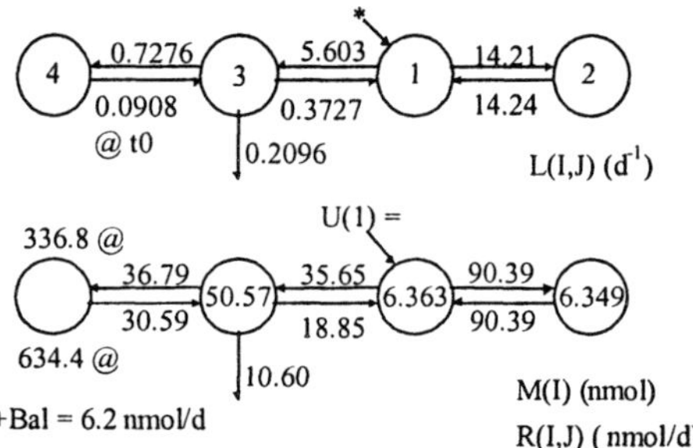

Figure 8. Non-steady state solution to the four-compartment model for iron-deficient rat #4. The top panel shows the model-predicted fractional transfer coefficients [L(I,J)s]; the asterisk is the site of tracer input. The lower panel shows the model-predicted transfer rates [R(I,J)s] and compartment masses [M(I)]s at time 0 and 48 days; U(1) represents input of vitamin A.

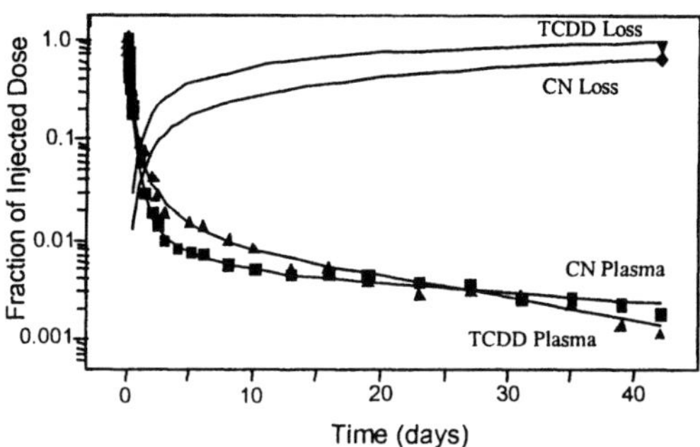

Figure 9. Plasma disappearance curves and tracer loss for a representative control (CN) and TCDD-treated rat (TCDD) from the study by Kelley et al. (1998). Symbols are observed data and lines are model-predicated values for the four-compartment steady state model.

fractional rate of loss of labeled vitamin A from the system and is consistent with the observed loss of liver vitamin A. Figure 10 shows the good fit of the TCDD-treated rat's tracer data for plasma and loss to the four-compartment model shown in Figure 2. When the plasma retinol pool size (33 nmol) and the site of vitamin A input [U(1)] were added to the SAAM deck, the steady state solution (Figure 11) predicted 725 nmol in the slowest turning-over vitamin A pool (compartment 4) and 442 nmol in compartment 3. Even if we accept the conventional wisdom that 80% of whole-body vitamin A is in the liver, these steady state predictions significantly overestimate the measured liver vitamin A level in TCDD-treated rat #2 at the end of the study (95 nmol). The model-predicted input and disposal rates were 67 nmol/day, compared to an average vitamin A intake of 74 nmol/day by TCDD-treated rats. This suggests an absorption efficiency of 92%, much higher than that measured previously in control rats (75%).

In view of the discrepancy between model-predicted and observed liver vitamin A levels and the unlikely prediction for vitamin A absorption, we decided to develop a non-steady state model for these data using the same strategy described earlier for the iron-deficient rat. Further justification for choosing L(3,4) (Figure 2) as the parameter most likely to be time variant came from another vitamin A/TCDD kinetic study by Kelley et al. (2000). There, we

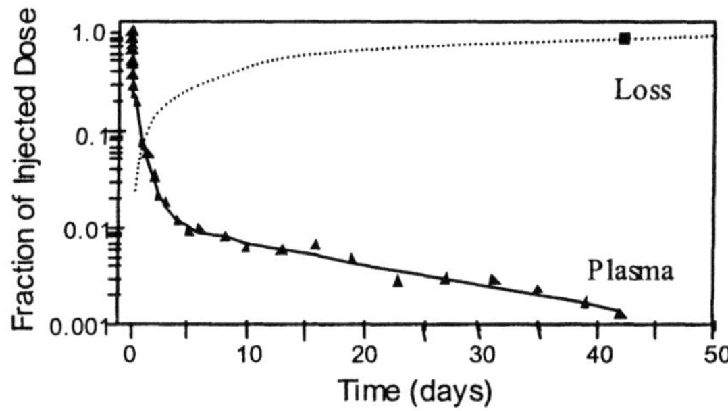

Figure 10. Plasma tracer disappearance curve and tracer loss for a TCDD-treated rat (TCDD #2) from the study by Kelley et al. (1998). Symbols are observed data and lines are model-predicted values for the four-compartment steady state model.

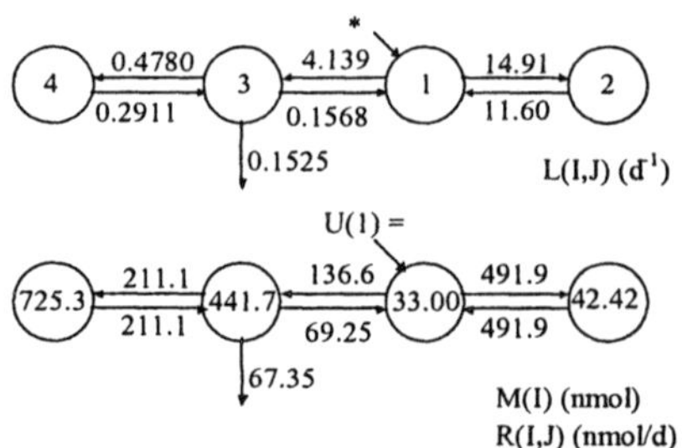

Figure 11. Steady state solution to the four-compartment model for TCDD-treated rat #2. The top panel shows the model-predicted fractional transfer coefficients [L(I,J)s]; the asterisk is the site of tracer input. The lower panel shows the model-predicted transfer rates [R(I,J)s] and compartment masses [M(I)]s; U(1) represents input of vitamin A.

injected control rats with [^{3}H]retinol-labeled plasma and "perturbed" the vitamin A system with TCDD 22 days later, when changes in plasma tracer concentration were in a terminal slope. TCDD administration caused an increase in plasma tracer concentrations. The model developed to fit those data predicted that L(3,4) had to increase a few days after TCDD administration to result in the increase in plasma tracer concentration. We concluded that an increase in vitamin A mobilization precedes the increased disposal of vitamin A.

To develop a non-steady state model for the TCDD-treated rat, we began by setting up a parallel model for tracee using the same L(I,J)s as in the tracer model. L(3,4) was set equal to the initial estimate for R(3,4) and then made function dependent by dividing it by M(4). In the case of the TCDD-treated rat, M(4) decreased with time, L(3,4) / M(4) increased, and R(3,4) remained constant. Our initial assumptions were that compartments 3 and 4 represented mainly liver vitamin A and that TCDD caused vitamin A depletion from both pools. We therefore set the sum of these compartments to be ~1000 nmol at the beginning of the kinetic study and tried to develop a model in which this pool would decrease to ~95 nmol 42 days later. Such a model was not compatible with the plasma tracer data since there was not enough tracer in compartments 3 + 4 to support the observed tail of the plasma tracer disappearance curve. That is, the model underestimated the plasma tracer data after ~ 25 days. This finding prompted us to rethink the model structure and hypothesize that compartment 3, rather than representing mainly liver vitamin A, might correspond to extrahepatic vitamin A pools. We hypothesized that this pool supported the later plasma tracer concentration. In the new model, the effect of TCDD on liver vitamin A was to mobilize vitamin A from the slowest turning-over retinyl esters in liver perisinusoidal stellate cells. Thus, it was the size of compartment 4 which decreased from ~1000 to 95 nmol during the kinetic study. After numerous adjustments in L(I,J)s, M(I)s, U(1), and the R(I,J)s, we arrived at a non-steady state model that fit both the tracer data for plasma and loss (Figure 12) and the tracee data (liver vitamin A) (Figure 13). Note how the non-steady state model better fits the contour of the terminal portion of the plasma tracer data (Figure 12 *vs.* Figure 10). As with the non-steady state model for the iron-deficient rat, we hypothesized, based on the time invariant plasma retinol concentration, that compartments 1, 2, and 3 were in a steady state, while compartment 4 depleted over time at a rate of ~26 nmol/day. As evident in Figure 13, our working hypothesis/non-steady state model is compatible with that assumption. Figure 14 shows the time variant change in L(3,4) / M(4).

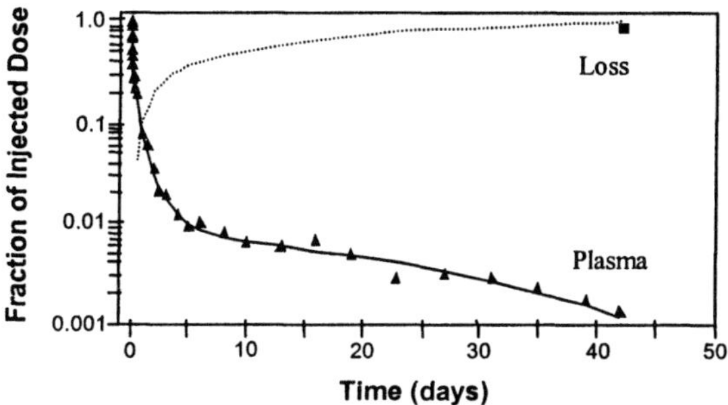

Figure 12. Fit of the non-steady state model to the working hypothesis/four-compartment model for TCDD-treated rat #2. Symbols are observed data and lines are model-predicted values.

The non-steady state model for the TCDD-treated rat is shown in Figure 15. The model predicts that the initial size of compartment 4 (predicted to be 1171 nmol) depleted to 87 nmol (92% decrease) by day 42, a depletion rate of 26 nmol/day ([day 42 liver vitamin A – day 0 liver vitamin A] / 42 days). The model also predicts an input rate of 55 nmol/day, which is more physiologically reasonable (absorption efficiency, 74%). Further, it predicts a disposal rate of 81 nmol/day; that is 68% higher than the control rats in this study (controls were in a positive vitamin A balance of 32 nmol/day). As in the case of the iron-deficient rat, the input and output rate predicted by the steady state model is the mean of the non-steady state rates.

Based on the non-steady state model, we hypothesize that TCDD caused an increase in mobilization of vitamin A stores, leading to an increase in plasma retinol concentration and an increase in vitamin A disposal rate. When the higher plasma retinol steady state was reached, the disposal rate remained elevated, and the depletion of liver vitamin A continued. The model overpredicts the observed irreversible loss by 10% (Figure 12), indicating that the disposal rate may have been decreasing toward the end of the study. Another interesting prediction of the model is that compartment 3 contains most of the system tracer and tracee by the end of the experiment. Since our current model predicts that compartment 3 is not

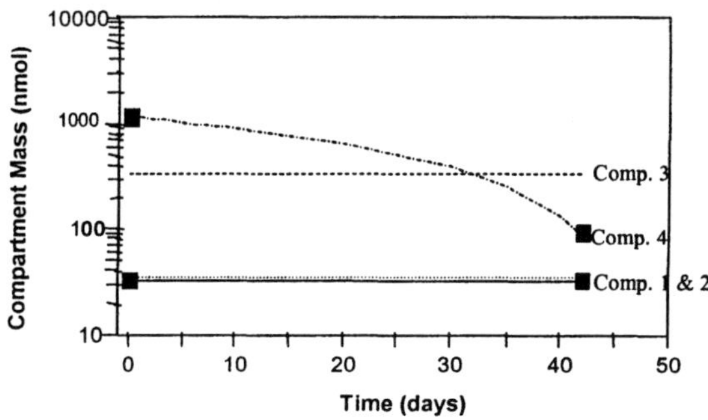

Figure 13. Predictions of the non-steady state model for compartment masses *vs.* time in TCDD- treated rat #2.

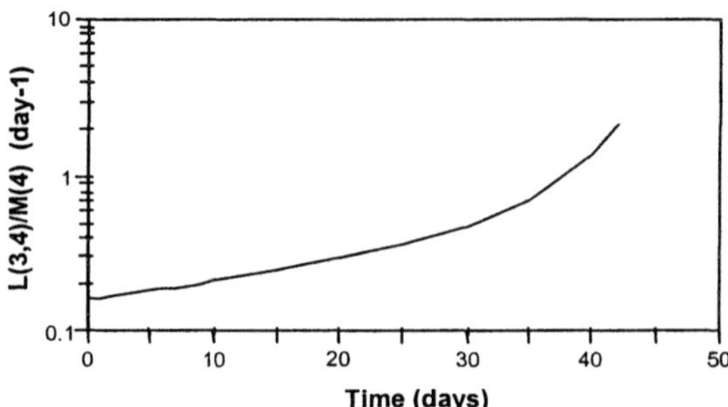

Figure 14. Model-predicted time variance in L(3,4) / M(4) for the non-steady state model for TCDD-treated rat #2.

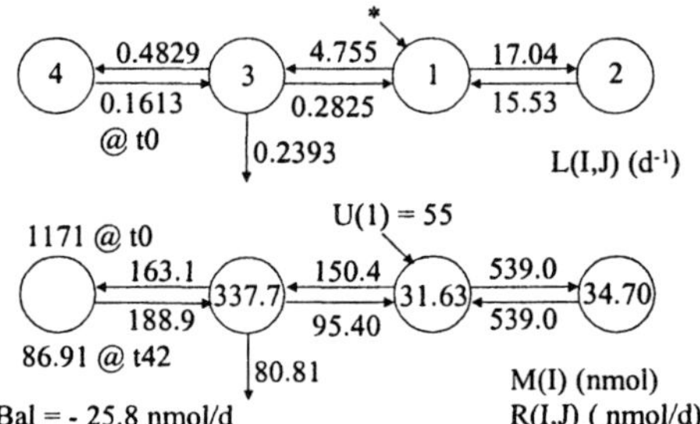

Figure 15. Non-steady state solution to the four-compartment model for TCDD-treated rat #2. The top panel shows the model-predicted fractional transfer coefficients [L(I,J)s]; the asterisk is the site of tracer input. The lower panel shows the model-predicted transfer rates [R(I,J)s] and compartment masses [M(I)]s at time 0 and 42 days; U(1) represents input of vitamin A.

primarily liver vitamin A, the model is compatible with the terminal tracer data that showed that only 1.6% of the injected dose was present in the liver *vs.* 15% in the carcass.   An interesting next step in model development would be to make the model mammillary, having compartment 4 exchange with compartment 1 rather than with compartment 3 since the latter appears to be extrahepatic. Predictions of the non-steady state model agree with our previous compartmental models describing the effects of TCDD administration on the biodynamics of vitamin A (Kelley et al., 1998; Kelley et al., 2000), in that all of the models predict an increase in net hydrolysis of liver vitamin A in stellate cells. This increase could be due to increases in known retinyl ester hydrolases or nonspecific carboxyl ester hydrolases or to other physiological processes involved in vitamin A mobilization from the liver. To date, results of studies on effects of TCDD on such hydrolases are inconclusive as noted by Kelley et al. (2000). Further work is needed to identify which hydrolase activity, if any, is responsible for the increased vitamin A mobilization following TCDD exposure.

CONCLUSIONS

To summarize, we note that non-steady state models such as we developed here are able to predict the homeostatic control over the biodynamics of the vitamin A system while liver vitamin A levels are increasing or decreasing. These models also reinforce the steady state models in hypothesizing that a likely site of perturbation in iron deficiency and TCDD exposure may be the enzyme(s) involved in vitamin A mobilization from liver retinyl ester stores (i.e., retinyl ester hydrolases or non-specific carboxyl ester hydrolases). In the case of iron-deficient rats, the net hydrolysis is reduced, while after TCDD treatment, net hydrolysis is increased. In agreement with previous models of vitamin A metabolism in rats (Kelley and Green, 1998), the non-steady state models predict a reduction in vitamin A utilization (disposal rate) compared with control rats when plasma retinol levels are lowered as in iron deficiency and an increase compared with control rats when plasma retinol is elevated as by TCDD treatment. Besides identifying which hydrolases may be affected by these two treatments, it would be worthwhile to determine whether there are effects on the metabolism of cellular retinol-binding protein, the protein involved in the intracellular metabolism of retinol.

We believe that continued use of model-based compartmental analysis will provide a powerful tool to quantify and pinpoint likely sites of perturbations in metabolism of vitamin A and other nutrients due to xenobiotics, diseases, or nutrient deficiencies. Although biological modelers have traditionally been challenged by systems that do not follow steady state kinetics, we have shown here that it is technically possible with WinSAAM, as well as intellectually fruitful and satisfying, to accommodate temporal changes in a compartment that has significant impact on system dynamics (such as the liver in the case of vitamin A).

ACKNOWLEDGMENTS

We acknowledge the contributions of collaborators Jing-Tsz Jang and John L. Beard (Penn State University; iron study) and Sean K. Kelley, Charlotte B. Nilsson and Helen Håkansson (Karolinska Institute, Stockholm, Sweden; TCDD study). The iron study was funded by a grant to J.-T. J. from the Inter-College Graduate Program in Nutrition (Penn State) and the TCDD study by a grant from the Swedish Environmental Protection Agency as well as Karolinska Institute Visiting Scientist Awards (to M.H.G and J.B.G) and a Fulbright Fellowship (to S.K.K.)

CORRESPONDENCE

Please address all correspondence to:
Michael H. Green
Pennsylvania State University
Nutrition Department, S-126 Henderson South
University Park, PA 16802
mhg@psu.edu

REFERENCES

Adams, W.R., Smith, J.E., and Green, M.H., 1995, Effects of N-(4-hydroxyphenyl)retinamide on vitamin A metabolism in rats, *Proc. Soc. Exp. Biol. Med.* 208:178-185.

Berman, M., and Weiss, M.F., 1978, *SAAM Manual,* DHEW Publ. #78-180, U.S. Government Printing Office, Washington, DC.

Blaner, W.S., and Olson, J.A., 1994, Retinol and retinoic acid metabolism, in: *The Retinoids: Biology, Chemistry, and Medicine, 2ⁿᵈ edn.,* M.B. Sporn, A.B. Roberts, and D.S. Goodman, eds., Raven Press, New York.

Blomhoff, R., Green, M.H., Green, J.B., Berg, T., and Norum, K.R., 1991, Vitamin A metabolism: new perspectives on absorption, transport, and storage, *Physiol. Rev.* 71:951-990.

Green, M.H., Uhl, L., and Green, J.B., 1985, A multicompartmental model of vitamin A kinetics in rats with marginal liver vitamin A stores, *J. Lipid Res.* 26:806-818.

Green, M.H., and Green, J.B., 1990a, Experimental and kinetic methods for studying vitamin A dynamics in vivo, *Meth. Enzymol.* 190:304-317.

Green, M.H., and Green, J.B., 1990b, The application of compartmental analysis to research in nutrition, *Ann. Rev. Nutr.* 10:41-61.

Green, M.H., and Green, J.B., 1994a, Vitamin A intake and status influence retinol balance, utilization and dynamics in rats, *J. Nutr.* 124:2477-2485.

Green, M.H., and Green, J.B., 1994b, Dynamics and control of plasma retinol, in: *Vitamin A in Health and Disease,* R. Blomhoff, ed., Marcel Dekker, New York.

Green, M.H., and Green, J.B., 1996, Quantitative and conceptual contributions of mathematical modeling to current views on vitamin A metabolism, biochemistry, and nutrition, *Adv. Food Nutr. Res.* 40:3-24.

Jang, J.-T., Green, J.B., Beard, J.L., and Green, M.H., 2000, Kinetic analysis shows that iron deficiency decreases liver vitamin A mobilization in rats, *J. Nutr.* 130:1291-1296.

Kelley, S.K., and Green, M.H., 1998, Plasma retinol is a major determinant of vitamin A utilization in rats, *J. Nutr.* 128:1767-1773.

Kelley, S.K., Nilsson, C.B., Green, M.H., Green, J.B., and Håkansson, H., 1998, Use of model-based compartmental analysis to study effects of 2,3,7,8-tetrachlorodibenzo-*p*-dioxin on vitamin A kinetics in rats, *Toxicol. Sci.* 44:1-13.

Kelley, S.K., Nilsson, C.B., Green, M.H., Green, J.B., and Håkansson, H., 2000, Mobilization of vitamin A stores in rats after administration of 2,3,7,8-tetrachlorodibenzo-*p*-dioxin: a kinetic analysis, *Toxicol. Sci.* 55:478-484.

Landaw, E.M., and DiStefano III, J.J., 1984, Multiexponential, multicompartmental and noncompartmental modeling. II. Data analysis and statistical considerations, *Am. J. Physiol.* 246:R665-R677.

Lewis, K.C., Green, M.H., Green, J.B., and Zech, L.A., 1990, Retinol metabolism in rats with low vitamin A status: a multicompartmental model, *J. Lipid Res.* 31:1535-1548.

Novotny, J.A., Dueker, S.R., Zech, L.A., and Clifford, A.J., 1995, Compartmental analysis of the dynamics of ß-carotene metabolism in an adult volunteer, *J. Lipid Res.* 36:1825-1838.

v Reinersdorff, D., Green, M.H., and Green, J.B., 1998, Development of a compartmental model describing the dynamics of vitamin A metabolism in men, in: *Mathematical Modeling in Experimental Nutrition,* A. J. Clifford and H.-J. Müller, eds., Plenum Press, New York.

Wastney, M.E., Patterson, B.H., Linares, O.A., Greif, P.C., and Boston, R.C., 1999, WinSAAM, in: *Investigating Biological Systems Using Modeling: Strategies and Software,* Academic Press, San Diego.

MODELING SHORT (7 HOUR)- AND LONG (6 WEEK)-TERM KINETICS OF VITAMIN B-6 METABOLISM WITH STABLE ISOTOPES IN HUMANS

Stephen P. Coburn, Douglas W. Townsend, Karen L. Ericson, Robert D. Reynolds, Paula J. Ziegler, David L. Costill, J. Dennis Mahuren, Wayne E. Schaltenbrand, Thomas A. Pauly, Yao Wang, William J. Fink, David R. Pearson, and David L. Hachey[*]

INTRODUCTION

The significance and relevance of the topics being discussed in this text is underscored by Westerhoff's (2000) suggestion that biochemistry is poised to make major advances in this century, similar to those made in physics in the 20th century. He anticipates that, to understand the nonlinear kinetic interactions between molecules and interpret the vast quantities of data becoming available, biochemistry may soon become as computation-intensive as elementary particle physics. The papers in this book should contribute to that process.

A review of much of the literature on the kinetic aspects of vitamin B-6 metabolism is available (Coburn, 1996) and will not be repeated here. This paper deals primarily with features of our current model. Because one of the goals of the 7th Conference on Mathematical Modeling in Nutrition and the Health Sciences (Penn State University, July, 2000) was to provide instruction in model development, we have included comments on the techniques used to adapt the model to WinSAAM notation (Wastney et al.,1998).

Our interest in vitamin B-6 was initially stimulated by reports that the response of persons with Down syndrome to administration of tryptophan (O'Brien and Groshek,

* Stephen P. Coburn, J. Dennis Mahuren, Wayne E. Schaltenbrand, Thomas A. Pauly, Fort Wayne State Developmental Center, Fort Wayne, IN 46835. Douglas W. Townsend, Karen L. Ericson, Indiana University-Purdue University in Fort Wayne, Fort Wayne IN 46805. Robert D. Reynolds, Yao Wang, University of Illinois at Chicago, Chicago, IL 60612. Paula J. Ziegler, David L. Costill, William J. Fink, David R. Pearson, Ball State University, Muncie, IN 47306. David L. Hachey, Vanderbilt University, Nashville, TN.

1962), pyridoxine (Gershoff et al., 1959), and 4'-deoxypyridoxine (McCoy et al., 1969) was consistent with a deficiency or increased turnover of vitamin B-6. More recently, vitamin B-6 supplementation has been proposed as a means of correcting low serotonin concentrations in Down syndrome (Coleman, 1988), and abnormalities in tryptophan metabolism continue to be found (Baran et al., 1996). Pyridoxal kinase activity was increased in the tongue in Down syndrome (Coburn et al., 1991a). Down syndrome accounts for about 10% of the mentally retarded population, making it one of the two most common genetic causes of mental retardation. Down syndrome results from a trisomy of chromosome 21. Therefore, persons with Down syndrome have a 50% excess of the genes carried on chromosome 21. The resulting biochemical disorders would presumably be manifested by increased rates of many reactions rather than the blockage of a single reaction as is observed in single gene disorders. Therefore, we decided to examine how the kinetics of vitamin B-6 metabolism in Down syndrome differ from normal. Because our ultimate goal is to understand the alterations in Down syndrome, from the very beginning we have attempted to examine the *in vivo* control of vitamin B-6 metabolism. Therefore, rather than focusing on the steady state, we have attempted to build a model which will respond appropriately to change. This introduces major challenges to the model building process. The results presented here reflect the current status of our work.

KEY FEATURES OF VITAMIN B-6 METABOLISM

Several key observations which should be considered when modeling vitamin B-6 metabolism are listed below.

- The metabolism of vitamin B-6 involves interconversion among 7 major forms (Figure 1). Additional forms, such as 5-pyridoxic acid and its lactone, may become significant at vitamin B-6 intakes in excess of 200 mg/d (Mahuren et al., 1991). There are variations between tissues and between species regarding which of the reactions can be performed. In some species including humans, erythrocytes can phosphorylate pyridoxine and oxidize it to pyridoxal 5'-phosphate. The relative importance of the erythrocytes in whole body vitamin B-6 kinetics remains to be established.

- Using radioisotope tracers and total body analysis in rats (Coburn et al., 1988b), guinea pigs (Coburn et al., 1984), and miniature swine (Coburn et al., 1985), we demonstrated that about 70-80% of the body pool of vitamin B-6 is located in muscle, and about 10% is in the liver. The remainder is distributed among the other organs.

- Although vitamin B-6 is a water soluble vitamin, the muscle pool is efficiently conserved during short periods of low intake. In vitamin B-6 deficient rats, vitamin B-6 was not released from the muscle until the deficiency became so severe that animals started losing weight (Black et al., 1978). In men consuming only about 0.4 mg vitamin B-6/d for 6 wk, the vitamin B-6 content of the muscle was not signifi- cantly altered (Coburn et al., 1991b). The urinary excretion dropped to approximate the intake within 2 wk, with a net loss of about 4% of the body pool, thus establishing a new steady state which would minimize further losses of vitamin.

- The sole source of pyridoxal 5'-phosphate in plasma appears to be secretion from the liver in association with plasma proteins. The half-life of this vitamer is about 2.5 h (Lumeng et al., 1974).
- 4-Pyridoxic acid is probably formed primarily in the liver (Korte and Bannuscher, 1958).
- Because pyridoxal 5'-phosphate is a charged molecule which cannot readily cross membranes, the plasma concentration is controlled in part by alkaline phosphatase. This relationship is illustrated by the marked increases in plasma pyridoxal 5'-phosphate concentrations associated with the lack of alkaline phosphatase in hypophosphatasia in humans (Whyte et al., 1985) and in knockout mice (Waymire et al., 1995). The role of alkaline phosphatase in controlling plasma concentrations of pyridoxal 5'-phosphate is further illustrated by the decline in pyridoxal 5'-phosphate during pregnancy in women with hypophosphatasia as placental alkaline phosphatase increases (Whyte et al., 1995). The situation is further complicated by the fact that physiological concentrations of inorganic phosphate can act as a competitive inhibitor of alkaline phosphatase (Coburn et al., 1998).
- Examination of the urinary excretion of tracer after a single dose of radioactive pyridoxine in rats and humans produced a curve which could be modeled by assuming a large, slow turnover pool in equilibrium with a small, rapid turnover pool (Johansson et al., 1966).
- The estimates of the total body pool ranged from 107 (Tillotson et al., 1967) - 190 μmol (Johansson et al., 1966) following oral administration of label (Johansson et al., 1966) to 345 - 725 μmol following intravenous administration (Johansson et al., 1966; Tillotson et al., 1967; Shane, 1970). These values are less than the 1,000 μmol

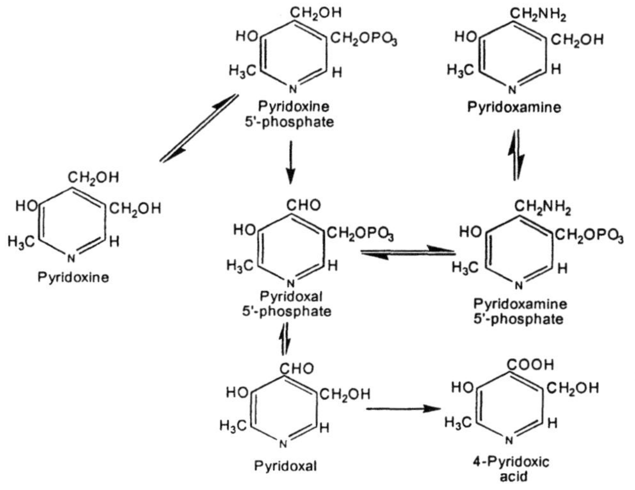

Figure 1. Normal metabolic pathways of vitamin B-6 metabolism.

pool we estimated from muscle biopsies (Coburn et al., 1988a). Our results suggest that a portion of the pool was not detected in the earlier studies, presumably because of the very slow turnover.

- Dietary pyridoxine and pyridoxamine are normally phosphorylated and oxidized to pyridoxal 5'-phosphate. The pyridoxal 5'-phosphate then must be hydrolyzed back to pyridoxal before it can be oxidized to 4-pyridoxic acid for excretion. In contrast, ingested pyridoxal can be oxidized directly to 4-pyridoxic acid. Therefore, the model must allow pyridoxal to enter at a different point than do pyridoxine and pyridoxamine.

COMPARTMENTAL MODEL OF VITAMIN B-6 METABOLISM

In view of the points just presented, we proposed the model shown in Figure 2 as the simplest one that might account for most of these key observations. This model expands our previous model (Coburn, 1996) by including plasma and erythrocyte vitamers. While blood contains only about 0.2% of the body vitamin B-6 pool, plasma concentrations are probably the most commonly utilized indicators of vitamin B-6 status. Plasma concentrations are also more readily measured than are tissue pools. Therefore, we felt it was important to include blood in our model, even though it is a minuscule fraction of the body's vitamin B-6 pool. We recognize that this model is not identifiable (i.e., does not have a unique solution) with the limited data currently available. However, the primary goal of this work is not to identify the precise rates of individual reactions but rather to predict the basic behavior of the system under a variety of conditions. This will allow us to better understand how the various parameters reflect changes in vitamin B-6 status and ultimately to improve our estimates of vitamin B-6 requirements. We hope that this model provides a physiologically-based system which can readily be refined as additional data become available.

The model shows pyridoxine and pyridoxamine in the gastrointestinal tract entering a

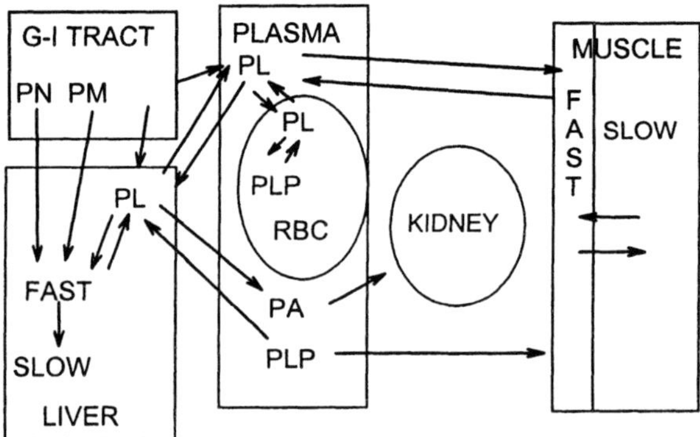

Figure 2. Compartmental model of vitamin B-6 metabolism. Items connected by arrows represent compartments in the model. PN, pyridoxine; PM, pyridoxamine; PL, pyridoxal; PLP, pyridoxal 5'-phosphate; PA, 4-pyridoxic acid.

rapidly exchangeable pool. This pool includes the phosphorylation and oxidation reactions. Although it is labeled as liver, at least some of this processing probably occurs in the intestinal wall. From the liver pool, pyridoxal may be oxidized to 4-pyridoxic acid, transferred to the plasma, and excreted in the urine. Alternatively, liver pyridoxal may enter the plasma or be phosphorylated and enter the fast liver pool. The pyridoxal 5'-phosphate and pyridoxamine 5'-phosphate produced from any of the vitamers may enter the slow turnover pool in the liver, be secreted into the plasma bound to protein, or be hydrolyzed back to pyridoxal. Pyridoxal 5'-phosphate in the plasma may be taken up by peripheral tissues, mainly muscle, or returned to the liver pyridoxal pool. We originally attempted to incorporate alkaline phosphatase activity into the rate function for the removal of pyridoxal 5'-phosphate from plasma, but for the time being, we have simply fixed the turnover of plasma pyridoxal 5'-phosphate to a half-life of 2.5 h as reported by Lumeng et al. (1974). Plasma pyridoxal may be taken up by erythrocytes, peripheral tissues, or the liver. In erythrocytes, the pyridoxal may be converted to pyridoxal 5'-phosphate, or it may be returned to the plasma. In peripheral tissues, material in the rapid turnover pool can be transferred to the slow turnover pool or returned to plasma. The slow turnover pools presumably represent protein-bound vitamers. The rapid turnover pools probably include non-phosphorylated vitamers, free phosphorylated vitamers, and rapidly exchangeable bound forms.

In order to account for the ability of the tissues to conserve vitamin B-6 when vitamin B-6 intake is restricted, we utilized equations for binding kinetics (Coburn, 1996). These were set up in a form that allowed the concentration of the free vitamer to be calculated from the total binding sites, the total ligand, and the dissociation constants. These are quantities which may eventually be measurable. For the moment, we have made arbitrary assumptions. The resulting equations are quadratic or, in the case of liver which was assigned to have 2 sets of binding sites, they are cubic:

$$S_F^3 + S_F^2(P_{1T} + P_{2T} - S_T + K_1 + K_2) + S_F(P_{1T}K_2 + P_{2T}K_1 - S_TK_2 - S_TK_1 + K_1K_2)$$
$$- S_TK_1K_2 = 0 \tag{1}$$

where S_F = free ligand (μmol), P_{1T} = total binding sites available for proteins other than plasma proteins (μmol/h), P_{2T} = total binding sites available for plasma protein (μmol/h), K_1 = dissociation constant for sites other than plasma proteins, and K_2 = dissociation constant for plasma protein sites. The roots of the cubic equation can be obtained by standard algebraic techniques (Hodgman, 1952). The model deck includes adjustments for the fact that WinSAAM has no arccosine function and calculates angles in radians. From the formulas for the coefficients for this cubic equation, it is possible to show that the equations obtained from the data sets used here have three real roots, exactly one of which is positive. Thus far, the root selected for use in the model has proved to be the positive root.

Once the concentration of free ligand is calculated, the amount bound (P_{XB}) to each protein (μmol/h) can be estimated from the equation

$$P_{1B} = S_F \, P_{1T} / (\, K_1 + S_F) \tag{2}$$

The fractional turnover rate [L(i,j)] from pool j [F(j)] resulting from the loss of P_{1B} is

$$L(i,j) = P_{1B} / F(j) \qquad (3)$$

The ability of the tissues to conserve vitamin B-6 can be explained by two mechanisms. Proteins may not degrade until replacement cofactor is available, or they may degrade continuously but then efficiently recapture the released cofactor. We utilized the second approach in the model for two reasons. First, the mathematics are simpler. Second, the assumption that there is no recycling of cofactor would result in protein half-lives of several years under conditions of low vitamin B-6 intakes. Since the rate of recycling between the slow and fast muscle pools might vary over a considerable range without affecting other parts of the system, and we had no data to guide the computer, we fixed the fractional turnover of the slow muscle pool at 0.0025/h or 6%/d. However, the amount of cofactor which returns to the plasma from the fast muscle pool is controlled by the binding calculations. Therefore, the model predicts that the permanent loss of material from the slow muscle pool will be dependent on intake up to a maximum of 6%/d. It is interesting to note that the turnover estimates of 1-4%/d (Johansson et al., 1966; Tillotson et al., 1967) were obtained in men consuming normal diets which provide about 1% of the body pool of vitamin B-6 per day. We felt that allowing a maximum of 6%/d would accommodate the experimental conditions used in the present studies.

Because we have conducted triple label experiments to monitor the relative rates of metabolism of the three forms of vitamin B-6, we set up four parallel versions of our model. The first 14 pools monitor endogenous, unlabeled vitamers; then three parallel models, each with14 pools, monitor the metabolism of $^2H_2$-pyridoxamine, $^2H_3$-pyridoxal, and $^2H_5$-pyridoxine, respectively. Each of the parallel systems uses identical rate equations. Because we are particularly interested in the response of the system to changes, it is important that the system starts at a steady state. Changes in parameters produced during iterations may cause changes in the steady state pool sizes. In order to insure that the system is at steady state when the program processes the data for our experiments, the first time change in the WinSAAM deck [TC(0) in WinSAAM notation] simulates 748 h with a constant, normal vitamin B-6 intake to establish the steady state pool sizes. We selected a constant intestinal infusion of 0.375 μmol/h, which is equivalent to 9 μmol/d (about 1.8 mg/d). It is possible to divide the intake and administer it as boluses to simulate meals. That would change the results slightly for the first few days, but it would have little influence at later times. Because dividing the dose markedly increases calculation time, we chose the continuous infusion mode for routine use. We assumed complete absorption since fecal excretion accounted for only 2-3% of orally administered tracer (Tillotson et al., 1967). This section of the deck also simulates administration of a minuscule amount of tracer into the plasma pyridoxal 5'-phosphate pool. That pool can then be monitored to verify that the half-life of plasma pyridoxal 5'-phosphate remains close to the experimentally observed value of 2.5 h (Lumeng et al., 1974).

The second section of the deck [TC(1)] contains observed data on the concentrations of B-6 vitamers in plasma and erythrocytes as well as on the appearance of tracer as 4-pyridoxic acid in urine during the 7 h after simultaneous oral administration of 5 μmol each of $^2H_2$-pyridoxamine dihydrochloride, $^2H_3$-pyridoxal hydrochloride, and $^2H_5$-pyridoxine hydrochloride to nonpregnant, pregnant, and lactating women. The third section [TC(2)] includes data from the only subject who received tracer orally in the study of Johansson et al. (1966). This study generated data for 10 d under reasonably steady state conditions. The last set of data presented here [TC(3)] is from a 6 wk study of men receiving a low vitamin B-6 intake of about 0.4 mg/d (Coburn et al., 1991b).

METHODS

The protocols were approved by the review boards at the participating institutions, and informed consent was obtained.

The labeled vitamers were synthesized by published procedures (Coburn et al., 1982). Nonpregnant (n=17), pregnant (n=17), and lactating women (n=16) consumed a daily multivitamin supplement containing 2 mg pyridoxine HCl plus a self-selected diet for 4 weeks prior to the study date. On the study day, subjects consumed a low vitamin B-6 meal at 7 a.m. Two hours later, they ingested 100 mL of an aqueous solution containing 5 μmol each of $^2H_2$-pyridoxamine 2HCl, $^2H_3$-pyridoxal HCl, and $^2H_5$-pyridoxine HCl. Blood and urine samples were collected over the next 7 h. After each voiding, subjects consumed 100 mL of 5% glucose polymer (Polycose®, Ross Product Div, Abbott Laboratories, Columbus, OH) as the sole source of nutrition during the collection period. The use of a glucose polymer under these conditions has been shown not to influence pyridoxal 5'-phosphate concentrations in plasma (Hofmann et al., 1991).

Because of the low concentration of B-6 vitamers in plasma, tracer data could not be obtained routinely. In order to obtain information on the tracers in plasma, a 200 mL blood sample was obtained from a male volunteer starting 30 min after ingestion of 5 μmol each of deuterated pyridoxine, pyridoxamine, and pyridoxal as described above. Pyridoxal and 4-pyridoxic acid were isolated from plasma. Pyridoxal was oxidized to 4-pyridoxic acid for mass spectrometry.

The protocol for the 6 wk low intake study has been described previously (Coburn et al., 1991b). Briefly, healthy men were fed a low vitamin B-6 diet based on rice and vitamin-free casein which provided about 0.81 μmol vitamin B-6/d. This was supplemented with 0.95 μmol $^2H_5$-pyridoxine HCl/d. Blood and urine samples were collected at intervals throughout the study.

B-6 vitamers in plasma and urine were determined by cation exchange HPLC (Mahuren and Coburn, 1990). 4-Pyridoxic acid was isolated from urine using Dowex 50 followed by purification with cation exchange HPLC. Isotope ratios were determined by mass spectrometry of the acetylated derivative as previously described (Hachey et al., 1985).

The model was fit to the data from each subject individually. Then, all of the subjects from a given portion of the study were processed as a group using the project utility in WinSAAM (version 2.2.1, Laboratory of Experimental and Computational Biology, National Cancer Institute, Bethesda, MD). The data from the women were fit using the steady state equilibration [TC(0)], the 7 h data [TC(1)], and the 10 d data [TC(2)]. The 6 wk low intake data were fit starting with the population means obtained from the nonpregnant women and utilizing the steady state equilibration [TC(0)] plus data from the 7 h [TC(1)] and 10 d studies [TC(2)] as well as the 6 wk data [TC(3)].

RESULTS AND DISCUSSION

The nonpregnant women excreted about 40% of the pyridoxal dose as pyridoxic acid within 7 h while the pregnant and lactating women excreted 20-30% (Figure 3). The increased retention in the pregnant and lactating women presumably resulted from uptake by the fetus and associated tissues in the pregnant women and by mammary tissue and milk in the lactating women. For longer-term studies, it would be essential to add an extra route for loss to the model for pregnant and lactating women to account for such processes. However, for the present model, we retained the material within the system. All three groups excreted only about 10% of the label administered as pyridoxine or pyridoxamine.

The model-predicted steady state rate constants and pool sizes are summarized in Tables 1 and 2. The large errors associated with many of the values reflect both the limited data available and the large individual variation among subjects. Therefore, these values should be taken as preliminary estimates subject to modification as additional data become available. Since we have no direct measurements of absorption rates to guide the calculations, the program tends to increase the rate to the available upper limit. Therefore, the rates shown indicate rapid absorption rather than absolute rates. We found, when administering pyridoxine to pigs via gastrostomy, that it was difficult to take blood samples quickly enough to catch the peak in the portal plasma (Coburn et al., unpublished data). We initially connected the incoming pyridoxal only to the liver pool

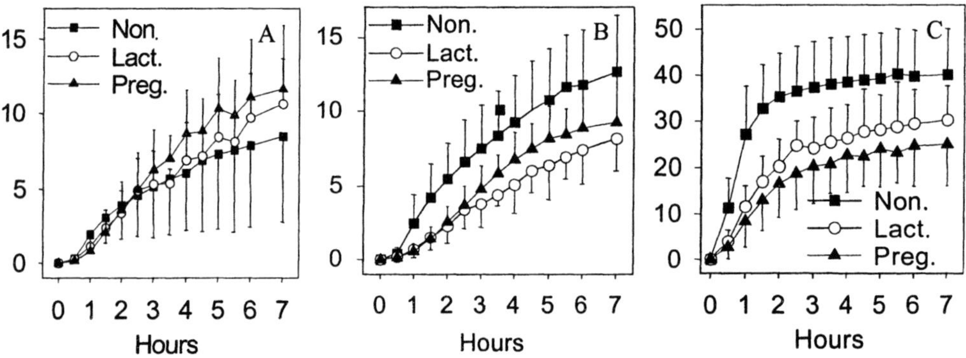

Figure 3. Cumulative excretion of tracer from pyridoxine (A), pyridoxamine (B), or pyridoxal (C) during the 7 h after ingestion of 5 µmol of each by nonpregnant (Non; n=17), lactating (Lact; n=16), and pregnant (Preg; n=17) women. Data are means ± SD.

but found that an additional connection to plasma was needed to fit the data. This area needs further attention since it is unlikely that the intestinal uptake rates for pyridoxal, pyridoxine, and pyridoxamine should differ as much as predicted by the current model.

The data allowed us to calculate the renal clearances of 4-pyridoxic acid independently of the model (Coburn et al., 2000). If we assume a plasma volume of 4 L, the mean observed clearances of 232, 337, and 215 mL/min for the nonpregnant, pregnant, and lactating women, respectively, would be comparable to fractional turnover rates of 3.5, 5.1, and 3.2/h compared with the rates of 3.2, 4.3, and 2.9/h predicted by the model. Since 4-pyridoxic acid is an end product of metabolism that is efficiently eliminated, it is reasonable that the model predicts that the rate of entry into the plasma 4-pyridoxic acid pool would be reduced in pregnant and lactating women compared to men and nonpregnant women. Remember that, except for the parameters which were fixed (Table

Table 1. Steady state rate constants (fractional rate/h) generated by the WinSAAM population utility

Parameter	Non-pregnant	Pregnant	Lactating	Low intake study
PL uptake from intestine to liver	1.2 ± 0.4	1.1 ± 1.4	1.5 ± 1.0	1.5 ± 0.4
PL uptake from intestine to plasma	1.0 ± 0.4	0.7 ± 0.5	0.5 ± 0.2	0.8 ± 0.08
PN uptake from intestine to fast liver	25^2	25^2	25^2	25^2
PM uptake from intestine to fast liver	23 ± 6	10 ± 16	13 ± 53	11 ± 40
Liver PL to fast liver PLP	19 ± 12	3.2 ± 2.2	7.8 ± 9	13 ± 29
Liver PL to plasma PL	7.8 ± 10	0.7 ± 2.0	1.3 ± 5.6	0.8 ± 5.6
Liver PL to plasma PA	18 ± 11	1.4 ± 2.2	2.9 ± 4.3	13 ± 29
Plasma PA to urine	3.2 ± 1.0	4.3 ± 1.4	2.9 ± 0.5	3.5 ± 0.2
Plasma PL to liver PL	3.2 ± 3.0	5.2 ± 3.3	3.1 ± 2.3	1.8 ± 0.3
Plasma PL to erythrocyte PL	23 ± 15	33 ± 17	23 ± 42	32 ± 45
Plasma PL to fast muscle	6.8 ± 3.4	18 ± 9	3.2 ± 2.4	14 ± 3
Fast liver to plasma PLP	0.02^3	0.001^3	0.003^3	0.006^3
Fast liver to slow liver	0.4^3	0.3^3	0.2^3	0.4^3
Fast liver to liver PL	0.07^3	0.1^3	0.1^3	0.07^3
Slow liver to fast liver	0.06 ± 0.02	0.05 ± 0.03	0.04 ± 0.02	0.03 ± 0.01
Plasma PLP to fast muscle	$.025^2$	0.25^2	0.25^2	0.25^2
Plasma PLP to liver PL	0.025^2	0.025^2	0.025^2	0.025^2
Fast muscle to slow muscle	0.075^3	0.073^3	0.065^3	0.059^3
Slow muscle to fast muscle	0.0025^2	0.0025^2	0.0025^2	0.0025^2
Fast muscle to plasma PL	0.03^3	0.04^3	0.01^3	0.05^3
Erythrocyte PL to PLP	8.2 ± 7.4	5.2 ± 6.1	8.8 ± 6.9	5.2 ± 8.5
Erythrocyte PLP to PL	3.7 ± 3.3	1.9 ± 1.8	4.3 ± 3.1	3.3 ± 5.1
Erythrocyte PL to plasma PL	41 ± 16	33 ± 15	40 ± 19	41 ± 56

[1]Abbreviations: PN, pyridoxine; PM, pyridoxamine; PL, pyridoxal; PLP, pyridoxal phosphate; PA, pyridoxic acid.

[2]These values were fixed.

[3]These values are based on binding equations. While the program calculates error estimates for the individual parameters, it does not estimate errors for the fractional rate constant calculated using those parameters.

1), WinSAAM determined rates which best fit the data. A similar change occurred in the rate of transfer from liver pyridoxal to the fast liver pool. Those are the most significant rate differences between the groups. As noted above, in this short-term study, the amount of material retained by the pregnant and lactating women is a sufficiently small portion of the total body pool that major changes in other rates would not be necessary. One of our next modifications will be to add an additional pool exiting from plasma pyridoxal and/or pyridoxal 5'-phosphate as an alternative route to account for the increased retention of label. This may allow the model to accommodate the increased retention of label without a major change in the release of label from liver pyridoxal to plasma pyridoxal 5'-phosphate.

We knew from the quantitative measurements of urinary 4-pyridoxic acid that the men lost about 4% of their body pool during the 6 weeks of low vitamin B-6 intake (Coburn et al., 1991b). The model predicted that the body pool declined 3.7% from 1011 μmol to 973 μmol. Most of the loss came from the two muscle pools, with a smaller amount from the fast liver pool. As emphasized earlier, this loss is strictly the result of the model responding to the change in intake and the computer fitting the model to the data. There were no restrictions on the changes in pool sizes built in to the model. The fact that the model controlled the losses under these conditions is encouraging.As noted earlier, we fixed the rate of loss from the slow muscle pool because the lack of kinetic data meant that the program would have no guidance toward a reasonable value. However, we allowed the rate from the slow liver pool to vary since that pool was a relatively small fraction of the total and because its location in the liver might have more

Table 2. Pool sizes (μmol) obtained by using the fractional rates from Table 1[1]

Pool	Nonpregnant	Pregnant	Lactating	6 wk low intake study	
				Beginning	End
Liver PL	0.02	0.31	0.15	0.03	0.007
Fast liver	10.9	12.7	14.4	10.9	5.6
Slow liver	72	70	81	167	165
Plasma PA	0.13	0.10	0.15	0.12	0.03
Plasma PL	0.09	0.06	0.10	0.08	0.02
Plasma PLP	0.16	0.05	0.16	0.23	0.07
Erythrocyte PL	0.05	0.05	0.05	0.06	0.02
Erythrocyte PLP	0.11	0.13	0.12	0.09	0.03
Fast muscle	27	27	30	28	9
Slow muscle	813	813	814	805	793
Total body pool	923	923	940	1011	973

[1]Pool sizes for the nonpregnant, pregnant, and lactating groups are the values obtained using the mean parameters for each group of subjects and a constant intake of 9 μmol/d for a simulated time of 748 h to achieve a steady state in TC(0). The values for the 6 wk low intake study are those predicted for the beginning and end of TC(3) using the mean parameters from the low intake study with an intake of 9 μmol/d for TC(0) to establish a steady state. At the start of TC(3), the intake was reduced to 0.81 μmol unlabeled pyridoxine/d plus 0.95 μmol $^2H_5$-pyridoxine/d. Since only one calculation was done using the mean parameters for each group of subjects, no error terms are available for the pool sizes.

impact on fitting the data. At this time, we have not determined whether the altered rate and consequent increase in the slow liver pool in men was an important aspect of fitting the data.

As discussed earlier, even though the slow muscle pool is set to turn over at 6%/d, much of the material released is recycled, with the residence time being dependent on intake. This can be examined in WinSAAM by using the QO mechanism to reset pool sizes at specified times. We used this technique to simulate introduction of a small amount of tracer into the slow muscle pool of one of the three systems not used by the endogenous vitamers. Then we determined the half-life. With intakes of 1.76, 9, or 18 μmol/d, the half-life of tracer in the slow muscle pool was 1,250, 292 and 188 d, respectively. Dual label studies in which protein and vitamin B-6 are labeled could be used to examine the relative turnover rates of the protein and pyridoxal 5'-phosphate cofactor. This simulation also demonstrated that, in the current model, the slow muscle pool takes about 3,000 h to reach steady state conditions.

Figures 4-15 illustrate how the model predictions using the mean parameters from the WinSAAM population analysis fit the data. There is obviously a lot of individual variation among the participants. In some cases, individual fits might be improved by adjusting the zero time conditions of the model to agree more closely with the observed values determined for individual subjects. However, overall, the model provides a reasonable approximation of most of the data.One of the intriguing observations in lactating women was that the concentration of tracer in the milk was too low to be

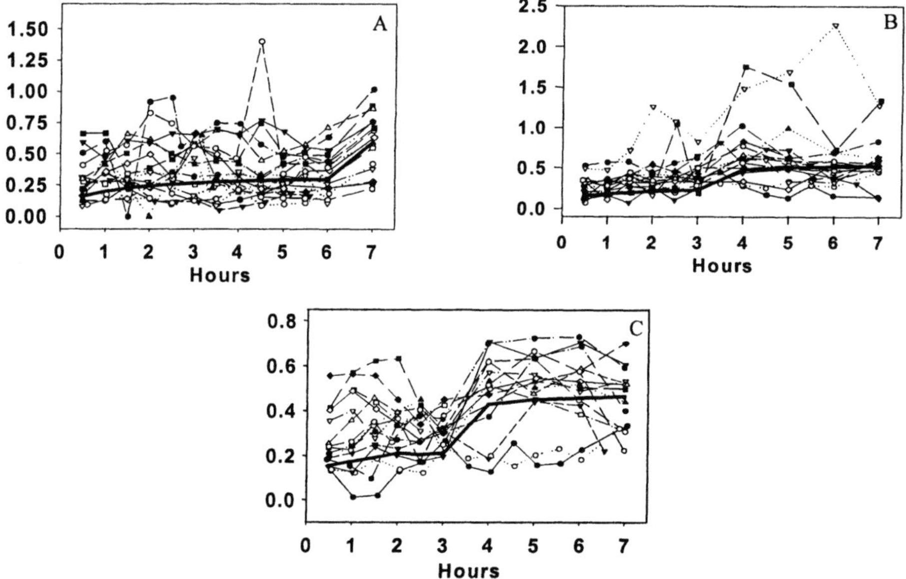

Figure 4. Excretion of endogenous 4-pyridoxic acid in nonpregnant (A), pregnant (B), and lactating (C) women during 7 h after oral administration of 5 μmol each of deuterated pyridoxine, pyridoxamine, and pyridoxal. The light lines are individual subjects. The heavy line is the model prediction using parameters from the WinSAAM population analysis.

measured accurately. In the urine, the peak concentration of labeled 4-pyridoxic acid originating from pyridoxal was about twice the concentration of unlabeled 4-pyridoxic acid at about 0.5-1 h. Plasma pyridoxal peaked about the same time (Figure 9). The model predicts that the plasma ratio of $^2H_3$-pyridoxal:endogenous pyridoxal should also be about 2:1. However, in a single large (200 mL) blood sample obtained from a man 30 min after ingesting the same tracer mixture ingested by the women, the ratio of $^2H_3$-pyridoxic acid (from pyridoxal): endogenous 4-pyridoxic acid was only 0.4, and the ratio of $^2H_3$-pyridoxal: endogenous pyridoxal was only 0.077, even though the plasma concentrations of 4-pyridoxic acid and pyridoxal were markedly elevated (Table 3). The low tracer content of plasma pyridoxal is consistent with finding only small amounts of tracer in milk. It also suggests that the plasma 4-pyridoxic acid and pyridoxal arise from different pools, and the model needs to be modified to account for this. The increase in unlabeled plasma pyridoxal might be another example of flushing molecules from binding sites. The increased sensitivity of accelerator mass spectrometry raises the potential for obtaining more detailed data from both plasma and tissues to clarify these observations. The metabolism of large doses of vitamin B-6 is a treatment which we have not yet incorporated into the model. That will require building in some saturation kinetics and alternate metabolic pathways.

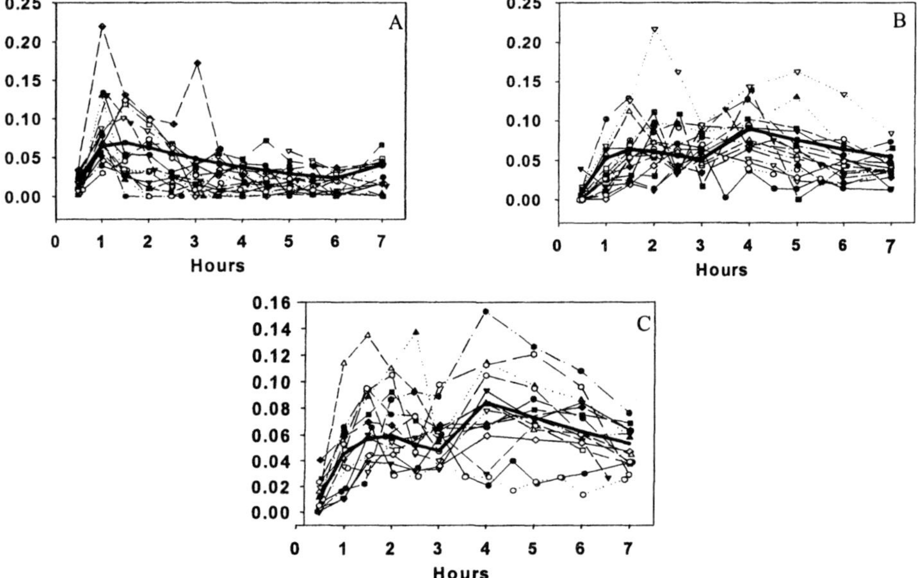

Figure 5. Excretion of 4-pyridoxic acid from pyridoxine in nonpregnant (A), pregnant (B), and lactating women (C) during 7 h after oral administration of 5 µmol each of deuterated pyridoxine, pyridoxamine, and pyridoxal. The light lines are individual subjects. The heavy line is the model prediction using parameters from the WinSAAM population analysis.

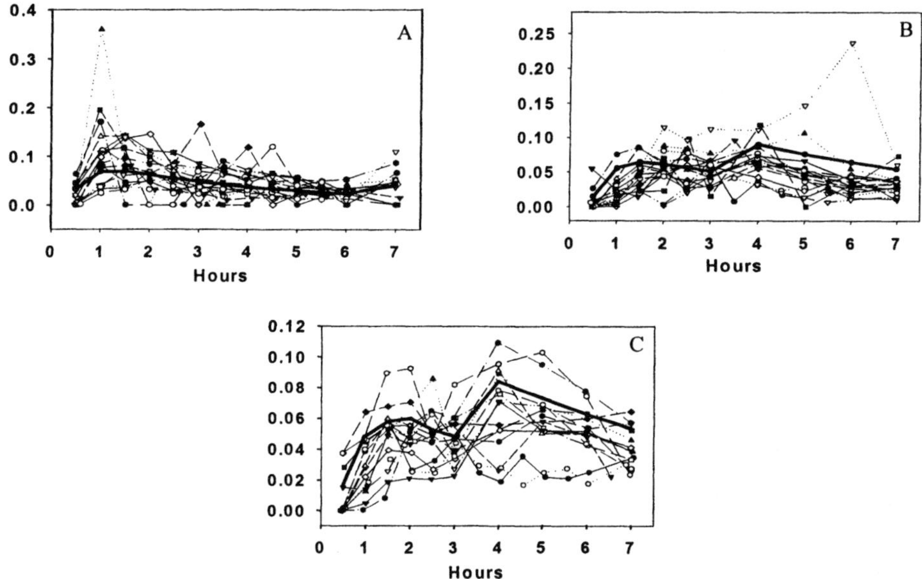

Figure 6. Excretion of 4-pyridoxic acid from pyridoxamine in nonpregnant (A), pregnant (B), and lactating women (C) during 7 h after oral administration of 5 μmol each.

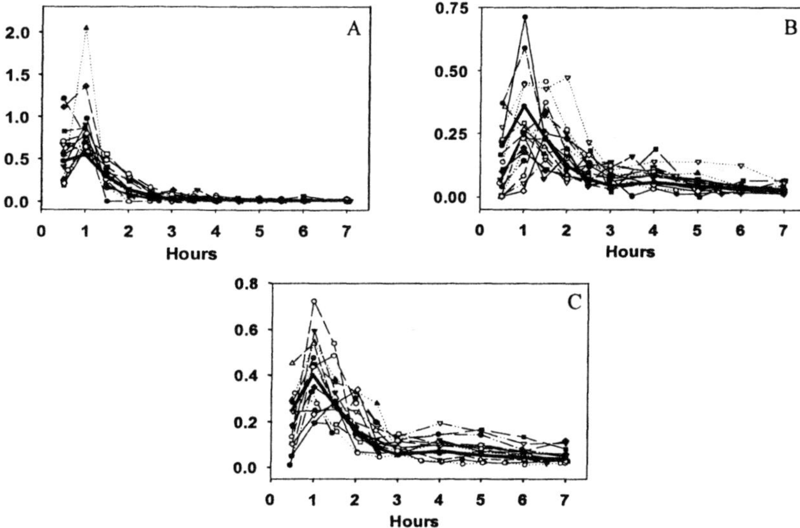

Figure 7. Excretion of 4-pyridoxic acid from pyridoxal in nonpregnant (A), pregnant (B), and lactating women (C) during 7 h after oral administration of 5 μmol each of deuterated pyridoxine, pyridoxamine, and pyridoxal. The light lines are individual subjects. The heavy line is the model prediction using parameters from the WinSAAM population analysis.

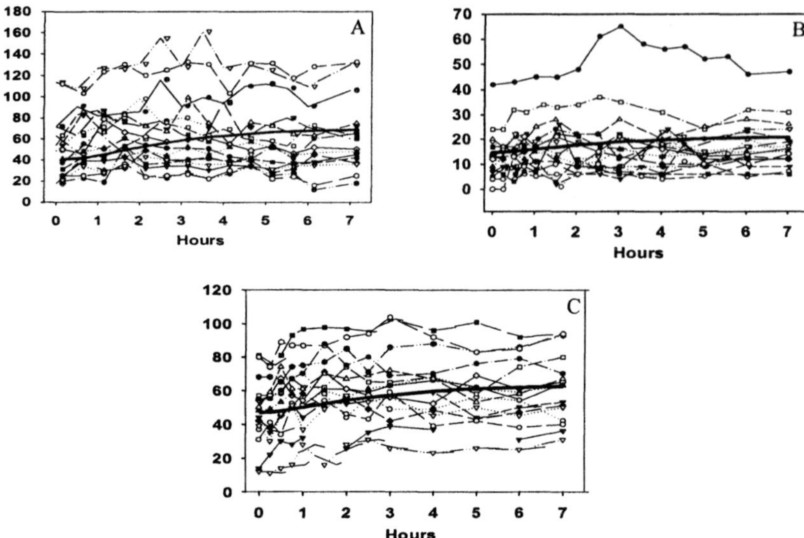

Figure 8. Plasma pyridoxal 5'-phosphate in nonpregnant (A), pregnant (B), and lactating women (C) during 7 h after oral administration of 5 μmol each of deuterated pyridoxine, pyridoxamine, and pyridoxal. The light lines are individual subjects. The heavy line is the model prediction using parameters from the WinSAAM population analysis.

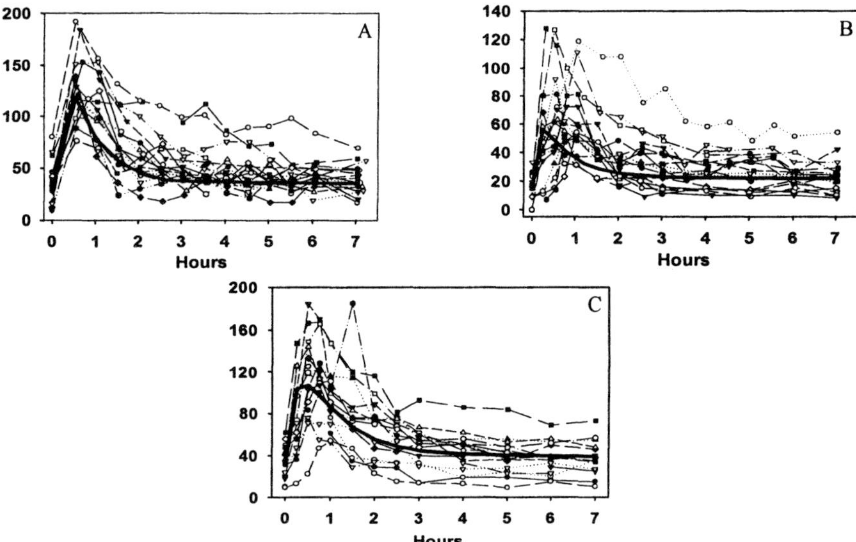

Figure 9. Plasma pyridoxal in nonpregnant (A), pregnant (B), and lactating women (C) during 7 h after oral administration of 5 μmol each of deuterated pyridoxine, pyridoxamine, and pyridoxal. The light lines are individual subjects. The heavy line is the model prediction using parameters from the WinSAAM population analysis.

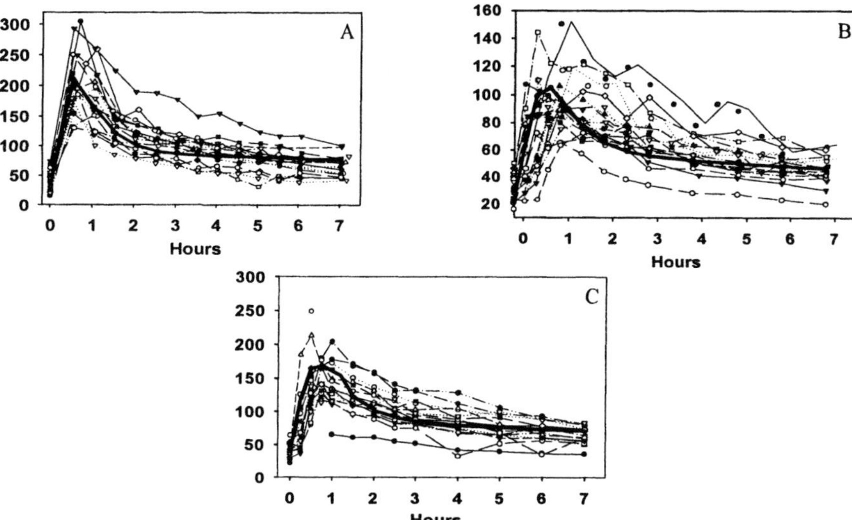

Figure 10. Plasma 4-pyridoxic acid in nonpregnant (A), pregnant (B), and lactating women (C) during 7 h after oral administration of 5 µmol each of deuterated pyridoxine, pyridoxamine, and pyridoxal. The light lines are individual subjects. The heavy line is the model prediction using parameters from the WinSAAM population analysis.

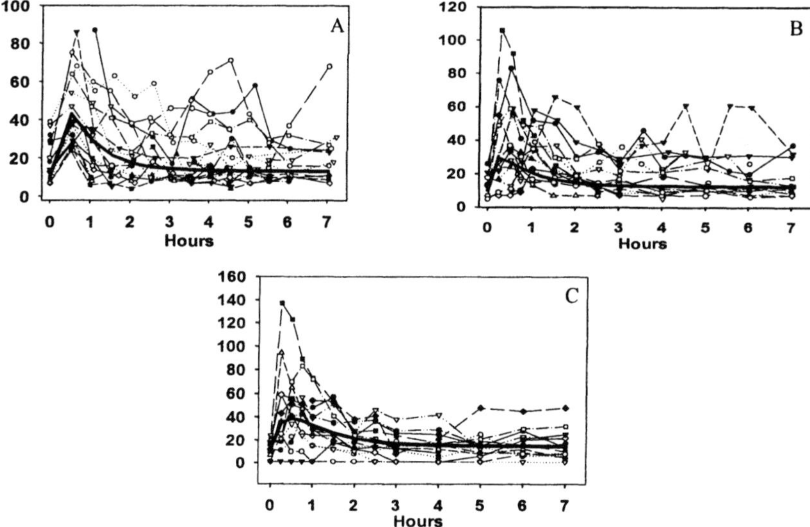

Figure 11. Erythrocyte pyridoxal in nonpregnant (A), pregnant (B), and lactating women (C) during 7 h after oral administration of 5 µmol each of deuterated pyridoxine, pyridoxamine, and pyridoxal. The light lines are individual subjects. The heavy line is the model prediction using parameters from the WinSAAM population analysis.

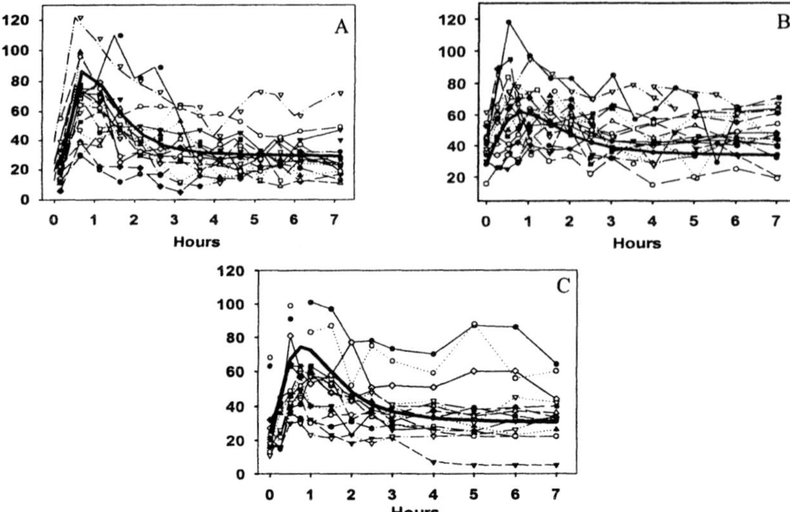

Figure 12. Erythrocyte pyridoxal 5'-phosphate in nonpregnant (A), pregnant (B), and lactating women (C) during 7 h after oral administration of 5 µmol each of deuterated pyridoxine, pyridoxamine, and pyridoxal. The light lines are individual subjects. The heavy line is the model prediction using parameters from the WinSAAM population analysis.

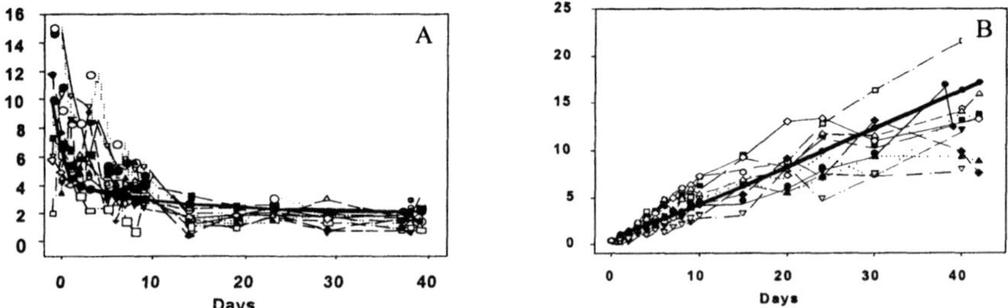

Figure 13. Total urinary 4-pyridoxic acid excretion (A) and isotope ratio in 4-pyridoxic acid (B) in men receiving a dietary vitamin B-6 intake of 0.81 µmol/d supplemented with 0.95 µmol ^{2}H$_5$-pyridoxine HCl/d. The light lines are individual subjects. The heavy line is the model prediction using parameters from the WinSAAM population analysis.

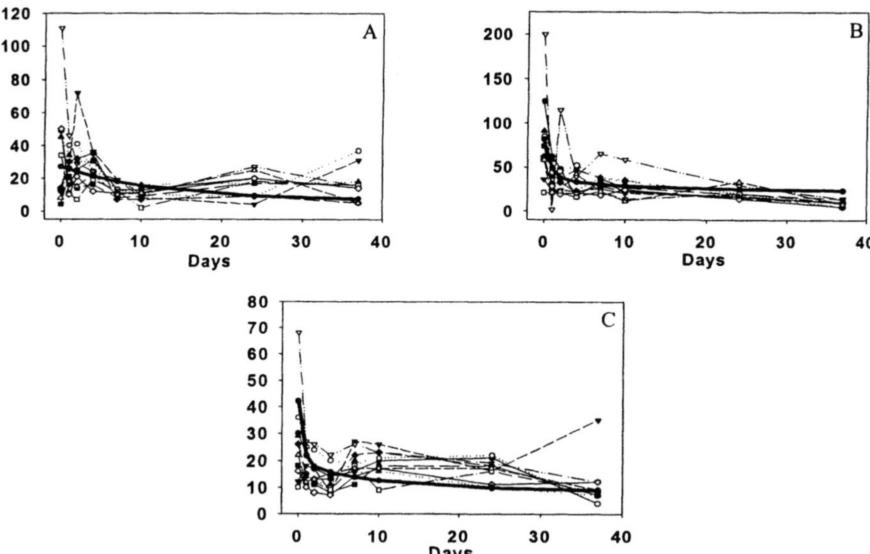

Figure 14. Plasma pyridoxal (A), pyridoxal 5'-phosphate (B), and 4-pyridoxic acid (C) in men receiving a low vitamin B-6 intake. The light lines are individual subjects. The heavy line is the model prediction using parameters from the WinSAAM population analysis.

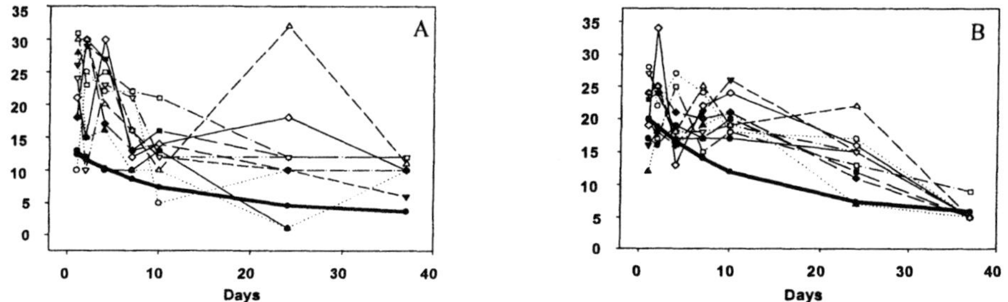

Figure 15. Erythrocyte pyridoxal (A) and pyridoxal 5'-phosphate (B) in men receiving a low vitamin B-6 intake. The light lines are individual subjects. The heavy line is the model prediction using parameters from the WinSAAM population analysis.

Table 3. Tracer ratios (^{2}H/^{1}H $*$ 100) in pyridoxal and 4-pyridoxic acid isolated from plasma of a man 30 min after ingesting 5 μmol each of ^{2}H$_2$-pyridoxamine 2HCl, ^{2}H$_3$-pyridoxal HCl, and ^{2}H$_5$-pyridoxine HCl

Vitamer	Plasma Concentration (nmol/L)	Source of tracer		
		Pyridoxamine	Pyridoxal	Pyridoxine
Pyridoxal	169	11.2	7.7	5.7
4-Pyridoxic acid	331	11.2	40.9	9.9

CONCLUSIONS

Six main conclusions can be drawn from the work presented here.

- About 40% of orally administered pyridoxal is converted to 4-pyridoxic acid and excreted within 7 h, compared to less than 15% of pyridoxine or pyridoxamine.
- Pregnant and lactating women retain more vitamin B-6 from pyridoxal than nonpregnant women.
- Only traces of label appeared in the milk. In the only plasma sample analyzed for tracer, the concentration of tracer from pyridoxal was about 4-fold greater in 4-pyridoxic acid than in pyridoxal. This suggests that, although both pyridoxal and pyridoxic acid increase in plasma after ingestion of pyridoxal, most of the pyridoxal arises from a different source (possibly peripheral tissues) than the 4-pyridoxic acid.
- In men ingesting only 1.76 μmol vitamin B-6/d, a new steady state was reached within 2 wk with a loss of about 4% of the body pool of vitamin B-6.
- A 14-pool model provides a reasonable approximation of the above observations. The model includes major plasma and erythrocyte pools of B-6 vitamers and uses protein binding kinetics to stabilize the large pools during variations in vitamin B-6 intake. This model assumes that much of the pyridoxal 5'-phosphate released from degraded protein will be reutilized either before or after removal of the phosphate group. The model predicts that the amount of recycling will decrease as vitamin B-6 intake increases.
- The model makes testable predictions that can be used to obtain further refinements.

ACKNOWLEDGMENTS

This work was supported in part by a series of grants from the USDA/NRICGP, the most recent of which was 95-37200-1703. The authors greatly appreciate the assistance provided by Loren Zech, who introduced us to SAAM, as well as Ray Boston, Peter

Greif, and David Foster, who assisted us with various aspects of modeling. We also acknowledge the technical assistance of R. Frederick and data entry by T. Weber.

CORRESPONDENCE

Please address all correspondence to:
Stephen P. Coburn
Fort Wayne State Developmental Center
4900 St. Joe Road
Fort Wayne, IN 46835
Coburn@ipfw.edu

REFERENCES

Baran, H., Cairns, N., Lubec, B., and Lubec, G., 1996, Increased kynurenic acid levels and decreased brain kynurenine aminotransferase I in patients with Down syndrome, *Life Sci.* 58:1891-1899.

Black, A.L., Guirard, B.M., and Snell, E.E., 1978, The behavior of muscle phosphorylase as a reservoir for vitamin B-6 in the rat, *J. Nutr.* 108:670-677.

Coburn, S.P., 1996, Modeling vitamin B-6 metabolism, *Adv. Food Nutr. Res.* 40:107-132.

Coburn, S.P., Lin, C.C., Schaltenbrand, W.E., and Mahuren, J.D., 1982, Synthesis of deuterated vitamin B-6 compounds, *J. Labelled Comp. Radiopharm.* 19:703-716.

Coburn, S.P., Mahuren, J.D., Erbelding, W.F., Townsend, D.W., Hachey, D.L., and Klein, P.D., 1984, Measurement of vitamin B-6 kinetics in vivo using chronic administration of labelled pyridoxine, in: *Chemical and Biological Aspects of Vitamin B-6 Catalysis, Part A,* A.E. Evangelopoulos, ed., Alan R. Liss, Inc., New York.

Coburn, S.P., Mahuren, J.D., Szadkowska, Z., Schaltenbrand, W.E., and Townsend, D.W., 1985, Kinetics of vitamin B-6 metabolism examined in miniature swine by continuous administration of labelled pyridoxine, in: *Mathematical Models in Experimental Nutrition,* N.L. Canolty, and T.P. Cain, eds., University of Georgia, Athens.

Coburn, S.P., Lewis, D.N., Fink, W.J., Mahuren, J.D., Schaltenbrand, W.E., and Costill, D.L., 1988a, Human vitamin B-6 pools estimated through muscle biopsies, *Am. J. Clin. Nutr.* 48:291-294.

Coburn, S.P., Mahuren, J.D., Kennedy, M.S., Schaltenbrand, W.E., Sampson, D.A., O'Connor, D.K., Snyder, D.L., and Wostmann, B.S., 1988b, B-6 vitamer content of rat tissues measured by isotope tracer and chromatographic methods, *BioFactors* 1:307-312.

Coburn, S.P., Mahuren, J.D., and Schaltenbrand, W.E., 1991a, Increased activity of pyridoxal kinase in tongue in Down's syndrome, *J. Ment. Def. Res.* 35:543-547.

Coburn, S.P., Ziegler, P.J., Costill, D.L., Mahuren, J.D., Fink, W.J., Schaltenbrand, W.E., Pauly, T.A., Pearson, D.R., Conn, P.S., and Guilarte, T.R., 1991b, Response of vitamin-B-6 content of muscle to changes in vitamin B-6 intake in men, *Am. J. Clin. Nutr.* 53:1436-1442.

Coburn, S.P., Mahuren, J.D., Jain, M., Zubovic, Y., and Wortsman, J., 1998, Alkaline phosphatase (EC 3.1.3.1) in serum is inhibited by physiological concentrations of inorganic phosphate, *J. Clin. Endocrinol. Metab.* 83:3951-3957.

Coburn, S.P., Reynolds, R.D., Mahuren, J.D., Schaltenbrand, W.E., Wang, Y., Ericson, K., Whyte, M.P., and Wortsman, J., 2000, Renal clearance of pyridoxic acid involves tubular secretion, *FASEB J.* 14:A242.

Coleman, M., 1988, Studies of the administration of pyridoxine to children with Down's syndrome, in: *Clinical and Physiological Applications of Vitamin B-6,* J.E. Leklem and R.D. Reynolds, eds., Alan R. Liss, Inc., New York.

Gershoff, S.N., Mayer, A.L., and Kulczycki, L.L., 1959, Effect of pyridoxine administration on the urinary excretion of oxalic acid, pyridoxine and related compounds in mongoloids and non-mongoloids, *Am. J. Clin. Nutr.* 7:76-79.

Hachey, D.L., Coburn, S.P., Brown, L.T., Erbelding, W.F., DeMark, B., and Klein, P.D., 1985, Quantitation of vitamin B-6 in biological samples by isotope dilution mass spectrometry, *Anal. Biochem.* 151:159-168.

Hodgman, C.D., 1952, *Mathematical Tables, 9th Edition*, Chemical Rubber Publishing Co., Cleveland.

Hofmann, A., Reynolds, R.D., Smoak, B.L., Villanueva, V.G., and Deuster, P.A., 1991, Plasma pyridoxal and pyridoxal 5'-phosphate concentrations in response to ingestion of water or glucose polymer during a 2-h run, *Am. J. Clin. Nutr.* 53:84-89.

Johansson, S., Lindstedt, S., Register, U., and Wadstrom, L., 1966, Studies on the metabolism of labelled pyridoxine in man, *Am. J. Clin. Nutr.* 18:185-196.

Korte, F., and Bannuscher, H., 1958, Bildung von 4-pyridoxinsäure und 4-pyridoxinsäurelaction aus pyridoxal-5'-phosphat in Leberhomogenat, *Biochem. Z.* 329:451-457.

Lumeng, L., Brashear, R.E., and Li, T.K., 1974, Pyridoxal 5'-phosphate in plasma: source, protein-binding, and cellular transport, *J. Lab. Clin. Med.* 84:334-343.

Mahuren, J.D., and Coburn, S.P., 1990, B-6 vitamers: cation exchange HPLC, *J. Nutr. Biochem.* 1:659-663.

Mahuren, J.D., Pauly, T.A., and Coburn, S.P., 1991, Identification of 5-pyridoxic acid and 5-pyridoxic acid lactone as metabolites of vitamin B-6 in humans, *J. Nutr. Biochem.* 2:449-453.

McCoy, E.E., Colombini, C., and Ebadi, M., 1969, Metabolism of vitamin B-6 in Down's syndrome, *Ann. N. Y. Acad. Sci.* 166:116-125.

O'Brien, D., and Groshek, A., 1962, The abnormality of tryptophane metabolism in children with mongolism, *Arch. Dis. Child.* 37:17-20.

Shane, B., 1970, *Metabolism of Vitamin B-6 in Pregnancy*, Ph.D. thesis, University of London, London.

Tillotson, J.A., Sauberlich, H.E., Baker, E.M., and Canham, J.E., 1967, Use of ^{14}C-labeled vitamins in human nutrition studies. Pyridoxine, *Proc. 7th Int. Congr. Nutr. 1966* 5:554-557.

Wastney, M.E., Patterson, B.H., Linares, O.A., Greif, P.C., and Boston, R.C., 1998, *Investigating Biological Systems Using Modeling: Strategies and Software*, Academic Press, New York.

Waymire, K.G., Mahuren, J.D., Jaje, J.M., Guilarte, T.R., Coburn, S.P., and MacGregor, G.R., 1995, Mice lacking tissue non-specific alkaline phosphatase die from seizures associated with defective metabolism of vitamin B-6, *Nature Genet.* 11:45-51.

Westerhoff, H.V., 2000, Live perspectives of biochemistry, *The Biochemist* 22:43-47.

Whyte, M.P., Landt, M., Ryan, L.M., Mulivor, R.A., Henthorn, P.S., Fedde, K.N., and Coburn, S.P., 1995, Alkaline phosphatase: placental and tissue nonspecific isoenzymes hydrolyze phosphoethanolamine, inorganic pyrophosphate, and pyridoxal 5'-phosphate (substrate accumulation in carriers of hypophosphatasia corrects during pregnancy), *J. Clin. Invest.* 95:1440-1445.

Whyte, M.P., Mahuren, J.D., Vrabel, L.A., and Coburn, S.P., 1985, Markedly increased circulating pyridoxal 5'-phosphate levels in hypophosphatasia, *J. Clin. Invest.* 76:752-756.

CALCIUM UTILIZATION IN YOUNG WOMEN: NEW INSIGHTS FROM MODELING

Meryl E. Wastney, Berdine R. Martin, Rebecca J. Bryant, and Connie M. Weaver[*]

INTRODUCTION

Low bone mass is associated with increased risk of fracture or osteoporosis (Figure 1). Bone mass increases with age to a peak that, depending on skeletal site, ranges from 16 years to the second or third decade of life. Subsequently, bone mass plateaus and then decreases gradually throughout life at rates of approximately 1%/year. Because bone mass is largely established during adolescence, osteoporosis has been termed a pediatric disease (Chesnut, 1988). One way to focus on improving bone health is to increase our understanding of calcium metabolism in young subjects.

The most significant factor that determines peak bone mass in an individual is genetic potential; this may be modified by exercise and diet (Matkovic et al., 1990). Specifically, calcium intake during periods of rapid bone growth is an important factor in determining whether subjects achieve their maximum peak bone mass. Two questions that are of interest are: how does calcium metabolism differ in subjects during rapid *vs.* slow bone accretion, and how do higher levels of calcium intake improve bone retention? To address these questions, we performed calcium balance experiments (Weaver et al., 1995; Jackman et al., 1997) in combination with tracer kinetic studies. Here we will describe how compartmental modeling was used to analyze kinetic data from those studies. We will review two experiments on calcium kinetics in girls *vs.* women (Wastney et al., 1996) and in girls on a low *vs.* high calcium intake (Wastney et al., 2000). Then we will show how, by calculating rates of calcium absorption, distribution, excretion, and bone turnover, we identified differences between girls and women as well as processes that promote increases in bone retention. Modeling challenges that were encountered in analyzing the data will also be discussed.

---

[*] Meryl E. Wastney, Metabolic Modeling Services Ltd., Hamilton 2030, New Zealand. Berdine R. Martin, Rebecca J. Bryant, and Connie M. Weaver, Department of Food and Nutrition, Purdue University, West Lafayette, IN 47907.

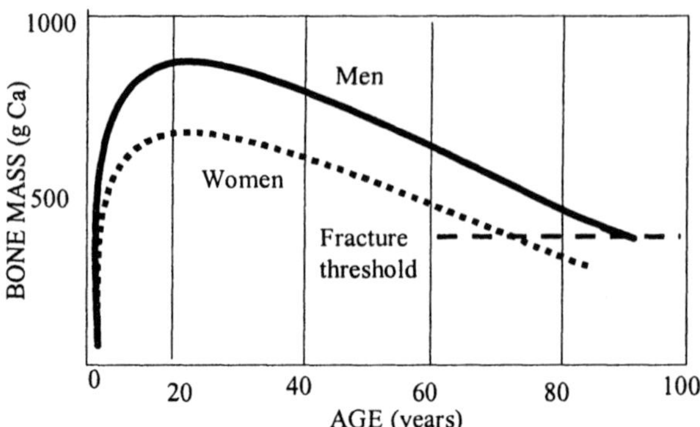

Figure 1. Schematic of change in bone mass with age showing the increase in bone mass during adolescence and the higher peak bone mass achieved in men than women. It also shows the similar rates of bone loss with age in both genders and how women reach a low bone mass, or 'fracture threshold,' at an earlier age than men. Adapted with permission from Grundy (1994).

METHODS

Protocol for Study 1: Calcium Kinetics in Girls *vs.* Young Women

To identify differences in bone retention between subjects with rapid and slow bone accretion, calcium kinetics were studied in white girls [13 ± 1 yr (mean ± SD), n=14] and young women (22 ± 4 yr, n=11). Subjects were studied while they consumed a controlled diet containing 1330 mg Ca/d during a 21-day camp conducted on the Purdue University campus. The diet was chosen so that calcium intake was close to the recommended daily allowance of 1200 mg/d for young women. Calcium was consumed in three meals: 250 mg with breakfast and 500-600 mg each at lunch and dinner. The ratio of polyethylene glycol (PEG, a non-absorbed marker) to calcium in feces did not change after one week, indicating that subjects had adjusted to the diet after 7 days (Weaver et al., 1995). On the eighth day of camp, subjects were admitted to the General Clinical Research Center at Indiana University Medical Center. They were given two stable isotope tracers (Oak Ridge National Laboratory, TN): the oral tracer (^{44}Ca) was administered as $CaCO_3$ in solid form (~38 mg Ca) with 250 mg Ca at breakfast. One h later, subjects were given intravenous tracer (^{42}Ca) administered as $CaCl_2$ (~40 mg Ca) in 10 mL saline through an intravenous catheter. The catheter was flushed with saline, and patency was maintained with saline and heparin. Subjects remained in the Clinical Research Center for 6 h of sampling and then returned to the Purdue campus. Frequent blood samples and complete collections of urine and feces were obtained for the 14 days after tracer administration. Samples were analyzed for calcium by atomic absorption spectrophotometry and for isotope enrichment by fast atom bombardment mass spectrometry. Further experimental details, including diet composition, are included in Weaver et al. (1995).

Protocol for Study 2: Calcium Kinetics During Low and High Calcium Intake

To identify the processes by which high calcium intake promotes bone retention, calcium kinetics were studied in white girls (12 ± 1 yr, n=10) consuming two calcium intakes: low (860 mg/d) and high (1900 mg/d). The studies lasted for 21 days and were conducted in the same format as described for Study 1. Each girl was studied twice in a randomized crossover design, with the two studies being separated by one month (Wastney et al., 2000). The high calcium intake was achieved by including fruit-flavored beverages containing calcium citrate malate (Procter and Gamble, Cincinnati, OH). The ratio of PEG to calcium in feces did not change for either the low or high calcium diets after one week, indicating that the subjects adjusted to the diet in the initial 7 days (Jackman et al., 1997). Tracers were administered as described in Study 1, except that the oral tracer was administered with one-third of the daily calcium intake. The sampling protocol and analytical procedures were the same as described for Study 1.

Compartmental Analysis

Stable isotope enrichment was converted to mg of isotope and expressed as percentage of dose. Stable isotope data from serum, urine, and feces were analyzed using the WinSAAM software (Boston et al., 1998) and compartmental modeling (Wastney et. al, 1996, 2000). The following notation is used. Fractional transfer coefficients [L(I,J), fraction/d] are defined as the fraction of compartment J which flows into compartment I per unit time. Steady state parameters are compartment masses [M(I), mg] and transport rates [R(I,J), mg/d]; rates are defined as the product of fractional transfer [L(I,J)] and compartment mass [M(J), g or mg]: R(I,J) = L(I,J) * M(J).

A model developed by Neer et al. (1967) and employed by many groups for analyzing calcium metabolism (O'Brien et al., 1998; Specker et al., 1994) was used (Figure 2).

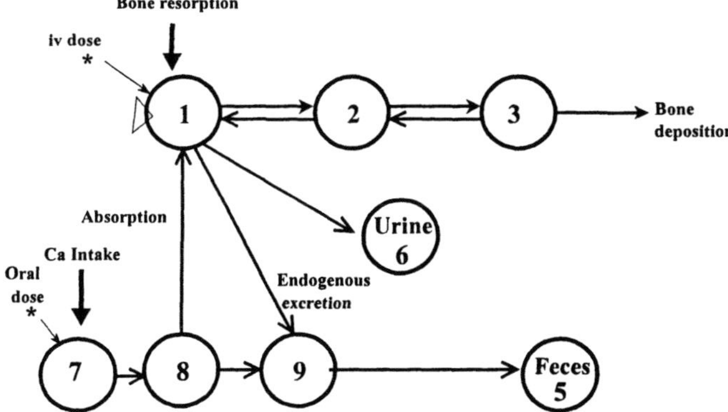

Figure 2. Model for calcium metabolism. Circles represent compartments, numbers in circles represent compartment number, thin arrows represent movement between compartments, heavy arrows represent entry of calcium via the diet or bone resorption, and asterisks represent sites of isotope administration. Calcium loss occurs by excretion into urine, endogenous excretion into feces, and by deposition into bone. Compartment 1 includes serum, compartment 2 is considered to be soft tissue, and compartment 3 represents exchangeable calcium on bone.

Compartments 1, 2, and 3 do not correspond directly to anatomic spaces but are considered to represent calcium in blood and extracellular fluid, soft tissues, and an exchangeable compartment on bone, respectively. Compartment 5 is feces, and compartment 6 is urine. Losses occur by excretion into urine, endogenous excretion into feces (via compartment 9), and deposition into bone. Compartment 7 is a short delay in the intestine prior to absorption from compartment 8. Compartment 9 is a delay before calcium is excreted in feces. Bone contains 99% of body calcium, and bone turnover is much longer than the length of the study. Therefore, calcium deposition into bone was represented as a loss pathway, and calcium re-entering the system by bone resorption was treated as an input of calcium into compartment 1. Calcium also enters serum by absorption from the diet [i.e., via L(1,8)].

Data from each subject were fit individually by the compartmental model (Figure 2) using as starting values the parameters determined by Neer et al. (1967) for healthy adults. These values were adjusted until the model fit the data for each subject. The fitting of a set of data is described in detail by Wastney et al. (1999a). Once the model was fit to the tracer data, pool sizes were calculated in WinSAAM from the initial volume of distribution and the serum concentration of calcium. This information, together with the parameter values and calcium intake, defined the other pool sizes and transport rates, including the rate of bone resorption. Goodness of fit was assessed graphically and statistically by the sum of squares of the errors between the calculated and observed values and by how well parameters were determined (based on their standard deviation).

For Study 1, after the data from each subject had been fit to the model, data from all subjects were submitted to Extended Multiple Studies Analysis (EMSA) in WinSAAM (Lyne et al., 1992; Wastney al., 1999b) to determine values for the population of girls *vs.* women. Fractional transfers, pool sizes, and transport rates in girls were compared to women using t-tests. In Study 2, the model was fit to the data for each subject on the high and low intakes simultaneously, and then the fractional transfers, pools sizes, and transport rates were compared using paired t-tests.

Modeling Challenges

Challenges in modeling kinetic data from Study 1 (girls *vs.* women) were to:

- fit data for two tracers (oral and iv). This was accomplished by using identical models, one with an initial condition in the intestine (representing the oral tracer) and the other with an initial condition in serum (representing the iv tracer). Oral and iv tracer data were then fit simultaneously using two models (one representing the iv experiment and another for the oral experiment). The pathways and their values were the same for both models as it was assumed that once absorbed, tracer calcium would be metabolized in the same way as calcium administered iv. Only the initial conditions or the site into which tracer was introduced into the system differed between the models. (Note that, using WinSAAM, the data could have been fitted alternatively by using one model and the T-interrupt feature. This can be used to reset time and initial conditions.)

- fit tracer data from three sites (serum, urine, and feces). This was accomplished by first fitting the iv data (for distribution of calcium in the body and endogenous fecal excretion) and then the oral data (for defining the absorption pathway).
- resolve a difference in appearance of tracer and tracee data. One problem was encountered in fitting the tracer and tracee data in that the value for absorption for tracer data predicted that too little calcium was excreted in feces. As this occurred in all the subjects, we allowed tracer absorption to differ from tracee absorption. We considered that absorption of tracer (which was administered after an overnight fast) might be higher than absorption of calcium from meals served later in the day. This phenomenon has been demonstrated by Birge et al. (1969).
- compare the results from two populations. This was accomplished by using EMSA (Lyne et al., 1992; Wastney al., 1999b). This option is available on the menu in WinSAAM. The user specifies the subjects to be included in the population study, and the program calculates population values and their statistical error.

Challenges in fitting kinetic data from girls studied at two levels of calcium intake (Study 2) were to:

- simulate two studies per subject when experiments were performed in a randomized order. Both studies were modeled simultaneously using the time-change facility in WinSAAM. The time-change was used to reset time at the start of the second study back to zero while retaining the tracer values in each compartment.
- determine carryover of tracer between studies. With the time-change approach, the model was used to determine the amount of tracer remaining in the body at the start of the second study.
- identify parameters that were different when calcium intake was higher. Parameters were identified by fitting data from the first study and then adjusting the minimum number of parameters required to fit the kinetics observed in the second study. In this way, only the minimum number of parameters that were necessary and sufficient to explain the kinetics on the two intakes differed.

RESULTS

Data from serum, urine, and feces (Figure 3) showed differences in calcium kinetics between girls and women (Study 1). Notably, the girls lost tracer more rapidly from serum and excreted less calcium in urine than women. The volume of distribution was 10 L based on dilution of the iv dose (~10 %/L; Figure 3). Absorption of tracer did not differ for girls *vs.* women (49 *vs.* 44%), but absorption of calcium from the diet was significantly higher (38 ± 18 *vs.* 22 ± 9%) (Wastney et al., 1996). As a result, girls absorbed 74% more calcium than women when consuming 1330 mg calcium/d (494 *vs.* 283 mg Ca/d; Table 1).

The kinetics showed that, in girls, parameter values for transfer from bone compartments back to serum [L(2,3) and L(1,2)] were lower than in women, indicating that calcium movement was toward the bone compartments (Figure 4). In contrast, in women, calcium was moved away from the bone compartments to blood, and from

3 (A)

SERUM
SERUM ⁴⁴Ca (%dose/L)
women
girls
TIME (days)
SERUM ⁴²Ca (%dose/L)
women
girls
TIME (days)
SERUM ⁴²Ca (%dose/L)
women
girls
TIME (days)

3 (B)

URINE
URINE ⁴⁴Ca (% of dose)
women
girls
URINE ⁴²Ca (% of dose)
women
girls
URINE Ca (mg)
women
girls
TIME (days)

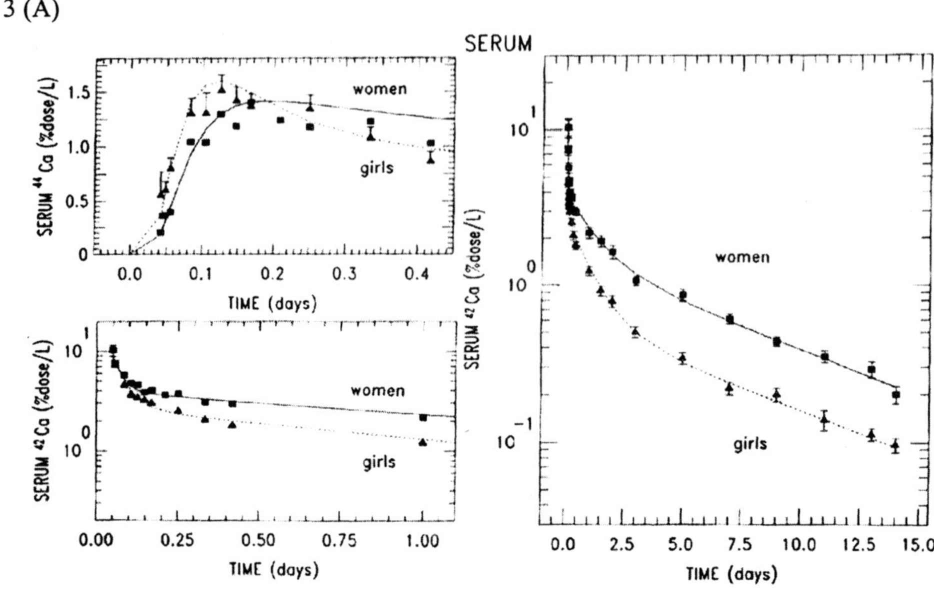

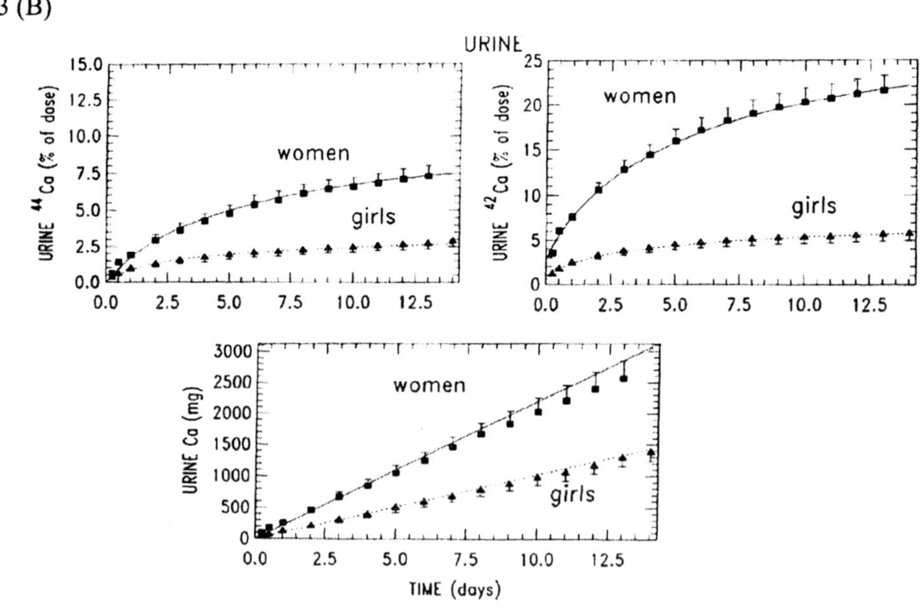

3 (C)

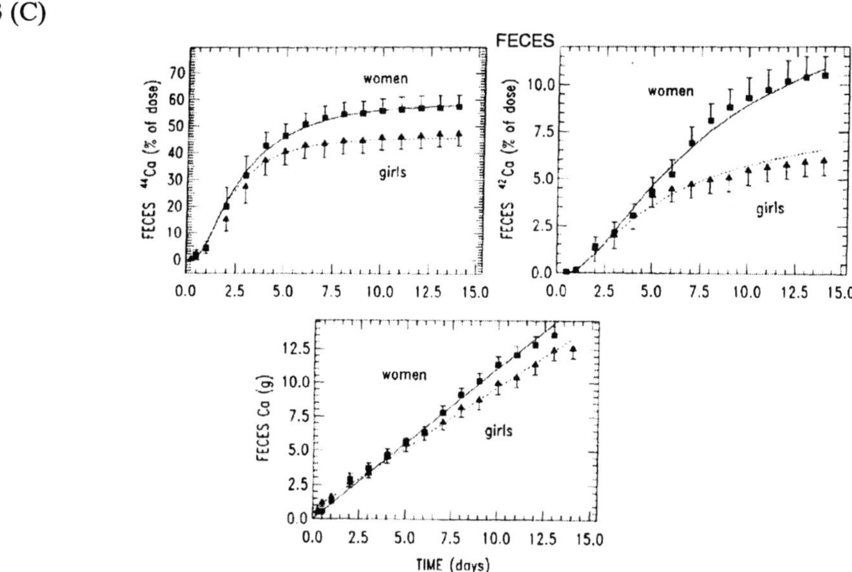

Figure 3. Study 1. Stable isotope and calcium data in girls (triangles) and women (boxes) after oral administration of ^{44}Ca orally and ^{42}Ca iv. Error bars (mean ± SE) are shown (but in some cases, they do not extend beyond the symbols). (A) Serum over 0.5 days (top left), 1 day (lower left), and the whole period of the study (right). (B) Cumulative excretion in urine of oral tracer (top left), iv tracer (top right), and total calcium (lower). (C) Cumulative excretion in feces of oral tracer (top left), iv tracer (top right), and total calcium (lower). Reprinted with permission from Wastney et al. (1996).

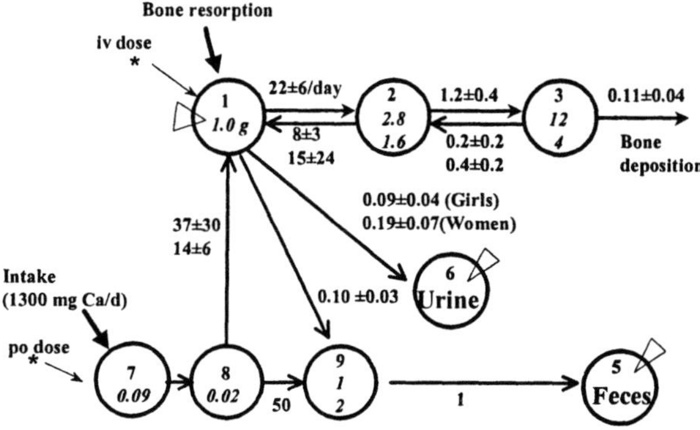

Figure 4. Model for calcium kinetics with parameter values (mean ± SD) for girls (top values) and women (bottom values). Single values indicate that differences were not significant (P>0.05, F test) between girls and women, and the value for girls is shown. Values next to the arrows without an error term indicate that the parameter value was fixed for girls and women. Values in circles are pool sizes (mg), values next to arrows are fractional transfer coefficients (fraction/d), and the triangles identify compartments that were sampled. Reprinted with permission from Wastney et al. (1996).

there, it was twice as likely to be excreted in urine than it was in girls (0.19/d *vs.* 0.09/d; Figure 4). Even though the fraction deposited in bone [L(0,3)] did not differ between girls and women (0.11/d), the rate of bone deposition was higher in girls (1459 *vs.* 501 mg/d; Table 1) due to the three-fold larger mass of compartment 3 (Figure 4). Girls had higher rates of bone turnover, but they retained 280 mg Ca/d *vs.* 0 in women (Table 1). When rates of bone deposition were plotted as a function of post-menarcheal age, it was found that the rate of bone deposition decreased exponentially with age (Figure 5). In summary, when consuming 1330 mg Ca/d, girls absorbed 74% more calcium than women, excreted 50% less calcium in urine (Figure 3B), and retained 280 mg calcium/d *vs.* 0 in young women.

Having established that 13 yr-old girls have a higher capacity to absorb and retain calcium than young women (Table 1), we then investigated the processes by which high calcium intake promotes bone retention (Study 2). To identify these processes, we compared calcium kinetics in girls while they consumed different calcium intakes: 860 mg (low) *vs.* 1900 mg (high) Ca per day (Wastney et al., 2000). It was possible to fit the kinetic data by allowing only one parameter, fractional urinary excretion, to change for the low *vs.* high calcium intakes (Figure 6). Even though other fractional transfers did not change, girls absorbed twice as much calcium on the high intake. Bone retention was increased by 450 mg/d compared to the low calcium intake (Table 1). Pool sizes did not change, as the total flow of calcium into and out of compartment 1 remained similar (Figure 6). Additional calcium absorbed on the high calcium intake was apparently matched by a reduction in bone resorption. This result was consistent with the qualitative changes in bone markers which showed no change in those associated with bone deposition but a decrease in urinary hydroxyproline, which is associated with bone resorption (Wastney et al., 2000). In summary, adolescent girls on 1900 *vs.* 860 mg calcium/d absorbed twice as much calcium and excreted 35% more calcium in urine but retained four-fold more calcium (Figure 6 and Table 1).

Table 1. Calcium utilization (mg/d) in girls and young women[1]

Age (yr)	(n)	Ca Intake	Absorption	Bone Deposition	Bone Resorption	Retention
13	14	860	320 ± 100	1528 ± 576	1400 ± 560	130 ± 144
13	14	1300	494 ± 232	1459 ± 542	1177 ± 436	282 ± 269
13	14	1900	784 ± 300[a]	1536 ± 552	960 ± 388[a]	580 ± 356[a]
20	11	1300	283 ± 122[b]	501 ± 129[b]	542 ± 212[b]	-41 ± 165[b]

[1]Data are means ± SD.

[a]Differ significantly (P<0.05, paired t-test) from girls on 860 mg/d (Wastney et al., 2000).

[b]Differ significantly (P<0.05, t-test) from girls on 1300 mg/d (Wastney et al., 1996).

DISCUSSION

Methodology

Tracer studies of calcium kinetics have been performed since the 1950s, initially in patients (Bellin and Laszlo, 1953; Blau et al., 1957) and later in healthy adults (Neer et al., 1967; Birge et al., 1969; Phang et al., 1969). Kinetic studies using stable isotopes have been performed in additional populations, including infants and children (Yergey et al., 1990; Abrams et al., 1999), adolescents (Abrams et al. 1995, 1996), pregnant women (Heaney and Skillman, 1971), lactating women (Specker et al., 1994), and astronauts (Smith et al., 1999).

Studies have varied in the type of data collected, number of data points, the length of study, and the extent of dietary control. The type of data obtained depends on the aspect of metabolism being studied. For clinical determination of absorption, a single serum sample (Heaney and Recker, 1985) or a 24-h urine sample (Yergey et al., 1994) have been used. For broader description of calcium metabolism, samples have been obtained from additional sites including serum, urine and feces, saliva as a 'window' on serum (Smith et al., 1996), and sweat (Charles et al., 1983). In terms of length of study, absorption studies have typically lasted 24 hours or less. Bone deposition has been determined in studies of up to 6 days in children (Abrams et al., 1996), but it has been proposed that, for adults, studies should be at least 20 days (Jung et al., 1978).

Studies have also varied in terms of diet and lifestyle control - from those performed in clinical units where food intake was totally regulated and activity minimized (Phang et al., 1969) to community-based studies. Our experiments were designed so the subjects were free-living (i.e., exposed to sunlight, exercise, and other factors that affect calcium metabolism) but under dietary control. Some studies have been cross-sectional (Abrams et al., 1996) or, as described for Study 2 here, longitudinal with experiments being repeated in the same individual. The latter approach limits individual variability for assessing dietary effects. O'Brien et al. (1998) performed an intergenerational study, limiting genetic variability, to determine differences related to incidence of osteoporosis.

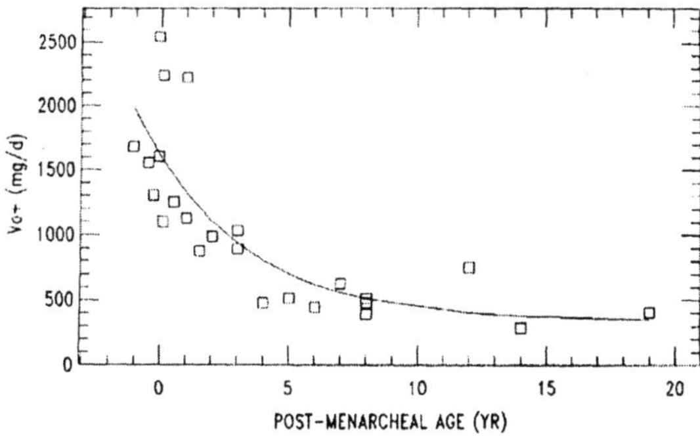

Figure 5. Change in bone deposition with post-menarcheal age. Reprinted with permission from Wastney et al. (1996).

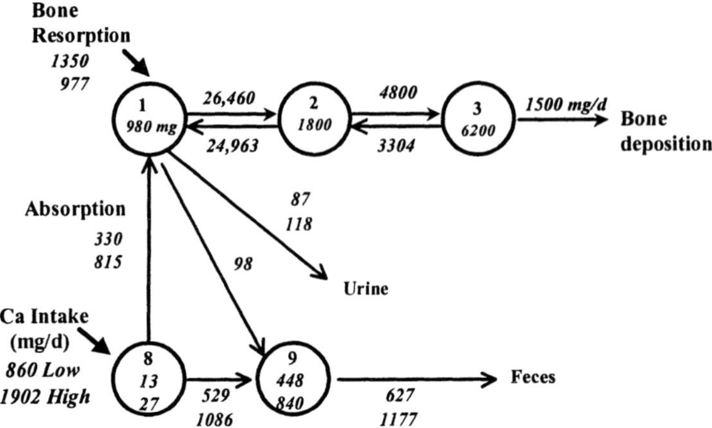

Figure 6. Study 2. Rates of calcium metabolism (mg/d) and pool sizes (mg) for 13 yr-old girls while on low (860 mg, upper values) or high (1900 mg, lower values) calcium intake per day. Reprinted with permission from Wastney et al. (2000). Note: these values were determined using the average parameter values for the population, and they differ slightly from those determined by averaging the rates calculated for each subject (Table 1).

Several multicompartmental models have been proposed for calcium metabolism. These range in complexity from 2 to 5 compartments (reviewed by Aubert et al. 1963), and they have various interpretations (Neer et al., 1967; Weiss et al., 1994). In addition, non-compartmental models (e.g., power functions) have been used to analyze kinetic data. Results of compartmental *vs.* non-compartmental analysis have been described (see Heaney, 1963; Jung et al., 1978).

Insights and Implications

Our kinetic studies showed that 13 year-old girls *vs.* 22 year-old women absorb and retain more calcium, and that bone deposition decreases rapidly within 2 years after menarche. Bronner and Abrams (1998) reported this sharp decrease after puberty in a cross-sectional study of girls aged 5-16 years and also showed how the rate of deposition increased as girls approached the age of puberty.

Our studies identified specific pathways between pools that differed between girls who are in a rapid phase of bone accretion and young women in whom growth is slowed. The flows of calcium from the bone-exchanging pools back to serum were slower in girls. The reasons for this may relate to one or more hormones associated with puberty (e.g., growth hormones). By identifying which hormones influence these pathways, perhaps through animal studies, new approaches may be suggested for modifying the pathways and, in this way, bone retention.

While many studies have shown that higher calcium intake improves calcium balance (Jackman et al., 1997) and bone health (Heaney, 2000), kinetic studies have shown that the increased retention in girls is due to a higher rate of calcium absorption and suppression of bone resorption (Wastney et al., 2000). A possible mechanism for the decrease in bone resorption may relate to the bone remodeling transient. Heaney (2000) states that calcium supplementation in adults increases parathyroid hormone (PTH), and this reduces the rate at which new bone resorption sites are established. The increase in bone mass with increased calcium intake is therefore due to an imbalance between the numbers of bone resorption *vs.* deposition sites. Heaney suggests that changes in bone mass can only be evaluated after a year of supplementation (or after 3-6 months in adolescents) once the balance is restored in number of remodeling sites. In our studies, PTH did not change in response to the higher calcium intake (Wastney et al., 2000). Further, in children, especially those in a state of rapid growth as were our subjects, the impact of the remodeling transient related to changing intake is overshadowed by their rapidly changing rates of calcium accretion. Unlike the situation in adults, bone remodeling in growing females is not in balance even without a change in intake.

With regard to long-term retention, in prepubertal children, increases in bone mass induced by high calcium intake were maintained for 3 years, although no differences in bone mass were observed in pubertal children (Johnston et al., 1992). There is evidence that, in children, the increased bone mass is not always retained with cessation of higher calcium intake (Slemenda et al., 1997). However, epidemiological studies have shown that higher calcium intake in childhood is associated with higher bone mass in adulthood (Nieves et al., 1995; Sandler et al., 1985). Further, lower bone density and higher proximal femur fracture rates were reported in one district in Yugoslavia compared to another that had a two-fold higher calcium intake (Matkovic et al., 1979). Kinetic studies in subjects supplemented for longer periods could add to our understanding of the individual variability and mechanisms involved in maintaining higher retention.

ACKNOWLEDGEMENTS

These studies were supported by NIH grant R01-AR-40553.

CORRESPONDENCE

Please address all correspondence to:
Meryl E. Wastney
Metabolic Modeling Services, Ltd
PO Box 23008
Dalesford
Hamilton 2030
New Zealand
Wastneym@netscape.net

REFERENCES

Abrams, S.A., O'Brien, K.O., Liang, L.K., and Stuff, J.E., 1995, Differences in calcium absorption and kinetics between black and white girls aged 5-16 years, *J. Bone Min. Res.* 10:829-833.

Abrams, S.A., O'Brien, K.O., and Stuff, J.E., 1996, Changes in calcium kinetics associated with menarche, *J. Clin. Endocrinol. Metab.* 81:2017-2020.

Abrams, S.A., Copeland, K.C., Gunn, S.K., Stuff, J.E., Clarke, L.L., and Ellis, K.J., 1999, Calcium absorption and kinetics are similar in 7- and 8-year-old Mexican-American and Caucasian girls despite hormonal differences, *J. Nutr.* 129: 666-671.

Aubert, J.-P., Bronner, F., and Richelle, L. J., 1963, Quantitation of calcium metabolism theory, *J. Clin. Invest.* 42:885-897.

Bellin, J., and Laszlo, D., 1953, Metabolism and removal of Ca^{45} in man, *Science* 117:331-334.

Birge, S.J., Peck, W.A., Berman, M., and Whedon, G.D., 1969, Study of calcium absorption in man: a kinetic analysis and physiologic model, *J. Clin. Invest.* 48:1705-1713.

Blau, M., Spencer, H., Swernov, J., Greenberg, J., and Laszlo, D., 1957, Effect of intake level on the utilization and intestinal excretion of calcium in man, *J. Nutr.* 61:507-521.

Boston, R.C., Greif, P., Wastney, M., and Linares, O., 1998, Balancing needs, efficiency and functionality in the provision of modeling software: a perspective of the NIH WinSAAM project, *Adv. Exp. Med. Biol.* 445:3-20.

Bronner, F., and Abrams, S.A., 1998, Development and regulation of calcium metabolism in healthy girls, *J. Nutr.* 128:1474-1480.

Charles, P., Taagehoj Jensen, F., Mosekilde, L., and Hvid Hansen, H., 1983, Calcium metabolism evaluated by ^{47}Ca kinetics: estimation of dermal calcium loss, *Clin. Sci.* 65:415-422.

Chestnut, C.H. III, 1988, Is osteoporosis a pediatric disease? Peak bone mass attainment in the adolescent female, *Public Health Reports* Suppl 50-54.

Grundy, G.R.,1994, Boning up on genes, *Nature* 367:216-217.

Heaney, R.P., 1963, Evaluation and interpretation of calcium-kinetic data in man, *Clin. Orthoped.* 31:153-183.

Heaney, R.P., 2000, Calcium, dairy products and osteoporosis. *J. Am. Coll. Nutr.* 19: 83S-99S.

Heaney, R.P., and Skillman, T.G., 1971, Calcium metabolism in normal human pregnancy, *J. Clin. Endocrinol.* 33:661-670.

Heaney, R.P., and Recker, R.R., 1985, Estimation of true calcium absorption. *Ann. Int. Med.* 103:516-521.

Jackman, L.A., Millane, S.S., Martin, B.R., Wood, O.B., McCabe, G.P., Peacock, M., and Weaver, C.M., 1997, Calcium retention in relation to calcium intake and postmenarcheal age in adolescent females, *Am. J. Clin. Nutr.* 66:327-333.

Johnston, C.C. Jr., Miller, J.Z., Slemenda, C.W., Reister, T.K., Hui, S., Christian, J.C., and Peacock, M., 1992, Calcium supplementation and increases in bone mineral density in children, *N. Engl. J. Med.* 327:82-87.

Jung, A., Bartholdi, P., Mermillod, B., Reeve, J., and Neer, R., 1978, Critical analysis of methods for analysing human calcium kinetics, *J. Theoret. Biol.* 73:131-157.

Lyne A., Boston, R., Pettigrew, K., and Zech, L., 1992, EMSA: a SAAM service for the estimation of population parameters based on model fits to identically replicated experiments, *Comput. Methods Progr. Biomed.* 38:117-151.

Matkovic, V., Kostial, K., Simonovic, I., Buzina, R., Brodarec, A., and Nordin, B.E.C., 1979, Bone status and fracture rates in two regions of Yugoslavia, *Am. J. Clin. Nutr.* 32:540-549.

Matkovic, V., Fontana, D., Tominac, C., Goel, P., and Chesnut, C. H. III, 1990, Factors that influence peak bone mass formation: a study of calcium balance and the inheritance of bone mass in adolescent females, *Am. J. Clin. Nutr.* 52:878-888.

Neer R., Berman, M., Fisher, L., and Rosenberg, L.E., 1967, Multicompartmental analysis of calcium kinetics in normal adults males, *J. Clin. Invest.* 46:1364-1379.

Nieves, J.W., Golden, A.L., Siris, E., Kelsey, J.L., and Lindsay, R., 1995, Teenage and current calcium intake are related to bone mineral density of the hip and forearm in women aged 30-39 years, *Am. J. Epidemiol.* 141:342-351.

O'Brien, K.O., Abrams, S.A., Liang, L.K., Ellis, K.J., and Gagel, R.F., 1998, Bone turnover response to changes in calcium intake is altered in girls and adult women in families with histories of osteoporosis, *J. Bone Min. Res.* 13:491-499.

Phang, J.M., Berman, M., Finerman, G.A., Neer, R.M., Rosenberg, L.E., Hahn, T.J., Fisher, L., and Granger, A., 1969, Dietary perturbation of calcium metabolism in normal man: compartmental analysis, *J. Clin. Invest.* 48:67-77.

Sandler, R.B., Slemenda, C.W., LaPorte, R.E., Cauley, J.A., Schramm, M.M., Barresi, M.L., and Kriska, A.M., 1985, Postmenopausal bone density and milk consumption in childhood and adolescence, *Am. J. Clin. Nutr.* 42:270-274.

Slemenda, C.W., Peacock, M, Hui, S., Zhao, L., and Johnston, C.C. Jr., 1997, Reduced rates of skeletal remodelling are associated with increased bone mineral density during the development of peak skeletal mass, *J. Bone Min. Res.* 12:676-682.

Smith, S.M., Nyquist, L.E., Shih, C.-Y., Wiesmann, H., Nillen, J.L., Wastney, M.E., and Lane, H.W., 1996, Calcium kinetics using microgram stable isotope doses and saliva sampling, *J. Mass Spect.* 31:1265-1270.

Smith, S.M., Wastney, M.E., Morukov, B.V., Larina, I.M., Nyquist, L.E., Abrams, S.A., Taran, E.N., Chih, C.-Y., Nillen, J.L., Davis-Street, J.E., Rice, B.L., and Lane, H.W., 1999, Calcium metabolism before, during, and after a 3-month space flight: kinetic and biochemical changes, *Am. J. Physiol.* 46:R1-R10.

Specker, B.L., Vieira, N.E., O'Brien, K.O., Ho, M.L., Heubi, J.E., Abrams, S.A., and Yergey, A.L., 1994, Calcium kinetics in lactating women with low and high calcium intakes, *Am. J. Clin. Nutr.* 59:593-599.

Wastney, M.E., Ng, J., Smith, D., Martin, B.R., Peacock, M., and Weaver, C.M., 1996, Differences in calcium kinetics between adolescent girls and young women, *Am. J. Physiol.* 40:R208-R216.

Wastney, M.E., Patterson, B.H., Linares, O.A., Greif, P.C., and Boston, R.C., 1999a, Starting modeling and developing a model, in: *Investigating Biological Systems Using Modeling: Strategies and Software,* p. 223-236, Academic Press, New York.

Wastney, M.E., Patterson, B.H., Linares, O.A., Greif, P.C., and Boston, R.C., 1999b, Multiple studies, in: *Investigating Biological Systems Using Modeling: Strategies and Software,* p. 257-274, Academic Press, New York.

Wastney, M.E., Martin, B.R., Ng, J., Peacock, M., Smith, D., Jiang, X-Y., and Weaver, C.M., 2000, Changes in calcium kinetics in adolescent girls induced by high calcium intake, *J. Clin. Endocrinol. Metab.* 85:4470-4475.

Weaver, C.M., Martin, B.R., Plawecki, K.L., Peacock, M., Wood, O.B., Smith, D.L., and Wastney, M.E., 1995, Differences in calcium metabolism between adolescent and adult females, *Am. J. Clin. Nutr.* 61:577-581.

Weiss, G.H., Goans, R.E., Gitterman, M., Abrams, S.A., Vieira, N.E., and Yergey, A.L., 1994, A non-Markovian model for calcium kinetics in the body, *J. Pharm. Biopharm.* 22:367-379.

Yergey, A.L., Abrams, S.A., Vieira, N.E., Aldroubi, A., Marini, J., and Sidbury, J.B., 1994, Determination of fractional absorption of dietary calcium in humans, *J. Nutr.* 124:674-682.

Yergey, A.L., Abrams, S.A., Vieira, N.E., Eastell, R., Hillman, L.S., and Covell, D.G., 1990, Recent studies of human calcium metabolism using stable isotopic tracers, *Can. J. Physiol. Pharmacol.* 68:973-976.

MODELING CHOLESTEROL IN HUMANS: UPDATE AND DEALING WITH THE PROBLEM OF EXCHANGE *IN VIVO* USING THE BLOOD CELL-LIPOPROTEIN PARADIGM

Charles C. Schwartz, Julie M. VandenBroek, and Patricia S. Cooper[*]

INTRODUCTION: AN UPDATE ON CHOLESTEROL METABOLISM

The cholesterol molecule, like some other lipids, is very weakly soluble in water. It is present in all cells of humans, mainly in membranes and especially in specialized areas of the plasma membrane such as caveolae and rafts. These areas are rich in sphingomyelin and carry out important functions like signaling. Cholesterol is also present in plasma and bile, but it is nearly absent from normal urine and spinal fluid. In plasma, cholesterol is localized in lipoprotein particles; in bile, it is in vesicles and micelles. The common denominators at these locations are phosphatidylcholine and sphingomyelin. Cholesterol is very mobile in this phospholipid environment; it flip-flops across the bilayer with a $t_{1/2}$ of seconds, diffuses laterally with similar speed, and exchanges/transfers between membranes in proximity by aqueous diffusion or with the facilitation of soluble lipid transfer proteins. The reasons why lipoproteins, bile vesicles, and areas of the plasma membrane are cholesterol enriched are active areas of investigation.

Cholesterol can undergo several transformations. It can be reversibly esterified by long-chain fatty acids (such as oleic and linoleic acids) at its 3β-OH position. Cholesteryl ester is very hydrophobic, insoluble in water, and resides in membrane-lined droplets in cells and the inner core of plasma lipoprotein particles. In the skin, cholesterol is often sulfated; this serves a barrier function. Furthermore, two degradation products of cholesterol are vital. These are bile acids and steroid hormones. About 500 and 50 mg, respectively, are formed daily. Adequate amounts of cholesterol can be synthesized in the tissues of adolescents and adults to meet all needs in the absence of

[*] Charles C. Schwartz, Julie M. VandenBroek, and Patricia S. Cooper, Department of Medicine, Medical College of Virginia, Virginia Commonwealth University, Richmond, VA 23298.

dietary cholesterol intake.   Therefore, cholesterol is not considered a nutrient or micronutrient, although this issue is not resolved in infants.

The methods of sterol balance, input-output analysis, and the 3-pool whole-body model used in the 1960s and '70s gave accurate estimates of cholesterol synthesis and absorption in humans.   Our previous criticism (Schwartz, 1982)--that synthesis measured with exogenous isotopes may be underestimated by 20% due to direct biliary secretion of 20% of newly synthesized cholesterol--was retracted.   We found that ~95% of synthesized cholesterol mixes with hepatic and plasma pools (Schwartz et al., 1993).   In normal subjects, <5% of newly synthesized cholesterol is secreted directly into bile or used for bile acid synthesis.   The major tissue sites of synthesis in humans remain an area of contention.   The traditional belief is that all tissues have the capacity for cholesterol synthesis from acetate but that synthesis is normally suppressed in all except liver and intestine.   Work by Dietschy and colleagues in several non-human primates has shown that hepatic synthesis is minor and that tissues such as muscle make significant amounts of cholesterol (Spady and Dietschy, 1983).   This paper focuses on whole-body aspects of cholesterol metabolism in humans, but we cannot omit recent work at the molecular level by Brown, Goldstein, and colleagues.   They have elucidated the regulation of cholesterol synthesis by a family of membrane-bound transcription factors (SREBPs) (Dorn et al., 1998).   This work builds on contributions they made ~20 years ago on the regulation of cholesterol synthesis by LDL receptor endocytosis.

The use of stable isotopes to study cholesterol metabolism (mainly synthesis) in humans became more popular in the mid-1990s, especially in the labs of Schoeller, Jones, Hellerstein, Hachey, Klein, and Schaefer (Wong et al., 1991; Faix et al., 1993; Jones et al., 1994; Lee et al., 1994).   The application of modeling to analysis of the raw data was limited.   The potential for stable isotopes to study human cholesterol metabolism is great but barely realized.

Transport of cholesterol within cells has been a major area of research during the past 15 years and the subject of several reviews (see, for example, Liscum and Underwood, 1995).   In Tangier disease, there is defective efflux to plasma from intracellular cholesterol stores (Francis et al., 1995).   In Niemann-Pick C disease, there is defective transport of low density lipoprotein (LDL)-derived cholesterol to the plasma membrane from lysosomes, whereas newly synthesized and high density lipoprotein (HDL)-derived cholesterol bypass lysosomes and are transported in a normal fashion within the cell (Shamburek et al., 1997).   The plasma membrane is the interface between intracellular and extracellular cholesterol pools.   Over 95% of cholesterol molecules are in unesterified (free) form in normal cells and over 80% reside in the plasma membrane. Cholesterol-rich HDL particles dock on or near the outer leaf of the plasma membrane, promoting rapid one-for-one exchange with membrane cholesterol.   In addition, there may be net transport to (or from) the plasma membrane from (or to) various HDL subclasses.   Transport of intracellular cholesterol is rapid.   Cholesterol that is either synthesized in the cell interior, derived from hydrolysis of cytoplasmic ester droplets, or captured by lysosomes moves to the plasma membrane with a $t_{1/2}$ of 15-30 minutes.

Most recently, genes encoding ATP-binding cassette transporters have been found to participate in cellular cholesterol traffic.   They are ABCA1 (or ABC1) in Tangier disease (Brooks-Wilson et al., 1999) and ABCG5/ABCG8 in sitosterolemia (Berge et al., 2000).

These membrane-bound proteins seem to regulate cell cholesterol efflux and intestinal cholesterol absorption, respectively, among other possibilities.

This update reinforces the concept that all cholesterol in humans is dynamic, especially unesterified (free) cholesterol. There are many pools, many transport and transformation pathways, and many sites at which exchange occurs. This complexity demands carefully designed experiments and comprehensive modeling techniques to study metabolism/kinetics *in vivo*.

THE PROBLEM: EXCHANGE—AN EXAMPLE OF HOW TO QUANTIFY CHOLESTEROL EXCHANGE AND SEARCH FOR NET TRANSPORT ON A BACKGROUND OF EXCHANGE

Exchange is defined as one-for-one transfer of cholesterol molecules between two compartments: $R(i,j) = R(j,i)$. It is the same as bidirectional transport with zero net flux. If there is bidirectional transport between two compartments in addition to net flux to one from the other, then we say that there is net transport on a background of exchange. The background of exchange between any two cholesterol compartments may be large relative to net transport to one from the other, making accurate measurement of net transport (net flux) especially difficult.

The term 'isotope exchange' is ambiguous. It could mean that the isotope is not secure and dissociates from the cholesterol molecule such as occurred with some of the original techniques for ^{3}H-labeling. This is no longer a problem. It could also imply that tracer exchanges, but tracee does not. This is not the case. Considerable evidence has shown that cholesterol tracers fulfill their function. Proof comes from complete recovery of exogenous tracers and findings that endogenous tracers, including precursors of all types, give the same results as exogenous radioactive tracers. Thus, the tracers are tracing the exchange of unlabeled cholesterol molecules.

Exchange has been a thorn in the side of many investigators in the cholesterol field. Studies of absorption using isotopic cholesterol were criticized because of the potential for exchange between ingested tracer in the intestinal lumen and unlabeled mucosal cholesterol. This might give the appearance of absorption (net uptake) when there was none. Grundy and colleagues used dual administration of isotopes among other techniques to clarify this problem (Grundy, 1982). Exchange between slowly- and rapidly miscible pools was estimated in the classic three-pool whole-body cholesterol model. In *in vitro* studies, cholesterol exchange and net transport are estimated by a combination of isotope and mass measurements, usually in a non-steady state.

We have previously described the central role of HDL in free cholesterol transport to liver, blood cells, and other lipoproteins (Schwartz et al., 1982b). Some studies were done with forcing functions (Foster and Boston, 1983) so that transport was determined in only one direction and exchange was possibly overlooked. Based on the experience of others and on insights gained from our previous studies, here we simultaneously administered isotopes in different compartments and then rapidly separated α lipoprotein (HDL), β lipoprotein (VLDL + LDL), and blood cells to prevent *ex vivo* cholesterol movement. Bile was collected as a surrogate for liver radioactivity. The raw data were subjected to multicompartmental analysis and sensitivity testing (Zech, 1982) without forcing functions. Our goal was to quantify cholesterol exchange and net transport involving blood cells.

METHODS

Four ambulatory, healthy volunteer subjects were studied in the Clinical Research Center at the Medical College of Virginia. Informed consent was obtained from each subject. Each had a normal physical examination, blood chemistry panel, and lipid profile. Body weight varied <0.5 kg from 10 days before and through the end of the study. Subjects 3 and 4 were female (post menopausal). Age was 46-65 years, weight 60-78 kg, and BMI 24-26.

Each subject was administered two labeled compounds simultaneously by vein at ~11 a.m. over a 1 min period. The subjects fasted except for water for 14 hours before and 8-10 hours after label administration. Labels were prepared and assayed for purity as described previously (Schwartz et al., 1982b; Schwartz et al., 1993). All administered lipoproteins were autologous. Subject 1 received LDL [4-^{14}C]cholesterol (1.6E+6 dpm) and VLDL [1,2-^{3}H]cholesterol (33E+6 dpm). Subject 2 received LDL [4-^{14}C]cholesterol (1.8E+6 dpm) and HDL [1,2-^{3}H]cholesterol (33E+6 dpm). Subject 3 received VLDL + LDL [1,2-^{3}H]cholesterol (47E+6 dpm) and HDL [4-^{14}C]cholesterol (1.9E+6 dpm). Subject 4 received HDL [1,2-^{3}H]cholesterol (20E+6 dpm) and DL[2-^{14}C]mevalonic acid (76E+6 dpm).

After label administration, blood and bile samples were obtained at 15-20 min intervals for the initial three hours and then less often. Samples were processed as described previously (Schwartz et al., 1982b; Schwartz et al., 1993). Note that blood samples were collected in heparinized tubes and immediately centrifuged to separate plasma from blood cells. The plasma was then immediately treated with heparin-MnCl$_2$ to precipitate β (VLDL + LDL) lipoproteins. The purpose of rapid isolation in the cold is to prevent *ex vivo* transfer/exchange of cholesterol as occurs in the ultracentrifuge (Schwartz et al., 1982b).

Samples were extracted using 2:1 chloroform:methanol. Cholesterol and cholesteryl ester were isolated by column chromatography and/or TLC, mass was determined by HPLC or GC, and ^{3}H and ^{14}C were analyzed by scintillation counting. Final data were expressed as specific activity (dpm/μmol).

Kinetic Analysis

The model was developed by subjecting the experimentally observed mass and specific activity data to multicompartmental analysis using the Simulation, Analysis and Modeling programs (SAAM/CONSAM; Berman and Weiss, 1978; Berman et al., 1983). In each subject, the concentration of cholesterol in blood cells and in α (HDL) and β lipoprotein was constant throughout the experiment, as was that of cholesteryl ester in α and β lipoprotein, supporting a steady state. The standard error of the mean for each concentration was <4% of the mean (n=10-18). Hematocrit was used in the calculation of blood cell cholesterol concentration in plasma. Pool size [M(I)] of the individual plasma compartments was calculated from the observed mean plasma concentration and estimated plasma volume [4.5% of body weight (kg)]. The final solution for each pool size was allowed to vary only ± 5% from the calculated value during iterations. The model was constrained by the simultaneous consideration of all ^{3}H and ^{14}C specific activity data in each subject. This imposes rigid constraints on the topology of the model and narrows the confidence limits on the final parameter solutions since the same kinetic parameters were required to fit both the ^{3}H and ^{14}C data that were obtained after administration of tracers in separate compartments.

RESULTS

Mass Data

See Table 1 for blood cell and lipoprotein cholesterol concentrations.

Specific Activity Data

Figures 1 and 2 show the ^{3}H and ^{14}C specific activities of cholesterol in the various components of blood as a function of time following administration of labeled compounds to subjects 2 and 4, respectively. All normal subjects studied in our lab have followed these patterns. Following administration of unesterified ^{3}H-cholesterol in HDL (Figures 1A and 2A), the HDL (α) cholesterol specific activity rapidly declined for ~40 min at which time an abrupt slowing occurred. The β (VLDL + LDL) cholesterol activity rapidly increased and reached that of HDL (α) within 30-40 min. After 40 min, the β cholesterol specific activity remained slightly higher than α, in a manner compatible with a precursor (α) / product (β) relationship. The α and β activity relationship was quite different after administration of LDL labeled with unesterified ^{14}C-cholesterol (Figure 1B). The β cholesterol specific activity initially declined more slowly, and the α peak specific activity did not intersect that of β cholesterol.

Blood cell cholesterol specific activity increased rapidly during the initial 30 min when isotopic cholesterol was administered in HDL (Figures 1A and 2A) and lagged behind when administered in LDL (Figure 1B). However, in each subject, the specific activities of cholesterol in blood cells and each lipoprotein were equalized by 400-500 min, after which they all declined with blood cell activity slightly higher.

Figure 2B shows cholesterol specific activities following administration of [2-^{14}C]mevalonic acid. The rapid rise in α (HDL) cholesterol activity and the short lag in β cholesterol activity during the initial two hours were previously reported (Schwartz et al., 1982b). After three hours, they declined in parallel, with the α cholesterol activity slightly lower (~5%) than β cholesterol for at least one day. Blood cell cholesterol activity reached that of lipoproteins in ~500 min.

We have previously shown that cholesterol in LDL is metabolized in an identical manner as cholesterol in VLDL (Schwartz et al., 1993). This was confirmed in subject 1, who was simultaneously administered VLDL labeled with ^{3}H cholesterol and LDL with ^{14}C cholesterol. The fraction of ^{3}H administered was identical to that of ^{14}C in bile, α cholesterol, β cholesterol, blood cells, α cholesteryl ester, and β cholesteryl ester in all

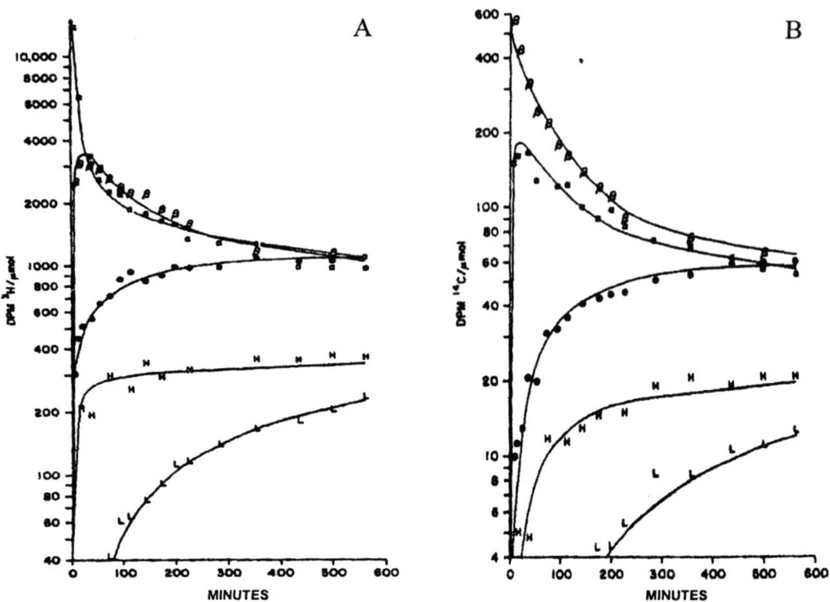

Figure 1. Specific activity-time courses. Subject 2, a male with a bile fistula, was simultaneously administered HDL [1,2-^{3}H]cholesterol (panel A) and LDL [4-^{14}C]cholesterol (panel B). α, α (HDL) cholesterol; β, β (VLDL + LDL) cholesterol; •, blood cell cholesterol; H, α (HDL) cholesteryl ester; L, β (VLDL + LDL) cholesteryl ester. Each line represents the computer simulation using the model shown in Figure 3. All rate constants were adjustable except that L(5,3) was dependent on L(3,5) so that R(3,5) = R(5,3) and R(3,4) = R(4,3) as discussed in the text.

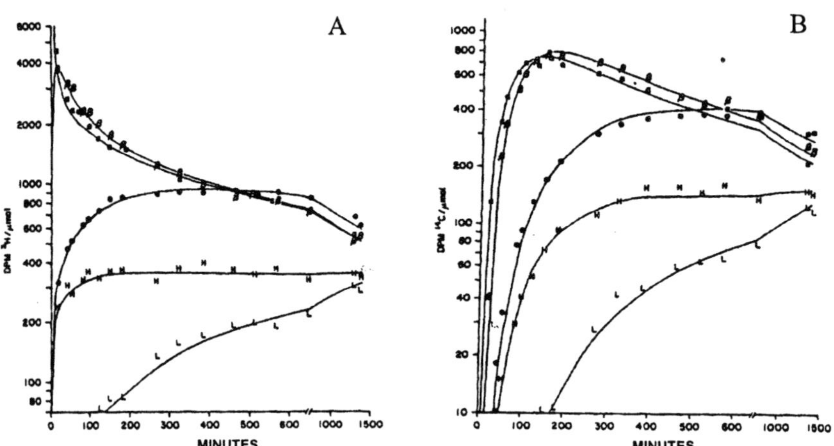

Figure 2. Specific activity-time courses. Subject 4, a normal female, was simultaneously administered HDL [1,2-^{3}H]cholesterol (panel A) and [2-^{14}C] mevalonic acid (panel B). Symbols, lines, and model parameters are as described in the legend to Figure 1.

samples. Thus, LDL and VLDL can be combined as β cholesterol (compartment 5 in the model) for the purpose of kinetic analysis.

The Model

Figure 3 shows the compartmental model. The structure and pathways have been identified previously (Schwartz et al., 1982a; Schwartz et al., 1993) and the current studies conform. During model development, we adhered to the minimal structure principle. The ^{3}H and ^{14}C specific activities were fit using a time-interrupt or by constructing two models running simultaneously, one for ^{3}H and one for ^{14}C, with all parameters of the ^{3}H model equal to those of the ^{14}C model. Both approaches gave identical solutions.

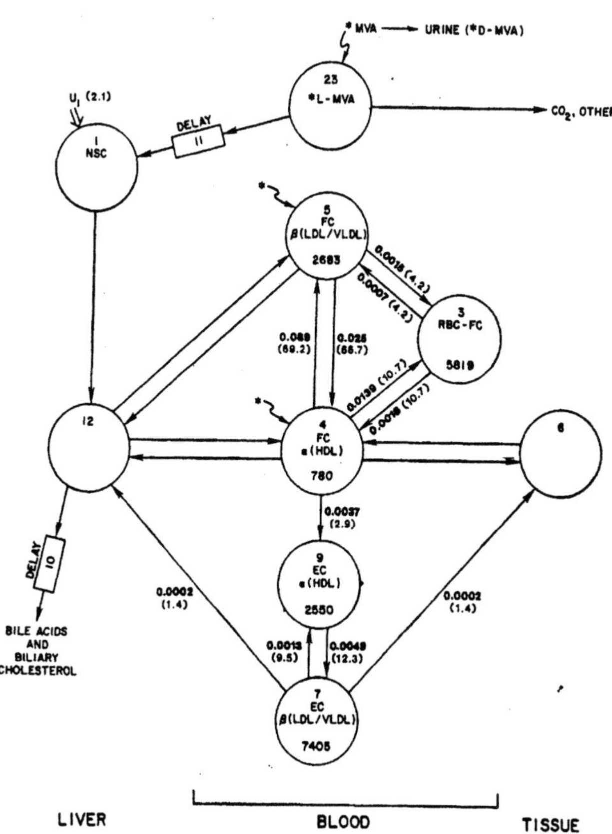

Figure 3. The model. Parameter values shown are from subject 4. Rate constants [L(i,j), min⁻¹] are adjacent to arrows; transports [R(i,j), µmol/min] are in parentheses adjacent to arrows; pool sizes (µmol) are inside the compartments. Pool size was calculated from the mean plasma concentration * plasma volume estimate (4.5% of body weight) and then constrained ± 5% in the solution. U₁ is the input of cholesterol (µmol/min) into compartment 1. NSC, newly synthesized cholesterol; MVA, mevalonic acid; *, labeled compounds administered in this paper; RBC, blood cells; FC, free cholesterol; EC, esterified cholesterol.

BLOOD CELL CHOLESTEROL MODELING AND DISCUSSION

Subjects 2 and 3, in whom isotopic cholesterol in VLDL + LDL and in HDL was simultaneously administered, gave the most rigorous information on simulation of blood cell specific activity data. This was not a surprise in view of the two lipoprotein compartments initially labeled and since blood cells are exposed extensively to these two lipoproteins. Input of cholesterol via nascent cells and loss in senescent cells (0.03-0.42 μmol/min assuming a life span as long as 120 days and as short as 10 days) could not be identified. These processes were negligible relative to the solutions for transport to and from lipoproteins.

An attempt was initially made to simulate the blood cell ^{3}H and ^{14}C specific activity observations using the model shown in Figure 3 with both L(3,5) and L(5,3) fixed at 0. The transport pathways between HDL and blood cells [L(4,3) and L(3,4)] were allowed to adjust during the iterative process as were all other rate constants. Under these conditions, it was not possible in either subject 2 or 3 to simultaneously fit the blood cell ^{3}H and ^{14}C observations with the same set of model parameters. The rate constants could be varied to minimize the difference between calculated and observed values for either ^{3}H observations *or* ^{14}C observations, but no adjustments could be found that resulted in simultaneous minimization of the difference for ^{3}H *and* ^{14}C observations. This dilemma could not be resolved either by introducing subpools of blood cell cholesterol into the topology of the model or by allowing L(5,3) to adjust with L(3,5) fixed at 0. An attempt was then made to simulate the blood cell ^{3}H and ^{14}C specific activity observations by fixing both L(3,4) and L(4,3) at 0 while allowing the transport pathways between VLDL + LDL and blood cells [L(3,5) and L(5,3)] to adjust. This led to the same general inconsistency: when the ^{3}H observations were simulated with minimal error, the ^{14}C observations were not and vice versa. The introduction of subpools of blood cell cholesterol and adjustment of L(4,3) did not resolve the inconsistency.

The next hypothesis tested was that blood cell cholesterol is derived from both HDL and VLDL + LDL cholesterol. All rate constants for pathways shown in Figure 3 were allowed to adjust. Under these conditions, a simultaneous minimum between observed and calculated values for ^{3}H and ^{14}C was obtained in each of the four studies. Pathways L(3,4), L(3,5), and L(4,3) were well defined in all studies and were especially well defined (FSD = 7-18%) in subjects 2 and 3 which offered the most rigorous experimental design to test the blood cell pathways. Pathway L(5,3) was not well defined.

With all rate constants adjustable, the solution for L(5,3) could range from 0 to approximately the same magnitude as L(4,3) without affecting the visible fit or minimal error. Thus, pathway L(5,3) was not adequately defined and would traditionally be eliminated according to the principle of minimal structure. However, L(5,3) was not eliminated for two reasons. First, it has been shown *in vitro* that blood cell cholesterol can be transported directly to LDL (Quarfort and Hilderman, 1970). Second, if L(5,3) was eliminated, our results would be interpreted to show that HDL promotes net transport of cholesterol from bloods cells and/or that VLDL + LDL promotes net transport of cholesterol to blood cells. Therefore, we tested whether it can be positively concluded that HDL promotes net transport from blood cells. It was postulated that blood cell cholesterol transport to and from individual lipoproteins is purely bidirectional with zero net flux. To disprove this postulate, it would be necessary to show that the fit of the data significantly improves when $R(4,3) \neq R(3,4)$. In other words, does the overall fit significantly improve when the ratio $R(4,3) / R(3,4)$ is less or greater than 1.0? To answer this question, the observed ^{3}H and ^{14}C data from each subject were simulated with all rate constants adjustable except $L(4,3)$. $L(4,3)$ was initially made dependent as a function of $L(3,4)$ so that $R(4,3) / R(3,4) = 1.0$. This was feasible in part because pool sizes were tightly constrained to observed values. Solutions were then obtained in each subject by systematically changing the function to yield a range of ratios for $R(4,3) / R(3,4)$. $L(4,3) = fx * L(3,4) * M(4) / M(3)$ where fx is the desired ratio. The absolute values of $L(3,4)$, $L(4,3)$, $R(3,4)$, and $R(4,3)$ were not fixed. The total error (sum of squares) for each solution was divided by the minimal total error for that subject [all $L(i,j)$ adjustable] to give total relative error. The results are shown in Figure 4. Also shown in Figure 4 are the absolute values for $R(4,3)$ and $R(3,4)$ from each solution in subject 3.

In subjects 2 and 3, minimal error was achieved and total relative error was 1.0 when $R(4,3) = R(3,4)$ and therefore when $R(5,3) = R(3,5)$. In subjects 1 and 4, total relative error was only 1.04 and 1.06 fold, respectively, above minimal error when $R(4,3) = R(3,4)$. These small elevations in total error are of doubtful significance to disprove the postulate. As anticipated, identical results were obtained when all rate constants were adjustable except $L(5,3)$ which was made dependent on $L(3,5)$ to vary the ratio $R(5,3) / R(3,5)$. It was concluded that the present experiments do not contain sufficient evidence to show that HDL promotes net transport of cholesterol from blood cells. The postulate could not be disproved. Therefore, $L(5,3)$ was included in the model but made dependent on $L(3,5)$ so that $R(5,3) = R(3,5)$ and $R(4,3) = R(3,4)$.

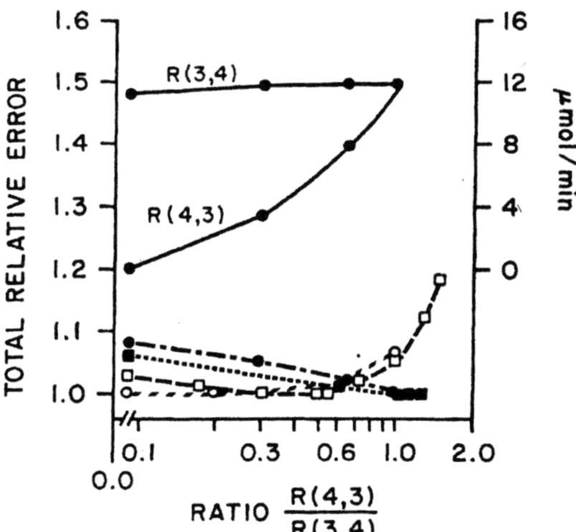

Figure 4. Sensitivity test showing evidence for cholesterol exchange between blood cells and HDL with zero net transport. Total relative error is total error (sum of squares) of each solution with $L(4,3)$ dependent ÷ minimal total error with all $L(i,j)$ adjustable. At a ratio of 1.0, $R(4,3) = R(3,4)$ and $R(5,3) = R(3,5)$, and there is only exchange of cholesterol. □, subject 1; ■, subject 2; ●, subject 3; o, subject 4. Absolute values for $R(4,3)$ and $R(3,4)$ are shown with scale on the right axis for subject 3.

In all above solutions from the four subjects, it was found that cholesterol transfer to blood cells from HDL [R(3,4)] was larger than from VLDL + LDL [R(3,5)]. This occurred even though blood cells were exposed to plasma containing 2.7-6.0 times more cholesterol in VLDL + LDL than in HDL (see Table 1). To evaluate this quandary, we tested the sensitivity of the data to exchange of cholesterol between blood cells and HDL *vs.* blood cells and VLDL + LDL. The observed ^{3}H and ^{14}C data were simulated first with all rate constants adjustable except L(5,3) which was always dependent on L(3,5) [i.e., L(5,3) = L(3,5) * M(5) / M(3)] so that R(5,3) = R(3,5) and R(4,3) = R(3,4) in all solutions. Second, L(3,5) was made dependent as a function of L(3,4) [i.e., L(3,5) = fx * L(3,4) * M(4) / M(5)] to obtain in each subject a series of solutions with a broad range of R(3,5) / R(3,4). The absolute values of L(3,5), L(3,4), R(3,5), and R(3,4) were not fixed. For each solution, the total relative error was plotted against the ratio R(3,5) / R(3,4).

These results, shown in Figure 5, indicate that total relative error in all subjects was lower when blood cell cholesterol exchange with HDL was greater than exchange with VLDL + LDL [when R(3,5) / R(3,4) < 1.0]. At a ratio of 1.0, total error was 1.1-1.3 fold greater than minimal error in the four subjects. At a ratio corresponding to the VLDL + LDL to HDL cholesterol concentration, total error was 1.3-2.5 fold greater. As predicted from the experimental design, data for subjects 2 and 3 were most sensitive to exchange of blood cell cholesterol with HDL *vs.* VLDL + LDL. This is shown by the rapid increase in total relative error as R(3,5) / R(3,4) changed in either direction from the solution with minimal error. The best solutions for subjects 2 and 3 were at ratios of 0.3 and 0.6, respectively. If exchange of blood cell cholesterol with HDL and with VLDL +

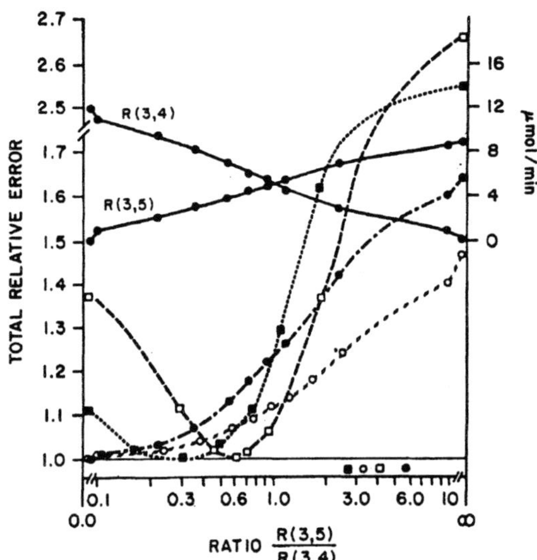

Figure 5. Sensitivity test showing preferential exchange of cholesterol between blood cells and HDL rather than VLDL + LDL. ●, subject 1; ■, subject 2; □, subject 3; o, subject 4. L(5,3) was dependent on L(3,5) so that R(5,3) = R(3,5) and R(4,3) = R(3,4) in all solutions. At a ratio of 0, R(3,5) and R(5,3) = 0; at a ratio of ∞, R(3,4) and R(4,3) = 0. In the lower right, isolated symbols represent the mass ratio of VLDL + LDL cholesterol to HDL cholesterol. Absolute values for R(3,4) and R(3,5) are shown with scale on the right axis for subject 1.

LDL was based on the concentrations of HDL cholesterol and VLDL + LDL cholesterol, the corresponding ratios would have been 2.7 in subject 2 and 4.4 in subject 3 as shown in Table 1 and Figure 5.

CONCLUSIONS

Exchange of many types of molecules such as cholesterol or glucose is a notorious and frequently dreaded problem for investigators of metabolic systems. The term should be used to denote one-for-one transfer of molecules (both tracee *and* tracer) between two compartments (i.e., bidirectional transport with zero net). Here, multicompartmental analysis of blood cell cholesterol in humans *in vivo* is described. Our purpose was to demonstrate examples of sensitivity testing as an approach to quantification of cholesterol exchange and detection of net transport on a background of exchange.

Sensitivity testing was done by allowing all $L(i,j)$ in the model to adjust during the iterative process except for one that was involved in the putative exchange process. This one exception was made dependent as a function of another $L(i,j)$ in the exchange process and the sum of squares determined. The function was varied over a range above and below the function that gave bidirectional transport with zero net. With the blood cell cholesterol paradigm, the sum of squares was at or very close to the minimum possible when there was considerable exchange of cholesterol between blood cells and HDL with zero net transport.

Success in detecting net transport on a large exchange background *in vivo* is predicated on rigorous experimental design that would include frequent observations of radioactivity in all compartments and accurate observations of pool sizes. Also helpful would be simultaneous administration of tracers in two or more compartments. The experimental design described here could be improved by administration of autologous blood cells containing isotopic cholesterol.

The results of our modeling are interpreted to show that, in humans *in vivo*, only one pool of cholesterol is present in blood cells. This implies that cholesterol in erythrocytes is localized almost exclusively to the plasma membrane in which it flip-flops with a $t_{1/2}$ <15 min. The turnover time of cholesterol in blood cells was ~6.5 h, compared to ~1.5 h in liver and ~40 h in other tissues (Schwartz et al., 1993). Furthermore, results of the

Table 1. Mean cholesterol concentrations and the ratio of β (VLDL + LDL) free cholesterol to α (HDL) free cholesterol[1]

Subject	Blood Cell	α(HDL) FC	β(VLDL+LDL) FC	β FC÷ α FC	α(HDL) EC	β(VLDL+LDL) EC
1	164	18.0	107.5	6.0	52.3	270
2	230	35.2	94.1	2.7	105.9	230
3	191	17.6	75.8	4.4	43.1	138
4	179	24.0	82.5	3.4	78.4	228

[1]Cholesterol concentrations are μmol/100 ml of plasma. The SEM of each concentration was <4% of the mean (n=10-18 for each). FC = free cholesterol; EC = esterified cholesterol.

sensitivity testing are in agreement with many *in vitro* studies showing the relative ease of cholesterol exchange between membranes and HDL compared to VLDL + LDL. No evidence was found for net cholesterol transport to HDL or to VLDL + LDL from blood cells.

ACKNOWLEDGMENT

This paper is presented with gratitude for the contributions of Loren Zech (1942-1997).

CORRESPONDENCE

Please address all correspondence to:
Charles C. Schwartz
12130 Springcreek Drive
Midlothian, VA 23113
schwartz@hsc.vcu.edu

REFERENCES

Berge, K.E., Tian, H., Graf, G.A., Yu, L., Grishin, N.V., Schultz, J., Kwiterovich, P., Shan, B., Barnes, R., and Hobbs, H.H., 2000, Accumulation of dietary cholesterol in sitosterolemia caused by mutations in adjacent ABC transporters, *Science* 290:1771-1775.

Berman, M., Beltz, W.F., Greif, P.C., Chabay, R., and Boston, R.C., 1983, *CONSAM User's Guide,* Publication No. 1983-421-132:3279, U.S. Government Printing Office, Washington, DC.

Berman, M., and Weiss, M.F., 1978, *SAAM Manual,* DHEW Publ. #78-180, U.S. Government Printing Office, Washington, DC.

Brooks-Wilson, A., Marcil, M., Clee, S.M., Zhang, L.-H., Roomp, K., van Dam, M., Yu, L., Brewer, C., Collins, J.A., Molhuizen, H.O.F., Loubser, O., Ouelette, B.F.F., Fichter, K., Ashbourne-Excoffon, K.J.D., Sensen, C.W., Scherer, S., Mott, S., Denis, M., Martindale, D., Frohlich, J., Morgan, K., Koop, B., Pimstone, S., Kastelein, J.J.P., Genest, J. Jr., and Hayden, M.R., 1999, Mutations in ABC1 in Tangier disease and familial high density lipoprotein deficiency, *Nature Genet.* 22:336-345.

Faix, D., Neese, R., Kletke, C., Wolden, S., Cesar, D., Coutlangus, M., Shackleton, C.H.L., and Hellerstein, M.K., 1993, Quantification of menstrual and diurnal periodicities in rates of cholesterol and fat synthesis in humans, *J. Lipid Res* 34:2063-2075.

Foster, D.M., and Boston, R.C., 1983, The use of computers in compartmental analysis: The SAAM and CONSAM programs, in: *Compartmental Distribution of Radiotracers,* J.S. Robertson, ed., CRC Press Inc., Boca Raton.

Francis, G.A., Knopp, R.H., and Oram, J.F., 1995, Defective removal of cellular cholesterol and phospholipids by apo A-1 in Tangier disease, *J. Clin. Invest.* 96:78-87.

Grundy, S.M., 1982, Role of isotopes for determining absorption of cholesterol in man, in: *Lipoprotein Kinetics and Modeling,* M. Berman, S.M. Grundy, and B.V. Howard, eds., Academic Press, New York.

Jones, P.J.H., Lichtenstein, A.H., and Schaefer, E.J., 1994, Interaction of dietary fat saturation and cholesterol level on cholesterol synthesis measured using deuterium incorporation, *J. Lipid Res.* 35:1093-1101.

Korn, B.S., Shimomura, I., Bashmakov, Y., Hammer, R.E., Horton, J.D., Goldstein, J.L., and Brown, M.S., 1998, Blunted feedback suppression of SREBP processing by dietary cholesterol in transgenic mice expressing sterol-resistant SCAP (D443N), *J. Clin. Invest.* 102:2050-2060.

Lee, W.N.P., Bassilian, S., Guo, Z., Schoeller, D., Edmond, J., Bergner, E.A., and Byerley, L.O., 1994, Measurement of fractional lipid synthesis using deuterated water (^{2}H$_2$0) and mass isotopomer analysis, *Am. J. Physiol.* 266:E372-E383.

Liscum, L., and Underwood, K.W., 1995, Intracellular cholesterol transport and compartmentation, *J. Biol. Chem.* 270:15443-15446.

Quarfort, S.H., and Hilderman, H.L., 1970, Quantitation of the in vitro free cholesterol exchange between human red cells and lipoproteins, *J. Lipid Res.* 11:528-535.

Schwartz, C., 1982, Cholesterol kinetics and modeling: introduction, in: *Lipoprotein Kinetics and Modeling,* M. Berman, S.M. Grundy, and B.V. Howard, eds., Academic Press, New York.

Schwartz, C.C., Berman, M., Vlahcevic, Z.R., and Swell, L., 1982a, Multicompartmental analysis of cholesterol metabolism in man: quantitative kinetic evaluation of precursor sources and turnover of high density lipoprotein cholesterol esters, *J. Clin. Invest.* 70:863-876.

Schwartz, C.C., Vlahcevic, Z.R., Berman, M., Meadows, J.G., Nisman, R.M., and Swell, L., 1982b, Central role of high density lipoprotein in plasma free cholesterol metabolism, *J. Clin. Invest.* 70:105-116.

Schwartz, C.C., Zech, L.A., VandenBroek, J.M., and Cooper, P.S., 1993, Cholesterol kinetics in subjects with bile fistula, *J. Clin. Invest.* 91:923-938.

Shamburek, R.D., Pentchev, P.G., Zech, L.A., Blanchette-Mackie, J., Carstea, E.D., VandenBroek, J.M., Cooper, P.S., Neufeld, E.B., Phair, R.D., Brewer, H.B., Brady, R.O., and Schwartz, C.C., 1997, Intracellular trafficking of the free cholesterol derived from LDL cholesteryl ester is defective in vivo in Niemann-Pick C disease: insights on normal metabolism of HDL and LDL gained from the NP-C mutation, *J. Lipid Res.* 38:2422-2435.

Spady, D.K., and Dietschy J.M., 1983, Sterol synthesis in vivo in 18 tissues of the squirrel monkey, guinea pig, rabbit, hamster, and rat, *J. Lipid Res.* 24:303-315.

Wong, W.W., Hachey, D.L., Feste, A., Leggitt, J., Clarke, L.L., Pond, W.G., and Klein, P.D., 1991, Measurement of in vivo cholesterol synthesis from ^{2}H$_2$0: a rapid procedure for the isolation, combustion, and isotopic assay of erythrocyte cholesterol, *J. Lipid Res.* 32:1049-1056.

Zech, L.A., 1982, Sensitivity in compartmental models, in: *Lipoprotein Kinetics and Modeling,* M. Berman, S.M. Grundy, and B.V. Howard, eds, Academic Press, New York.

PART VI

MODELING OF PROTEIN METABOLISM, ENERGY, AND GROWTH

CHALLENGING THE ASSUMPTIONS IN ESTIMATING PROTEIN FRACTIONAL SYNTHESIS RATE USING A MODEL OF RODENT PROTEIN TURNOVER

Heidi A. Johnson, Chris C. Calvert, and Kirk C. Klasing[*]

INTRODUCTION

Understanding the effects of physiological changes on protein synthesis and degradation is limited by our ability to quantify protein metabolism. Current methods for estimating the rate of protein synthesis, such as the flooding dose, continuous infusion, and pulse dose, are based on measuring the rate of incorporation of a radiolabeled amino acid into protein in a particular tissue or in the whole body. Each method is limited by approximations and assumptions that must be made in order to summarize radioactivity data into a single rate estimate. In the flooding dose method, a large bolus of unlabeled amino acid is injected with the radiolabeled amino acid to expand free amino acid pools (extracellular and intracellular tRNA pools) so that their specific radioactivities ($\mu Ci/\mu mol$) are equivalent (Garlick et al., 1980). In the continuous infusion method, a constant amount of radiolabeled amino acid is continuously infused so that specific radioactivities of plasma and free amino acid pools reach a plateau or become equal (Waterlow et al., 1978). The pulse dose method involves injection of a small amount of radiolabeled amino acid and collection of blood for determination of free amino acid specific radioactivities over a specified period of time (Peters and Peters, 1972). In all three methods, when free amino acid pool specific radioactivities reach a plateau or a predesignated time, specific radioactivity of the protein is measured. Then the ratio of free amino acid specific radioactivity to protein specific radioactivity is calculated to estimate protein fractional synthesis rate (FSR).

Each method uses the specific radioactivity of one of the free amino acid pools rather than the true precursor pool (aminoacyl tRNA), and each assumes that specific radioactivity of the free amino acid pool represents specific radioactivity of aminoacyl tRNA. To calculate FSR, the specific radioactivity of the free amino acid pool is essentially averaged over the experimental time (in the flooding dose and continuous

---

[*] Heidi A. Johnson, Chris C. Calvert, and Kirk C. Klasing, Department of Animal Science, University of California, Davis, CA 95616.

infusion methods) or described mathematically by fitting the curve (pulse dose method). Protein specific radioactivity is averaged over time (pulse dose) or measured at the end of the incorporation period (flooding dose and continuous infusion methods). Inherent in these methods are the assumptions that (1) free amino acid pools are homogeneous, with specific radioactivities being equivalent for all free amino acid pools; (2) recycling of amino acids from protein degradation to protein synthesis does not occur or does not affect free amino acid pool specific radioactivities; (3) the protein pool is homogeneous and will not change over the experimental time; and (4) growth has an insignificant impact on pool specific radioactivities over the experimental time. These assumptions largely ignore current developments in our knowledge of the process of protein synthesis and degradation. In this paper, we examine the assumptions in light of a mathematical model for protein and amino acid kinetics in rodents.

CURRENT UNDERSTANDING OF PROTEIN SYNTHESIS AND DEGRADATION

Protein turnover is a complex and dynamic process that may involve (1) several sources of free amino acid for aminoacyl tRNA charging; (2) a variety of initiation, elongation and termination factors; and (3) a link to cell structure. Regarding the first point, Schneible and Young (1984) found that they could not force leucyl-tRNA specific radioactivities to a level equal to that of either intracellular or extracellular amino acid specific radioactivities in cultured chick muscle cells. Using the flooding dose method, Barnes et al. (1994) found that specific radioactivities of aminoacyl tRNA in chick heart and liver were significantly different from intracellular and extracellular specific radioactivities. Only in pectoralis and thigh did specific radioactivities of intracellular and aminoacyl tRNA pools equilibrate. The traditional amino acid sampling pool for specific radioactivity (plasma or extracellular fluid) never approximated aminoacyl tRNA specific radioactivities. In addition, the source of amino acids for protein synthesis may vary among cell types as well as in different nutritional and physiological states. The specific radioactivity of aminoacyl tRNA changes when cells and tissues are exposed to regulatory hormones or processes which impact protein accretion. For example, the specific radioactivity of aminoacyl tRNA changed by a factor of 2 during the anabolic phase that accompanies the recovery of rat hind limb muscles from immobilization (Morrison et al., 1987). Kipnis et al. (1961) described direct incorporation of extracellular amino acids into rat muscle protein without mixing of extracellular pools with intracellular pools. Their work indirectly indicated that the specific radioactivity of aminoacyl tRNA deviates from intracellular amino acid specific radioactivity as a result of growth hormone treatment.

The linkage between protein turnover and cell structure has been implied from several observations:

- the tethering of mRNA, ribosomes, and both initiation and elongation factors to specific locations on the cytoskeletal network (Jansen, 1999)

- the supramolecular organization of the protein translation machinery including tRNA synthetases, elongation factor-1α (EF-1α), and the ribosome (Barbarese et al., 1995)
- a physical association of EF-1α with tRNA synthetase and tRNA (Negrutskii and Budkevich, 1996; Petrushenko et al., 1997)
- the inability of exogenous aminoacyl-tRNAs to serve as precursors for protein synthesis in intact cells (Negrutskii and Deutscher, 1991)
- protein synthesis in intact cells with close spatial association of all factors involved in translation occurs at forty times the rate observed in cell-free systems (Negrutskii et al., 1994)

The sum of this evidence suggests that aminoacyl-tRNAs are recycling from aminoacyl-tRNA synthetase by EF-1α *directly* to the ribosome without dissociation into the fluid cytoplasm. Then the deacylated tRNA is returned by EF-1α to the synthetase complex. All three methods for estimating FSR assume that recycling of amino acids from protein degradation to protein synthesis does not occur or is not sufficient to affect estimates of protein synthesis during the short times of measurement. This assumption may be false and lead to underestimates of protein synthesis rates.

Because protein synthesis and degradation are multifaceted processes, it is difficult to estimate changes in FSR and protein turnover based on a single, static measurement in time. Using methods in which specific radioactivities of free amino acid (source) pools can be dissimilar during the measurement period, and different across nutritional or physiological states, adds to the difficulty of estimating FSR. In order to address effects of these assumptions, an analytical, dynamic, mathematical model of protein synthesis and degradation in the rodent was created (Johnson et al., 1999a). The purpose of the rodent model was twofold: to interpret time course specific radioactivity data and estimate FSR independent of method and to quantitatively assess the impacts of method assumptions on estimates of FSR. Here we use existing time course specific radioactivity data and sensitivity analyses with the model to test the assumptions of the methods (flooding dose, continuous infusion, and pulse dose) for estimating FSR.

THE RODENT MODEL

The rodent model is a dynamic representation of leucine (leu) or phenylalanine (phe) kinetics in a nongrowing rodent. It has been used for interpreting specific radioactivity data for the whole body (Johnson et al., 1999a, b) and for tissues (e.g., brain, liver, and muscle) (Johnson et al., 2000, 2001). The structure of the tissue models is identical to the structure for the whole body. The model traces the turnover of leu or phe from the free amino acid pools (plasma + extracellular, intracellular, and aminoacyl tRNA) to oxidation or to the protein pool as shown in Figure 1. Amino acids for protein synthesis [i.e., aminoacyl tRNA (T)] can arise from several sources: the intracellular pool (I), the extracellular pool (E), or protein (P). Leu or phe can be incorporated into protein (TP) or released through protein degradation to either be converted back into protein (PT) or released into the intracellular pool (PI). In the intracellular pool, leu or phe can be oxidized (IO), remain as a reserve, or exchange with the extracellular pool (EI and IE).

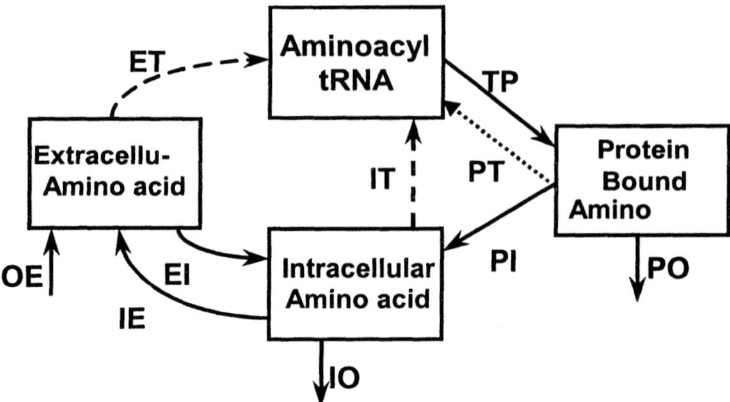

Figure 1. The model structure for leu or phe kinetics in whole body and tissues. E, I, and T are free leu or phe in the extracellular (plus plasma), intracellular, and aminoacyl tRNA pools, respectively; P is leu or phe in the protein pool; EI is the flux of leu or phe from the extracellular pool to the intracellular pool; ET is the flux of leu or phe from the extracellular pool to the aminoacyl tRNA pool; IE is the flux of leu or phe from the intracellular pool to the extracellular pool; IO is the flux of leu or phe oxidized from the intracellular pool; IT is the flux of leu or phe from the intracellular pool to the aminoacyl tRNA pool; OE is the intake flux of leu or phe to the extracellular pool; PI is the flux of leu or phe from protein to the intracellular pool; PO is the flux of leu or phe from protein exported out of the tissue (liver and gut models); PT is the flux of leu or phe from protein degradation to the aminoacyl tRNA pool (recycling); TP is the flux of leu or phe from the aminoacyl tRNA pool to protein.

Unlabeled leu or phe intake and radiolabeled leucine are flooded, pulsed, or continuously infused into the extracellular pool (OE). For export tissues (liver and gut), protein export (PO) is included as shown in Figure 1. The primary differences between the whole-body and tissue models are in the estimates of initial pool sizes and in the flux equations. Conservation of mass principles give differential equations for the models for unlabeled and radiolabeled amino acids (Table 1).

DATA SETS

The data required for the model to estimate FSR must include changes over time in specific radioactivity in protein and in one or more free amino acid pools. Four data sets were available, covering two of the dosing methods: two whole-body data sets using the flooding dose method, a set of tissue data using the flooding dose method (gastrocnemius, gut, heart, kidney, liver, soleus, and tibialis), and a data set from a pulse dose experiment (brain, liver, and muscle). The first whole body data set was from Bernier and Calvert (1987). Thirty-gram mice were injected with 3 µCi ^{14}C leu/30 µmol leu interperitoneally, and specific radioactivity measurements were taken at 2, 5, 10, 15, 20, and 30 min. The specific radioactivity of protein and a combined plasma, extracellular, intracellular, and aminoacyl tRNA pool were measured. The second data set was from Obled et al. (1991) using 70 g rats injected with 0.21 µCi ^{14}C leu/140 µmol leu via a caudal vein. Specific radioactivity measurements were obtained at 5, 7, 9, 11, 13, and 15 min. Specific radioactivities of protein, plasma, and combined extracellular, intracellular, and aminoacyl tRNA pools were measured. The third data set for tissues (Lajtha, 1959) followed a whole-body pulse dose of 0.2 µCi U ^{14}C leu /0.025 µmol leu given to 20 g (100-d old) mice. Specific radioactivities were measured at 3, 5, 10, 20, 30, 45, 60, and 120 min in plasma, brain, liver and muscle. Six mice were pooled to obtain data at each time point. Plasma, protein, and combined tissue extracellular, intracellular, and aminoacyl tRNA specific radioactivities were measured. The fourth data set was from a flooding dose experiment by Goldspink and Kelly (1984), Lewis et al. (1984), Goldspink et al. (1984), and Kelly et al. (1984). A flooding dose of 65 µCi L-4 ^{3}H phe /150 µmol phe /100 g body weight was given to 200 g rats. Tissues were collected at 2.5, 5, 10, 15, 20, and 30 min after injection, and data from four rats at each time point were pooled. Specific radioactivities were determined in plasma, protein, and combined extracellular, intracellular, and aminoacyl tRNA for liver, kidney, heart, tibialis anterior, soleus, gastrocnemius, and gut. Suitable data for the continuous infusion method are not available since time course specific radioactivity data in protein (which is required by the model) are not collected.

In order to determine the quantitative impact of the four assumptions on estimates of FSR, a combination of sensitivity analyses and results of parameter fitting to the model for the four data sets were used. Data were fit using a generalized reduced gradient in ACSL Optimize (AEgis, 2000). ACSL Optimize uses the method of maximum likelihood to estimate parameters based on data input into a model. Model parameters are chosen to maximize the likelihood of fitting the experimental data. Errors are assumed to be normally distributed and statistically independent. For all models, leu and phe intake, ^{14}C leu and ^{3}H phe dose to the tissue, and the model-predicted FSRs were estimated based on fitted specific radioactivities.

RESULTS AND DISCUSSION

Assumption 1

The first assumption is that free amino acid pools are homogeneous, with specific radioactivities being equivalent for all free amino acid pools and aminoacyl tRNA and that sampling pool specific radioactivities can be manipulated to be equivalent.

Results from fitting the specific radioactivity data in the whole-body flooding dose and tissue pulse dose experiments are shown in Figures 2 and 3. Specific radioactivities of the source pools (extracellular free leu and intracellular free leu) decreased in both the flooding and pulse dose experiments. However, the model predicted that source pool sizes were also changing throughout the course of the experiments. Pool size changes did not necessarily reflect the change in specific radioactivities. The I and E pool sizes decreased, but I and E specific radioactivities (SRI and SRE, respectively) increased in the flooding dose experiment (Figure 2). In the pulse dose experiment (Figure 3), both E and I pool sizes increased but their specific radioactivities (SRE and SRI, respectively) decreased.

Table 1. Equations for the whole-body (WBM) and tissue models (TM) for protein turnover in rodents[1]

Leu or Phe Concentration (μmol/L)	
Extracellular Pool	$C_E = E/vol_E$
Intracellular Pool	$C_I = I/vol_I$
OE (μmol/min)	
WBM	$OE = [0.5 \cdot bw^{0.67}/(24 \cdot 60)]^2 \cdot 0.012^3/0.00011^4$
TM	$OE = [0.5 \cdot bw^{0.67}/(24 \cdot 60)]^2 \cdot 0.012^3/0.00011^4 \cdot vol_E/vol_{WB}$
IO (μmol/min) WBM & TM	$IO = 3 \cdot OE^5 / [1 + (km_{IO}{}^6 / C_I)]$
IT (μmol/min)	
WBM	$IT = (1 - PC) \cdot PI$
TM; no export	$IT = (1 - PC) \cdot PI$
TM; liver, gut	$IT = (1 - PC) \cdot (PI + PO)$
ET (μmol/min)	
WBM & TM	$ET = PC \cdot PI$
TM; liver, gut	$ET = PC \cdot (PI + PO)$
EI (μmol/min)	
WBM	$EI = Vm_{EI}{}^7 \cdot twt / [1+(Km_{EI}{}^8 / C_E)]$
TM	$EI = Vm_{EI}{}^9 \cdot twt / [1+(Km_{EI}{}^{10} /C_E)]$
IE (μmol/min)	
WBM	$IE = Kd_{IE}{}^{11} \cdot C_I \cdot bw$
TM	$IE = Kd_{IE}{}^{12} \cdot C_I \cdot twt$
PT (recycling) (μmol/min)	$PT = PR \cdot (TP - PO)$
PI (μmol/min)	$PI = (1 - PR) \cdot (TP - PO)$
PO (μmol/min) (export)	$PO = 0.60^{13} \cdot TP$
TP (μmol/min)	$TP = K_S \cdot P / 1440$
Differential Equations	
E	$dE /dt = IE + OE - EI - ET$
SE	$dSE /dt = SRI \cdot IE - SRE \cdot (EI + ET)$
I	$dI /dt = EI + PI - IO - IE - IT$
SI	$dSI /dt = SRE \cdot EI + SRP \cdot PI - SRI \cdot (IO + IE + IT)$
T	$dT /dt = ET + IT + PT - TP$
ST	$dST /dt = SRE \cdot ET + SRI \cdot IT + SRP \cdot PT - SRI \cdot TP$
P (no export)	$dP /dt = TP - PI - PT$
P (export)	$dP /dt = TP - PI - PT - PO$
SP (no export)	$dSP /dt = SRT \cdot TP - SRP \cdot (PI + PT)$
SP (export)	$dSP /dt = SRI \cdot TP - SRP \cdot (PI + PT + PO)$
SR (μCi/μmol)	
SRE	SE / E
SRI	SI / I
SRT	ST / T
SRP	SP / P

[1] Besides the abbreviations given in Figure 1, bw is body weight (g); C_E is the concentration of leu or phe in the extracellular (plasma) pool; C_I is the concentration of leu or phe in the intracellular pool; Kd_{IE} is the diffusion constant for leu or phe from the intracellular to the extracellular pool $(ml \cdot min^{-1} \cdot g^{-1})$, Km_{EI} is the concentration of extracellular leu or phe at half maximal velocity of flux of leucine or phenylalanine from the extracellular to the intracellular pool (μmol/L), Km_{IO} is the concentration of intracellular leu or phe at half maximal velocity of leu or phe oxidized (μmol/L), K_S is the protein synthesis rate (%/d), PC is the percent from E exclusive of recycling, PR is the percent recycling, E is leu or phe in the extracellular pool (μmol), SE is ^{14}C leu or ^{3}H phe in the extracellular pool (μCi), I is leu or phe in the intracellular pool (μmol), SI is ^{14}C leu or ^{3}H phe in the intracellular pool (μCi), P is leu or phe in the protein pool (μmol), SP is ^{14}C leu or ^{3}H phe in protein (μCi), T is leu or phe in the aminoacyl tRNA pool (μmol), ST is ^{14}C leu or ^{3}H phe in the aminoacyl tRNA pool (μCi), SRE is radioactivity in the extracellular pool (μCi ^{14}C leu/μmol leu or μCi ^{3}H phe/μmol phe), SRI is specific radioactivity in the intracellular pool (μCi ^{14}C leu/μmol leu or μCi ^{3}H phe/μmol phe), SRP is specific radioactivity in the protein pool (μCi ^{14}C leu/μmol leu or μCi ^{3}H phe/μmol phe), SRT is the specific radioactivity in the aminoacyl tRNA pool (μCi ^{14}C leu/μmol leu or μCi ^{3}H phe/μmol phe), t is time (min), twt is weight of tissue (g), Vm_{EI} is maximal velocity of flux from the extracellular to the intracellular pool $(mol \cdot min^{-1} \cdot L^{-1})$, vol_E is volume of the extracellular pool (L), vol_I is the volume of the intracellular pool (L), and vol_{WB} is the volume of the whole body (L).

[2] Based on a food intake 0.3785 μmol leu/min for a 20 g mouse or 1.46 μmol phe/min for a 200 g rat (John and Bell, 1976).

[3] Leu absorption of 1.2% of total intake, phe absorption of 1.4%, or total intake (John and Bell, 1976).

[4] Molecular mass of leu or phe in protein (g/μmol).

[5] Assumes a high maximal velocity (Vm_{IO}) so the rate of the equation is set by the concentration of leu or phe in Q_I and Km_{IO} is set so the intake of leu or phe (F_{OE}) equals the oxidation of leu or phe (F_{IO}).

[6] $Km_{IO} = 38962$ μmol/L.

[7] $Vm_{EI} = 23 mol \cdot min^{-1} \cdot L^{-1}$ (Stevens et al., 1984).

[8] $Km_{EI} = 500 mol/L$ (Oxender and Christensen, 1963).

[9] $Vm_{EI} = 0.005$ $mol \cdot min^{-1} \cdot L^{-1}$ (Stevens et al., 1984).

[10] $Km_{EI} = 0.0007$ mol/L (Stevens et al., 1984).

[11] $Kd_{IE} = 1.2$ $ml \cdot g^{-1} \cdot min^{-1}$ (Stevens et al., 1984).

[12] $Kd_{IE} = 0.045$ $ml \cdot g^{-1} \cdot min^{-1}$ for phe (Miller et al., 1985) or $Kd_{IE} = 1.2$ $ml \cdot g^{-1} \cdot min^{-1}$ for leu (Stevens et al., 1984).

[13] Assumes 60% of the protein synthesized in tissue is exported.

A single pool may not be the exclusive source of amino acids for aminoacyl tRNA. Table 2 shows the relative proportions of amino acids from each source pool that contribute to protein synthesis. Values were predicted from fitting tissue flooding dose specific radioactivity data to two models. The first model separated extracellular pools (E) and plasma (Pls) from intracellular pools (I) and the second separated E and I from Pls. The models predict that heart, kidney, and liver had more than one free amino acid pool contributing to aminoacyl tRNA.

Table 2 also indicates that, in some tissues (gastrocnemius, soleus, and tibialis), there may be a large amount of recycling. Therefore, all of the source pools (including protein) could be contributing to aminoacyl tRNA. If several pool specific radioactivies are contributing to protein synthesis, and their pool sizes are changing during the experimental time period, it becomes more difficult to choose which source pool specific radioactivity truly reflects aminoacyl tRNA specific radioactivity. Figure 4 shows how specific radioactivity of aminoacyl tRNA is expected to change with increasing contributions of amino acids from E (instead of I). Using parameters estimated from the Bernier and Calvert data (1987), simulations were run with different proportions of E or I supplying amino acids for aminoacyl tRNA charging. Therefore, each point represents

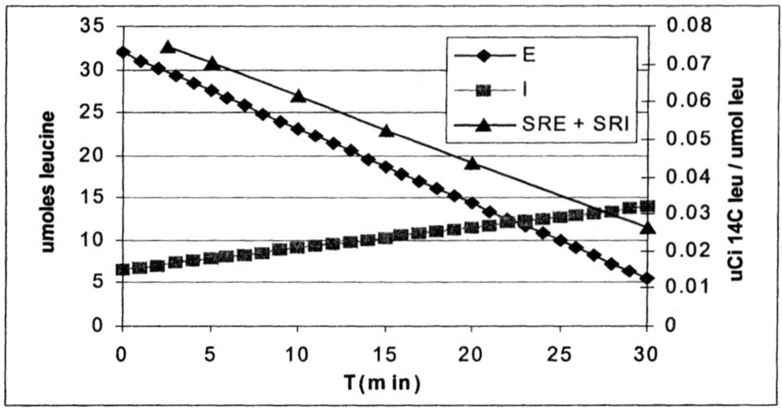

Figure 2. Changes in extracellular (E) and intracellular (I) leu pool sizes and specific radioactivity of E + I (SRE + SRI) over time. Data are from the whole-body flooding dose experiment of Bernier and Calvert (1987). Specific radioactivities of free amino acids and protein in the whole body were fit to the model.

the specific radioactivity of the pool at the end of the experiment for each method (flooding dose, 30 min; pulse dose, 60 min; and continuous infusion, 180 min).

The pulse dose method is the only method in which specific radioactivity differences between free amino acid pools are less than differences due to different amino acid sources. E or I specific radioactivity could be used to approximate aminoacyl tRNA specific radioactivity regardless of the relative contributions of E *vs.* I. For the flooding dose and continuous infusion methods, the specific radioactivity of aminoacyl tRNA is different depending on which pool is the predominant source of amino acids. If both E

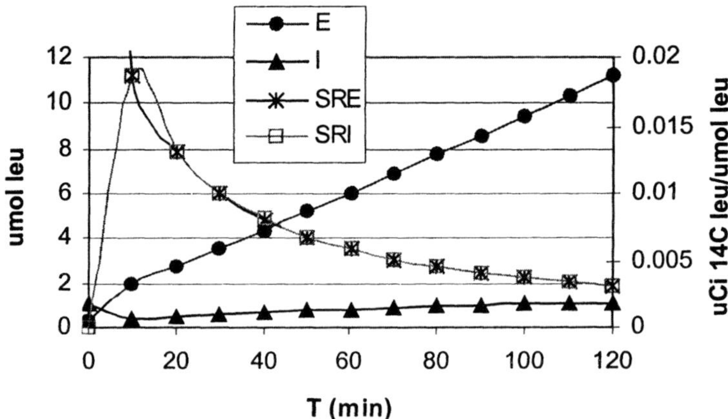

Figure 3. Changes in extracellular (E) and intracellular (I) leu pool sizes and specific radioactivities of E (SRE) and I (SRI) over time. The data are from a tissue pulse dose experiment (Lajtha, 1959). Specific radioactivities of free amino acids and protein in muscle were fit to the model.

and I contribute equally (50% E), aminoacyl tRNA specific radioactivity is not reflected by either pool. Therefore, the flooding dose and continuous infusion methods are not able to manipulate free amino acid pool specific radioactivities to consistently reflect aminoacyl tRNA specific radioactivity.

The effects of the failure of the flooding dose method to manipulate E or I specific radioactivity to reflect aminoacyl tRNA specific radioactivity on estimates of FSR are shown in Table 3. FSR estimates vary from +5.5%/d to −3.1%/d around the actual protein synthesis rate (40.8 %/d).

Assumption 2

The second assumption is that recycling of amino acids from protein degradation to protein synthesis does not affect free amino acid pool specific radioactivities.

Results from fitting tissue flooding dose data (Table 2) and tissue pulse dose data (Johnson et al., 2001) indicate that protein degradation could be contributing amino acids to aminoacyl tRNA. However, it is generally assumed that, within short experimental time periods, recycling does not affect estimates of FSR (Waterlow et al., 1978). This assumption was tested by simulating flooding, pulse, and continuous infusion experiment specific radioactivities (similar to Figure 4) at increasing rates of recycling (Figure 5). Specific radioactivities of E and I increased slightly while aminoacyl tRNA specific radioactivities decreased as recycling increased in the flooding and continuous infusion methods [panels (A) and (C), respectively]. However, with the pulse dose method [panel (B)], aminoacyl tRNA specific radioactivity was not affected by recycling until recycling was high (greater than 60%). For all methods, the greater the percent of amino acids supplied by recycling, the lower the specific radioactivity of aminoacyl tRNA.

Although specific radioactivities of E and I were only affected slightly by recycling, protein specific radioactivity and therefore estimates of FSR changed. Figure 6 shows the effects of increased recycling on estimates of FSR with the flooding and pulse dose methods. As recycling increased, specific radioactivity of protein decreased and therefore FSR decreased.

Assumption 3

The third assumption is that the protein pool is homogeneous and will not change over the experimental time.

All three methods used to estimate FSR depend on protein specific radioactivity remaining the same over time and between experiments. However, if recycling is occurring, protein specific radioactivity will plateau and then decrease. Also, protein synthesis rate is not homogeneous in the whole body or in a given tissue. Since FSR is an aggregate of all protein synthetic processes, and these proceed at different rates, the length of the experiment can have an effect on the resulting estimated FSR. Tables 4 and 5 show how estimates of FSR change with longer experimental times and with increased protein synthesis rates in a portion of the total protein pool. In the flooding dose method, fractional synthesis rate decreased as the length of the experiment progressed, and it increased with an increase in protein synthesis rate. Using an experimental time period of 15 *vs.* 30 min decreased FSR estimates by 6% (Table 4).

Table 2. Percent phe for aminoacyl tRNA from different sources relative to amount of protein synthesized as predicted by fitting ^{3}H phe flooding dose specific radioactivity data[1]

	% Pls	% E+Pls	% I	% E+I	% P
GASTROCNEMIUS					
E+Pls		5.5 (12)	0		94.5 (0.42)
E+I	5.5 (14)			0	94.5 (0.038)
GUT					
E+Pls		100 (11)	0		0
E+I	94.2 (12)			0	5.80 (3.7)
HEART					
E+Pls		45.3 (5.5)	54.7 (9.0)		0
E+I	36.8 (8.2)			63.2 (16)	0
KIDNEY					
E+Pls		69.8 (47)	0		30.2 (5.3)
E+I	16.5 (1.6)			74.6(0.068)	8.89 (0.41)
LIVER					
E+Pls		44.5(0.00065)	53.6(0.0013)		1.88 (0.12)
E+I	34.1(0.66)			65.9(0.042)	0
SOLEUS					
E+Pls		7.40 (8.5)	0		92.6 (11)
E+I	7.40 (10)			0	92.6 (0.20)
TIBIALIS					
E+Pls		5.10 (6.0)	0		94.9 (0.39)
E+I	5.10 (7.8)			0	94.9 (0.30)

[1]Data from Goldspink et al. (1984), Goldspink and Kelly (1984), Kelly et al. (1984), and Lewis et al. (1984) as predicted by the rat tissue models. Standard deviations are in parentheses following the values. E is extracellular, E + I is combined E and I phe, E + Pls is combined E and Pls phe, I is intracellular, Pls is plasma, and P is protein.

Increasing protein synthesis in 12% of the protein pool (representing fast protein turnover by the gut and liver) from 104%/d to 154%/d increased FSR by approximately 7%/d (data not shown). FSR did not match the protein synthesis rate until 120 min. Estimates of FSR using the pulse dose method were much more sensitive to experimental time length (Table 5). A change from 15 to 30 min tripled the estimated FSR. The pulse dose method usually relies on measurement of protein specific radioactivity when it reaches a peak. If the peak is missed, FSR estimates are affected.

Assumption 4

Assumption 4 is that growth has an insignificant impact on pool specific radioactivities over the experimental time.

Table 3. Effect of different sampling pools used to calculate FSR (%/d) from whole-body flooding dose simulations [1]

% P	% E	% I	FSR I	FSR T	FSR E	True FSR
0	100	0	47.1	45.2	45.5	40.8
0	0	100	44.7	44.7	43.4	40.8
12	88	0	47.0	45.2	45.3	40.8
12	0	88	44.6	44.7	43.3	40.8

[1] FSR I is calculated using intracellular specific radioactivity as the precursor pool, FSR T uses aminoacyl tRNA specific radioactivity, FSR E uses extracellular (or plasma) specific radioactivity as the precursor specific radioactivity, and True FSR is the μmol leu incorporated into protein/μmol leu in protein.

Since most experiments to estimate FSR last 5 min to 6 hr, depending on the method, it is assumed that the animal does not grow enough in that time to change FSR (Waterlow et al., 1978). Figures 7 and 8 show the change in FSR with increasing protein degradation rate (Kd). With both the flooding dose and pulse dose methods, estimated FSR does not change if aminoacyl tRNA specific radioactivity is used in the calculation. Growth over the experimental time period has no effect on FSR. However, if protein specific radioactivities in the intracellular or extracellular pools are used, estimates of FSR are lower with increasing growth using either method.

SUMMARY

Published estimates of protein fractional synthetic rate vary widely (Johnson et al., 1999a). Contributing to the large standard deviation for FSR are physiological and methodological differences that do not account for changes in specific radioactivities of I, E, T, and P. Current methods for estimating FSR are based on four assumptions which may not be valid. The first assumption, that the free amino acid pool is homogenous and reflects the specific radioactivity of the true precursor pool (aminoacyl tRNA), can cause FSR estimates to increase by up to 8%/d. The second assumption, that recycling has an insignificant effect on FSR estimates, could result in decreases in estimates of FSR from 10 to 20%/d. The third assumption, that the protein pool is homogeneous and will not change over time, results in a 4-10%/d change using the flooding dose method. The fourth assumption, that growth will not affect estimated FSR over a short experimental time, is true if aminoacyl tRNA specific radioactivity is used to estimate FSR. Otherwise, estimates can vary 4-5%/d. Although specific radioactivity of aminoacyl tRNA is difficult to measure, the first and fourth assumptions are valid if aminoacyl tRNA specific radioactivity is used. Using a model of protein turnover, as described in this paper, to interpret specific radioactivity data allows the inclusion of all four assumptions and the potential to better quantify changes in FSR under different physiological conditions.

(A) Flooding Dose

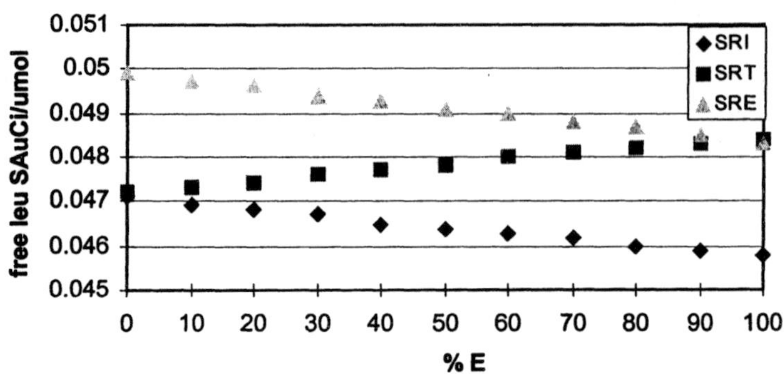

(B) Pulse Dose

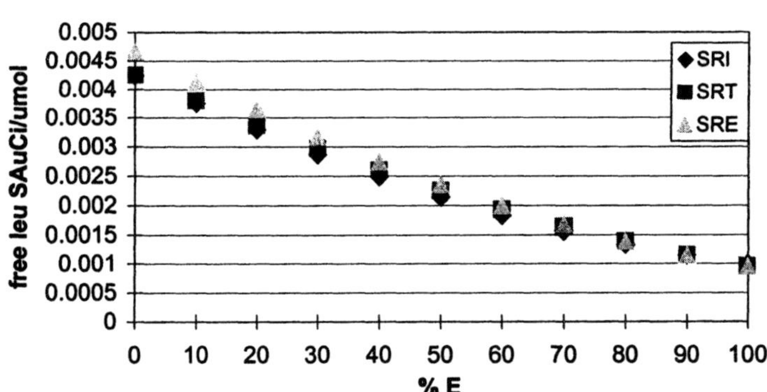

(C) Continuous Infusion

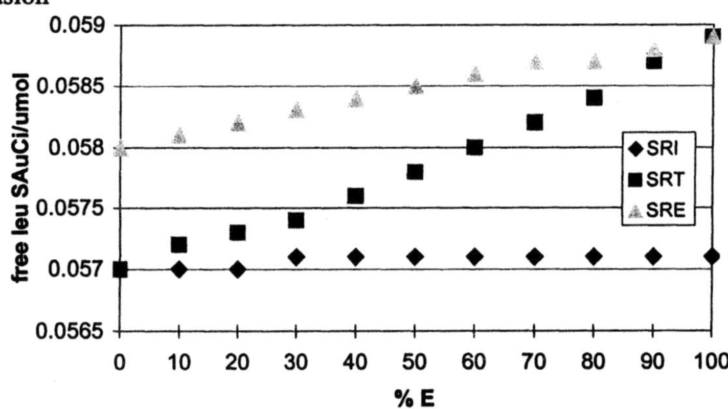

Figure 4. Simulations of changes in free amino acid pool specific radioactivities (whole body) with increasing aminoacyl tRNA charging from extracellular sources for the flooding dose method (A), the pulse dose method (B), and the continuous infusion method (C). Each point represents specific radioactivities at the end of an experiment: flooding dose, 30 min; pulse dose, 60 min, and continuous infusion, 180 min. SRI is the specific radioactivity of the intracellular pool, SRT is the specific radioactivity of aminoacyl tRNA, and SRE is the specific radioactivity of the extracellular pool.

(A) Flooding dose

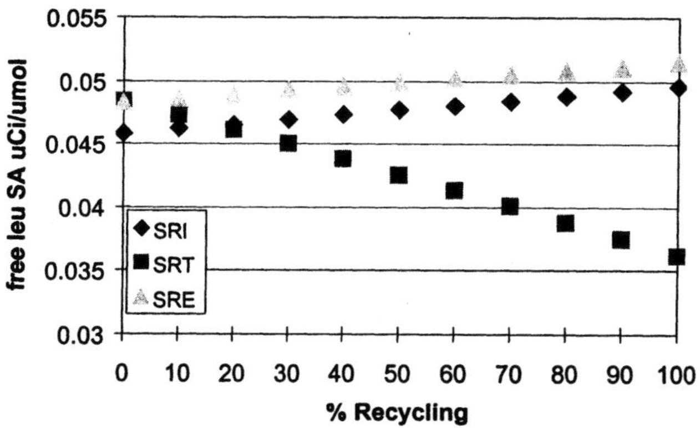

(B) Pulse dose

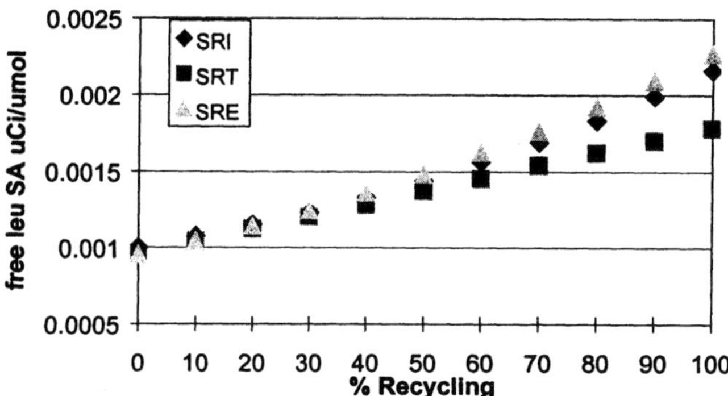

(C) Continuous infusion

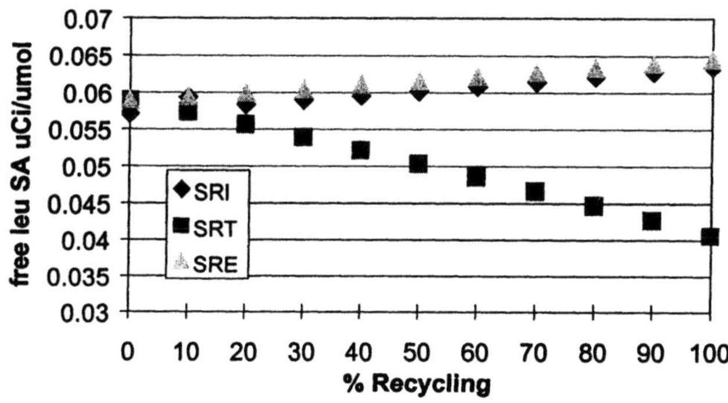

Figure 5. Simulations of changes in free amino acid pool specific radioactivities (whole body) with increasing aminoacyl tRNA charging from protein (recycling) for the flooding dose method (A), the pulse dose method (B), and the continuous infusion method (C). Each point represents specific radioactivities at the end of an experiment: flooding dose, 30 min; pulse dose, 60 min; and continuous infusion, 180 min. SRI is the specific radioactivity of the intracellular pool, SRT is the specific radioactivity of aminoacyl tRNA, and SRE is the specific radioactivity of the extracellular pool.

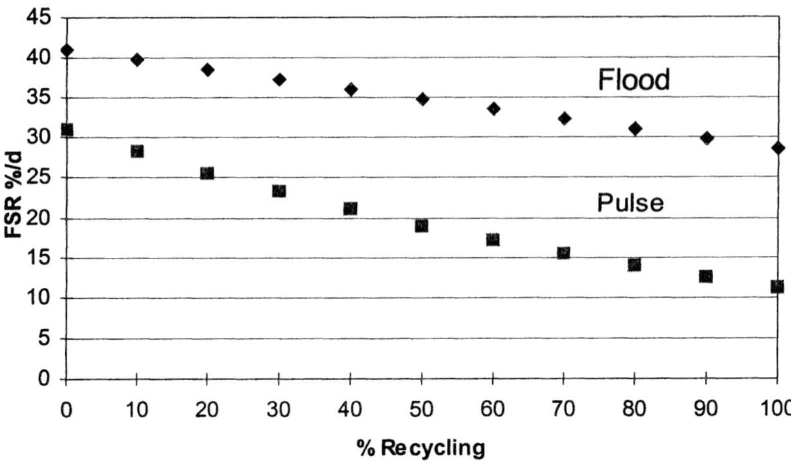

Figure 6. Simulations of the effect of increasing recycling on estimates of whole-body FSR for the flooding and pulse dose methods.

Table 4. Estimates of whole-body FSR from the flooding dose method as affected by length of the experiment

Length of Experiment (min)	FSR (%/d) eFSR[1] = 41	FSR (%/d) eFSR = 48
5	101	116
15	51.1	58.6
30	45.4	52.1
60	42.9	49.2
120	41.3	47.5
180	41.1	46.6

[1]eFSR is the FSR reported in the experiment.

Table 5. Estimates of tissue FSR over time with the pulse dose method [based on simulations of data from Lajtha (1959)] as affected by length of the experiment

Length of Experiment (min)	Liver FSR (%/d) eFSR[1] = 125	Muscle FSR (%/d) eFSR = 9.80	Brain FSR (%/d) eFSR = 13.7
5	0.308	0.00540	0.00739
15	6.09	0.459	2.39
30	24.4	1.93	7.69
60	106	8.95	24.0
120	481	43.9	92.1
180	1161	113	236

[1]eFSR is the FSR reported in the experiment.

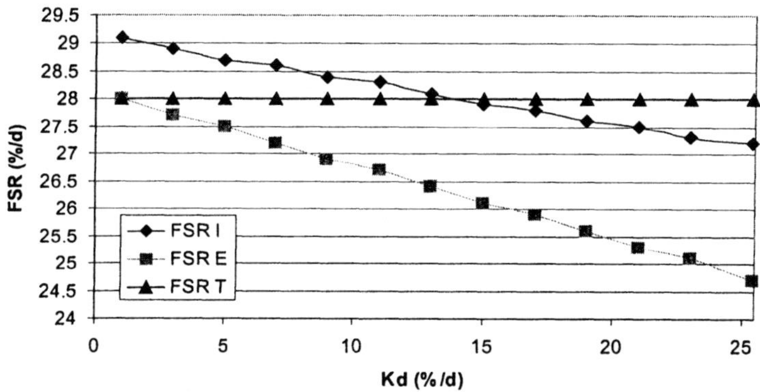

Figure 7. Changes in whole-body FSR with increasing degradation rate (flooding dose method). Each point represents FSR estimated at the end of a 30 min simulated experiment. Kd is protein degradation rate.

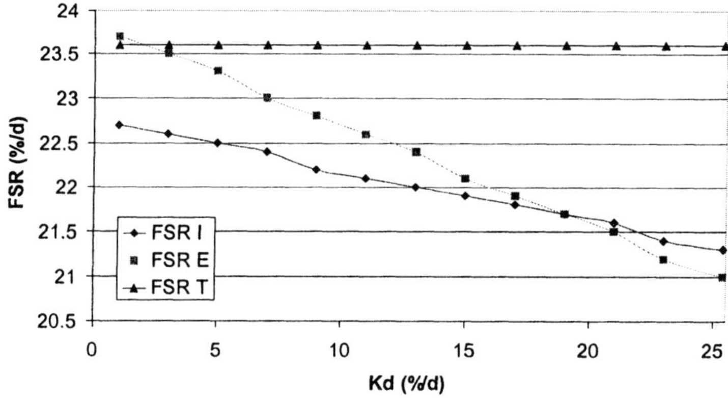

Figure 8. Change in tissue FSR with increasing degradation rate (pulse dose method). Each point represents FSR estimated at the end of a 60 min simulated experiment. Kd is protein degradation rate.

CORRESPONDENCE

Please address all correspondence to:
Heidi A. Johnson
University of California, Davis
Department of Animal Science
One Shields Avenue
Davis, CA 95616
HAJohnson@UCDavis.edu

REFERENCES

AEgis Technologies Group Inc., ACSL Sim ®, 2000, *Advanced Continuous Simulation Language,* Huntsville.

Barbarese, E., Koppel, D.E., Deutscher, M.P., Smith, C.L., Ainger, K., Morgan, F., and Carson, J.H., 1995, Protein translocation components are colocalized in granules in oligodendrocytes, *J. Cell Sci.* 108:2781-2790.

Barnes, D.M., Calvert, C.C., and Klasing, K.C., 1994, Source of amino acids for tRNA acylation in growing chicks, *Amino Acids* 7:267-278.

Bernier, J.F., and Calvert, C.C., 1987, Effect of a major gene for growth on protein synthesis in mice, *J. Anim. Sci.* 65:982-995.

Garlick, P.J., McNurlan, M.A., and Preedy, V.R., 1980, A rapid and convenient technique for measuring the rate of protein synthesis in tissues by injection of [^{3}H] phenylalanine, *Biochem. J.* 192:719-723.

Goldspink, D.F., and Kelly, F.J., 1984, Protein turnover and growth in the whole body, liver and kidney of the rat from foetus to senility, *Biochem. J.* 217:507-516.

Goldspink, D.F., Lewis, S.E.M., and Kelly, F.J., 1984, Protein synthesis during the developmental growth of the small and large intestine of the rat, *Biochem. J.* 217:527-534.

Jansen, R.P., 1999, RNA-cytoskeletal associations [see comments], *FASEB J.* 13:455-466.

John, A., and Bell, J.M., 1976, Amino acid requirements of the growing mouse, *J. Nutr.* 106:1361-1367.

Johnson, H.A., Baldwin, R.L., France, J., and Calvert, C.C., 1999a, A model of whole body protein turnover based on leucine kinetics in rodents, *J. Nutr.* 129:728-739.

Johnson, H.A., Baldwin, R.L., France, J., and Calvert, C.C., 1999b, Recycling, channeling and heterogeneous protein turnover estimation using a model of whole body protein turnover based on leucine kinetics in rodents, *J. Nutr.* 129:740-750.

Johnson, H.A., Baldwin, R.L., Klasing, K.C., France, J., and Calvert, C.C., 2000, Designing an experiment for measuring the rates of channeling, recycling and protein synthesis using a rodent model of protein turnover, *J. Nutr.* 130:3097-3102.

Johnson, H.A., Baldwin, R.L., Klasing, K.C., and Calvert, C.C., 2001, Impact of separating amino acids between plasma, extracellular and intracellular compartments on estimating protein synthesis in rodents, *Amino Acids.* 20:389-400.

Kelly, F.J., Lewis, S.E.M., Anderson, P., and Goldspink, D.F., 1984, Pre and postnatal growth and protein turnover in four muscles of the rat, *Muscle Nerve* 7:235-242.

Kipnis, D.M., Reiss, E., and Helmreich, E., 1961, Functional heterogeneity of the intracellular amino acid pool in mammalian cells, *Biochim. Biophys. Acta* 51:519-524.

Lajtha, A., 1959, Amino acid and protein metabolism of the brain – V. turnover of leucine in mouse tissues, *J. Neurochem.* 3:358-365.

Lewis, S.E.M., Kelly, F.J., and Goldspink, D.F., 1984, Pre and postnatal growth and protein turnover in smooth muscle, heart and slow and fast twitch skeletal muscles of the rat, *Biochem. J.* 217:517-526.

Miller, L.P., Pardridge, W.M., Braun, L.D. and Oldendorf, W.H., 1985, Kinetic constants for blood-brain barrier amino acid transport in conscious rats, *J. Neurochem.* 45:1427-1432.

Morrison, P.R., Muller, G.W., and Booth, F.W., 1987, Actin synthesis rate and mRNA level increase during early recovery of atrophied muscle, *Am. J. Physiol.* 253:C205-C209.

Negrutskii, B.S., and Budkevich, T.V., 1996, Rabbit translation elongation factor 1 alpha stimulates the activity of homologous aminoacyl-tRNA synthetase, *FEBS Lett.* 382:18-20.

Negrutskii, B.S., and Deutscher, M.P., 1991, Channeling of aminoacyl tRNA for protein synthesis in vivo, *Proc. Natl. Acad. Sci. USA* 88:4991-4995.

Negrutskii, B.S., Stapulionis, R., and Deutscher, M.P., 1994, Supramolecular organization of the mammalian translation system, *Proc. Natl. Acad. Sci. USA* 91:964-968.

Obled, C., Barre, F., and Arnal, M., 1991, Flooding-dose of various amino acids for measurement of whole-body protein synthesis in the rat, *Amino Acids* 1:17-27.

Oxender, D.L., and Christensen, H.N., 1963, Distinct mediating systems for the transport of neutral amino acids by the Erlich cell, *J. Biol. Chem.* 238:3686-3699.

Peters, T., and Peters, J.C., 1972, The biosynthesis of rat serum albumin, *J. Biol. Chem.* 247:3858-3863.

Petrushenko, Z.M., and Negrutskii, B.S., 1997, Evidence for the formation of an unusual ternary complex of rabbit liver EF-1alpha with GDP and deacylated tRNA, *FEBS Lett.* 407:13-7.

Schneible, P.A., and Young, R.B., 1984, Leucine pools in normal and dystrophic chicken skeletal muscle cells in culture, *J. Biol. Chem.* 259:1436-1440.

Stevens, B.R., Kaunitz, J.D., and Wright, E.M., 1984, Intestinal transport of amino acids and sugars: advances using membrane vesicles, *Ann. Rev. Physiol.* 46:417-433.

Waterlow, J.C., Garlick, P.J., and Millward, D.J., 1978, *Protein Turnover in Mammalian Tissues and in the Whole Body,* North-Holland Press, Amsterdam.

SIMULATING PATTERNS OF CHANGE IN RATES OF SECRETION OF PROTEIN INTO MILK

John Cant[*]

INTRODUCTION

Protein is a very important component of cow's milk, both nutritionally for the consumer and economically for the dairy farmer. In order to manage milk protein production at the farm level, it would be advantageous to have an understanding of its biological regulation that was adequate to predict daily yield. Most of our information regarding regulation of milk protein secretion is derived from arteriovenous difference methodology in which the balance of milk precursors and products is measured across the udder of lactating cows. The purpose of the models presented here is to study the connection between arterial concentrations of milk precursors and the composition of milk produced by bovine mammary glands. More specifically, the models should predict the secretion rate of milk protein by dairy cows, the percentage of milk volume that is protein (this is a factor in many milk pricing schemes), and the arteriovenous differences of milk precursors across the mammary glands (the intermediate pieces of information upon which the knowledge base has been built).

Although protein secretion by mammary glands is the main subject of this paper, synthesis of fat and lactose in the mammary glands cannot be ignored. Lactose is synthesized from glucose in the same Golgi vesicles that secrete proteins into the alveolar lumen. As lactose accumulates, the vesicles swell with osmotic movement of water, and it is the emptying of vesicle contents into the alveolus that is the main route of water secretion into milk (Linzell and Peaker, 1971). Water represents 88% of bovine milk volume, so lactose yield is the primary determinant of milk yield. Protein secretion rates relative to lactose secretion rates essentially determine milk protein percentage. By this mechanism, changes in protein percentage are often mimicked by changes in fat percentage.

---

* John Cant, Department of Animal and Poultry Science, University of Guelph, Guelph, Ontario, Canada N1G2W1.

PATTERNS OBSERVED

There are several physiological and nutritional effects on milk protein secretion that a successful model should be able to simulate.   One is the temporal lactation curve following parturition (Figure 1A).   For the first 50 days or so, daily milk production increases, while fat and protein percentages drop. (Note that fat and protein yields increase during this time.)   After peak yield, total milk production declines, and fat and protein percentages rise.   This mirror image of yield and percentage (i.e., when one increases, the other declines) is also observed in the seasonal curve (Figure 1B).   As daily photoperiod increases and temperatures rise in spring, milk protein and fat yields increase, while their percentages drop. Following a summer peak, the reverse occurs.

Nutritionally, some effects of arterial metabolite concentrations have been established. When fat was added to dairy cow diets at 3.5% of dry matter intake, arterial long-chain fatty acid concentrations and uptakes by mammary tissue increased (Cant et al., 1993b).   A drop in mammary blood flow was associated with fat feeding, presumably through a link to metabolic activity (Cant and McBride, 1995), which reduced net glucose and amino acid uptakes by the mammary glands. Because fatty acid uptakes were elevated on the high-fat treatment, fat secretion into milk increased. However, the decline in blood flow rate prevented amino acid uptakes from increasing concomitantly, and milk protein

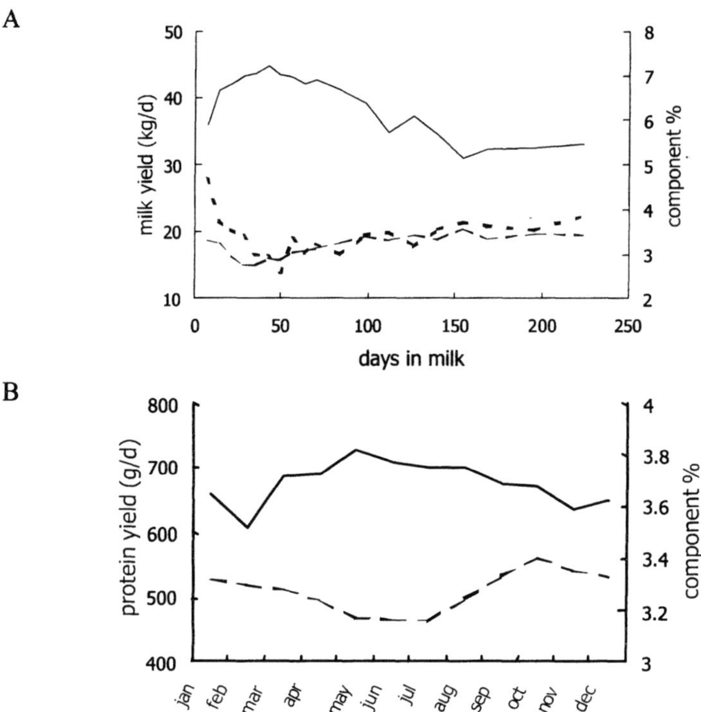

Figure 1.   Changes in composition of bovine milk (A) during lactation (one cow from University of Guelph research herd) and (B) with season (provincial means from Sargeant et al., 1998).   Yields are shown with a solid line, fat percentages with a dotted line, and protein percentages with a dashed line.

percentage dropped. Similarly, infusion of glucose into the bloodstream of cows to increase plasma glucose concentration from 3.63 to 6.37 mM caused a greater uptake of glucose into mammary tissue (Cant et al., submitted for publication). Again, mammary blood flow declined, thereby decreasing fatty acid uptakes (amino acid uptakes were not measured). The net result of elevated glucose concentration was an increase in lactose yield and total milk production, and a drop in fat and protein percentages.

Glucose and fatty acid uptakes into mammary tissue are related to their respective arterial concentrations, and the rate of synthesis of products follows suit (Miller et al., 1991; Cant et al., submitted for publication). The same cannot be said of amino acids and milk protein. A summary plot of data from 16 different publications (Figure 2A) shows a significant correlation between arteriovenous concentration difference of potentially limiting amino acids (met, phe, tyr and his) and their arterial concentrations. However, milk protein yield was not related to these arterial concentrations (Figure 2B), even when each experiment was plotted individually to remove between-experiment variance (Figure 2C). To further emphasize the lack of relationship, when a complete profile of 17 amino acids was infused into the arterial supply of the bovine udder, causing plasma group I amino acid concentrations to increase 3-fold, milk protein yield only increased from 842 to 991 g/day (Figure 2D). The net reaction of protein synthesis can be considered as

$$\text{amino acids} + \text{mRNA} + \text{ATP} \longrightarrow \text{protein}$$

where the "enzyme" facilitating the reaction is a mixture of ribosomal subunits plus initiation, elongation, and termination factors. The small milk protein yield response to amino acid concentration may have been due to a limitation in the concentration of the other substrates, mRNA, and ATP. However, we have previously calculated that a 12% increase in mammary glucose uptake (a relatively small change) would provide the ATP necessary for an additional 600 g/day of protein to be secreted into milk, even with degradation of 50% of the synthesized protein. Such numbers make it difficult to expect an ATP limitation to milk protein secretion rate. There is little experimental evidence regarding effects of mRNA concentration on milk protein secretion, but McClenaghan et al. (1995) inserted multiple copies of sheep ?-lactoglobulin gene into mice. The transgenic mice secreted sheep protein at 7 to 33 mg/mL of milk, while total protein content remained constant at 80 mg/mL. These results suggest that translational machinery, not mRNA concentration, limited protein secretion. Because there is little kinetic response in milk protein secretion to any of the substrates, it was suggested that the process was operating very near to its V_{max}.

In summary, a model of milk protein secretion by bovine mammary glands should be able to simulate the fat-induced milk protein depression, the decline in milk fat and protein percentage when arterial glucose concentration increases, the small response to arterial amino acid concentrations, the link between mammary blood flow rate and metabolic activity of the glands, and the reciprocal relationship between milk yield and milk fat and protein percentages.

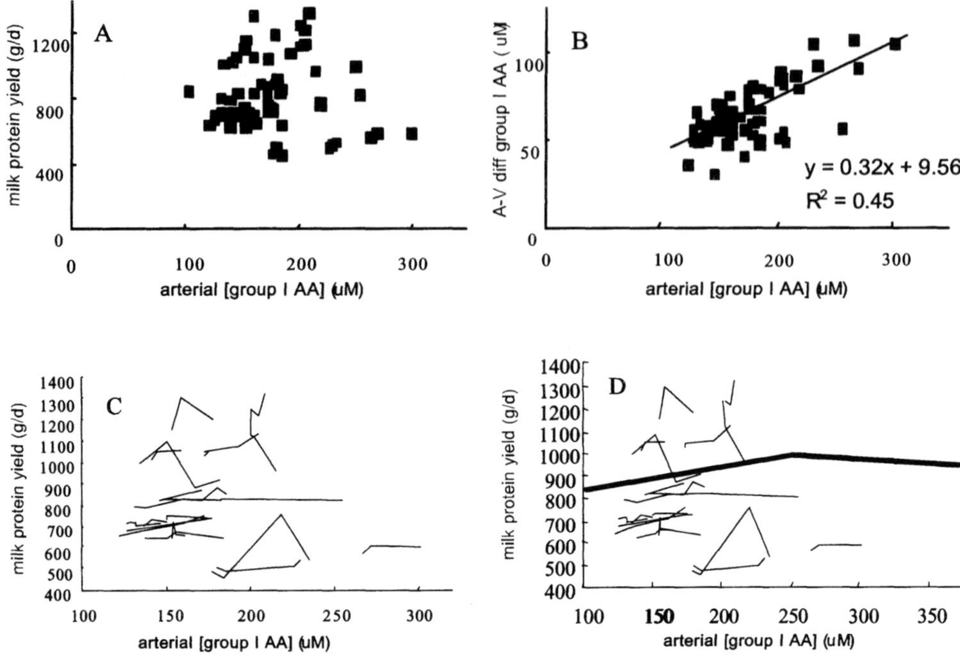

Figure 2. Relationship of (A) arteriovenous difference of group I essential amino acids (met, phe, tyr + his) and (B, C) milk protein yield to arterial concentration of group I essential amino acids from 16 different experiments (Bickerstaffe et al., 1974; Kellaway et al., 1974; Cant et al., 1993a; Coomer et al., 1993; Baker et al., 1996; Metcalf et al., 1996; Chan et al., 1997; Guinard and Rulquin, 1994a and b; Guinard and Rulquin, 1995; Lykos and Varga, 1997; Nichols et al., 1998; Vanhatalo et al., 1999; Varvikko et al., 1999; and Mackle et al., 2000). Relationships within each experiment are shown in panel C. The response to arterial infusion of 17 amino acids (Cant et al., 2001) is presented as a thick line in panel D.

SIMULATION STRATEGY 1

A model was developed to explore the hypothesis that mammary blood flow is controlled locally by vasodilatory mechanisms to maintain ATP production in secretory cells equal to ATP utilization (Cant and McBride, 1995). The model assumes that net uptake of milk precursors from blood into secretory cells is determined by a first order rate constant k describing capillary exchange. The concentration of substrate S flowing out of the venous end of a single capillary is then dependent on the transit time of blood through the capillary:

$$[S]_V = [S]_A \, e^{-k \frac{Vol_{cap}}{BF_{cap}}} \qquad (1)$$

where BF is blood flow. Assuming all capillaries in the udder are the same size and receive the same flow rate of blood, uptake is calculated for the whole udder as

$$\text{uptake } ? [S]_A (1 ? e^{? k \frac{Vol_{cap} N_{cap}}{BF_{tot}}}) BF_{tot} \tag{2}$$

The number of open capillaries (N_{cap}) and total blood flow rate (BF_{tot}) are modulated with time to maintain ATP balance.

In order to simulate a lactation curve with the model, a setpoint for milk production was introduced using the equation of Rook et al. (1993):

$$\text{setpoint } ? 2000e^{?0.0026t}(1 ? 0.5e^{0.08t}) \tag{3}$$

where t = day of lactation. Parameters were set for a reference cow producing 11,000 kg milk over a 305-day lactation which peaked at 47 kg/day. For the same reference cow, maximum rates of protein, fat, and lactose secretion were linear functions of the milk production setpoint:

$$U_{Aa,AaPt} ? \frac{0.632\,mmol/h ? setpoint}{1 ? \dfrac{1.6\,mM}{[Aa]_I}} \tag{4}$$

$$U_{Ac,AcFa} ? \frac{1.610\,mmol/h ? setpoint}{1 ? \dfrac{2.5\,mM}{[Ac]_I} ? \dfrac{[Fa]_I}{0.2\,mM}} \tag{5}$$

$$U_{Gl,GlLc} ? \frac{0.656\,mmol/h ? setpoint}{?1? \dfrac{0.2\,mM}{[Gl]_I} ??1? \dfrac{1.6\,mM}{[Aa]_I} ?} \tag{6}$$

where $U_{Aa,\,AaPt}$, $U_{ac,\,AcFa}$, and $U_{gl,\,GlLc}$ are the rates of utilization of amino acids, acetate, and glucose in the three processes, respectively. Interstitial concentrations of the four metabolites were calculated as logarithmic averages of their respective predicted arteriovenous differences:

$$[S]_I ? \frac{[S]_A ? [S]_V}{k\,Vol_{cap}N_{cap}/\,blood\,flow} \tag{7}$$

As in the original model (Cant and McBride, 1995), milk precursors taken into mammary tissue and not used for synthetic purposes were considered oxidized (Figure 3). The equations were solved numerically using Advanced Continuous Simulation Language software (ACSL, 1999). Milk yields plotted in Figure 4 were calculated from the synthesis rate of lactose ($U_{Gl, GlLc}$), assuming a constant proportion of lactose in milk.

The simulated ratio of mammary blood flow to milk yield (Figure 4) followed the same pattern as that derived from the measurements of Linzell (1960) and Kronfeld et al. (1968). Also, fat and protein percentages were predicted to drop in early lactation and then gradually increase, just as is observed in cows (Figure 1A). Although maximum capacities for lactose, fat, and protein secretion were all modified to the same extent by the temporal setpoint curve so that one might expect the predicted percentage composition of milk to remain unchanged throughout lactation, the change in blood flow had a greater impact on uptakes of metabolites removed inefficiently from blood. This is because capillary transit time ($Vol_{cap}N_{cap}$/blood flow) changed in the same direction as blood flow. When blood flow decreased, predicted capillary transit time fell, and the lower the k, the greater the impact on arteriovenous difference and uptake [Eq. (2)]. Glucose is the least efficiently extracted milk precursor of the four (Cant et al., 1993b), so that rates of lactose secretion and milk yield were affected by the change in blood flow more than were fat and protein secretion rates- hence, the fall and rise of flow/yield ratio and milk fat and protein contents. This lactation curve simulation depended on the precursor uptake rate being a major determinant of component secretion rate.

The model also adequately simulated the milk protein depression on fat feeding (Table 1). Yields of all milk components were underpredicted by about 10%, but the ratios between them (reflected in protein and fat percentages) were very close to observed values. However, milk protein percentage was much too sensitive to arterial amino acid concentrations to conclude that the model appropriately simulated the patterns of milk protein secretion normally observed (Figure 5).

Table 1. Predictions with simulation strategy 1 of mammary effects of dietary fat using arterial plasma concentrations of milk precursors as input[1]

	control	5% added fat observed	5% added fat predicted
arterial [AA] (mM)	1.77		1.77
arterial [acetate] (mM)	2.2	2.2	
arterial [LCFA] (mM)	0.784		1.154
arterial [glucose] (mM)	3.71		3.72
mammary blood flow (L/h)	725	687	685
lactose yield (g/d)	1183	1226	1142
protein yield (g/d)	718	746	677
protein %	2.97	2.87	2.89
fat yield (g/d)	770	950	843
fat %	3.18	3 .66	3.6

[1] Observations are from Gachuiri (1993). Equation parameters were set to match prediction with observation on the control diet.

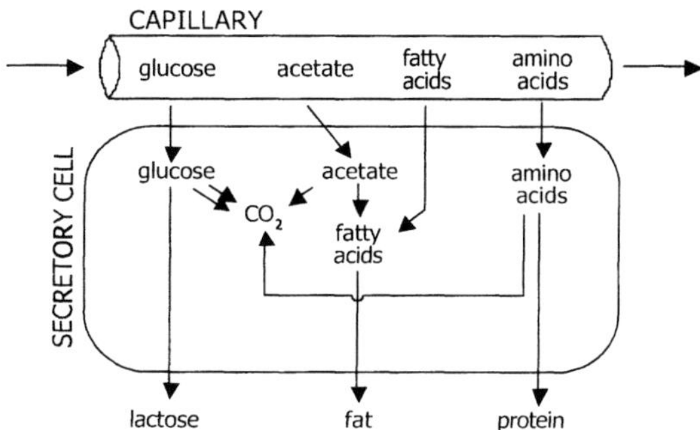

Figure 3. Mammary uptake and metabolism of milk precursors considered in simulation strategy 1.

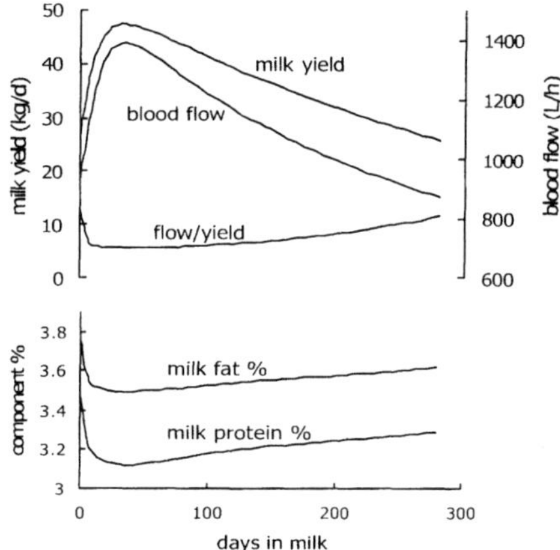

Figure 4. Milk yield, blood flow, and milk composition over the course of a lactation predicted using simulation strategy 1 (see text for details).

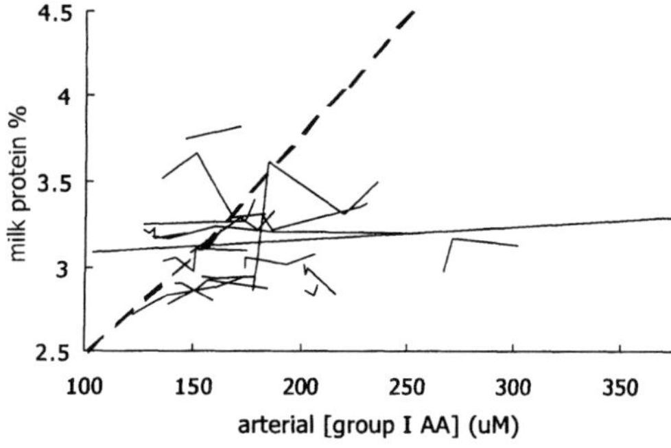

Figure 5. Simulation strategy 1. Predicted relationship between milk protein percentage and arterial concentration of group I essential amino acids (dashed line) overlayed on observed data from Figure 2.

SIMULATION STRATEGY 2

The determinant of milk protein yield in the first simulation strategy was the rate constant k for net capillary exchange of amino acids. Because of the failure of this approach, it was decided to let the V_{max} of a Michaelis-Menten-type relationship be the determinant of protein secretion rate. In this scenario, net amino acid uptake would be dependent on milk protein synthesis rate instead of arterial amino acid concentrations. Additional amino acids taken out of the blood would not necessarily end up in milk protein if the process was at saturation, so an escape route for excess amino acid would be needed. That escape route is known to be reflux back out of the cell into venous drainage (Bequette et al., 2000). Bidirectional flux across the capillary wall during flow through the vascular space is difficult to simulate with ordinary differential equations, because there are changes with time and distance occurring simultaneously. Additionally, we have found that the short-term venous dilution of multiple indicators injected into the arterial supply of a cow's udder is adequately simulated with an assumption of instantaneous mixing of milk precursors between vascular and extracellular compartments (Qiao, F. and J.P. Cant, unpublished data). That same assumption was used by Hanigan et al. (1998b) in calculation of rate constants for the net uptake of amino acids into mammary tissue. Their model was essentially that shown in Figure 6A, where a well-mixed extracellular pool receives substrate from arterial blood and cell reflux, and it loses substrate to venous drainage and cellular influx. The intracellular substrate is used for biosynthetic and catabolic processes in our model according to Michaelis-Menten saturation kinetics.

Mammary metabolism of glucose, acetate, and amino acids were each simulated with the two-compartment model (Figure 6B). Cherepanov et al. (2000) used the same approach. Blood flow was regulated as in strategy 1 to maintain intracellular ATP balance but, to avoid overprediction of blood flow responses, the V_{max} for catabolism in each two-pool model was also modified by the same ATP balance equations.

This second model simulated the drop in mammary blood flow observed during arterial glucose infusion (Table 2). A lactation curve was again simulated with setpoint modification [Eq. (3)] of the respective V_{max} for protein, fat, and lactose secretion. With the proper set of parameters describing uptake and metabolism of the milk precursors, blood flow as a proportion of milk yield dropped in early lactation and then increased by almost two-fold (Figure 7). Mabjeesh et al. (2000) observed a two-fold increase in flow/yield from day 80 to 233 of lactation in goats and suggested that the effect was contradictory to the concept of flow regulation by energy metabolism. In fact, our simulation of the same increase was based solely on an energy balance mechanism. This example highlights the value of mechanistic modeling to correct or focus intuitive interpretation of experimental observations. Milk fat percentage was much more sensitive to the changes in V_{max} than was protein percentage, but this is often the case in cows (Figure 2A). Most importantly, the second simulation strategy was insensitive to elevated concentrations of amino acids (Figure 8).

Table 2. Predictions with simulation strategy 2 of mammary effects of arterial glucose infusion using arterial plasma concentrations of milk precursors as input[1]

	control	infused glucose observed	infused glucose predicted
arterial [AA] (mM)	2	2	
arterial [acetate] (mM)	2.92	2.85	
arterial [LCFA] (mM)	0.732	0.641	
arterial [glucose] (mM)	3.63	6.37	
mammary blood flow (L/h)	597	351	363
lactose yield (g/d)	635	758	1023
protein yield (g/d)	497	502	494
protein %	3.45	3.14	2.13
fat yield (g/d)	619	610	543

[1]Observations are from Cant et al. (submitted for publication). Equation parameters were set to match predictions with observations on the control treatment.

CONCLUSION

If milk protein yields cannot be predicted from circulating amino acid concentrations, then what should a management model use as the predictor? Subnel et al. (1994) found that the efficiency of milk protein production was highly correlated with protein/energy ratios in the diet of cows. Hanigan et al. (1998a) reported that dietary energy intake explained more of the variance in milk protein yield than did dietary protein intake. In the 16 experiments used to generate Figure 2, which shows no relationship between protein yields and amino acid concentrations, net energy intakes by the cows explained 75% of the variance in milk protein yield (Figure 9). The residuals were regressed against individual and summed amino acid concentrations, and the best fit was obtained with variables and parameters shown in Table 3. Amino acid concentrations explained 68% of the residual variance and, together with net energy intake (Figure 9), 92% of the variance in milk protein yield was accounted for (Table 3). We suggest that hormonal signals integrating information on dietary nutrient supply are primary determinants of milk protein yield, with much smaller effects of the concentrations of amino acids and potentially limiting amino acids in blood. Thus, an accurate and biologically consistent mammary model will have to consider the hormonal control of milk synthesis.

The second simulation strategy was able to simulate the major patterns of change in milk protein production by bovine mammary glands. Specifically, these were the fat-induced milk protein depression, the decline in milk fat and protein percentage when arterial glucose concentration increases (Table 2), the small response to arterial amino acid concentrations (Figure 8), the link between mammary blood flow rate and metabolic activity of the glands (Table 2 and Figure 7), and the reciprocal relationship between milk yield and milk fat and protein percentages (Figure 7). Future work with the model should

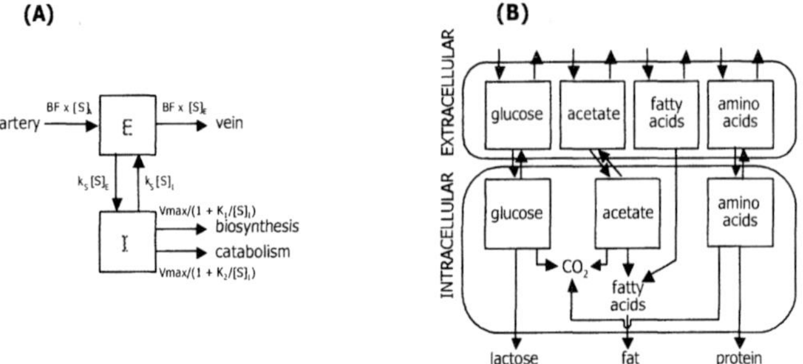

Figure 6. Flow diagrams of (A) extracellular and intracellular metabolite pools and (B) mammary uptake and metabolism of milk precursors considered in simulation strategy 2.

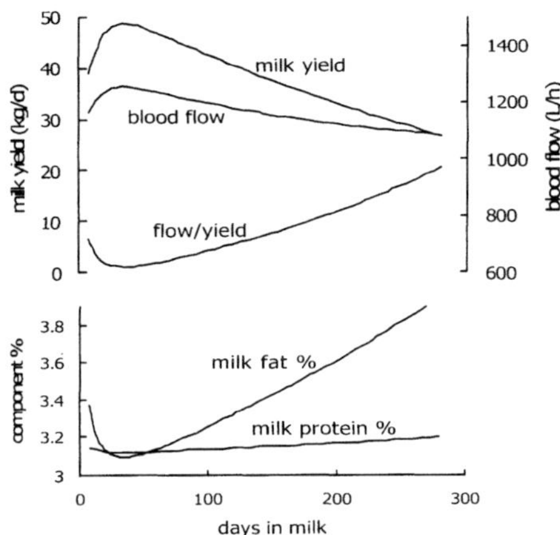

Figure 7. Milk yield, blood flow, and milk composition over the course of a lactation predicted using simulation strategy 2 (see text for details).

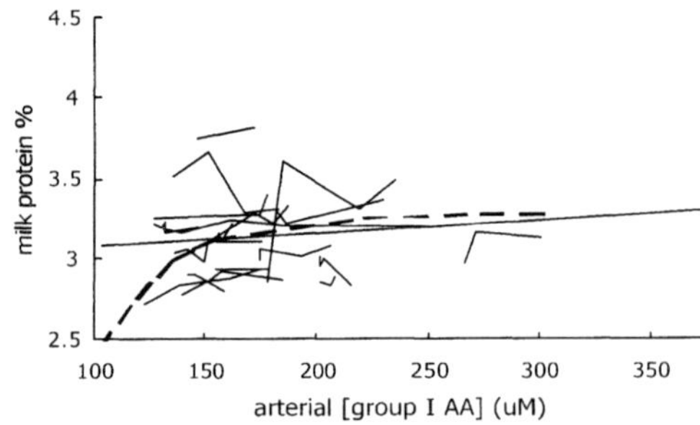

Figure 8. Simulation strategy 2. Predicted relationship between milk protein percentage and arterial concentration of group I essential amino acids (dashed line) overlayed on observed data from Figure 2.

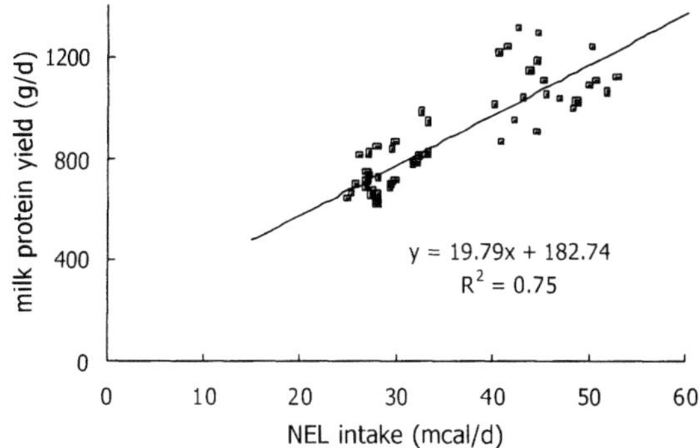

Figure 9. Relationship of milk protein yield to dietary intake of net energy for lactation (NEL) from 16 different experiments (Bickerstaffe et al., 1974; Kellaway et al., 1974; Cant et al., 1993a; Coomer et al., 1993; Guinard and Rulquin, 1994a and b; Guinard and Rulquin, 1995; Baker et al., 1996; Metcalf et al., 1996; Chan et al., 1997; Lykos and Varga, 1997; Nichols et al., 1998; Vanhatalo et al., 1999; Varvikko et al., 1999; and Mackle et al., 2000).

Table 3. Results of linear regression of 51 observed milk protein yields on net energy intake and amino acid concentrations in blood from 16 different experiments[1]

variable	parameter	SE	r^2root	MSE
dependent variable = residual after milk protein yield vs. NE_L intake				
intercept (g/d)	-359	85.9		
[total amino acids]$_{artery}$ (mM)	0.158	0.051		
[met+phe+tyr+his]$_{artery - vein}$ (mM)	5.87	1.31		
[his]$_{artery}$ (mM)	2.6	0.64		
[leu]$_{artery - vein}$ (mM)	-7.62	0.99	0.68	60
dependent variable = milk protein yield				
intercept (g/d)	-190.8	86.5		
NE_L intake (Mcal/d)	20.78	1.3		
[total amino acids]$_{artery}$ (mM)	0.154	0.051		
[met+phe+tyr+his]$_{artery - vein}$ (mM)	5.87	1.32		
[his]$_{artery}$ (mM)	2.51	0.67		
[leu]$_{artery - vein}$ (mM)	-7.88	1.09	0.92	60

[1] Data are from Bickerstaffe et al., 1974; Kellaway et al., 1974; Cant et al., 1993a; Coomer et al., 1993; Guinard and Rulquin, 1994a and b; Guinard and Rulquin, 1995; Baker et al., 1996; Metcalf et al., 1996; Chan et al., 1997; Lykos and Varga, 1997; Nichols et al., 1998; Vanhatalo et al., 1999; Varvikko et al., 1999; and Mackle et al., 2000.

include tests of the accuracy of predictions in real situations and experimentation to identify signal and response participants in the regulation of V_{max} for milk protein secretion. Additionally, mechanisms of mammary blood flow changes with amino acid supply should be investigated. Such effects have been observed (Bequette et al., 2000; Rulquin et al., 2000) but are not understood well enough to simulate mechanistically.

ACKNOWLEDGEMENT

This work was supported by funding from NSERC Canada.

CORRESPONDENCE

Please address all correspondence to:
John Cant
Department of Animal and Poultry Science
University of Guelph
Guelph, Ontario, Canada N1G 2W1
jcant@aps.uoguelph.ca

REFERENCES

Advanced Continuous Simulation Language (ACSL) Reference Manual, 1999, AEgis Simulation, Inc., Huntsville.

Baker, M.J., Amos, H.E., Nelson, A., Williams, C.C., and Froetschel, M.A., 1996, Undegraded intake protein: effects on milk production and amino acid utilization by cows fed wheat silage, *Canad. J. Anim. Sci.* 76:367-376.

Bequette, B.J., Hanigan, M.D., Calder, A.G., Reynolds, C.K., Lobley, G.E., and MacRae, J.C., 2000, Amino acid exchange by the mammary gland of lactating goats when histidine limits milk production, *J. Dairy Sci.* 83:765-775.

Bickerstaffe, R., Annison, E.F., and Linzell, J.L., 1974, The metabolism of glucose, acetate, lipids and amino acids in lactating dairy cows, *J. Agric. Sci.* 82:71-85.

Cant, J.P., and McBride, B.W., 1995, Mathematical analysis of the relationship between blood flow and uptake of nutrients in the mammary glands of a lactating cow, *J. Dairy Res.* 62:405-422.

Cant, J.P., DePeters, E.J., and Baldwin, R.L., 1993a, Mammary amino acid utilization in dairy cows fed fat and its relationship to milk protein depression, *J. Dairy Sci.* 76:762-774.

Cant, J.P., DePeters, E.J., and Baldwin, R.L., 1993b, Mammary uptake of energy metabolites in dairy cows fed fat and its relationship to milk protein depression, *J. Dairy Sci.* 76:2254-2265.

Cant, J.P., Trout, D.R., Qiao, F., and McBride, B.W., 2001, Milk composition responses to unilateral arterial infusion of complete and histidine-lacking amino acid mixtures to the mammary glands of cows, *J. Dairy Sci.* 84:1192-1200.

Chan, S.C., Huber, J.T., Theurer, C.B., Wu, Z., Chen, K.H., and Simas, J.M., 1997, Effects of supplemental fat and protein source on ruminal fermentation and nutrient flow to the duodenum in dairy cows, *J. Dairy Sci.* 80:152-159.

Cherepanov, G.G., Danfaer, A., and Cant, J.P., 2000, Simulation analysis of substrate utilization in the mammary gland of lactating cows, *J. Dairy Res.* 67:171-188.

Coomer, J.C., Amos, H.E., Williams, C.C., and Wheeler, J.G., 1993, Response of early lactation cows to fat supplementaton in diets with different nonstructural carbohydrate concentrations, *J. Dairy Sci.* 76:3747-3754.

Gachuiri, C.K., 1993, *Effects of Supplemental Dietary Fat on Rumen Fermentation, Nutrient Flow to the Duodenum, and Milk Yield and Composition of Lactating Dairy Cows,* Ph.D. dissertation, University of California, Davis.

Guinard, J., and Rulquin, H., 1994a, Effect of graded levels of duodenal infusions of casein on mammary uptake in lactating cows. 2. Individual amino acids, *J. Dairy Sci.* 77:3304-3315.

Guinard, J., and Rulquin, H., 1994b, Effects of graded amounts of duodenal infusions of lysine on the mammary uptake of major milk precursors in dairy cows, *J. Dairy Sci.* 77:3565-3576.

Guinard, J., and Rulquin, H., 1995, Effects of graded amounts of duodenal infusions of methionine on the mammary uptake of major milk precursors in dairy cows, *J. Dairy Sci.* 78:2196-2207.

Hanigan, M.D., Cant, J.P., Weakley, D.C., and Beckett, J.L., 1998a, An evaluation of postabsorptive protein and amino acid metabolism in the lactating dairy cow, *J. Dairy Sci.* 81:3385-3401.

Hanigan, M.D., France, J., Wray-Cahen, D., Beever, D.E., Lobley, G.E., Reutzel, L., and Smith, N.E., 1998b, Alternative models for analyses of liver and mammary transorgan metabolite extraction data, *Brit. J. Nutr.* 79:63-78.

Kellaway, R.C., Ranawana, S.S.E., Buchanan, J.H., and Smart, L.D., 1974, The effect of nitrogen source in the diet on milk production and amino-acid uptake by the udder, *J. Dairy Res.* 41:305-315.

Kronfeld, D.S., Raggi, F., and Ramberg, C.F., Jr., 1968, Mammary blood flow and ketone body metabolism in normal, fasted, and ketotic cows, *Am. J. Physiol.* 215:218-227.

Linzell, J.L., 1960, Mammary-gland blood flow and oxygen, glucose and volatile fatty acid uptake in the conscious goat, *J. Physiol.* 153:492-509.

Linzell, J.L., and Peaker, M., 1971, Mechanism of milk secretion, *Physiol. Rev.* 51:564-597.

Lykos, T., and Varga, G.A., 1997, Varying degradation rates of total nonstructural carbohydrates: effects on nutrient uptake and utilization by the mammary gland in high producing Holstein cows, *J. Dairy Sci.* 80:3356-3367.

Mabjeesh, S.J., Kyle, C.E., MacRae, J.C., and Bequette, B.J., 2000, Lysine metabolism by the mammary gland of lactating goats at two stages of lactation, *J. Dairy Sci.* 83:996-1003.

Mackle, T.R., Dwyer, D.A., Ingvartsen, K.L., Chouinard, P.Y., Ross, D.A., and Bauman, D.E., 2000, Effects of insulin and postruminal supply of protein on use of amino acids by the mammary gland for milk protein synthesis, *J. Dairy Sci.* 83:93-105.

McClenaghan, M., Springbett, A., Wallace, R.M., Wilde, C.J., and Clark, A.J., 1995, Secretory proteins compete for production in the mammary gland of transgenic mice, *Biochem. J.* 310:637-641.

Metcalf, J.A., Wray-Cahen, D., Chettle, E.E., Sutton, J.D., Beever, D.E., Crompton, L.A., MacRae, J.C., Bequette, B.J., and Backwell, F.R.C., 1996, The effect of dietary crude protein as protected soybean meal on mammary metabolism in the lactating dairy cow, *J. Dairy Sci.* 79:603-611.

Miller, P.S., Reis, B.L., Calvert, C.C., DePeters, E.J., and Baldwin, R.L., 1991, Patterns of nutrient uptake by the mammary glands of lactating dairy cows, *J. Dairy Sci.* 74:3791-3799.

Nichols, J.R., Schingoethe, D.J., Maiga, H.A., Brouk, M.J., and Piepenbrink, M.S., 1998, Evaluation of corn distillers grains and ruminally protected lysine and methionine for lactating dairy cows, *J. Dairy Sci.* 81:482-491.

Rook, A.J., France, J., and Dhanoa, M.S., 1993, On the mathematical description of lactation curves, *J. Agric. Sci.* 121:97-102.

Rulquin, H., and Pisulewski, P.M., 2000, Effects of duodenal infusion of graded amounts of His on mammary uptake and metabolism in dairy cows, *J. Dairy Sci.* 83 (Suppl 1):164.

Sargeant, J.M., Leslie, K.E., Shoukri, M.M., Martin, S.W., and Lissemore, K.D., 1998, Trends in milk component production in dairy herds in Ontario: 1985-1994, *Canad. J. Anim. Sci.* 78:413-420.

Subnel, A.P.J., Meijer, R.G.M., van Straalen, W.M., and Tamminga, S., 1994, Efficiency of milk protein production in the DVE protein evaluation system, *Livest. Prod. Sci.* 40:215-224.

Vanhatalo, A., Huhtanen, P., Toivonen, V., and Varvikko, T., 1999, Response of dairy cows fed grass silage diets to abomasal infusions of histidine alone or in combinations with methionine and lysine, *J. Dairy Sci.* 82:2674-2685.

Varvikko, T., Vanhatalo, A., Jalava, T., and Huhtanen, P., 1999, Lactation and metabolic responses to graded abomasal doses of methionine or lysine in cows fed grass silage diets, *J. Dairy Sci.* 82:2659-2773.

MECHANISTIC EQUATIONS TO REPRESENT DIGESTION AND FERMENTATION

Richard A. Kohn[*]

INTRODUCTION

Current mathematical models use different forms of equations to represent feed degradation and microbial fermentation in the rumen. For example, the Cornell Net Carbohydrate and Protein System (CNCPS; Sniffen et al., 1992) assumes that feed digestion rate is first order with respect to substrate, but the models of Baldwin et al. (1987) and Dijkstra et al. (1992) assume Michaelis-Menten kinetics. A model proposed by Kohn and Boston (2000) suggests an alternative form of equations that incorporates the Second Law of Thermodynamics to represent fermentation processes in the rumen. The purpose of this paper is to explore the differences among these approaches, including the form of the equations, the theoretical basis and assumptions, and the behavior. This analysis will identify a number of important research questions to pursue.

Ruminal fermentation initially results in the degradation of macromolecules such as starch, fiber, and protein to short-term intermediates such as sugars and amino acids. The products of this initial degradation are readily metabolized to microbial mass and carbon dioxide (CO_2), methane (CH_4), ammonia (NH_3), and volatile fatty acids (VFA). The latter are primarily acetate, propionate, butyrate, and, to a lesser degree, branched chain VFA and occasionally lactate (Prins, 1977). The control mechanisms, and thus the appropriate forms of equations, are likely to differ for initial degradation reactions and the subsequent fermentation reactions. Degradation is characterized by an abundant supply of substrate (e.g., cellulose, protein) with very little product (e.g., glucose, amino acids). The substrate offers considerable resistance to degradation. Fermentation reactions are characterized by a low level of substrate (e.g., glucose, amino acids) and an accumulation of products (e.g., VFA, ammonia). The substrates are readily attacked, but there are limitations (absorption and passage) to removal of products. Because of these differences, in this paper, different models will be contrasted for both the degradation and fermentation reactions.

---

[*] Richard A. Kohn, Department of Animal and Avian Science, University of Maryland, College Park, MD 20742.

FIRST ORDER DEGRADATION

The term 'mass action kinetics' is routinely used to refer to a first-order reaction in which the rate of the reaction depends linearly on the amount of reactant. Plots representing first order degradation are shown in Figure 1. The disappearance of substrate in this way is referred to as exponential decay or disappearance, and the appearance of product is referred to as an inverted exponential. See Appendix 1 for a review of the mathematical basis of these equations. Variations of these first order equations are routinely used to fit experimental data representing degradation of fiber and protein in the rumen (Mertens, 1993). When nylon bags containing feed samples are suspended in the rumen of cattle or sheep, or when feeds are incubated with ruminal contents, the macromolecules disappear in a manner reflecting first order degradation. End products of fermentation (gases, VFA) accumulate during *in vitro* fermentation similar to the products shown in Figure 1.

However, there is a caveat. These equations are often altered slightly to improve the fit of the data observed. Especially for fiber, lag assumptions are frequently included (Mertens, 1993). The rate of disappearance of substrate and the rate of appearance of products are delayed or slowed initially. These changes to the simple exponential equations often improve the fit of the data, but they also raise a number of questions. Is the observed lag effect a part of the degradation process that needs to be included to describe how susceptible a feed is to degradation, is it an artifact of the method used to estimate degradation, or is it both? The answers to these questions are essential to determining how to use the data obtained from *in situ* or *in vitro* studies.

Consider an *in vitro* fermentation in which ruminal microbes are incubated with buffer and feed samples. The initial degradation may be slowed by the reduced concentration of microbes initially or by their initial inactivity due to the shock of the

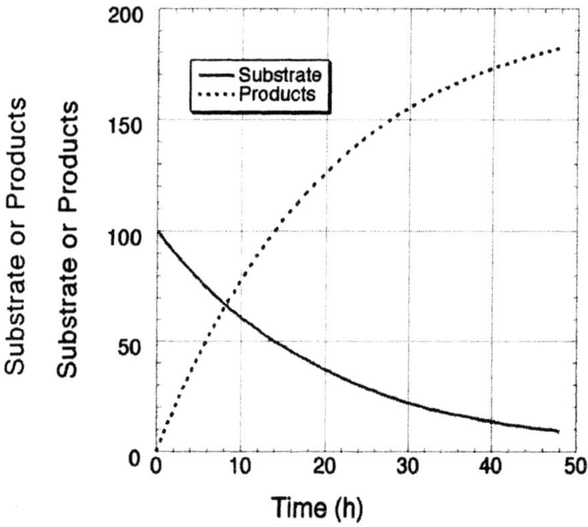

Figure 1. Exponential decay and inverted exponential reactions.

transition from the rumen to the lab. Thus, this would be an artificial situation that should be accounted for and removed from models of digestion of these feeds. One would fit the data using a lag function simply to improve the estimate of the degradation rate. Once the digestion rate is determined, the data describing the lag function would be discarded when developing a model of feed degradation because the lag would not be relevant to the *in vivo* data. In practice, all of the models of feed degradation known to this author disregard data related to lag.

However, what if the lag is a real part of the degradation process? For example, it may represent the time required to hydrate the feed before it can be attacked by microbes. It may also represent the time needed for microbes to attach to the feed. Or, in the case of cellulose, it may represent the way in which the cellulose microfibrils unravel and become more available to degradation once the process of degradation begins. In these cases, fitting a lag function to improve prediction of the rates and then discarding the data describing the lag would make the overall predictions from the digestion model less accurate than fitting the simple exponential. Allen and Mertens (1988) point out, however, that, for some predictions, the lag descriptors cancel out and thus do not need to be included.

The CNCPS is a static kinetic model describing digestion in the rumen and other aspects of cattle nutrition. A graphical depiction of the CNCPS rumen model is shown in Figure 2. The particular rates represented in the CNCPS are the rate of passage and the rate of degradation for feed fractions. Each feed fraction (e.g., soluble protein and soluble carbohydrate) for each feed has a different fractional rate constant for passage and digestion. The change in the amount of feed in the rumen with respect to time (dR/dt) equals the intake (F) minus the disappearance from digestion ($k_d R$) and passage ($k_p R$), where R is the amount of feed in the rumen. That is,

$$dR/dt = F - k_d R - k_p R \qquad (1)$$

In addition to assuming first order kinetics of digestion and passage, the CNCPS also assumes steady state (i.e., that the inflow to the rumen over time equals the outflow). Thus, on average

$$dR/dt = 0 \qquad (2)$$

and Eq. (1) simplifies to

$$F = R (k_d + k_p) \qquad (3)$$

Furthermore, the fraction of feed degraded in the rumen (D) can be defined as

$$D = k_d R / F \qquad (4)$$

Substituting Eq. (3) into Eq. (4) yields

$$D = k_d R / \{ R (k_d + k_p) \} \qquad (5)$$

which simplifies to

$$D = k_d \ / \ (k_d + k_p) \tag{6}$$

Thus, the CNCPS does not actually predict the amount of feed in the rumen, but it uses assumptions about rates of digestion and passage to predict the fraction of feed components digested in the rumen. These assumptions can be tested by calculating the amount of feed in the rumen that would be predicted from a given intake and the fractional rates.

The theoretical basis for assuming that digestion and passage are first order with respect to substrate has been considered (Van Soest, 1994). Any given macromolecule in the rumen is a candidate for either passage or digestion. Logically, the greater the amount of a given feed fraction (e.g., cellulose, starch, or protein), the more likely it will be digested or pass from the rumen. The lag effects that are observed *in vitro* or *in situ* are not included in this model. Also, the amount of microbial activity would not affect the rate of digestion or passage.

MICHAELIS-MENTEN KINETICS

In contrast to first order kinetics, Michaelis-Menten kinetics is responsive to changes in enzyme activity, or in the case of the rumen, microbial activity. One representation of the equation is

$$dP/dt = k_2 \, [E]_0 \, [S]_0 \ / \ \{ \ [S]_0 + K_m \ \} \tag{7}$$

where $[E]_0$ and $[S]_0$ represent initial enzyme and substrate concentrations, respectively, k_2 represents a rate constant, and K_m represents a constant for a given substrate. When $[S]_0 = K_m$, the reaction proceeds at half the maximal velocity (V_{max}). V_{max} is the rate of product formation when all enzyme is saturated by substrate ($V_{max} = k_2 \, [E]_0$). Derivation of the equation is provided in Appendix 2.

Several assumptions are made to derive the Michaelis-Menten equation. One assumption is that the rate of digestion must be limited by enzyme concentration (i.e., $[E]_0$

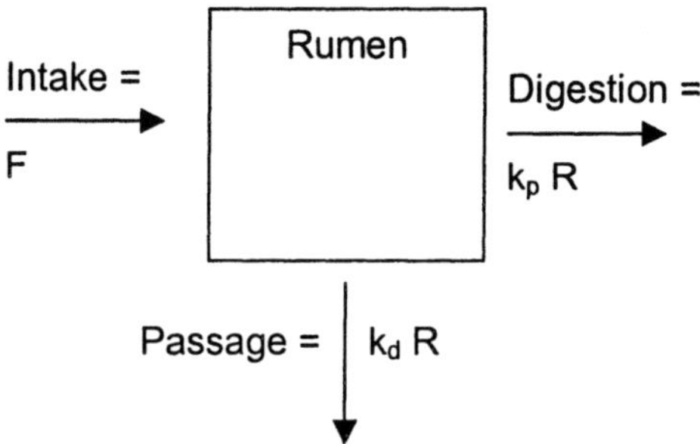

Figure 2. A schematic of the Cornell Net Carbohydrate and Protein System ruminal digestion model.

$< [S]_0$). Imagine that you have a small amount of enzyme and a moderate amount of substrate. In this case, the chance of a complex forming between substrate and enzyme depends on the substrate bumping into an enzyme. Increasing the amount of enzyme would increase the chance of complex formation and increase the rate of the reaction. Also, increasing the amount of substrate would increase the chance for a collision, up to a point. Once the enzyme sites are completely saturated by a high level of substrate, V_{max} is achieved, and there would be no advantage to further increasing substrate, although more enzyme would still increase the reaction velocity. Therefore, an assumption of the Michaelis-Menten equation is that there will come a time when increasing substrate will no longer continue to increase the rate of feed digestion. Models with Michaelis-Menten kinetics would therefore predict a decrease in fractional digestion rate due to higher rumen concentration of substrate. In contrast to both of these approaches, France et al. (1982) represented degradation as first order with respect to *both* substrate and microbe, and the microbes were allowed to increase by growth. Fitting this model yields the S-shaped logistic growth curve familiar to microbiologists. The growth of microbes or production of product begins slowly and then increases until the substrate disappears and the reaction slows. It is precisely an inverted exponential with a lag. In this case, degradation depends on both enzyme and substrate concentration, but the system can not be saturated.

The reaction considered for the Michaelis-Menten equation is as follows:

$$E + S \; \rightleftharpoons \; ES \longrightarrow P$$

But in the rumen, E is represented by microbes (M), and they are both the enzyme and a product of the reaction as shown here:

$$M + S \; \rightleftharpoons \; MS \longrightarrow M$$

Consider the rumen immediately after a meal or an *in vitro* system at the start of incubation. Microbes (enzyme) may be in very low concentration, and the system may be near the customary V_{max}. Increasing enzyme would increase degradation, but changing substrate concentration would have no effect. As microbes grow and substrate is degraded, degradation would increase, and the system would become dependent on substrate concentration according to a Michaelis-Menten type equation. As enzyme continues to increase while substrate decreases, there might be more enzyme sites than sites on the substrate for the enzyme to bind. Under these conditions, an alternative equation can be derived representing the reaction velocity:

$$dP/dt = k_2 \, [E]_0 \, [S]_0 \, / \, \{ \, [E]_0 + K_m \, \} \tag{8}$$

This equation is the other solution to the polynomial set up when deriving the Michaelis-Menten type equation. It is derived by making the assumption that $[E] > [S]$. If the enzyme concentration is very high, the velocity approaches

$$dP/dt = k_2 \, [S]_0 \tag{9}$$

Note that the rate of digestion in this case is a constant times initial substrate concentration as was assumed in Eq. (1), the primary assumption of the first order digestion rate. Therefore, there is a sound theoretical basis for the first order assumption if enzyme concentrations are high relative to substrate concentrations. But if substrate concentrations are high relative to enzyme concentrations, the Michaelis-Menten assumption is more appropriate. While one form of the equation may give way to another under different circumstances, these equations are clearly not the same in terms of their behavior. It would be appropriate here to compare how these different forms fit actual data, but that research is not trivial considering all the possible conditions that can occur in ruminants.

These different approaches attempt to explain degradation of macromolecules in the rumen mechanistically. However, the profiles of different VFA are typically predicted with empirical equations related to substrate type.

THERMODYNAMIC CONTROL

Chemical reactions are controlled either kinetically or thermodynamically (Chang, 1981). Both first order and Michaelis-Menten equations are examples of kinetic control of reactions. If the rates of utilization of a substrate control the pathways for product formation when all of the reactions are thermodynamically favorable, the reactions are kinetically controlled. These rates may depend on substrate and/or enzyme concentrations as described previously. In contrast, Kohn and Boston (2000) suggest that assuming thermodynamic control of glucose fermentation may lead to development of mechanistic equations to predict the profile of different VFA. If reactants are sufficiently limited relative to products, then the reactions cannot proceed according to the Second Law of Thermodynamics. Under these circumstances, thermodynamics controls which pathway branches are available, and net reaction rates depend on both substrate and product concentrations.

While thermodynamics may not directly affect the degradation of macromolecules, it may affect fermentation of glucose in the rumen. Consider a system in which key intermediates are found in very low concentrations relative to products. For example, one would not expect to find appreciable concentrations of glucose in the rumen where fermentation occurs rapidly. This expectation results from our understanding that glucose is rapidly metabolized to other compounds. If the very moment a glucose molecule is released into solution, it is a candidate for metabolism, whether that molecule becomes acetate, propionate, butyrate, or lactate would depend on the speed of the respective reactions or at least on how quickly the reactions can remove the molecule from solution. If the end products build up, however, VFA and especially H_2 may inhibit the reaction from occurring at all based on the Second Law of Thermodynamics. The Second Law of Thermodynamics requires that free energy (G) be released for a process to occur spontaneously. If the concentration of precursor becomes small enough relative to product, the forward reaction will not proceed. In this case, the precursor may be used to produce a different product, and there may be a shift from one VFA or gas to another.

Integration of the Second Law of Thermodynamics with the Ideal Gas Law provides the mathematical relationship that is necessary to develop a model incorporating thermodynamics:

$$\Delta G = \Delta G^\circ + RT \ln\{[\text{Products}] / [\text{Reactants}]\} \tag{10}$$

where [ ] represents activity (or pressure in atmospheres for ideal gases and molarity for ideal solutes), R is the gas constant and equals 8.314 J $^\circ K^{-1}$ mol^{-1}, T represents the temperature in degrees Kelvin, and ΔG° represents the change in free energy for the reaction under standard conditions. 'Standard conditions' are 278.15 $^\circ K$, one unit of activity for all solutes, and one atmosphere of pressure for all gases. If a reaction goes to equilibrium, ΔG is equal to zero. Therefore, ΔG° can be determined by measuring the concentration of products and reactants. Once this value is determined, it can be used with other similar values to determine the ΔG of other reactions contained within these reactions. The amount of free energy of a material is intrinsic to that material. Furthermore, once G is determined for potential reactants and products, the equilibrium constant for those reactions can be calculated easily. Setting ΔG to 0 in Eq. (10) represents equilibrium. Using book values for ΔG° enables determination of the equilibrium constant (K_{eq}) as follows:

$$K_{eq} = e^{-\Delta G/RT} \tag{11}$$

For reactions that are very fast and inexpensive to catalyze (i.e., no complex enzyme systems are needed), the reactions can approach equilibrium. However, complex biological pathways have built-in inefficiencies so that it is not feasible for the reactions to proceed all the way to equilibrium. The negative final ΔG of such reactions represents the free energy lost from the system and reflects the incomplete capture of free energy in the form of ATP generated. The efficiency of ATP synthesis can be determined from the concentrations of reactants and products. For example, Kohn and Boston (2000) calculated the efficiencies of some key reactions in the rumen under typical conditions. As expected, the efficiency for methanogenesis was higher (65%) than that for VFA production (56%). Methane production is a simpler process than VFA formation, so we should not be surprised to learn that less free energy is lost from carrying it out. Methanogenesis can occur at pH 7 and -0.33 V or more negative reducing potential (Wolfe and Higgins, 1979). An equivalent H_2 concentration would be observed at pH 6.5 and –0.30 V reducing potential. Thus, the conditions where methanogenesis can begin to occur are quite similar to the conditions in the rumen, and methanogenesis appears to be thermodynamically limited.

The observed efficiencies for acetate and propionate production approach the maximal efficiencies that are feasible, allowing for the necessary losses of heat from fermentation. Thus, the concentrations of acetate and propionate are probably limited by thermodynamics. On the other hand, maximal butyrate efficiency is probably higher than observed (43%). Therefore, this pathway may not approach the concentrations that would be infeasible, and it is probably not dependent on product concentration.

USE OF THERMODYNAMICS IN DYNAMIC MODELS

Judging from the previous analysis of the efficiencies of VFA production, it seems appropriate to incorporate thermodynamics into kinetic and dynamic models of fermentation. Production of acetate, propionate, methane, and lactate all seem to be limited by thermodynamics at some time. Dynamic models need to account for these limits so that infeasible concentrations are not predicted. Including thermodynamics will make the models more stable. Kohn and Boston (2000) calculated the efficiencies of free energy utilization for specific metabolite concentrations. Development of dynamic models using thermodynamics requires estimation of the maximum potential efficiencies. The actual efficiency may be lower unless the reactions approach thermodynamic limits. The ΔG at the maximum efficiency is called the threshold free energy (ΔG_T). It is the ΔG of a reaction when that reaction is as close as it can ever get to zero, but, for reactions with built-in inefficiency, it is always negative. The ΔG_T is equivalent to the unavoidable losses in free energy (i.e., heat loss from a system), and it can be calculated from a modification of Eq. (10):

$$\Delta G_T = \Delta G^\circ + RT \ln\{[\text{Products}]_T / [\text{Reactants}]_T\} \tag{12}$$

where $[\ X\ ]_T$ represents the concentration of reactant or product X when the system approaches its threshold. When a reaction approaches ΔG_T, the forward reaction rate equals the reverse reaction:

$$k_{rxn} [\text{Reactants}] = k_{-rxn} [\text{Products}] \tag{13}$$

where K_{rxn} and K_{-rxn} represent the fractional rate constants. Combining Eqs. (12) and (13) yields a ratio that is analogous to an equilibrium constant for reactions that never obtain equilibrium due to the inherent inefficiencies and complexities of the system:

$$k_{rxn} / k_{-rxn} = e^{(\Delta G^\circ - \Delta G^\circ T) / RT} \tag{14}$$

A dynamic model of glucose fermentation was developed (Kohn and Boston, 2000) assuming 56% efficiency of glucose fermentation to acetate, propionate, or butyrate, and 70% efficiency of methane production. These efficiencies were used to represent the highest efficiencies typically observed for the rumen. The ΔG_T was determined from these assumed efficiencies, and, from these values, the k_{rxn} / k_{-rxn} was determined for each reaction. Initial concentrations of each metabolite were set to approximate values that were similar to those observed in the rumen. Forward rate constants were set to values such that, when multiplied by each of the reactant concentrations, the resulting fluxes were realistic. The reverse rate constants were calculated as the k_{rxn} (as set previously) divided by k_{rxn} / k_{-rxn}. Glucose was assumed to be infused continuously into this system at a rate of 25 mmol h^{-1} L^{-1} and VFA were assumed to be removed at a fractional rate of 0.40 h^{-1} (Neal et al., 1992). As gas pressures exceeded 1 atm, the model simulated CO_2 and CH_4 escape in the proportions that they were predicted to be represented in the gas phase.

Table 1. Predicted steady state concentrations and fluxes for mathematical model incorporating thermodynamic and kinetic elements[1]

Metabolite	Concentration		
Glucose, mmol / L	0.15		
Acetate, mmol / L	59.1		
Propionate, mmol / L	38.6		
Butyrate, mmol / L	13.8		
CH_4, atm	0.31		
CO_2, atm	0.69		
H_2, atm	1.3×10^{-3}		
Acetate / Propionate	1.5		

Reaction	Flux (mmol substrate used per hour)		
	Forward	Reverse	Net
Glucose → 2 Acetate	15.2	3.6	11.6
Glucose → 2 Propionate	11.9	4.2	7.7
Glucose → Butyrate	5.4	10^{-16}	5.4
CO_2 → CH_4	14.4	3.6	10.8

[1] Model assumed 56% maximal free energy efficiency for acetate, propionate and butyrate formation, and 70% maximal efficiency for methane formation, pH 6.5, glucose infusion at 25 mmol L^{-1} h^{-1}, fatty acid removal at 0.40 / h, and gas removal as accumulated in proportion to partial pressures. From Kohn and Boston (2000).

The results of the simulation to steady state, shown in Table 1, appear to be a realistic starting point. While this model shows why a certain profile of VFA may form, it does not address why that profile may shift. For example, ionophores result in reduced methane, increased propionate, and higher pH. They permit ions to penetrate gram negative bacteria such as the acetate producers (Russell and Strobel, 1989). This effect may eventually be modeled by showing that ionophores increase the cost of acetate production by causing the gram negative organisms to expend ATP to repair internal ion concentrations. This cost would directly decrease the threshold ΔG (i.e., make it more negative) for acetate and would shift the equilibrium against acetate production. The glucose spared from this shift would further increase propionate production, which would decrease H_2 available for methane. In contrast, an equation to explain acetate production as first order with respect to substrate could not be used to model an effect of ionophores. Changes in microbial activity due to using ionophores would not affect the predictions made by first order equations. A Michaelis-Menten equation would be responsive to both microbial and substrate effects, and it may offer an alternative explanation. However, such a set of equations would risk exceeding thermodynamic boundaries.

Another example that needs to be explained is the effect of increasing energy density of a diet. As the energy density of the diet increases, acetate to propionate ratio and methane production decline while H_2 concentration increases. Because methane does not increase in response to the increased H_2, the calculated efficiency for methanogenesis would decrease. This change suggests that the ruminal conditions of high energy diets make it more energetically expensive to produce methane, causing the efficiency to decline. This change would favor propionate production over methane production. Therefore, thermodynamics suggest that methanogenesis is inhibited on high energy diets, thus causing the shift to propionate (Kohn and Boston, 1995). Recent research supports this hypothesis by suggesting that methanogenesis appears to be inhibited by low pH (Russell, 1998) which could result from rapid fermentation. If end products continue to build up in the rumen, propionate production may also be inhibited, and glucose and H_2 would accumulate until lactate production would become a thermodynamically feasible pathway.

CONCLUDING COMMENTS

The equations for modeling digestion and fermentation discussed in this paper raise many research questions and should help identify gaps in our understanding of digestion and fermentation. In particular, to what extent is digestion limited by enzyme availability and to what extent is it controlled by substrate? What role do the products of digestion and fermentation have in controlling pathway branches? A great deal more empirical research is needed to answer these questions and improve current theoretical models.

CORRESPONDENCE

Please address all correspondence to:
R.A. Kohn
University of Maryland
Department of Animal and Avian Sciences
College Park, MD 20842
rkohn@wam.umd.edu

REFERENCES

Allen, M.S., and Mertens, D.R., 1988, Evaluating constraints on fiber digestion by rumen microbes, *J. Nutr.* 118:261-270.

Baldwin, R.L., Thornley, J.H.M., and Beever, D.E., 1987, Metabolism of the lactating cow. II. Digestive elements of a mechanistic model, *J. Dairy Res.* 54:107-131.

Chang, R., 1981, The Second Law of Thermodynamics, in: *Physical Chemistry with Applications to Biological Systems, 2nd ed.,* Macmillian, New York.

Dijkstra, J., Neal, H.D.C., Beever, D.E., and France, J., 1992, Simulation of nutrient digestion, absorption and outflow in the rumen: model description, *J. Nutr.* 122:2239-2256.

France, J., Thornley, J.H.M., and Beever, D.E., 1982, A mathematical model of the rumen, *J. Agric. Sci. Camb.* 99:343-353.

Kohn, R.A., and Boston, R.C., 2000, The role of thermodynamics in controlling rumen metabolism, in: *Modelling Nutrient Utilization in Farm Animals,* J.P. McNamara, J. France, and D.E. Beever, eds, CABI Publishing, Wallingford.

Mertens, D.R., 1993, Rate and extent of digestion, in: *Quantitative Aspects of Ruminant Digestion and Metabolism,* J.M. Forbes and J. France, eds, CABI Publishing, Wallingford.

Neal, H.D.C., Dijkstra, J., and Gill, M., 1992, Simulation of nutrient digestion, absorption and outflow in the rumen: model evaluation, *J. Nutr.* 122:2257-2272.

Prins, R.A., 1977, Biochemical activities of gut micro-organisms, in: *Microbial Ecology of the Gut,* R.T.J. Clarke and T. Bauchop, eds, Academic Press, New York.

Russell, J.B., 1998, The importance of pH in the regulation of ruminal acetate to propionate ratio and methane production in vitro, *J. Dairy Sci.* 81:3222-3230.

Russell, J.B., and Strobel, H.J., 1989, Effect of ionophores on ruminal fermentation, *Appl. Environ. Microbiol.* 55:1-6.

Sniffen, C.J., O'Connor, J.D., Van Soest, P.J., Fox, D.G., and Russell, J.B., 1992, A net carbohydrate and protein system for evaluating cattle diets: II. Carbohydrate and protein availability, *J. Anim. Sci.* 70:3562-3577.

Van Soest, P.J., 1994, Mathematical applications: digestibility, in: *Nutritional Ecology of the Ruminant, 2nd ed.* Cornell University Press, Ithaca.

Wolfe, R.S., and Higgins, I.J., 1979, Microbial biochemistry of methane -- a study in contrasts. *Microbial Biochem.* 21:267-353.

APPENDIX 1. Derivation of Exponential (First Order) Decay Equation

The term 'mass action kinetics' is routinely used to refer to a first order reaction in which the rate of the reaction depends linearly on the amount of reactant. For an example, consider a process in which each mole of reactant R is degraded to produce a moles of product P:

$$R \rightarrow a\,P$$

If the reaction is first order with respect to the reactant, the differential equation representing the reaction is as follows:

$$dR/dt = -k\,[R] \qquad (1a)$$

where dR/dt represents the change (disappearance) of reactant over time, $-k$ represents the degradation rate, and $[R]$ represents the concentration of the reactant. If the reactant R is converted to a product P, an alternative equation can be used:

$$dP/dt = a\,k\,[R] \qquad (2a)$$

where a represents the number of moles of product produced per mole of reactant; other terms were defined previously. These two differential equations represent the change in reactant and product over time. To find the amount or concentration of reactant at a specific time requires taking the anti-derivative (integral) of Eq. (1a) which provides

$$[R]_t = [R]_0\,e^{-kt} \qquad (3a)$$

where $[R]_t$ represents the concentration of reactant at time t, and $[R]_0$ represents the concentration of reactant at the start. The concentration of product at any point in time equals the concentration of product at time t plus the change in concentration from degradation of reactant.

$$[P]_t = [P]_0 + a\,([R]_0 - [R]_t) \qquad (4a)$$

Equation (3a) can be substituted into Eq. (4a) to derive an equation that defines $[P]_t$ entirely by $[P]_0$ and $[R]_0$.

APPENDIX 2. Derivation of Michaelis-Menten Equation

Enzyme-catalyzed reactions are considered in two steps:

$$E + S \rightleftharpoons ES \longrightarrow P$$

where E represents enzyme and S represents substrate, ES represents the complex between enzyme and substrate, and P represents product. The formation of ES depends on the concentration of E and S, but, at the same time, the reverse reaction is considered dependent on ES. The formation of product is considered dependent on P. The change in concentration of ES (d[ES]/dt) is therefore

$$d[ES]/dt = k_1 ([E][S]) - (k_{-1} k_2) [ES] \tag{5a}$$

and

$$dP/dt = k_2 [ES] \tag{6a}$$

where k_1, k_{-1} and k_2 are fractional rate constants for forward (+) and reverse (-) reactions, and [ ] indicate concentrations. Once steady state is reached between the first forward and reverse reaction, if the first reaction is faster than the second, d[ES]/dt can be set equal to 0. Furthermore, the concentrations of E and S at steady state are equal to the initial amounts ($[E]_0$ and $[S]_0$) minus $[ES]_{ss}$. Thus,

$$d[ES]_{ss}/dt = 0 = k_1 ([E]_0 - [ES]_{ss})([S]_0 - [ES]_{ss}) - (k_{-1} k_2) [ES]_{ss} \tag{7a}$$

Another assumption is made to further simplify solution of the equation. That assumption is that the sites on the substrate where enzyme can attach greatly outnumber the sites on the enzyme available for substrate (i.e., [S] >>>[E]). Since there is little enzyme to remove substrate from solution,

$$[S]_0 - [ES]_{ss} \approx [S]_0 \tag{8a}$$

Thus, incorporating Eq. (8a) into Eq. (7a) yields

$$0 = k_1 ([E]_0 - [ES]_{ss})[S]_0 - (k_{-1} k_2) [ES]_{ss} \tag{9a}$$

which simplifies to

$$[ES] = [S]_0 [E]_0 / (k_{-1} k_2 / k_1) \tag{10a}$$

A definition is often introduced to simplify the constant portion of the equation for a given enzyme and substrate system:

$$K_m = (\, k_{-1}\, k_2 / k_1)$$
(11a)

Substituting Eqs. (10a) and (11a) into Eq. (6a) yields the Michaelis-Menten equation

$$dP/dt = k_2\, [E]_0\, [S]_0 / [S]_0 + K_m$$
(12a)

The term V_{max} is defined as the velocity of a reaction when enzyme systems are saturated and is equal to

$$V_{max} = k_2\, [E]_0$$
(13a)

MODELING GROWTH OF CATTLE FOR APPLICATION WITHIN THE STRUCTURE OF THE CORNELL NET CARBOHYDRATE AND PROTEIN SYSTEM

Danny G. Fox and Michael E. Van Amburgh[*]

INTRODUCTION

We use cattle around the world to convert forages, feed grains, and food byproducts into human food under production conditions that vary in soil, climate, feed composition, and genotype of cattle. Our goal over the past 20 years has been to develop a model of cattle nutrition that can be used to balance diets, thus converting these nutrients into human food as efficiently as possible. Our analyses indicate that accurate prediction of nutrient balance in each situation depends on the ability to predict cattle requirements and the supply of ruminally degraded carbohydrates, protein, microbial growth, total digestible nutrients (TDN), metabolizable energy (ME), net energy (NE), heat increment (HI), metabolizable protein (MP), net protein (NP), and essential amino acids (EAA). Formulating rations for varying cattle requirements thus requires an accounting system for the variables known to influence requirements and the dietary supply of energy and absorbed amino acids. In case studies and field experiences, we have found that the efficiency of cattle production can be improved using models to account for performance variation by accurately predicting requirements and feed utilization in individual production settings. The 'Cornell Net Carbohydrate and Protein System' (CNCPS), as described and validated by Fox et al. (1992), Russell et al. (1992), Sniffen et al. (1992), Ainslie et al. (1993), O'Connor et al. (1993), Tylutki et al. (1994), Fox et al. (1995), the National Research Council (NRC, 1996), Pitt et al. (1996), Fox et al. (2000a, b), and Tedeschi et al. (2000 a, b), was developed for that purpose. The CNCPS has been used as a teaching tool for students and consultants as well as to design and interpret experiments, apply research results, develop tables of nutrient requirements for any cattle type and production level, and evaluate and improve feeding programs on farms.

---

[*] Danny G. Fox and Michael E. Van Amburgh, Department of Animal Science, Cornell University, Ithaca, NY 14853.

This paper presents first a summary of the structure of the CNCPS, then describes the development of the growth model to be applied within the structure of the CNCPS and its application in predicting requirements of cattle in diverse production situations.

THE STRUCTURE OF THE CORNELL NET CARBOHYDRATE AND PROTEIN SYSTEM

The definition of a proposed model by Gill et al. (1989) describes the CNCPS: an integrated set of equations and transfer coefficients that represent the various physiological functions in cattle. Included are predictions of tissue requirements (for maintenance, growth, pregnancy, lactation, and tissue reserves) and nutrient supply to meet requirements (feed carbohydrate and protein fractions, their characteristic digestion and passage rates, microbial growth, intestinal digestion, and metabolism of absorbed nutrients). The purpose of a model is to describe mathematically the response of each compartment or several connected compartments to a variable or combination of variables. A model is considered to be mechanistic when it simulates behavior of a function through processes operating at a lower level (Gill et al., 1989). Most biological responses are integrated, nonlinear, and dynamic (Sauvant, 1991). A model that is totally mechanistic will accurately simulate whole animal metabolism under all conditions with little risk of use. However, such a model is beyond the capability of existing science (Gill et al., 1989). Further, ration formulation models are limited by the quality and availability of information about all of the model compartments and by the amount of data and work needed to test and validate the functions of such a model. Knowledge of the metabolism of nutrients is not as advanced as the prediction of ruminal fermentation because of the almost infinite metabolic routes connecting various tissue and metabolic compartments, the multiple nutrient interactions, and the sophisticated metabolic regulations that determine the partitioning of absorbed nutrients (homeorhesis and homeostasis) (Sauvant, 1991). Therefore, the CNCPS uses a combination of mechanistic and empirical approaches, assumes steady state conditions, and uses statistical representations of data that describe the aggregated response of whole compartments (Fox et al., 2000a).

A key challenge in modeling cattle nutrition is to determine the most appropriate level of aggregation of knowledge (whole animal, physiological function, cellular, subcellular, etc.). The most critical step is to describe the objective of the model and then to determine the appropriate mix of empirical and mechanistic representations of physiological functions. These decisions are made based on the availability of development and validation data, whether the needed inputs are typically available, and a risk-benefit analysis of the increased sensitivity (Fox et al., 1995). Because our CNCPS model is used on farms, it must be able to accurately predict animal requirements and the supply of nutrients from inputs that are routinely available on most farms, it must provide output that will help producers improve their feeding programs, and the risk of use must be small. To continue to be a tool to apply current scientific knowledge, it must have a structure that can be readily updated and refined as new knowledge becomes available.

The CNCPS model contains a biologically-based structure to evaluate diets for all classes of cattle (i.e., beef, dairy, and dual purpose) based on consideration of the existing animals, feeds, management, and environmental conditions. The approach taken and the level of aggregation of variables are based on our experience in working with farmers and consultants to diagnose animal performance problems and to develop more accurate feeding programs. Over 20 years, separate submodels that can be classified by physiological function have been developed and refined: (1) feed intake and composition, (2) rumen fermentation, (3) intestinal digestion, (4) metabolism, (5) maintenance, (6) growth, (7) pregnancy, (8) lactation, (9) reserves, and (10) nutrient excretion. New information has been periodically incorporated into these submodels. The user must have some nutritional knowledge to use the CNCPS because of the risks associated with not knowing how to choose inputs and interpret results. However, with training and experience, the CNCPS can be used to evaluate the interactions between animal type, production level, environment, feed composition, and management.

The following sections summarize how requirements and supply of nutrients are computed in the CNCPS.

Maintenance Requirement for Energy and Amino Acids

The maintenance energy requirement is predicted from metabolic body size and rate, with adjustments reflecting breed type, physiological state, previous nutritional treatment, activity, environment (temperature, wind velocity, and animal surface area and insulation), and heat gain or loss required to maintain normal body temperature (Fox and Tylutki, 1998). The proportion of the energy and protein intake needed for productive functions cannot be accurately determined until the proportion needed for maintenance has been estimated. The combined animal and diet heat production must thus be determined to assess energy balance in a particular environment, requiring the prediction of both ME and NE. The amino acid requirements for maintenance depend on the prediction of sloughed protein and net tissue turnover losses as predicted from metabolic fecal nitrogen, urinary nitrogen loss, and scurf protein.

Energy and Amino Acid Requirements for Tissue Deposition and Milk Synthesis

Expected mature weight, actual body weight, and rate of growth are used to predict the energy and protein content of growth. Prediction of amino acid requirements will not be accurate without accurate predictions of empty body gain or milk composition. Pregnancy requirements are predicted from uterine and conceptus demand based on expected birth weights and day of gestation. These become critical in the accuracy of ration formulation during the last 90 days of pregnancy because that is the period of highest conceptus weight accumulation and demand. Lactation requirements are computed for different days of lactation as well as levels and composition of milk. Body reserves are used to meet maintenance or lactation requirements when nutrient intake is inadequate. Reserves must be taken into account when evaluating ability to meet requirements, especially under environmental stress, feed shortage, or early lactation conditions. Visual appraisal is used to assign a body condition score, which in turn is used to predict body fat and energy reserves. The cycle of reserve depletion and replenishment during lactation and the dry period, respectively, is reflected by a predicted condition score change.

Prediction of Intake and Ruminal Degradation of Feed Carbohydrate and Protein Fractions and Microbial Growth

The ME and MP available to meet requirements depend on accurate determination of dry matter intake (DMI), ingredient content of carbohydrate and protein fractions, microbial growth on the fiber and non-fiber carbohydrates consumed, and the unique rates of digestion and passage of the individual feed carbohydrate and protein fractions that are being fed. A first limiting factor in the CNCPS is accurate determination of DMI. Where possible, we use as an input for this variable actual DMI values for the group of animals being evaluated. Then we use predicted DMI as a benchmark for diagnostic purposes or if DMI is not available (e.g., in grazing cattle). The interactions of DMI, digestion, and passage have several implications. First, the growth rate of each microbial pool that digests available carbohydrate fractions, and thus the absorbable microbial amino acids produced, will depend on the special characteristics and intake of the feeds being fed. This in turn determines the demand for the nitrogen source required by each pool (Russell et al., 1992; Sniffen et al., 1992). Second, the percentage of cell wall carbohydrate and protein that escapes digestion will change depending on level of intake, particle size, specific gravity, rumination, and degree of lignification (Sniffen et al., 1992). Third, the site of digestion and, depending on the rate of whole tract passage, the extent of digestion will be altered (Fox et al., 1992: Sniffen et al., 1992). Variable rates of digestion and passage have similar implications for protein fractions in feeds (Sniffen et al., 1992). Those readily available will be degraded in the rumen, while those more slowly degraded will be partially degraded in the rumen and partially degraded postruminally, the proportion depending on rates of digestion and passage of the protein fractions in the feeds (Sniffen et al., 1992).

Prediction of Intestinal Digestion and Excretion

Empirically-derived coefficients are used to predict intestinal digestibilities and fecal losses based on summaries of data in the literature. A more mechanistic approach is needed to incorporate the integration of digestion and passage to predict intestinal digestion. However, the accuracy of prediction of pool sizes digested depends on the accuracy of prediction of ruminal flows, and therefore it has second priority to the prediction of ruminal fermentation, particularly since, with most feeds, over 75% of total tract digestion occurs in the rumen. Until routine predictions of feed content of carbohydrate and protein fractions are available with accurate digestion and passage rates, the use of a more complex intestinal submodel could result in a multiplication of errors. In recent work, mineral excretion predictions needed for manure nutrient management planning (N and P) were nearly identical to that measured in manure applied to fields in a 500-cow dairy when volatilization losses of N and bedding contributions were considered (Fox et al., 2000b).

Prediction of Metabolism of Absorbed Energy and Amino Acids

We currently use empirically-derived equations and transfer coefficients for metabolism in this application-level model because of the limitations in predicting end products of ruminal fermentation as well as absorbed carbohydrate and amino acids. A mechanisitc model of metabolism must account for the large number of metabolic routes connecting the numerous tissue and metabolic compartments, the multiple nutrient interactions, and the sophisticated metabolic regulations which drive the partitioning of absorbed nutrients in various productive states. The equations used to predict ME from DE reflect the variation in methane produced across a wide range in diets. The equations used for lactating dairy cows to predict NE for lactation from ME reflect the energetic efficiency associated with the typical mix of metabolites in the ME, based on respiration chamber data (Moe, 1981) and validated using independent data (Roseler, 1994). The equations used for growing cattle to predict NE for maintenance and NE for growth reflect the wide variation in metabolites used in growing cattle and dry cows and validated with little bias across a wide range of ME contents (NRC, 1996). To predict the impact of diet on composition of tissue and milk, our metabolism submodel needs to be able to predict absorbed carbohydrates, volatile fatty acids, lipids, and amino acids available for various physiological functions and their metabolism with changes in productive states. Pitt et al. (1996) described the prediction of ruminal fermentation end products within the CNCPS structure as a first step.

DEVELOPMENT OF THE GROWTH MODEL

The objective of the growth model is to be able to predict the energy, protein, and mineral requirements at any stage of post-weaning growth for any mature size of cattle. Table 1 contains a list of abbreviations used in describing this model.

Net energy for gain (NE_g) is defined as the energy content of the tissue deposited, which is a function of the proportion of fat and protein in the empty body tissue gain (fat contains 9.367 kcal/g and nonfat organic matter contains an average of 5.686 kcal/g) (Garrett et al., 1959). Simpfendorfer (1974) summarized body composition data of growing cattle from birth to maturity and found that, within cattle of a similar mature size, 95.6 to 98.9% of the variation in the chemical components and empty body energy content was associated with the variation in weight. When energy does not limit growth, the empty body contains an increasingly smaller percentage of protein and an increasingly larger percentage of fat as BW increases. An animal reaches chemical maturity when additional weight gain contains little additional protein. The energy content of weight gain was described in equation form by Garrett (1980). Those equations were adapted by the Subcommittees on Beef and Dairy Nutrition of the National Research Council (1984, 1989). This data set included 72 comparative slaughter experiments conducted at the University of California between 1960 and 1980 on approximately 3,500 cattle (predominately Hereford steers and heifers, and some Hereford x Charolais crossbred steers) receiving various diets. The equation developed

Table 1. Definitions for abbreviations used in equations

$APSWG_1$ = target SWG after first pregnancy (kg/d)
$ACSWG_1$ = target SWG after first calving (kg/d)
$ACSWG_2$ = target SWG after second calving (kg/d)
$ACSWG_3$ = target SWG after third calving (kg/d)
AF = proportion of empty body fat
AP = proportion of empty body protein
BCS = body condition score
$BPSWG_1$ = target SWG before first pregnancy (kg/d)
BW = live full body weight (kg)
CI = calving interval (d)
CBW = calf birth weight (kg)
EBF = empty body fat (%)
EBP = empty body protein (%)
EQSBW = weight equivalent to the standard reference animal (kg)
EQEBW = equivalent empty body weight (kg)
EBW = empty body weight (kg)
EBW 5 = EBW at BCS 5 (kg)
EBG = empty body weight gain (kg/d)
ER = energy reserves (Mcal)
FBW = full body weight; assumed to be the same as BW (kg)
LWG = live weight gain, with no adjustment for gut fill (kg/d)
Mcal = megacalorie
NEFG = net energy available for weight gain (Mcal/d)
NEg = net energy required for growth (Mcal/d)
NPg = net protein required for growth (g/d)
RE = net energy retained (Mcal/d)
SBW = shrunk body weight; defined as 96% of full weight
SRW = weight of the standard reference animal at different stages of growth
(400, 435, 462, and 478 kg at 22, 25, 27, and 28% body fat, respectively)
SWG = shrunk body weight gain
t = days pregnant
TPW_1 = target first pregnant weight (kg)
TCW_1 = target weight post-calving for first lactation animals (kg)
TCW_2 = target weight post-calving for second lactation animals (kg)
TCW_3 = target weight post-calving for third lactation animals (kg)
TCW_4 = target weight post-calving for fourth lactation animals (kg)
TCA_1 = target calving age (d)
TPA_1 = target first pregnant age (d)
TF = total fat (kg)
TP = total protein (kg)

mathematically describes the relationship between retained energy (RE) and empty body weight gain (EBG) for a given empty body weight (EBW). A summary of the database indicates that EBW was 89.1% of shrunk body weight (SBW), and EBG was 95.6% of shrunk weight gain.

The weight at which cattle reach the same chemical composition differs depending on mature size and sex; hence, composition is different even when weight is the same (Fox and Black, 1984). The nutrient requirement systems that are used most widely in the world and have been revised in recent years use some type of size-scaling approach to adjust for differences in weight at a given composition. These include systems from the Institut National de la Recherche Agronomique (INRA, 1989), the Commonwealth Scientific and Industrial Research Organization (CSIRO, 1990), the Agricultural and Food Research Council (AFRC, 1993), and the National Research Council (NRC, 1996). The NRC (1996) adapted the size-scaling equation developed by Fox et al. (1992) with refinements published by Tylutki et al. (1994) [see Eq. (1) later] and by Fox et al. (1999) for dairy cattle. This is used to account for differences in mature size of cattle in Eq. (2). As in the CSIRO (1990) and INRA (1989) systems, this growth model assumes that various types of growing cattle have a similar chemical composition of growth at the same degree of maturity. Similar to the CSIRO system, the size-scaling equation in this model adjusts the body weights of cattle of various mature sizes to a weight at which they are equivalent in body composition to a standard reference animal (EQSBW).

This model uses several different body weight measurements including full (FBW), shrunk (SBW), or empty (EBW) body weight (BW). These different weight measurements are used because different body weights were used to derive the maintenance and growth equations. The NE_m requirement is computed with SBW, thus accounting for both fasting metabolism and maintaining the temperature of gut contents (NRC, 1984). The equations used to compute shrunk weight gain allowed by the diet net energy available for growth (NEFG) and to compute target ADG for herd replacements employ shrunk weight gain (SWG). SBW is defined as 96% of FBW and is generally equivalent to weight after an overnight fast without feed or water. The net energy required for a target SWG is computed from EBW (89.1% of SBW) and EBG (95.6% of SWG). Empty body weights were used in development of this equation because net energy requirements are a function of the proportion of fat and protein in the empty body tissue gain (Garrett et al., 1959).

The equations of Garrett (1980) adjusted for mature size [Eqs. (2) and (5)] are used to compute the energy content of gain at various stages of growth and rates of gain because they were developed from a large, robust data set (72 comparative slaughter studies; Garrett, 1980). They have been used with success in previous NRC publications (1984, 1989, 1996), and they accurately described the net energy and protein content of steers and heifers used for beef (NRC, 1996) and dairy replacement heifers (Fox et. al., 1999) when adjusted for mature body size as described in Eqs. (2) and (5). All body weight values used in the equations to predict growth requirements are measured in kilograms (kg).

$$EQSBW = SBW * (SRW/FSB + W) \qquad (1)$$

$$RE \ (Mcal) = 0.0635 * EQEBW^{0.75} * EBG^{1.097} \qquad (2)$$

where EQEBW is 0.891 * EQSBW, and EBG is 0.956 * SWG. EQSBW is the weight of the animal in question equivalent to the standard reference animal at the same stage of growth. The standard reference weight (SRW) is the weight of the standard reference animal at different stages of growth. These are 400, 435, 462, and 478 kg at 22, 25, 27, or 28% body fat, respectively. The FSBW is the weight of the animal in question at the stage of growth of interest. For breeding herd replacement heifers, 478 kg is the weight at which the standard reference breeding female is assumed to reach chemical maturity, and MSBW is the expected mature shrunk body weight of herd replacement heifers.

Given the relationship between energy retained and protein content of gain, protein content of SWG (retained protein, RP) is computed from the following equation (NRC, 1984, 1996):

$$RP \ (g/d) = SWG * \{268 - [29.4 * (RE/SWG)]\} \tag{3}$$

Equation (3) is adjusted for mature size because the RE used in that equation is based on EQEBW. The absorbed protein requirement is

$$MP_g = NP_g / [0.83 - (EQSBW * 0.00114)] \tag{4}$$

If EQSBW is > 478 kg, then an EQSBW of 478 kg is used in the equation.

To predict animal performance from a given diet, daily gain needs to be predicted from the diet being fed. To accomplish this, EQSBW is substituted for SBW, and the net energy available for growth (NEFG) is substituted for RE in the Garrett equation (1980) as shown in Eq. (5) to predict SWG.

$$SWG = 13.91 * NEFG^{0.9116} * EQSBW^{-0.6837} \tag{5}$$

Actual SWG and NEFG for RE can be substituted in Eq. (3) to compute the protein required for the observed SWG and NEFG. Then, Eq. (4) can be used to compute the MP required for the observed SWG to evaluate adequacy of the protein intake.

Table 2 shows the net energy requirements of dairy heifers of different mature sizes (650, 800, and 400 kg) growing at different rates (0.6, 0.8, and 1.0 kg/d) that were computed with our growth model (NRC, 2001). Several important relationships are shown in this table. First, as BW increases, the energy content of the gain increases and protein content of the gain decreases because more energy is deposited as fat. Second, as SWG increases, energy content of the gain increases and protein content of the gain decreases because the gain contains a higher proportion of fat as growth rate increases. Third, as animals increase in weight, metabolizable protein required does not decrease as rapidly as net protein required because the efficiency of protein absorption declines. As energy intake above maintenance increases, it is assumed that the rate of protein deposition becomes limiting, and excess energy is deposited as fat. Fat dilutes the body content of protein, ash, and water, which are deposited at nearly constant ratios to each other at a given age (Garrett, 1987).

We conducted an evaluation of the growth model with three different types of cattle commonly used for producing meat and milk: *Bos taurus* breeds of cattle used for beef production, *Bos indicus* breeds of cattle used for beef and dairy production, and Holstein

Table 2. Relationship between mature size and growth requirements[1]

Mature weight	Live Body weight during growth (kg)[2]						
650 kg Holstein	200	250	300	350	400	450	500
800 kg Holstein	246	308	369	431	493	554	616
400 kg Jersey	139	173	208	242	277	312	346
Daily gain, kg/day[3]	NE_g required (Mcal/d)[3]						
0.6	1.34	1.58	1.81	2.03	2.25	2.46	2.66
0.8	1.83	2.17	2.48	2.79	3.08	3.37	3.64
1.0	2.34	2.77	3.17	3.56	3.94	4.30	4.65
	Net protein required for growth (g/d)[4]						
0.6	122	114	108	101	95	89	83
0.8	161	151	141	132	124	115	107
1.0	199	187	175	163	152	142	131
	Metabolizable protein required for growth (g/d)[5]						
0.6	182	183	185	187	190	194	199
0.8	241	241	243	245	248	253	259
1.0	299	299	300	302	305	310	316

[1] National Research Council (2001).

[2] The body weights are full, not shrunk, weights. Weights in the same column are at the same stage of growth. Daily gain values are on a shrunk weight basis.

[3] NEg requirement is computed from Eq. (2): Retained energy (RE) $=0.0635 * EQEBW^{.75} * EBG^{1.097}$, where EQEBW is equivalent empty body weight and EBG is 0.956 * SWG.

[4] Net protein in the gain is computed from Eq. (3): RP (g/d) = SWG * {268 − [29.4 * (RE/SWG)]}.

[5] Metabolizable protein required is computed from Eq. (4): $MP_g = NP_g$ / [0.83 − (EQSBW * 0.00114)]; if EQSBW is > 478 kg, then an EQSBW of 478 kg is used in the equation.

heifers used in intensive dairy production. We used a defender-challenger approach, with the defender system being the 1984 and 1989 NRC systems and the challenger being the growth model developed for the CNCPS.

Evaluations of the Growth Model with *Bos taurus* Breeds of Cattle Used for Beef Production

Data set 1 included 82 pen observations (65 pens of steers and 17 pens of heifers) from published studies with body composition determined by the same procedures used by Garrett (1980) in developing the 1984 NRC system (Tylutki et al., 1994). Included were FSBW representative of the range in cattle fed in North America; all silage to all corn-based diets; no anabolic implant, estrogen only, or estrogen + TBA; and *Bos taurus* breed types representative of those fed in North America (British, European, Holstein, and their crosses). Data set 2 included 142 serially slaughtered (whole body chemical analysis by component; Fortin et al., 1980; Anrique et al., 1990) non-implanted steers, heifers, and bulls ranging widely in body size. A detailed description of these data sets, the validation procedures, and results were published by Tylutki et al. (1994).

In nearly every subclass, the CNCPS growth model presented here accounted for more of the variation and had less bias than did the defender systems. Nearly identical results were obtained between the 1984 NRC and the present systems when energy retained was used to predict SWG in Eq. (5); this equation is the one most commonly used to predict ADG. When all subclasses were combined, the CNCPS growth model accounted for 94% of the variation with a 2% over-prediction bias for retained energy and 91% of the variation in retained protein with a 2% under-prediction bias. It was concluded that the CNCPS growth model gives accurate predictions of net energy requirements for growth across wide variations in cattle breed, body size, implant, and nutritional management systems.

Evaluations of the Growth Model with *Bos indicus* Breeds of Cattle Used for Beef Production

An evaluation of the CNCPS growth model was conducted at the University of Sao Paulo at Piracaba, Brazil with *Bos indicus* breeds common to Latin America (Lanna et al., 1996). Included were ADG predicted from 63 different diets fed to a total of 687 *Bos indicus* bulls (Nellore breed) in which ME intake and body composition were determined; FSBW was determined from final EBW fat content. For the NRC 1984 and the CNCPS systems, the r^2 was 0.51 and 0.67, and the bias was -21% and -1%, respectively.

Evaluation of the Growth Model with Dairy Herd Replacement Heifer Data

Because they are the predominant breed used for dairying in North America, we conducted an evaluation to specifically determine the accuracy of the growth model in predicting growth in Holstein replacement heifers (Fox et al., 1999). The equations used for predicting energy and protein retained during growth were validated with the serially slaughtered non-implanted Holstein heifer data published by Fortin et al. (1980) and Anrique et al. (1990) as described by Fox et al. (1999). The $Y = X$ and residuals lines showed both Angus and Holstein heifer composition of gain data fit the same line (no breed effect using the size-scaling approach). The regression describing the Holstein heifers had the same slope and a similar intercept as the $Y = X$ line for both breeds. For RE in Holstein heifers, the r^2 was 0.86 for the 1989 National Research Council equations and 0.96 for the CNCPS growth model, with biases of -11% and -4%, respectively. For RP in Holstein heifers, the r^2 was 0.91 for the 1989 National Research Council equations and 0.71 for the model presented here, with biases of -13% and -10%, respectively. The bias for the 1989 National Research Council equations was non-uniform, with under-prediction of RP at lower body weights. These results indicate that the model described here can be used to predict RE and RP for dairy heifers. We suggest, however, that more research is needed to account for factors influencing RP, as indicated by the lower r^2 and higher bias in predicting RP compared to RE. Some of the bias might be due to previous plane of nutrition. An assumption is made that the composition of the empty body at the weight entered in the model is similar for all animals with the same mature weight; this might not be true. For example, if two groups of heifers with the same mature weight reached 200 kg either at 1 kg/d or 0.5 kg/d, the empty body fat content of the heifers reared at the higher rate of gain will be greater. Therefore, the energy required for a unit of gain will be different between the two groups. Further research is needed to refine the CNCPS growth model to account for this effect.

Information from 32 Holstein heifers fed alfalfa or corn silage diets at two rates of ADG (0.78 and 0.99 kg/d) from 181 to 334 kg of full body weight (FBW) (Waldo et al., 1997) was used to evaluate the prediction of empty body weight and empty body gain from shrunk weights. EBW averaged 89% of SBW compared to 89.1% in the Garrett database (1980) which was used to develop this model. EBG averaged 87.4% of SWG compared to 95.6% in the Garrett database (1980). In Holstein steers at the same stage of growth as the heifers in the trials conducted by Waldo et al. (1997) (< 400 kg SBW), EBW was 89% of SBW and EBG averaged 95.7% of SWG (Abdalla et al., 1988). These values are nearly identical to those used in this model.

Computing Target Growth Rates for Herd Replacement Heifers

Breeding herd replacement heifer growth rate affects economic returns on the dairy farm (Cady and Smith, 1996) and in beef production systems (NRC, 1996). Inadequate size at first parturition may limit milk production and conception during first lactation (Hoffman et al., 1996). Excess energy intake, however, can have negative effects on mammary development. For example, excessive energy intake has a negative effect on mammary parenchyma (ductular epithelial tissue) (Harrison et al., 1983; Foldager and Serjsen, 1987). However, adequate metabolizable protein supply has recently been shown to lessen the effects of higher energy intake on mammary development (Radcliff et al., 1997). Because puberty is associated with weight, parenchyma tissue growth, which is not linearly related to body growth, may be truncated before full ductal development as a result of excess energy intake before puberty (Van Amburgh et al., 1991). Excess energy intake, as evidenced by over-conditioning from 2 to 3 months of age until after conception, can reduce first lactation milk production (Van Amburgh et al., 1998b). Numerous data are available to support the concept of a genetically determined threshold age and weight at which heifers attain puberty (NRC, 1996). Joubert (1963) proposed that heifers would not attain puberty until they reached a given degree of physiological maturity; this is similar to the "target weight" concept proposed by Lamond (1970). Simply stated, the concept is to feed replacement heifers to attain a pre-selected or target weight at a given age to achieve optimum first lactation performance while controlling heifer rearing costs. In general, heifers of typical beef breeds are expected to attain puberty at about 60% of mature weight, and heifers of dual purpose or dairy breeds tend to attain puberty at a younger age and lower weight relative to mature weight (about 55% of mature weight) (NRC, 1996). Conversely, heifers of *Bos indicus* breeds (e.g., Brahman, Nellore, Sahiwal) generally attain puberty at older ages and heavier weight and at a slightly higher percentage of mature weight as compared to European beef breeds (NRC, 1996).

Recent research has identified optimum growth rates for dairy heifers that minimize replacement costs while maximizing first lactation milk production (Ferguson and Otto, 1989; Van Amburgh et al., 1998a, b). The 1996 NRC equations to predict target weights for beef herd replacement heifers, as modified and evaluated by Fox et al. (1999) for predicting target weights for dairy heifers, are used in the CNCPS growth model. The equations to predict target weights and rates of gain are as follows:

$$\text{Target weight first bred} = \text{Mature SBW} * 0.55 \tag{6}$$

$$\text{Target 1}^{\text{st}} \text{ pregnant age} = \text{Target first calving age} - 280 \tag{7}$$

$$\text{Target SWG before 1}^{\text{st}}\text{ pregnancy} = (\text{Target weight first bred} - \text{current SBW}) / \\ (\text{Target 1}^{\text{st}}\text{ pregnant age} - \text{current age}) \tag{8}$$

$$\text{Target 1}^{\text{st}}\text{ calving weight} = \text{Mature SBW} * 0.85 \tag{9}$$

$$\text{Target 2}^{\text{nd}}\text{ calving weight} = \text{Mature SBW} * 0.92 \tag{10}$$

$$\text{Target 3}^{\text{rd}}\text{ calving weight} = \text{Mature SBW} * 0.96 \tag{11}$$

$$\text{Target 4}^{\text{th}}\text{ calving weight} = \text{Mature SBW} * 1.00 \tag{12}$$

$$\text{First pregnancy SWG} = (\text{Target 1}^{\text{st}}\text{ calving weight} - \text{Target weight first bred}) / 280 \tag{13}$$

$$\text{1}^{\text{st}}\text{ lactation SWG} = (\text{Target 2}^{\text{nd}}\text{ calving weight} - \text{Target 1}^{\text{st}}\text{ calving weight}) \\ / \text{ Calving interval} \tag{14}$$

$$\text{2}^{\text{nd}}\text{ lactation SWG} = (\text{Target 3}^{\text{rd}}\text{ calving weight} - \text{Target 2}^{\text{nd}}\text{ calving weight}) \\ / \text{ Calving interval} \tag{15}$$

$$\text{3}^{\text{rd}}\text{ lactation SWG} = (\text{Target 4}^{\text{th}}\text{ calving weight} - \text{Target 3}^{\text{rd}}\text{ calving weight}) \\ / \text{ Calving interval} \tag{16}$$

where calving interval is in days.

For all target rates of gain, Eq. (2) is used to compute the NE_g requirement, and Eqs. (3) and (4) are used to compute the protein requirement for growth. Observed weights can be substituted for previous target weight and divided by days left to reach the next target weight in order to determine SWG required to reach the next target weight. The NE_g required to reach the target weight can then be calculated.

For pregnant animals, weight gain due to growth of the gravid uterus should be added to predicted daily SWG as follows:

$$\text{ADG}_{\text{preg}} = \text{CBW} * [18.28 * (0.02 - 0.0000286 * t)] * e^{[(0.0200 \, * \, t) - (0.0000143 \, * \, t*t)]} \tag{17}$$

where CBW = expected calf birth weight (kg), e is the base of natural logarithms, and t is days pregnant.

For pregnant heifers, weight of fetal and associated uterine tissue should be subtracted from SBW to compute growth requirements. The conceptus weight (CW) can be calculated as follows:

$$\text{CW} = \text{CBW} * (0.01828) * e^{[(0.02 \, * \, t) - (0.0000143 \, * \, t*t)]} \tag{18}$$

Evaluation of Target Weight Equations

Data from the study published by Van Amburgh et al. (1998a, b) will be used to demonstrate how the growth model can be applied to calculate NE_g required for growth of replacement dairy heifers. This study involved 273 Holstein heifers fed from an average of 77 d of age through the first lactation. The average mature weight of the herd was 641 kg. The average post-weaning and transition weight of calves was 84 kg. Average age at first calving was 687 d, and calving interval was 431 d. Average body weights and SWG observed in this experiment compared well with model-predicted values. The SWG before first calving averaged 0.82 kg/d *vs.* a target of 0.87 kg/d; weight at first pregnancy was 370 kg *vs.* a target of 352 kg; the SWG during first pregnancy averaged 0.63 kg/d *vs.* a target of 0.69 kg/d; weight after first calving averaged 533 kg *vs.* a target of 545 kg; first lactation SWG averaged 0.136 kg/d *vs.* a target of 0.104; and second post-calving weight was 592 kg (projected from SBW and SWG during the first 40 weeks of lactation) *vs.* a target of 590 kg.

CHANGES IN COMPOSITION OF BODY TISSUE AFTER MATURITY

Once mature, body stores of energy and protein become reserves that allow mobilization of energy and protein from body tissue when dietary intake is insufficient to meet animal requirements. These reserves must then be replaced by providing a dietary intake that exceeds the animal's requirements. Optimum management of energy reserves is critical to economic success with dairy and beef cows. Whether too fat or thin, cows at either extreme are at risk of metabolic problems and diseases, decreased milk yield, low conception rates, and difficult calving (Ferguson and Otto, 1989; NRC, 1996). Over-conditoning is expensive and can lead to calving problems and lower dry matter intake during early lactation. Conversely, thin cows may not have sufficient reserves for maximum milk production and will not likely rebreed on schedule. Both post-calving cow condition score and energy balance control ovulation (Wright et al., 1992). Ovulation occurs in dairy cattle 7 to 14 days after the energy balance nadir is reached during early lactation (Butler and Canfield, 1989). If the cow is too fat, intake will be lower and reserves will be used longer during early lactation, resulting in an extended time to nadir (point of maximum negative energy balance). Even if thin cows consume enough to meet energy requirements, a feedback mechanism mediated through hormonal changes seems to inhibit ovulation if body condition is inadequate (Roche et al., 1992). Energy reserves are managed on farms by observing body condition changes and by using body condition scores (BCS) to describe energy reserves. Body condition score is highly related to body fat and energy content (NRC, 1996).

We used the equations developed for the National Research Council body reserves model (NRC, 1996) to predict body composition from BCS for cows varying in breed type, mature body weight, and body condition score, as refined and evaluated for use with dairy cattle by Fox et al. (1999). These equations were developed from data on chemical body composition and body condition scores (1 to 9 scoring system) from 106 mature cows of diverse breed types, mature weights, and body condition scores (NRC, 1996). The resulting best fit equations to describe relationships between BCS and empty body percentage of fat, protein, water, and ash were linear. A zero intercept model was used with individual animal data to describe the relationship between percent empty body fat and BCS. BCS accounted for 65, 52, and 66% of the variation in body fat, body protein, and body energy, respectively, among individual animals. The SBW was 85.1% of EBW.

The mean SBW change associated with a BCS change (1 to 9 scale) was 6.85% of mean SBW.

The following equations summarize this model as modified for use with dairy cattle by Fox et al. (1999). Modifications include using the 1 to 5 dairy condition scoring system (Wildman et al., 1982) and using the SBW change per condition score to develop a method for computing the weight change associated with any mature body size and BCS.

$$1996 \text{ NRC 1 to 9 scale BCS} = [(\text{dairy body condition score} - 1) * 2] + 1 \qquad (19)$$

$$EBW = 0.851 * SBW \qquad (20)$$

Using the database value of 6.85% of SBW change per BCS, EBW at BCS 5 is calculated as

$$EBW\ 5 = (\text{current EBW/adjustment factor}) \qquad (21)$$

where the adjustment factor for a BCS of 1 to 9 is 0.726, 0.7945, 0.863, 0.9315, 1, 1.0685, 1.137, 1.2055, and 1.274, respectively. Then EBW is computed for each BCS using these factors applied to EBW at BCS 5. Next, body composition is computed for each BCS as follows:

$$\text{Proportion of empty body fat} = -0.00021 + 0.037683 * BCS \qquad (22)$$

$$\text{Proportion of empty body protein} = 0.200886 - 0.0066762 * BCS \qquad (23)$$

$$\text{Total fat (kg)} = \text{Proportion of empty body fat} * EBW \qquad (24)$$

$$\text{Total protein (kg)} = \text{Proportion of empty body protein} * EBW \qquad (25)$$

$$\text{Total body energy (Mcal)} = (9.4 * \text{Total fat}) + (5.7 * \text{Total protein}) \qquad (26)$$

Total body energy for the next lower and higher BCS is subtracted from the current BCS to compute energy and protein gain or loss to reach the next BCS. During mobilization, 1 Mcal of energy reserves (ER) substitutes for 0.82 Mcal of diet NE_L; 1 Mcal diet NE_L will provide 0.75/0.64 = 1.17 Mcal of ER. In this equation, 0.75 is the efficiency of use of ME for reserves in lactating cows, and 0.64 is the efficiency of use of ME for lactation (Moe, 1981).

Evaluation of Equations to Predict Changes in Body Reserves from Body Condition Score

This model was validated for beef cattle by the NRC (1996) and for dairy cattle by Fox et al. (1999) with the data of Otto et al. (1991). In this study, body composition and BCS of 56 Holstein cows selected to represent the range in dairy body condition scores of 1 to 5 were determined. Body fat at a particular condition score in Holstein cows was predicted with an R^2 of 0.95 and a bias of -1.6%. The relationship between body weight change and BCS in these Holstein cows was 85 kg/BCS ($R^2 = 0.96$), which compared to 80 kg predicted by the model and 82 kg in the data summarized by the National Research Council (1996). The intent of this validation was to test the model data containing both chemical composition of cows and BCS representing the range in dairy BCS, with the BCS obtained consistent with the way they are applied on dairy farms. Although the validation strongly supports the use of this model, further validation with other data sets should be conducted.

Table 3 shows the percentage composition and SBW change associated with each BCS computed with this model. Body fat changes 7.8 percentage units per dairy condition score. This model predicts energy reserves to be 5.98 Mcal/kg live weight loss at a BCS of 3.0 compared to the INRA (1989) and the 1989 National Research Council values of 6.0. Predicted energy content of weight loss ranges from 4.36 Mcal/kg at BCS 1.5 to 7.59 Mcal/kg at BCS 4.5 compared to CSIRO (1990) values of 3.0 and 7.1, respectively. Protein in the weight loss from a BCS 3 to a BCS 2 is predicted to be 83 g/kg.

These results (Tables 3 and 4) agree with available data indicating that the composition of live weight loss at maturity is approximately equal to the composition of live weight gain in animals (CSIRO, 1990; AFRC, 1993; NRC, 1996). Table 2 shows that 1 kg of SWG contains 5.86 Mcal and 93 g protein after a heifer reaches mature weight, compared to the Table 3 value of 5.98 Mcal and 83 g protein/kg at BCS 3. Table 4 shows the Mcal mobilized in moving to the next lower score or required to move from the next lower score to the one being considered for cows with different mature sizes.

Diet NE_L replaced by mobilized reserves or required to replenish reserves is computed by assuming 1 Mcal of mobilized tissue will replace 0.82 Mcal of diet NE_L, and 1 Mcal of diet ME will provide 0.75 Mcal of tissue NE if lactating or 0.60 Mcal if dry, based on Moe (1981) and the National Research Council (1996, 2001). For example, a 600 kg cow at BCS 3 will mobilize 470 Mcal NE in declining to BCS 2. If NE_L intake is deficient by 3 Mcal/day, this cow will lose 1 BCS in (470 * 0.82) / 3 = 125 days. If it is consuming 3 Mcal NE_L above daily requirements, this cow will move back to a BCS 3 in 470 / [(3 / 0.64) * 0.75] = 134 days.

Table 3. Empty body chemical composition at different body condition scores

Dairy BCS[1]	% in empty body				SBW, % of BCS 3	Mcal/kg SBW change[2]
	Fat	Protein	Ash	Water		
1.0	3.77	19.42	7.46	69.35	72.6	---
1.5	7.54	18.75	7.02	66.69	79.5	4.36
2.0	11.30	18.09	6.58	64.03	86.3	4.90
2.5	15.07	17.42	6.15	61.36	93.2	5.44
3.0	18.84	16.75	5.71	58.70	100.0	5.98
3.5	22.61	16.08	5.27	56.04	106.9	6.51
4.0	26.38	15.42	4.83	53.37	113.7	7.05
4.5	30.15	14.75	4.39	50.71	120.6	7.59
5.0	33.91	14.08	3.96	48.05	127.4	8.13

[1] BCS is body condition score and SBW is shrunk body weight.
[2] Weight change is the difference between the body weight at the current BCS and the body weight at the next lower or higher BCS. The weight loss from BCS 3 to BCS 2 contains 8.3% protein.

Table 4. Energy reserves for dairy cows with different body size and condition scores (BCS)

| Dairy BCS | \\multicolumn{8}{c}{Shrunk body weight, kg} |

Dairy BCS	400	450	500	550	600	650	700	750
	\\multicolumn{8}{c}{Mcal NE required or provided for each half BCS[1]}
1.5	120	135	150	165	180	195	210	225
2.0	134	151	168	184	201	218	235	251
2.5	149	168	186	205	224	242	261	280
3.0	164	184	205	225	246	266	287	307
3.5	179	201	223	246	268	290	312	335
4.0	193	217	242	266	290	314	338	362
4.5	209	235	261	287	313	339	365	391
5.0	222	250	278	305	333	361	389	417

[1] Represents the energy mobilized in moving to the next lower score or required to move from the next lower score to this one.

APPLICATIONS OF THE CNCPS GROWTH MODEL

The CNCPS growth model has been incorporated into several applications software programs, as shown in the following list. In addition, because the model is published, it may be that others have used it as well.

- CNCPS version 4.0 for evaluating whole herd nutrient management
- CNCPS UK, an application of CNCPS version 4 for the United Kingdom
- CPM Dairy, a computer program designed by three cooperating academic institutions (Cornell University, University of Pennsylvania, and Miner Institute) to formulate diets for dairy cattle
- DALEX, a commercial sector computer program for applications by the feed industry to formulate diets for beef and dairy cattle
- National Research Council Nutrient Requirements of Beef Cattle model (2000)
- National Research Council Nutrient Requirements of Dairy Cattle model (2001)
- Cornell Cattle Systems 5, a computer program to predict growth, feed requirements, and profits for growing cattle
- BEEF TECH, a commercial computer program to predict growth, feed requirements, and profits for growing cattle
- VALUE DISCOVERY, a computer program to predict individual animal feed requirements from the animal's own body size and growth rate when fed in groups
- ACCUTRAC, a commercial computer program developed by Micro Beef Technologies for allocating feeds for individual animals fed in groups in commercial feedlots to market individual animals at their optimum point of profits with minimum waste fat

In our experience, uses of the growth model in applications beyond those implied by each program definition listed here include

- as a teaching tool to improve skills in evaluating the interactions of feed composition, feeding management, and animal requirements in varying farm conditions;
- to design and interpret experiments;
- to apply experimental results;
- to develop tables of net energy and metabolizable protein requirements and adjustment factors that can extend and refine the use of conventional diet formulation programs;
- as a tool for extending research results to varying farm conditions;
- as a diagnostic tool to evaluate feeding programs and to account for more of the variation in performance in a specific production setting, and
- to provide information that can be used to improve whole-farm nutrient management planning. Included are reduction of imported nutrients and reduction of manure nutrients and crop production that matches herd and soil productivity (Fox et al., 2000).

Although not well documented, use of the CNCPS growth model to more accurately formulate diets in specific production situations has been observed to improve performance while reducing cost of growth. Results to date indicate the potential impact of applying this growth model. The ability to accurately predict growth and feed requirements of individual growing animals so that each individual can be harvested at the optimum point in growth has the potential for doubling profits from growing and finishing cattle (Perry and Fox, 1997). The ability to formulate diets that allow dairy replacement heifers to grow at optimum rates would have a large impact on first-lactation milk production and rearing costs of replacement heifers (Van Amburgh et al., 1998).

CORRESPONDENCE

Please address all correspondence to:
Danny Fox
Cornell University
Department of Animal Science
130 Morrison Hall
Ithaca, NY 14853
dgf4@cornell.edu

REFERENCES

Abdalla, H., Fox, D., and Thonney, M., 1988, Compensatory gain by Holstein calves after underfeeding protein, *J. Anim. Sci.* 66:2687-2695.

Ainslie, S., Fox, D., Perry, T., Ketchen, D., and Barry, M., 1993, Predicting metabolizable protein and amino acid adequacy of diets fed to lightweight Holstein steers, *J. Anim. Sci.* 71:1312.

Agricultural and Food Research Council, 1993, *Energy and Protein Requirements of Ruminants*, CABI, Wallingford.

Anrique, R., Thonney, M., and Ayala, H., 1990, Dietary energy losses of cattle influenced by body type, size, sex and intake, *Anim. Prod.* 50:467-474.

Butler, W., and Canfield, R., 1989, Interrelationships between energy balance and postpartum reproduction, in: *Proceedings of the 1989 Cornell Nutrition Conference*, Department of Animal Science, Cornell University, Ithaca.

Cady, R., and Smith, T., 1996, Economics of heifer raising programs, in: *Proceedings of the Calves, Heifers and Dairy Profitability National Conference*, Northeast Regional Agricultural Engineering Service No. 74, Cornell University, Ithaca.

Commonwealth Scientific and Industrial Research Organization, 1990, *Feeding Standards for Australian Livestock: Ruminants*, CSIRO Publications, East Melbourne, Victoria.

Ferguson, J., and Otto, K., 1989, Managing body condition in cows, in: *Proceedings of the 1989 Cornell Nutrition Conference*, Department of Animal Science, Cornell University, Ithaca.

Foldager, J., and Serjsen, K., 1987, *Research in Cattle Production: Danish Status and Perspectives*, Landhusholdningsselskabets Forlag, Tryk.

Fortin, A., Simpfendorfer, S., Reid, J., Ayala, H., Anrique, R., and Kertz, A., 1980, Effect of level of energy intake and influence of breed and sex on the chemical composition of cattle, *J. Anim. Sci.* 51:604-614.

Fox, D., and Black, J., 1984, A system for predicting body composition and performance of growing cattle, *J. Anim. Sci.* 58:725-739.

Fox, D., and Tylutki, T., 1998, Accounting for the effects of environment on the nutrient requirements of dairy cattle, *J. Dairy Sci.* 81:3085-3095.

Fox, D., Sniffen, C., O'Connor, J., Russell, J., and Van Soest, P., 1992, A net carbohydrate and protein system for evaluating cattle diets. III. Cattle requirements and diet adequacy, *J. Anim. Sci.* 70:3578-3596.

Fox, D., Van Amburgh, M., and Tylutki, T., 1999, Predicting requirements for growth, maturity, and body reserves in dairy cattle, *J. Dairy Sci.* 82:1968-1977.

Fox, D., Tylutki, T., Van Amburgh, M., Chase, L., Pell, A., Overton T., Tedeschi, L., Rasmussen, C., and Durbal, V., 2000a, The Net Carbohydrate and Protein System for evaluating herd nutrition and nutrient excretion, *Cornell Animal Science Department Mimeo 213*, Cornell University, Ithaca.

Fox, D., Tylutki, T., Czymmek, K., Rasmussen, C., and Durbal, V., 2000b, Development and application of the Cornell University Nutrient Management Planning System, in: *Proceedings of the 2000 Cornell Nutrition Conference*, Department of Animal Science, Cornell University, Ithaca.

Garrett, W., 1980, Energy utilization by growing cattle as determined in 72 comparative slaughter experiments, *Proceedings of the 1980 Energy Metabolism Symposium* 26:3-7.

Garrett, W., 1987, Relationship between energy metabolism and the amounts of protein and fat deposited in growing cattle, *Proceedings of the 1987 Energy Metabolism Symposium* 32:98-101.

Garrett, W., Meyer, J., and Lofgreen, G., 1959, The comparative energy requirements of sheep and cattle for maintenance and gain, *J. Anim. Sci.* 18:528-547.

Gill, M., Beever, D., and France, J., 1989, Biochemical bases needed for the mathematical representation of whole animal metabolism, *Nutr. Abst. Rev.* 2:181.

Harrison, R., Reynolds, I., and Little, W., 1983, A quantitative analysis of mammary glands of dairy heifers reared at different rates of live weight gain, *J. Dairy Res.* 50:405-412.

Hoffman, P., Brehm, N., Price, S., and Prill-Adams, A., 1996, Effect of accelerated post-pubertal growth and early calving on lactation performance of primiparous Holstein heifers, *J. Dairy Sci.* 79:2024-2031.

Institut National de la Recherche Agronomique, 1989, *Ruminant Nutrition*, Libbey Eurotext, Montrouge.

Joubert, D., 1963, Puberty in female farm animals, *Anim. Breed. Abstr.* 31:295-306.

Keele, J., Williams, C., and Bennett, G., 1992, A computer model to predict the effects of level of nutrition on composition of empty body gain in beef cattle: I. Theory and development, *J. Anim. Sci.* 70:841-857.

Lamond, D., 1970, The influence of undernutrition on reproduction in the cow, *Anim. Breed. Abst.* 38:359-372.

Lanna, D., Fox, D., Boin, C., Traxler, M., and Barry, M., 1996, Validation of the Cornell Net Carbohydrate and Protein System estimates of nutrient requirements of growing and lactating zebu germplasm in tropical conditions, *J. Anim. Sci.* 74 (Suppl. 1):287.

Moe, P., 1981, Energy metabolism of dairy cattle, *J. Dairy Sci.* 64:1120-1139.

National Research Council, 1984, *Nutrient Requirements of Beef Cattle, Sixth Revised Edition,* National Academy Press, Washington, DC.

National Research Council, 1989, *Nutrient Requirements of Dairy Cattle, Sixth Revised Edition, Update,* National Academy Press, Washington, DC.

National Research Council, 1996, *Nutrient Requirements of Beef Cattle, Seventh Revised Edition,* National Academy Press, Washington, DC.

National Research Council, 2001, *Nutrient Requirements of Dairy Cattle, Seventh Revised Edition,* National Academy Press, Washington, DC.

O'Connor, J., Sniffen, C., Fox D., and Chalupa, W., 1993, A Net Carbohydrate and Protein System for evaluating cattle diets. IV. Predicting amino acid adequacy, *J. Anim. Sci.* 71:1298-1311.

Otto, K., Ferguson J., Fox, D., and Sniffen, C., 1991, Relationship between body condition score and composition of 9-10-11 rib tissue in Holstein dairy cows, *J. Dairy Sci.* 74:852-859.

Perry, T., and Fox, D., 1997, Predicting carcass composition and individual feed requirement in live cattle widely varying in body size, *J. Anim. Sci.* 75:300-307.

Pitt, R., Van Kessel, J., Fox, D., Barry, M., and Van Soest, P., 1996, Prediction of ruminal volatile fatty acids and pH within the Net Carbohydrate and Protein System, *J. Anim. Sci.* 74:226-244.

Radcliff, R., Vandehaar, M., Skidmore, A., Chapin, L., Radke, B., Lloyd, J., Stanisiewski, E., and Tucker, H., 1997, Effects of diet and bovine somatotropin on heifer growth and mammary development, *J. Dairy Sci.* 80:1966-2003.

Roche, J., Corwe, M., and Boland, M., 1992, Postpartum anestrus in dairy and beef cows, *Anim. Reprod. Sci.* 28:371-378.

Roseler, D., 1994, *Development and Evaluation of Feed Intake and Energy Balance Prediction Models for Lactating Dairy Cows,* Ph.D. dissertation, Cornell University, Ithaca.

Russell, J., O'Connor, J., Fox, D., Van Soest, P., and Sniffen, C., 1992, A Net Carbohydrate and Protein System for evaluating cattle diets. I. Ruminal fermentation, *J. Anim. Sci.* 70:3551-3561.

Sauvant, D., 1991, The use of modelling to predict animal responses to diet, *Proc. Ralston Purina Int. Scientific Advisory Board Mtg.*, Paris.

Simpfendorfer, S., 1974, *Relationship of Body Type, Size, Sex, and Energy Intake to the Body Composition of Cattle,* Ph.D. dissertation, Cornell University, Ithaca.

Sniffen, C., O'Connor, J., Van Soest, P., Fox, D., and Russell, J., 1992, A Net Carbohydrate and Protein System for evaluating cattle diets. II. Carbohydrate and protein availability, *J. Anim. Sci.* 70:3562-3577.

Tedeschi, L., Fox, D., and Russell, J., 2000a, Accounting for the effects of a ruminal nitrogen deficiency within the structure of the Cornell Net Carbohydrate and Protein System, *J. Anim. Sci.* 78:1648-1658.

Tedeschi, L., Fox, D., Chase, L., and Wang, S., 2000b, Whole herd optimization with the Cornell Net Carbohydrate and Protein System. I. Predicting feed biological values for diet optimization with linear programming, *J. Dairy Sci.* 83:2139-2148.

Tylutki, T., Fox, D., and Anrique, R., 1994, Predicting net energy and protein requirements for growth of implanted and nonimplanted heifers and steers and nonimplanted bulls varying in body size, *J. Anim. Sci.* 72:1806-1813.

Van Amburgh, M., Galton, D., Fox, D., and Bauman, D., 1991, Optimizing heifer growth, *Proceedings of the 1991 Cornell Nutrition Conference,* Department of Animal Science, Cornell University, Ithaca.

Van Amburgh, M., Fox, D., Galton, D., Bauman, D., and Chase, L., 1998a, Evaluation of National Research Council and Cornell Net Carbohydrate and Protein Systems for predicting requirements of Holstein heifers, *J. Dairy Sci.* 81:509-526.

Van Amburgh, M., Galton, D., Bauman, D., Everett, R., Fox, D., Chase, L., and Erb, H., 1998b, Effects of three prepubertal body growth rates in Holstein heifers during first lactation performance, *J. Dairy Sci.* 81:527-538.

Waldo, D., Tyrrell, H., Capuco, A., and Rexroad, C. Jr., 1997, Components of growth in Holstein heifers fed either alfalfa or corn silage diets to produce two daily gains, *J. Dairy Sci.* 80:1674-1684.

Wildman, E., Jones, G., Wagner, P., Boman, R., Troutt, H., and Lesch, T., 1982, A dairy cow body condition scoring system and its relationship to selected production characteristics, *J. Dairy Sci.* 65:495-501.

Wright, I., Rhind, S., Whyte, T., and Smith, A., 1992, Effects of body condition at calving and feeding level after calving on LH profiles and the duration of the post-partum anoestrous period in beef cows, *Anim. Prod.* 55:41-46.

PART VII

MODELING TO EXPLORE DISEASE

MATHEMATICAL MODELS OF TUMOR GROWTH: FROM EMPIRICAL DESCRIPTION TO BIOLOGICAL MECHANISM

John A. Adam[*]

INTRODUCTION

Cancer is a complex phenomenon (Weinberg, 1998), and in order to attempt to model any of its different facets (such as avascular spheroid growth, angiogenesis and vascularization, invasion, or metastasis) in a reasonable mathematical manner, many simplifications are necessary. If the simplifications are reasonable, the model may be of considerable use, not only as a receptacle for what is already known but also for its predictive capabilities.

The question may arise as to what constitutes a *model* as opposed to a metaphor (or even a simile) for cancer. This is well illustrated in the paper by Glass (1973), discussed below, which is very suggestive of the fundamental phenomenon - uncontrolled cellular proliferation - yet clearly is too simple to account for more complex aspects of tumor growth and development. Several complementary levels of description are possible, and it is the purpose of this paper to provide a brief survey of some of the basic mathematical models that have been developed in the field of cancer biology over the last three decades. Further details can be found in the references listed at the end of the text.

An important question to be asked at the outset is: *what is a mathematical model?* One basic answer is that it is the formulation in mathematical terms of the assumptions and their consequences believed to underlie a particular "real world" problem. The aim of mathematical modeling is the practical application of mathematical models to help unravel the underlying mechanisms involved in biological (or other) processes. Common pitfalls of mathematical modeling include the indiscriminate, naïve, or uninformed use of models. However, when developed and interpreted thoughtfully, mathematical models can provide insight into the nature of a problem, be useful in interpreting data, and stimulate experiments. There is not necessarily a "right" model; obtaining results which are consistent with observations is only a first step and does not imply that the model is

---

[*] John A. Adam, Department of Mathematics and Statistics, Old Dominion University, Norfolk, VA 23529.

the only one that applies, or even that it is "correct." Furthermore, mathematical descriptions are not *explanations*, and never on their own can they provide a complete solution to the biological problem – often there may be complementary levels of description possible within the particular scientific paradigm. Collaboration with biologists (insert here whichever category of scientist is appropriate!) is needed for realism and help in modifying the model mechanisms to reflect the biology more accurately. On the other hand, workers in the biological sciences (for example) need to appreciate what mathematics (and its practitioners) can and cannot do! (The mathematician needs to do the educating here; good communication, as always, is necessary.)

The art of good modeling relies on (i) a sound understanding and appreciation of the biological problem, (ii) a realistic mathematical representation of the important biological phenomena, (iii) finding useful solutions, preferably quantitative ones, and (iv) biological interpretation of the mathematical results - insights, predictions, etc. The mathematics is dictated by the biology and not, in general, vice versa, however tempting that may be! Sometimes the mathematics used can be very simple. The usefulness of a mathematical model should not be judged by the sophistication of the mathematics but by different (and no less demanding) criteria. It should also be pointed out that, while the techniques of statistical analysis may frequently be used in portraying and interpreting data, the term "mathematical model" as used here refers to deterministic models rather than probabilistic or statistical ones. Note that many of the preceding comments also apply to these latter types of mathematical models.

We concentrate here on deterministic models of tumor growth, with the exception of a computer simulation (based on a probabilistic model for tumor growth) presented by Williams and Bjerknes (1972). They defined tumor growth as occurring when a single abnormal cell divides faster than normal cells by a factor called the "carcinogenic advantage." The simulations in two dimensions appear to have fractal-like boundaries, and this "metaphor" serves to define yet another complementary level of description in cancer biology (Adam, 1988).

It is to be hoped that mathematical models of tumor growth will be able eventually to incorporate as many of the following phenomena as possible (note that these are not mutually exclusive): mechanical/pressure effects; oxygen and nutrient distribution; growth inhibitor/activator distribution; destructive enzyme action; metabolic activity; blood vessel and capillary distribution; cell adhesiveness; the immune response; invasion; metastasis; and growth inhibition due to radiation, chemotherapy, or other treatment modalities.

SIMPLE MODELS OF TUMOR GROWTH

At the simplest level of description, a population (of cells, bacteria, or people, depending on the context) growing at a rate proportional to its present value increases in size exponentially. Historically, this is related to Malthusian growth, and it is described by a simple differential equation if the continuum assumption is made. This assumption is that, given a sufficiently large initial population, the rate of change of that population will be continuous in the standard mathematical sense. This assumption is not always appropriate, but when it is valid, it has the pleasing consequence that differential and integral calculus may be used to describe many of the characteristics of that population's growth.

One can develop a sequence of models, increasing in mathematical complexity, that describe the qualitative features of population growth in a given context. With enough free parameters in the model, it is possible to fit a growth curve to the empirical data and generate a reasonable fit. However, the questions to be asked of these models in the context of biological applications are: what are the biological mechanisms giving rise to the growth curve(s), and are they incorporated in the model?

The simplest models are almost purely phenomenological in character and by themselves do not offer any biological insight into the nature of the problem (e.g., in the study of tumor growth). As mentioned above, the simplest possible growth model (ignoring very trivial special cases) is that of exponential growth. Another simple model is that of limited growth: the population grows at a rate proportional to the *difference* between the present population and some limiting saturation value. This is a convenient model for describing the spread of information by mass media in a closed community, for example, or any other kind of saturation phenomenon. A model which combines both this saturation effect and the linear growth rate model mentioned above is the *logistic* differential equation which incorporates both a linear growth rate term and a quadratic "self-interaction" term. This model has been used to describe a variety of different growth situations, but it has the disadvantage that it is symmetric (i.e., the point of inflection occurs midway in the range of the function); this is atypical in practical terms. This problem can be overcome from an empirical point of view by generalizing the growth rate dN/dt in terms of a dimensionless parameter α (Adam, 1997). A family of empirical curves of growth that saturates either more slowly or more rapidly than the logistic solution can be derived from a generalization of the Verhulst equation (Adam, 1997). This formulation has the additional advantage that the logistic equation is recovered when $\alpha = 1$, and the Gompertz equation is recovered in the limit $\alpha \rightarrow 0$. The latter is useful in the modeling of some types of tumors in small animals (Laird, 1965).

OTHER BASIC MODELS OF TUMOR CELL POPULATION GROWTH: MULTICELL SPHEROIDS

Other models have been developed by Marusic and co-workers (1994) and Vaidya and Alexandro (1982). Many prevascular diffusion models of tumor growth predict growth curves which are qualitatively similar to the saturated-growth curves derived from the above models, but a major advantage of *these* models is that the governing differential (or integro-differential) equations of growth are based on plausible physical and biological assumptions. Thus, any comparison of model-generated curves of growth with experimental data can, at least in principle, provide some information on the appropriate parameter ranges necessary for consistency with the data.

As pointed out by Wheldon (1988), the value of modeling to a science will depend on the extent to which that science incorporates defined assumptions that lend themselves to quantitative expression. This is the basis of the model by Greenspan (1972) which gives the outer radius of the spheroid [and various inner radii (e.g., the radius of the necrotic core)] as a function of time. The first mathematical model we review, however, is that by Burton (1966).

Burton was the first to introduce diffusion into a model of spheroid growth, taking the model out of the realm of pure phenomenology into a more biologically-driven environment. Interestingly, by considering a diffusion problem alone, Burton was able to glean important information about the relative thickness of the viable layer (the outer rim of proliferating cells) without utilizing an evolution equation for the outer tumor radius. The main features of Burton's models are easily summarized: the assumptions include spherical symmetry and diffusive equilibrium (see below for a discussion of this concept). The oxygen consumption rate per unit volume of non-necrotic tissue is assumed to be constant, so the distribution of oxygen concentration satisfies a time-independent diffusion equation, subject to appropriate boundary conditions. The necrotic core is defined by a zero flux boundary condition at the location where the oxygen concentration reaches a critical value below which cells are assumed to die. Further, at the outer surface of the spheroid, the oxygen concentration is maintained at a constant value. Under these circumstances, a simple mathematical relationship between these quantities is readily established. This simple model has obvious limitations: there is in principle no limit to the size of the spheroid (clearly a major deficiency of the model!), but even here a useful result can be found for the relative size of the viable layer (the layer containing non-hypoxic cells, for example, wherein they still proliferate). It transpires that, within the confines of this two-layer model, the viable layer approaches a constant thickness, equal to 58% of the critical tumor radius (i.e., the radius at which necrosis first occurs).

DIFFUSION OF GROWTH INHIBITOR

In 1973, Glass published a one-dimensional model of growth inhibitor production in tissue; that model has proven to be seminal because of its simplicity and its usefulness. In a related fundamental paper, with more biologically realistic boundary conditions, Shymko and Glass (1976) dealt fully with the corresponding three-dimensional multicellular spheroid problem. The former paper is summarized here, with the understanding that the basic biological features and consequences are not significantly changed (apart from geometric factors) in the more realistic spherically symmetric case.

Consider a "slab" of slowly growing tissue of width L, centered on the origin, producing growth inhibitor at a rate P (molecules/unit volume/second) which is depleted or decays at a rate λ. This central tissue is embedded in an infinite expanse of non-active tissue. For simplicity, the diffusion coefficient D for the inhibitor is assumed to be constant everywhere. A fundamental assumption made is that of diffusive equilibrium: the timescale for significant tissue growth is considered large compared with the timescale for readjustment of the inhibitor concentration profile due to such growth (i.e., the system is essentially in a steady state). This is a reasonable assumption for central tissue sizes not in excess of, say, 1-2 mm (typical of multicellular spheroids), for which the typical diffusion time (for oxygen) is of the order of 10 minutes (Edelstein-Keshet, 1988). The most straightforward boundary conditions of interest to us here are that the concentration of growth inhibitor C (in units of molecules/unit volume) should be smoothly varying across the boundary between the "active" tissue (producing the inhibitor) and the "normal" tissue, with the obvious symmetry condition that $C'(0) = 0$ (i.e., the derivative of the concentration of inhibitor is zero in the center of the slab because there is no flux there). The governing mathematical description can be represented in terms of an ordinary differential equation (rather than by a partial differential equation, in view of diffusive equilibrium).

Now we suppose, following Glass, that there exists a switch-like mechanism such that if the concentration C equals or exceeds some critical concentration θ, growth ceases in that region, and if $C < \theta$, growth continues. It follows that, for growth to cease *throughout the tissue,* the concentration of growth inhibitor must satisfy $C \geq \theta$ at the edges (or boundaries) of the slab (or at the surface of the spheroid in the three-dimensional case). Two other important parameters are the inhibitor depletion or decay rate λ and the inhibitor production rate P (both are assumed constant here). In terms of the dimensionless parameter $n = P/\lambda\theta$, it can be readily shown that the stable limiting size of tissue is given by the formula

$$L_S = \sqrt{\frac{D}{\lambda}} \ln\left(\frac{n}{n-1}\right) \tag{1}$$

L_S is clearly defined for $n > 1$; under these circumstances, growth eventually ceases as the system size approaches L_s. Glass speculated that the "singular region" defined in parameter space by $n \leq 1$ corresponds to the converse of controlled stable growth, namely uncontrolled, unstable growth. This is a very basic definition of cancer: cellular proliferation defying the normal control mechanisms. While claiming in essence to be no more than a metaphor for cancer at this level of description (given the lack of biological input insofar as cellular replication is concerned), this approach has been and continues to be an intriguing and suggestive avenue to a complex phenomenon. It may well require many different but complementary levels of description to enable even a partial understanding to be obtained (Adam, 1988). Interestingly enough, a similar mathematical approach has recently been applied to the phenomenon of wound healing in bone, with particular reference to the so-called *critical size defect.* For further details, see Adam (1999) and Arnold and Adam (1999).

By introducing somewhat more biologically reasonable geometry and boundary conditions at the tissue boundary, Shymko and Glass (1976) were able to make comparisons between their model and experiments on multicellular spheroids. The authors delineated parameter regions for unstable (or unlimited) growth throughout the tissue, for unstable growth with mitosis confined to a peripheral region of the tissue, and for stable, limited growth. A very interesting discussion of the implications of the model and experimental data on cellular and geometric control of tissue growth is presented in their paper, and the reader is urged to consult it for further details. Subsequent developments of this model have focused on non-homogeneous source terms (Adam, 1986, 1987a, b; Swan, 1992) and incorporation of necrosis as a possible source of inhibitor [e.g., see Adam and Maggelakis (1989) and references therein].

TIME-EVOLUTIONARY DIFFUSION MODELS

As stated by Greenspan (1972), the objective of the type of diffusion model he developed is to infer the major internal process affecting tumor growth from the most easily obtained *in vitro* data. The data are assumed to be measurements of the outer nodule radius as a function of time and a cross-section of the final dormant state which provides the limiting radius and the limiting necrotic core radius of the final dormant *in vitro* state. In this model, there develops in general a three-layer structure consisting of a central necrotic core above which there is a layer of viable nonproliferating cells and finally an outer shell where all mitosis occurs. (No account is taken of the cell age or stage of cycle.) Greenspan lists many assumptions that are explicit or implicit in the model; the reader is referred to that paper for further details. See also Adam (1997). The model of Greenspan (1972) represents a fine example of mathematical modeling in biology. The stability of such a spheroid, initially spherically symmetric, to non-symmetric perturbations was considered elsewhere by Greenspan (1976), and an extension will be discussed below.

The basis of a growth equation describing the evolution of the outer nodule radius can be described in the following manner: $A = B + C - D - E$, where A = total volume of living cells at any time t; B = initial volume of living cells at time $t = 0$; C = total volume of cells produced in $t > 0$; D = total volume of necrotic debris at time t; and E = total volume lost in necrotic core for $t > 0$. One feature common to all these prevascular models is that the geometry is spherically symmetric: this is mathematically convenient, for in this ideal situation, the net resultant of any intercellular forces within the colony is radially directed. Under these circumstances, no more precise examination of local cell dynamics is needed. However, it is clear that, in practice, such spherical symmetry is likely to be very rare. Even if it is initially the case, subsequent growth of a small tumor cell colony is unlikely to maintain this symmetry. Chaplain (1993) has contributed to the understanding of this situation, based in part on earlier work by Greenspan (1976), and here we draw on the salient features of Chaplain's work.

There is one caveat, and it applies to all models in which it is assumed that gross internal forces may be characterized by a pressure distribution, non-uniformities of which affect cell motion. Recent experiments on multicellular spheroids (R.K. Jain, 1995, private communication) failed to measure any pressure at all within the spheroids prior to vascularization. Once a spheroid or metastasis has been vascularized, the resultant pressure distribution must reflect the external systemic blood pressure and, by regarding the vascularization to correspond in some sense to a spatially smooth pressure distribution [as in Adam and Noren (1993)], the analysis in these models is still useful. Another prevalent assumption, also already noted, is that internal cell adhesion, like molecular attraction, produces a "surface tension" at the outer boundary of the nodule that maintains compactness and counteracts internal expansive pressures. While it is known that such multicellular spheroids are held together by a variety of junctions [see Chaplain (1993)], this postulated quasi-balance of forces (and associated surface area/volume considerations) is not sufficient to explain the stable limiting size of such a spheroid *in vitro*. Experimental work by Freyer (1988) and others suggests that tumor growth inhibitors have at least some role to play in this regard. This issue is addressed to some extent by the previously mentioned work of Greenspan (1972, 1974) and its extensions [e.g., Maggelakis and Adam (1990); Adam and Maggelakis (1990)].

With this in mind, we are in a better position to appreciate the strengths and weaknesses which characterize models of growing tumor colonies and their stability. An excellent introduction to models of this type can be found in Jones and Sleeman (1983), who identify the basic assumptions of Greenspan's (1976) model and re-derive the governing equations. Thus formulated, the problem falls into the class known as moving boundary value problems, in which the outer surface is represented by the general expression $\Gamma(x, y, z, t) = 0$.

The interior pressure p and exterior relative nutrient concentration σ respectively satisfy the equations $\nabla^2 p = S$ inside $\Gamma = 0$ and $\nabla^2 \sigma = 0$ outside $\Gamma = 0$. S is a constant representing the rate of volume loss per unit volume within the cancer cell colony (due to necrosis and the freely permeable necrotic debris being replaced by previously living cells via the compaction mechanism postulated above). Various other boundary and initial data are necessary to complete the statement of the mathematical problem. A prime concern of the model is *instability*: do small perturbations to an initial equilibrium configuration continue to grow in time or do they stabilize (as far as a linear analysis is concerned)? The simplest case mathematically is that of initial spherical symmetry, for which the equation $\Gamma(x, y, z, t) = 0$ becomes the simpler expression $r - R(t) = 0$, where R(t) is the outer tumor radius and r is the distance of any interior point from the tumor center. Since the time variable enters the system only in determination of the tumor radius, the preceding system of equations can be solved explicitly. This enables a single ordinary differential equation for R(t) to be derived. On solving this equation, it is found that the tumor (cell colony) typically grows in a logistic-type manner and asymptotes to the steady state radius in a time scale regulated mainly by the rate of volume loss of necrotic debris.

Growth is judged unstable to infinitesimal perturbations if any such disturbance amplifies at an exponential rate [exceeding that of the radius R(t)]. In this circumstance, the instabilities radically alter the shape of the colony and can even lead to fracturing into two or more pieces. The tumor becomes unstable if and when it reaches a critical size beyond which surface tension is overcome by pressure forces. The precise criterion for this is obtained from the analysis of Greenspan (1976) and Chaplain (1993). See also Byrne and Chaplain (1995), who considered the growth of a spherical tumor (or cell colony) which is subject to small deviations that are always inevitably present in a real environment. Originally, for simplicity, Greenspan took the perturbations from complete sphericity to be axially symmetric [i.e., independent of the azimuthal angle φ in spherical polar coordinates (r, θ, φ)]. For completeness, Chaplain included the azimuthal dependence in his analysis. The equation of the moving surface is now written as $r - R(t) - \varepsilon\eta(\theta, \varphi, t) = 0$. The solutions of this equation are given in terms of associated Legendre polynomials. It can be shown that an initial disturbance amplifies or decays according to whether a certain complicated expression is positive or negative. The tumor development is unstable if small perturbations can amplify; otherwise it is stable. Under some circumstances, the tumor is stable during its entire growth and it attains its symmetric equilibrium configuration. It is possible, however, that the tumor becomes unstable at some definite time in its growth when small disturbances amplify and change both the structure and the shape of the tumor. In this case, the changing pressure distribution overcomes the surface tension before spherical equilibrium is reached.

The onset of stability in the simplest non-trivial mode is manifested as a pinch in the outer surface of the tumor around the equatorial region. As the tumor grows and increases in size, other modes become unstable and more radical changes in configuration may occur which can be used to model the tumor invading the surrounding tissues. Further work by Chaplain and Sleeman (1992, 1993) on tumor growth involves the use of results and techniques from nonlinear elasticity theory and differential geometry. Their mathematical models describe the growth of a solid tumor using membrane and thick-shell theory. A central feature of their analysis is the characterization of the material composition of the model through the use of a strain-energy function, thus permitting a mathematical description of the degree of differentiation of the tumor explicitly in the model. Conditions are given in terms of the strain-energy function for the processes of invasion and metastasis which frequently occur as the tumor evolves. These are interpreted as the modes of bifurcation of a spherical shell. The results are compared with actual experimental results and with the general behavior exhibited by both benign and malignant tumors. The authors also use their mathematical results in conjunction with aspects of surface morphogenesis in tumors (in particular, the Gaussian and mean curvatures of the surface of a solid tumor) in an attempt to produce a mathematical formulation and description of the important medical processes of staging and grading cancers.

TUMOR ANGIOGENESIS

Without the phenomenon of angiogenesis, solid tumors do not grow beyond a size of at most 2 mm in diameter (corresponding to approximately one million cells). Indeed, this is the motivation behind the search for angiogenesis inhibitors: if this mechanism can be suppressed, then a nodular carcinoma of this size is essentially harmless. Clearly, all the models discussed so far apply to the avascular or pre-vascular growth phase. The next level of biological and mathematical complexity requires an appropriate representation of vascularization, and to this end the work of Chaplain and Anderson (1997) is very important. See also a summary of this and other work in Panetta et al. (1998). The work of Chaplain and Anderson (1996) focused attention on three features essential to the process: tumor angiogenic factors (e.g., vascular endothelial growth factor), endothelial cells, and matrix macromolecules (e.g., fibronectin).

The events leading to angiogenesis are as follows. At the end of the avascular phase, the tumor cells secrete chemicals [collectively known as tumor angiogenesis factors (TAF)] which induce endothelial cells (EC) in their vicinity to secrete matrix- degrading enzymes (e.g., proteases) which degrade the vascular basement membrane. These cells are then free to migrate through the membrane toward the tumor. Next, the cells associated with the capillary sprout-tip region subsequently proliferate, leading to sprout growth, initially in parallel, but eventually they merge and form tip-to-tip and tip-to-sprout fusions and loop formations. Shortly after this, the first signs of circulation become evident, with new sprouts and buds repeatedly emerging, further extending the capillary network. Eventually the tumor is penetrated, leading to vascularization and a vastly increased supply of oxygen for the tumor cells in the interior of the nodule.

The mathematical model developed by Chaplain and Anderson consists of three coupled partial differential equations describing the response of the endothelial cells to angiogenic factors via *chemotaxis* (whereby the cells are sensitive to and move up gradients of TAF, from lower to higher concentrations) and to fibronectin via *haptotaxis*. (Fibronectin affects how endothelial cells adhere to collagen and to the underlying substratum in general.) In the following "word" equations, $n(x, y, t)$ is the endothelial cell density, $c(x, y, t)$ is the TAF concentration, and $f(x, y, t)$ is the fibronectin concentration. (The reader is referred to the original article for the full mathematical details.) The meanings of the governing equations are (i) the time rate of change of n = spatial changes due to diffusion, chemotaxis, and haptotaxis; (ii) the time rate of change of f = rate of production by EC - rate of uptake by EC; and the (iii) time rate of change of c = rate of uptake by EC. Chaplain and Anderson solved these equations numerically (in discretized form) with parameter values based on experimental data. They were able to generate realistic capillary network structures, and by simulating the movement of endothelial cells toward a small circular tumor (in two dimensions x and y), they found that several of the loops achieved anastomosis within 4.5 days; complete vascularization was achieved at about 15 days.

TUMOR PROGRESSION

In a recent paper (Greller et al., 1996), a conceptual model for tumor progression was presented. Such a model seeks to address various aspects of cancer including (i) the dynamic features of tumor cell heterogeneity, progression, and growth and (ii) phenotypic progression as distinct from growth (i.e., changes in tumor bulk). Clearly, it is important to define what is meant by "tumor" and "progression" in this context, since we are implicitly dealing with many different aspects of tumor phenomenology. We can address these topics at various (complementary) levels of description, and this is one reason why the subject of cancer and its therapy is such a complex one.

Question: What is a Tumor?

A tumor is an assembly of cellular subpopulations exhibiting diverse traits at several levels of biological organization, namely genetic, phenotypic, cellular, and physiological. A tumor is *heterogeneous* – at the genotypic, phenotypic, spatial, and temporal levels of description. To gain an understanding of tumor dynamics includes studying (among other things) variable patterns of gene expression, enzymatic activity, cell surface properties, metabolic control, hormonal dependencies, tissue invasiveness, metastatic competencies, host immune responses, and resistance to treatment modalities.

Question: What is Tumor Progression?

Tumor progression is rather vaguely defined in the literature, but in broad terms, it may be characterized in several ways that are not mutually exclusive. It is an aggregate phenomenological property, it is an irreversible qualitative change in one or more characters of the neoplastic cells, and it is "...different from a mere extension in space and time without qualitative change..." (Foulds 1954). Progression does not necessarily correlate uniformly with actual elapsed chronological time. Indeed, according to Foulds, progression indicates "development of a tumor by way of permanent, irreversible qualitative change in one or more of its characters." This leads to another question.

Question: What Characteristics "Define" Cancer as a Disease?

There are many facets in the "answer" to this question. Some of the more obvious ones are a tumor's exploitation of cellular heterogeneity vis-à-vis increasing growth autonomy, loss of proliferative constraints, invasion of neighboring tissues, and aquisition of metastatic potential.

It is the case that certain phenotypic traits in tumor cells appear to be primarily progression-driven, as opposed to growth-driven. Examples of progression-driven traits include drug resistance, genetic instability, invasiveness, and metastatic potential. By way of contrast, tumor mass, growth rate, and vascularization are probably more growth-driven. In reality, progression and growth are not independent features of tumor development, so any separation of the two is artificial. Nevertheless, it may still be a conceptually useful contrivance because, by defining progression from a cellular perspective, the issue of tumor bulk is avoided (Greller et al., 1996). By projecting an "object" in a higher-dimensional space (defined by a host of molecular, cellular, tissue, and host-related characteristics) onto a one-dimensional space ("progression value"), information is obviously lost, but there is no longer a need (or opportunity!) to focus on detailed mechanisms. In light of this, though, we may ask another question.

Question: Is Such a Univariate Approach Useful?

Although the phenotypic state of a cell cannot be adequately described by a single quantity (cf. IQ!), this approach does allow for the modeling (in principle) of a large collection of cooperative phenomena throughout a population of cells.

Question: How Can the "State of Progression" be Characterized?

This is essentially the subject of the paper by Greller et al. (1996); it is discussed below. Nevertheless there are various vague but roughly equivalent statements that are often made in connection with this state, namely "degree of malignancy," "degree of transformation," and "departure from normality." In the model of Greller et al., the progressive state of each cell is not static but is allowed to change over time; the resulting "rate of change" is termed *progression velocity*. Progression is treated as a continuous variable, and the progression velocity will "move" each tumor cell toward higher progression values. Furthermore, each type of tumor can have a different functional form for its progression velocity. The progression velocity essentially represents the changing phenotypic behavior of a tumor in a relatively straightforward manner.

With progression as a continuous variable, the model may not have distinct subpopulations; however, there will be continuous distributions of cells, each of which has different progression values. The overall progression of a tumor is averaged over all the constituent cells and is closely linked with pathological staging (insofar as both convey the cumulative extent of departure away from normality). However, this average behavior alone is not sufficient for understanding the characteristics of tumor behavior, since one or more clonal subpopulations may become metastatically competent (a serious clinical manifestation). The reader is referred to the paper by Greller et al. (1996) and the references therein for a detailed account of the potential applications of this mathematical model.

CATASTROPHE THEORY

Another level of description that may be useful in trying to understand some qualitative features of cancer is known as catastrophe theory; see Zeeman (1973), Woodcock (1974), and Thom (1989). In catastrophe theory (as in quantum mechanics), the phenomenon under study is assumed to be governed by a potential function of some kind. Stable states of the system are regarded as minima of this potential. If the potential function has multiple minima, more than one stable state may be accessible to the system (e.g., normal/cancerous, benign/malignant, good prognosis/poor prognosis). Changing the control parameters in a treatment modality may alter the form of the governing potential function so as to change the positions, relative heights, or total number of local minima. Thus the observed state of a system *may* change in a discontinuous way as the controls are changed smoothly. These observed discontinuous changes in state are called *catastrophes*.

Space does not permit a detailed discussion here, so we will focus on the so-called *cusp catastrophe* in order to illustrate the features that may be relevant here. This catastrophe surface arises from consideration of extrema (particularly minima) of a particular form of a quartic potential V(a, b; x). Related to this potential is the cusp catastrophe surface which is in the form of a folded "sheet" with a cusp; see Figure 2 in Adam (1996). *What has this to do with modeling of cancer?* Many things, in all likelihood. To be specific, a recent model of tumor/immune system interactions (Adam, 1996) involved a study of the cusp catastrophe. The model, in crude terms, describes both remission and rapid metastatic growth behaviors.

There are five interrelated qualitative features associated with this catastrophe surface. It is not necessarily the case that they will all be manifested if the model applies, but if none appear, then the model is surely vacuous! These features are *bimodality*, *discontinuity*, *hysteresis*, *divergence*, and *inaccessibility*. The properties refer to selected behaviors of the system which may not all occur under normal circumstances. If conditions are perturbed, however, the system may move to a configuration that exhibits other properties in this list. A system that evolves smoothly but irregularly to one or the other of two possible final states (bimodality and divergence) may jump catastrophically (discontinuously) between these states if an appropriate new perturbation is applied. *Bimodality* refers to situations in which observations tend to cluster around two statistical measures. Tumors, for example, can be benign or malignant (and prognosis can be either good or bad). Bimodality is basically a static property which we incorporate as a given feature. *Discontinuity* refers to a large behavioral change resulting from a small change in control variables, as has been discussed earlier. *Hysteresis* occurs when a system has a delayed response to a changing stimulus. A plot of response against stimulus will follow one path when the stimulus increases and another when it decreases. *Divergence* occurs when initial conditions which are in some sense "nearby" evolve to widely separated final states. Thus, two patients with similar histories of cancer and identical treatment modalities may respond in radically different fashion. Cell differentiation is another example (indeed, Thom originally developed the theory to study morphogenesis). *Inaccessibility* refers to the existence of the unstable middle sheet of the folded surface. This represents those unstable equilibria which are rarely, if ever, observed, and thus are completely destroyed when perturbed (e.g., as with a coin balanced on edge).

SOME COMMENTS ON MODELING USING CATASTROPHE THEORY

For most biological systems, we are not in a position to write down the equations that govern their evolution (where the word *evolution* is not necessarily meant in the temporal sense). There may be thousands or millions of equations! Catastrophe theory may assist us by allowing us to use observations directly to suggest the generic types of equations we are seeking. (It is much easier to work backward from the answer!) We do not have to know in advance what approximations we may make. Instead, we can determine which ones we *must* make if the model is to be consistent with observations. This may lead to a greater understanding of the way the system works, by connecting certain internal features of it to the observed phenomena in ways which otherwise might be difficult to discover. Of course, there is also the possibility that the model is useless to us, and in that situation we must cut our losses and look for new pastures.

Lest it should appear that, in the application of catastrophe theory to a system, we are trying to get something for nothing, it is perhaps worthwhile to add some further comments. Firstly, there is a great amount of work "behind the scenes" so to speak, as far as catastrophe theory is concerned. The theorems proved by Thom and others are very "deep" and have an austere beauty not only in their own right, but also in connection with the power, applicability, and relative simplicity of the seven elementary catastrophes. Secondly, a complete model of the system contains (usually) an enormous amount of information and so requires a great deal of input. In attempting to apply catastrophe theory to a particular system, we are in effect seeking the answer to one or two specific questions (e.g., is the prognosis good or bad for this patient?). A relatively small amount of information (which still may be a lot!) may suffice. In this sense, catastrophe theory is an efficient technique if it can tell us what we want to know with the minimum of data and standard theoretical background as input. Thirdly, in some applications (e.g., in sociology), the governing family of potential functions is used strictly as a phenomenological tool. However, when augmented by other arguments (be they statistical or otherwise in nature), catastrophe theory can be used as a predictive tool to account in a quantitative manner for the changes in a given system as the constraints on the system are changed. This is a distinct possibility in cancer biology.

ACKNOWLEDGMENT

This article is dedicated to the memory of my friend, colleague, and fellow adventurer in mathematical biology, Dr. Philip R. Wohl, who died of cancer in January 1996.

CORRESPONDENCE

Please address all correspondence to:
John A. Adam
Old Dominion University
Department of Mathematics and Statistics
Hampton Boulevard
Norfolk, VA 23529
jadam@odu.edu

REFERENCES

Adam, J.A., 1986, A simplified mathematical model of tumor growth, *Math. Biosci.* 81:229-244.

Adam, J.A., 1987, A mathematical model of tumor growth: II. Effects of geometry and spatial non-uniformity on stability, *Math. Biosci.* 86:183-211.

Adam, J.A., 1987, A mathematical model of tumor growth: III. Comparison with experiment, *Math. Biosci.* 86:213-227.

Adam, J.A., 1988, On complementary levels of description in applied mathematics. II. Mathematical models in cancer biology, *Int. J. Math. Ed. Sci.Tech.*19:519-535.

Adam, J.A., 1996, Mathematical models of prevascular spheroid development and catastrophe-theoretic description of rapid metastatic growth/tumor remission, *Invas. Metast.* 16:247-267.

Adam, J.A., 1997, General aspects of modeling tumor growth and immune response, in: *A Survey of Models for Tumor-Immune System Dynamics,* J.A. Adam and N. Bellomo, eds, Birkauser, Boston.

Adam, J.A., 1999, A simplified model of wound healing (with particular reference to the critical size defect), *Math. Comput. Model.* 30:23-32.

Adam, J.A., and Maggelakis, S., 1989, A mathematical model of tumor growth. IV. Effects of a necrotic core, *Math. Biosci.* 97:121-136.

Adam, J.A., and Maggelakis, S.A., 1990, Diffusion regulated growth characteristics of a prevascular carcinoma, *Bull. Math. Biol.* 52:549-582.

Adam, J.A., and Noren, R., 1993, Equilibrium model of a vascularized spherical carcinoma with central necrosis: some properties of the solution, *J. Math. Biol.* 31:735-745.

Arnold, J.A., and Adam, J.A., 1999, A simplified model of wound healing. II: The critical size defect in two dimensions, *Math. Comput. Model.* 30:47-60.

Burton, A.C., 1966, Rate of growth of solid tumors as a problem of diffusion, *Growth* 30:159-176.

Byrne, H.M., and Chaplain, M.A.J., 1995, Growth of non-necrotic tumors in the presence and absence of inhibitors, *Math. Biosci.* 130:151-181.

Chaplain, M.A.J., The development of a spatial pattern in a model for cancer growth, in: *Experimental and Theoretical Advances in Biological Pattern Formation,* H.G. Othmer, P.K. Maini, and J.D. Murray, eds, Plenum Press, New York.

Chaplain, M.A.J., and Sleeman, B.D., 1992, A mathematical model for the growth and classification of a solid tumor: a new approach via nonlinear elasticity theory using strain-energy functions, *Math. Biosci.* 111:169-215.

Chaplain, M.A.J., and Sleeman, B.D., 1993, Modelling the growth of solid tumors and incorporating a method for their classification using nonlinear elasticity theory, *J. Math. Biol.* 31:431-473.

Chaplain, M.A.J., and Anderson, A.R.A., 1996, Mathematical modelling, simulation and prediction of tumour-induced angiogenesis, *Invas. Metast.* 16:222-234.

Edelstein-Keshet, L., 1988, *Mathematical Models in Biology,* Random House, New York.

Freyer, J.P., 1988, The role of necrosis in regulating the growth saturation of multicellular spheroids, *Cancer Res.* 48:2432-2439.

Foulds, L., 1954, The experimental study of tumor progression: a review, *Cancer Res.* 14:327-339.

Glass, L., 1973, Instability and mitotic patterns in tissue growth, *J. Dyn. Syst. Meas. Contr.* 95:324-327.

Greller, L.D., Tobin, F.L., and Poste, G., 1996, Tumor heterogeneity and progression: conceptual foundations for modeling, *Invas. Metast.* 16:177-208.

Greenspan, H.P., 1972, Models for the growth of a solid tumor by diffusion, *Stud. Appl. Math.* 51:317-340.

Greenspan, H.P., 1974, On the self-inhibited growth of cell cultures, *Growth* 38:81-95.

Greenspan, H.P., 1976, On the growth and stability of cell cultures and solid tumors, *J. Theor. Biol.* 56:229-242.

Jones, D.S., and Sleeman, B.D., 1983, *Differential Equations and Mathematical Biology,* George Allen and Unwin, London.

Laird, A.K., 1965, Dynamics of tumor growth. Comparison of growth rates and extrapolation of growth curve to one cell, *Brit. J. Cancer* 19:278-291.

Maggelakis, S., and Adam, J.A., 1990, Mathematical model for prevascular growth of a spherical carcinoma, *Math. Comp. Model.* 13:23-38.

Marusic, M., Bajzer, Z., Vuk-Pavlovic, S., and Freyer, J.P., 1994, Tumor growth in-vivo and as multicellular spheroids compared by mathematical models, *Bull. Math. Biol,* 56:617-631.

Panetta, J.C., Chaplain, M.A.J., and Adam, J.A., 1998, The mathematical modelling of cancer: a review, in: *Mathematical Models in Medical and Health Science,* M.A. Horn, MA, G. Simonett, and G.F. Webb, eds, Vanderbilt University Press, Nashville.

Swan, G.W., 1992, The diffusion of inhibitor in a spherical tumor, *Math. Biosci.* 108:75-79.

Thom, R., 1989, *Structural Stability and Morphogenesis: An Outline of a General Theory of Models,* Addison-Wesley, New York.

Vaidya, V.G., and Alexandro, F.J., 1982, Evaluation of some mathematical models for tumor growth, *Int. J. Bio-Med. Comp.* 13:19-35.

Weinberg, R.A., 1998, *One Renegade Cell,* Basic Books, New York.

Wheldon, T.E., 1988, *Mathematical Models in Cancer Research*, Adam Hilger, Bristol.

Williams, T., and Bjerknes, R., 1972, Stochastic model for an abnormal clone spread through epithelial basal layer, *Nature* 236:19-21.

Zeeman, E.C., 1973, *Applications of Catastrophe Theory,* Tokyo University Press,Tokyo.

A SYSTEMS MODELING APPROACH TO THE STUDY OF RETINOID FUNCTION: IMPLICATIONS FOR EVALUATION OF RETINOIDS IN CANCER CHEMOPREVENTION AND/OR CHEMOTHERAPY

Kevin C. Lewis, James F. Hochadel, and Loren A. Zech Jr.[*]

INTRODUCTION

Vitamin A-related compounds (retinoids) play critical roles in cell growth and differentiation (Roberts and Sporn, 1984). Since unregulated cell growth and differentiation are factors associated with most types of cancer, there has been a great deal of effort put into understanding the complex relationship between retinoids and various carcinogenic processes. An emerging body of evidence supports the notion that important relationships exist between retinoids and different types of cancer. Critical roles for naturally occurring- and synthetic retinoids in the prevention and/or therapy of certain cancers have been demonstrated in a wide variety of epidemiological, laboratory, and clinical studies (Lotan, 1980; Roberts and Sporn, 1984; Hong and Itri, 1994; Moon et al., 1994). However, human trials using retinoids in chemoprevention and/or chemotherapy programs have yielded mixed results. The reasons for this are not entirely clear. However, one likely problem is the relative paucity of information in a number of areas related to basic retinoid functioning and interactions. The importance of delineating the basic functions and interactions of the native and synthetic variants of retinoids, preferably in the physiological milieux in which they would normally exist, is of paramount importance if we are to more clearly understand the relationship between retinoids and carcinogenic processes. This is essentially the rationale we have used to guide and focus our research efforts. To obtain basic information on retinoid function and interactions, we have made extensive use of mathematical/compartmental modeling techniques not only to obtain critical information in this area, but also to optimize our experimental design and resources and thereby help to direct our research efforts appropriately.

---

[*] Kevin C. Lewis and Loren A. Zech, Jr, Drug-Nutrient Interactions Group, Basic Research Laboratory, National Cancer Institute, National Institutes of Health, MD 21702. James F. Hochadel, Intramural Research Support Program-Science Applications International Corp., Frederick, MD 21702.

BACKGROUND

Despite the relatively widespread use of various retinoids as chemopreventive and/or chemotherapeutic agents in recent years, many aspects of the metabolism of native retinoids (i.e., those normally found in the human system and diet) remain to be delineated. These include primary and secondary control mechanisms that regulate plasma and tissue levels of native forms of vitamin A as well as their turnover, uptake, processing, and ultimate utilization by tissues. Moreover, even less is known of the fundamental interactions that occur among native and synthetic retinoids, other nutrients, hormonal factors, and chemotherapeutic drugs. A well defined role for retinoids (administered alone or in conjunction with other agents) in chemoprevention and/or chemotherapy has proven elusive. However, it has been our view that, given the essential functions of retinoids in cell growth and differentiation, it is critical that we more clearly understand the metabolism of native retinoids, as well as the numerous interactions of these compounds with other components in the system. It is unlikely that the full clinical potential of retinoids as chemopreventive and/or chemotherapeutic agents will be fully realized until we do so.

Accordingly, an important area of research in our laboratory has involved the identification and delineation of the metabolism of native retinoids. This in turn has led to a better understanding of how these compounds interact and respond to a variety of nutritional, physiological, and pharmacological perturbations. We obtain our data by conducting experiments in complimentary *in vivo* and *in vitro* systems that are optimized for studies of retinoid - nutrient - drug interactions. We use the information derived from these experiments to develop mathematical/compartmental models to describe how different retinoids function in the system under study.

FOCUS OF THIS PAPER

To demonstrate the usefulness, power, and potential clinical applicability of our approach, we discuss here our *in vivo* and *in vitro* experiments designed to study the effects of the synthetic retinoid *N*-(4-hydroxyphenyl)retinamide (4-HPR) on the metabolism and kinetics of native retinoids. The effectiveness of this compound in the inhibition of cancer in a variety of tissues, coupled with its relatively low pharmacological toxicity compared to other retinoids, has made it an attractive candidate for a number of trials in which it has been studied alone and in conjunction with other agents. In this paper, we describe our findings on the effects of 4-HPR on the normal, physiological metabolism of native retinoids in the rat as well as in two specific tissues, the prostate gland and the eyes. Portions of this work have been described earlier (Lewis et al., 1994; Lewis et al., 1996; Lewis and Hochadel, 1999).

Our work with 4-HPR illustrates how we use relatively simple mathematical/compartmental models to help direct our research efforts in both *in vivo* and *in vitro* systems. In addition, our work with this agent serves as an example of how we have translated our hypotheses concerning the actions of 4-HPR into explanations of the some of the mechanisms involved in the adverse side effects of 4-HPR that have been observed in clinical trials (Kaiser-Kupfer et al., 1986; Kingston et al., 1986; Costa et al., 1989; Formelli et al., 1989; Modiano et al., 1990; Rotmensz et al., 1991; Gross and Helfgott, 1992; Cobleigh et al., 1993; Decensi et al., 1993). We choose to study eyes because of the relatively high prevalence of problems with visual function associated with administration of 4-HPR (see references just cited). The prostate gland was of interest to us based on the suggestion that 4-HPR might be a chemopreventive agent for certain types of prostate cancer (Welsch et al., 1983; Pollard et al., 1991; Pienta et al., 1993). Most importantly, the examples we discuss here highlight the fact that it was only by way of our use of a mathematical/compartmental modeling approach to investigate retinoid functioning and interactions that we were able to identify and delineate some mechanisms involved in the metabolic alterations that we observed. We are not aware of other approaches that would allow us to obtain and appropriately interpret this critical information.

METHODOLOGY

In Vivo Studies

The basic approach we have used in *in vivo* studies involves the use of specially prepared "donor" rats to incorporate a radiolabeled (tritium) dose of retinol in its normal physiological plasma transport complex (i.e., retinol:retinol-binding protein:transthyretin). To carry out retinol turnover studies, plasma containing the physiologically labeled complex is collected from donor animals and small, non-perturbing aliquots are injected into "recipient" rats that have been fed either a control diet or the control diet plus an additional retinoid and/or drug of interest. The donor - recipient approach was developed in Barbara Underwood's laboratory years ago (Lewis et al., 1976) and, with modifications, it has been used successfully in other studies (e.g., Green et al., 1985; Green et al., 1987; Lewis et al., 1990; Lewis et al., 1994; Lewis et al., 1996; Lewis and Hochadel, 1999). A major advantage of this approach is that recipient rats receive intravenously a labeled dose of the vitamin in physiological form and amount. The labeled dose is of high enough specific activity to carry out long-term studies. It should be emphasized that what we are ultimately monitoring using this approach is the metabolism and kinetic behavior of vitamin A (retinol) as it normally occurs in the circulation. We can also study how these processes are affected by the administration of retinoids such as 4-HPR. A general description of the experimental protocols we used in several recent studies is outlined in the following paragraphs.

For the *in vivo* studies, groups of rats were injected with label and studied for various periods of time between approximately one-half hour and 40 - 45 days. Following dose administration, plasma samples were collected at geometrically increasing intervals. Optimal times for sampling and the length of studies were based on preliminary data and computer simulations of the expected plasma and tissue kinetics. At the termination of each study, tissues were collected including livers, kidneys, testes, prostates, lungs, eyes, adrenals, and epididymal and perinephric fat pads. For long-term groups, urine and feces were also collected during the study. Plasma retinoids were extracted using hexane:ethanol and tissue samples were extracted using hexane:isopropanol. Retinoids were quantitated by reverse-phase HPLC using methanol:water as mobile phase. Urine samples were diluted, solubilized in scintillation solution, and counted. Fecal samples were first combusted in a biological materials oxidizer, then solubilized in scintillation solution and counted.

We usually develop whole-body models of native retinoid metabolism in a step-wise fashion. For the work presented here, we used a Windows version (WinSAAM; kindly provided by Dr. Peter Greif) of the SAAM/CONSAM computer modeling programs (Berman and Weiss, 1978; Berman et al., 1983). We used WinSAAM to first estimate a number of kinetic parameters for traced vitamin A based on analysis of plasma retinol tracer and mass (tracee) over time (Rescigno and Segre, 1966; Shipley and Clark, 1972; Rescigno and Gurpide, 1973). Kinetic parameters calculated include *fractional catabolic rates* (the rate of irreversible utilization expressed as a fraction of a compartment of interest), *transit times* (amount of time spent in a particular compartment during one passage through that compartment), *residence times* (total amount of time spent in a compartment prior to irreversible loss), *disposal or utilization rates* (the amount of vitamin utilized per unit of time), *number of recyclings* (number of times a molecule of vitamin A recycles through a particular compartment), and *mean recycling times* (the amount of time a molecule of vitamin A spends outside the plasma compartment before cycling back to this compartment). These parameters are all essentially based on information derived from the area under the plasma tracer response curve, and they provide important and easily assessable information concerning the metabolism of the retinoid of interest in the system as a whole.

To focus on metabolism in individual tissues, we next develop subsystem models for individual tissues of interest. To describe vitamin A metabolism in tissues of control rats and 4-HPR-fed animals, we set up parallel models for the tissues of interest to fit both retinoid tracer and mass (tracee) data. We then use a forcing function approach (Berman and Weiss, 1978; Berman et al., 1983; Foster and Boston, 1983) to model individual tissue or subsystem kinetics. The assumption in this approach is that each of the tissues sampled exchanges vitamin A with plasma but not directly with other tissues. Thus, one is able to model vitamin A turnover in an individual tissue using a mathematical description of the plasma (i.e., plasma tracer data fit to a multiexponential equation in SAAM/CONSAM) along with the tracer response profile in the tissue of interest. In essence, a forcing function decouples a tissue (or subsystem) from the rest of the system and simulates input to the tissue, thus allowing different portions of the whole-body model to be developed individually. In the final stages of development of the whole-body model, once all of the individual components of the system have been modeled using the forcing function, the forcing function is removed, allowing the different parts of the system to interact with one another. The input to these components previously provided by the forcing function is then replaced by differential equations that, following some parameter adjustments, will ultimately be used to describe the kinetics of native retinoids in the system.

Among the advantages of using a forcing function approach to develop a large and complex model of the type required to describe whole-body vitamin A metabolism is that it is possible to focus on one subsystem at a time without having to consider potentially confounding influences from other subsystems. Moreover, it has been our experience in developing these types of models that, in the final stages of the modeling process, when the forcing function is removed, the subsystem parameters are usually in close agreement with those obtained in the whole-body model. What this all means in practical, working terms is that, for a particular tissue of interest, one is able to obtain a fairly reliable estimate of the model structure for that tissue as well as the kinetic parameters associated with the tissue without having to first build a model of the system in its entirety.

In Vitro Studies

Based primarily on our *in vivo* findings, we have initiated a number of *in vitro* experiments to allow us to investigate more easily and in more detail observations from our *in vivo* studies. We were interested in using *in vitro* systems that would parallel and complement the metabolism and kinetics of the native retinoids we had studied *in vivo*. Thus far, we have focused our attention on tissue culture systems for the prostate (normal human prostate epithelial cells [PrEC] and human adenocarcinoma prostate cells [LNCaP]) and the eyes (human retinal pigment epithelial cells [ARPE-19]). Clearly, inherent limitations with tissue culture systems in general preclude precise simulation of *in vivo* conditions. However, given the nature of our studies, which at their basis are designed to better understand normal *in vivo* retinoid metabolism, it was important that, to some degree, we be able to compare selected aspects of our *in vivo* and *in vitro* experiments. We reasoned that any tissue culture system should meet certain minimal requirements, including the presence of retinoid stores in the cells and the ability to process retinoids in a manner similar to that which we had observed *in vivo*.

EFFECTS OF ADMINISTRATION OF THE SYNTHETIC RETINOID 4-HPR ON *IN VIVO* METABOLISM OF NATIVE VITAMIN A

Mean plasma retinol levels in control and 4-HPR treated rats are presented in Figure 1. During the 41-d study, plasma retinol levels in rats treated with 4-HPR (782 mg/kg diet) were significantly lower than in controls. Figure 2 presents the total native retinoid levels in liver and kidneys during the course of the experimental period. In 4-HPR treated animals, liver levels of vitamin A, commonly used as a indicator of overall vitamin A nutriture, were significantly elevated over those of controls at the end of the study.

To obtain a general picture of how the administration of 4-HPR affected metabolism of native retinoids, we first estimated a number of kinetic parameters for the control and 4-HPR treated rats from the respective plasma tracer response curves (Figure 3). The

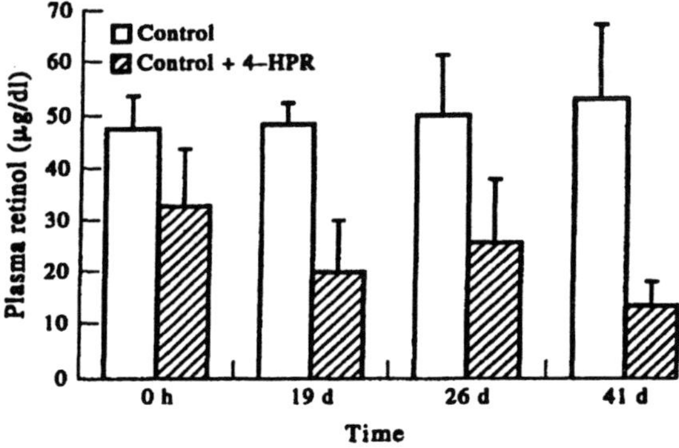

Figure 1. Plasma retinol levels in control (n=4) and 4-HPR treated rats (n=5) during a 41-d turnover study. Data are presented as means ± SD. Reprinted with permission from Lewis et al. (1996).

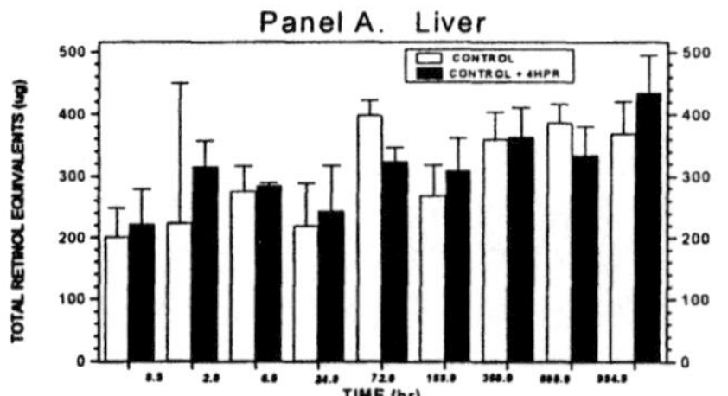
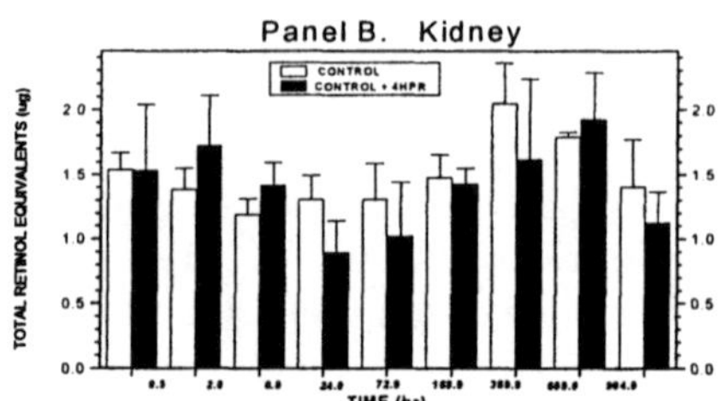

Figure 2. Total retinoid mass in liver (A) and kidneys (B) of control and 4-HPR treated rats. Data are means ± SD and are expressed as total μg retinol equivalents.

SAAM/CONSAM computer modeling programs were used to carry out graphical analysis of the plasma tracer data and to derive the kinetic parameters for each group. One should be aware of potential problems that may arise when dealing with models based on pooled or group average data as compared to data from individual animals or subjects. It has been suggested that, in general and if feasible, one should design experiments in such a way to allow for the development of models for individual animals or subjects (Coccheto et al., 1980; Jacquez, 1996). Cocchetto et al. (1980) reported that averaging individual subject data prior to doing compartmental analysis may result in an underestimation of rate constants. Jacquez (1996) also suggests modeling individual animal data, particularly for studies involving human subjects. However, he also recognizes that this is not always possible and points out that, in contrast to humans, estimates of kinetic parameters in inbred animal strains may have fairly tight distributions that allow for the development of models using pooled data. Thus, for our purposes, we reasoned that the highly inbred rats we have used were adequate to allow development of compartmental models from pooled or group average data.

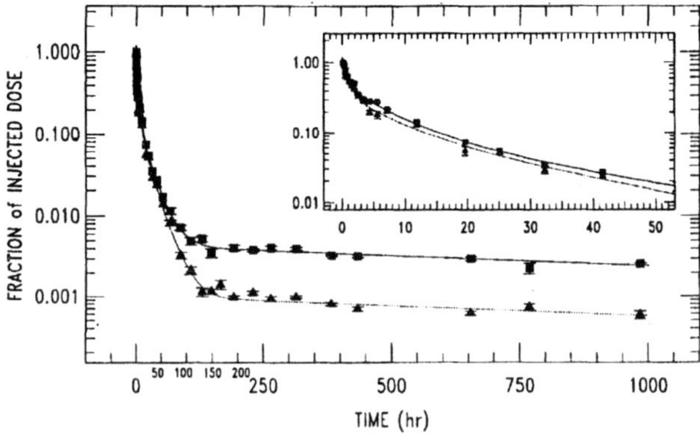

Figure 3. Plasma tracer response curve for retinol kinetics in control (■) and 4-HPR treated rats (▲). Data are group geometric mean (± SEM) fraction of injected dose in plasma *vs.* time. The inset shows data from the first 41.5 h on an expanded time scale. Reprinted with permission from Lewis et al. (1994).

Kinetic parameters for control and 4-HPR treated rats are shown in Table 1. Nearly all of the kinetic parameters calculated were affected by the administration of 4-HPR including the number of recyclings, the residence time, and the disposal rate. In contrast, the transit time was similar in both groups. The decreased utilization by 4-HPR treated rats was reflected in a decreased output of tracer in urine and feces over the course of the study (data not shown).

The overall picture that emerges from these initial kinetic findings is that, in both groups, vitamin A metabolism is, in general, a dynamic system. The native forms of the vitamin, reflected by the retinol tracer, move through the system - the circulation, the tissues - numerous times before they are utilized. Clearly, the rate at which native retinoid is turning over or moving through the system in both groups is much higher than that which is actually being utilized and ultimately lost from the system. However, with 4-HPR administration, the cycling of vitamin A through the plasma is reduced, native vitamin A spends less time there before it leaves irreversibly from this compartment, and on a daily basis, the apparent whole-body utilization of native vitamin A is reduced. Basic homeostatic regulatory mechanisms involved in the metabolism of native vitamin A are normally under fairly tight control (Keilson et al., 1979; Underwood et al., 1979), but these mechanisms are apparently altered significantly by the administration of 4-HPR.

Our next step was to focus on specific tissues to examine how the administration of 4-HPR might affect the metabolism and kinetics of native retinoids in these "subsystems." We found that, besides the liver, there were different effects of 4-HPR administration on vitamin A stores in a number of the tissues analyzed. For example, the prostates of the 4-HPR treated animals had depressed levels of vitamin A throughout most of the study (Figure 4). By 41 days, the native vitamin A stores in the prostates of the controls were more than four times

Table 1. Kinetic parameters derived from plasma tracer response curves for retinol metabolism in control and 4-HPR supplemented rats[1,2]

Group		FCR_p[3] (d^{-1})	t_p[4] (h)	T_p[5] (h)	RN_p[6]	DR[7] ($\mu g/d$)	SRT[8] (d)	RT[9] (d)
Control	Mean	2.00[a]	1.19[a]	12.86[a]	9.81[a]	16.31[a]	38.63[a]	3.99[a]
	±SD	0.68	0.033	3.42	2.83	2.47	9.62	0.53
4-HPR	Mean	3.61[b]	1.03[a]	6.70[b]	5.54[b]	11.01[b]	19.20[b]	3.57[a]
	±SD	0.49	0.06	0.79	0.82	3.1	7.13	1.57

[1]Reprinted with permission from Lewis et al. (1994). Calculation of kinetic parameters based on Rescigno and Segre (1966), Shipley and Clark (1972), and Rescigno and Gurpide (1973); see definitions in text and below.
[2]Means in the same column not having a common superscript letter are significantly different (P <0.05).
[3]Fractional catabolic rate: rate of irreversible utilization of vitamin A expressed as a fraction of the plasma pool.
[4]Mean transit time: amount of time, on average, that a vitamin A molecule spends in the plasma during a single transit.
[5]Mean residence time: total amount of time, on average, that a vitamin A molecule spends in the plasma prior to irreversibly leaving the plasma.
[6]Recycling number: number of times, on average, that a vitamin A molecule cycles back to the plasma before irreversibly leaving the plasma.
[7]Disposal rate: rate of irreversible utilization of vitamin A.
[8]Mean system residence time: total amount of time a molecule of vitamin A spends in the body before being irreversibly lost from the system.
[9]Mean recycling time: the amount of time, on average, a molecule of vitamin A spends outside the plasma before cycling back to the plasma.

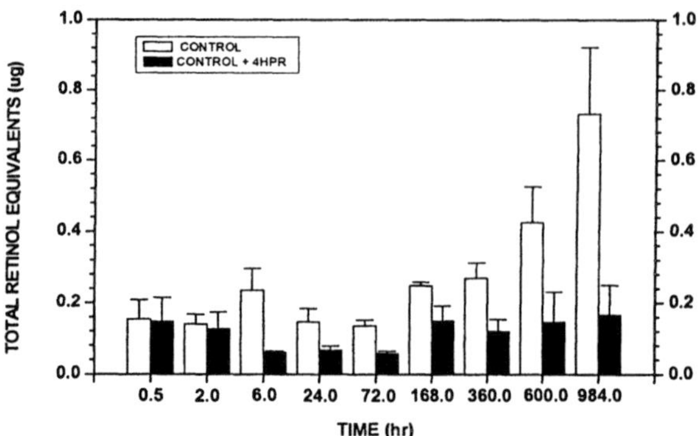

Figure 4. Mass of vitamin A in prostate glands of control and 4-HPR treated rats during the 41-d turnover study. Data are presented as means ± SD, n=3-5 rats/group/time.

higher than those of the 4-HPR treated group (0.732 ± 0.190 and 0.166 ± 0.0827 RE, respectively). Prostates in the 4-HPR treated rats had significantly lower vitamin A stores than control rats throughout the sampling period (α level of 0.05 or less). During the same time period, there was a gradual accumulation of native retinoids in prostates of control rats.

By applying the "forcing function" approach outlined earlier to prostate data, we examined consequences of the alterations observed in our analysis of the plasma tracer response curves. Models proposed to describe vitamin A metabolism in the prostate glands of control and 4-HPR treated rats, as well as corresponding tracer response curves, are shown in Figure 5. For both groups, a minimum of two compartments was required to fit the data obtained from the prostates. The fraction of vitamin A leaving the prostate (i.e., the model-derived fractional transfer coefficient) was similar in both groups [0.149 ± 0.103 and 0.155 ± 0.191 h^{-1} (mean ± fractional standard deviation or FSD) for control *vs.* 4-HPR, respectively]. The turnover of vitamin A from the prostates of 4-HPR treated rats was nearly three times less than that in controls (0.0885 and 0.243 $\mu g/d$, respectively). The model-predicted vitamin A mass in control prostates was 0.688 μg whereas that of the 4-HPR group was 0.123 μg. Since most of the increase in control prostate vitamin A stores occurred during the latter portion of the experimental period, and the corresponding 4-HPR group values remained consistently depressed throughout the study period (Figure 4), the chemically determined 41 d values for the two groups were closest to the model-predicted values, averaging 0.732 ± 0.189 and 0.166 ± 0.0826 μg, respectively.

Residence times, the amount of time on average that a vitamin A molecule spent in each of the prostate compartments before irreversible loss from the system, were obtained from the inverse matrix calculations derived by CONSAM (Berman et al., 1983). Residence times in the faster turning-over compartments were similar in both groups, averaging 0.28 d (6.7 h) for compartments 3 and 23. The estimated residence times for compartments 4 and 24 were 30.9 and 13.3 d for controls *vs.* 4-HPR treated rats, respectively, suggesting the possibility that these secondary compartments serve as longer-term storage pools in both groups. However, vitamin A molecules entering the secondary compartment of the 4-HPR treated prostates (compartment 24) remained there less than half the time as in control compartment 4. This is likely related to the decreased mass of vitamin A in the 4-HPR treated prostates.

The second tissue subsystem to be discussed here is the eyes. As was the case with the model we developed for the prostate, we found that, during the course of a 41-d experimental period during which 4-HPR was administered, there was a gradual depletion of vitamin A stores in the eyes of 4-HPR treated animals (Figure 6). Total vitamin A masses in the control group were significantly higher (α level of 0.05 or less) than in 4-HPR treated rats throughout most of the sampling period. By the terminal collection point, total ocular vitamin A levels in the control group (0.50 ± 0.053 RE; mean $\pm$ SD) were more than five times those of the 4-HPR treated group (0.098 ± 0.075 RE). The models we developed to describe vitamin A kinetics in the eyes of the control vs. 4-HPR treated animals are presented in Figure 7. Based on these models, a number of kinetic parameters were calculated. For example, the flow of vitamin A through the eyes of the 4-HPR treated animals was significantly reduced (0.0162 ± 0.101 μg/d) compared to the controls (0.0604 ± 0.0672 μg/d), and compensatory mechanisms that would normally act to conserve the depletion of ocular vitamin A stores were not apparent in the 4-HPR treated animals. Thus, our initial *in vivo* findings provide a possible explanation for the visual function problems that have been reported in human trials with 4-HPR. Moreover, our findings illustrate the practical applicability and translational potential of our approach. As will be seen in the sections which follow, we have used these initial observations to design experiments which have helped us to further clarify the mechanisms that are involved in the alterations we describe here.

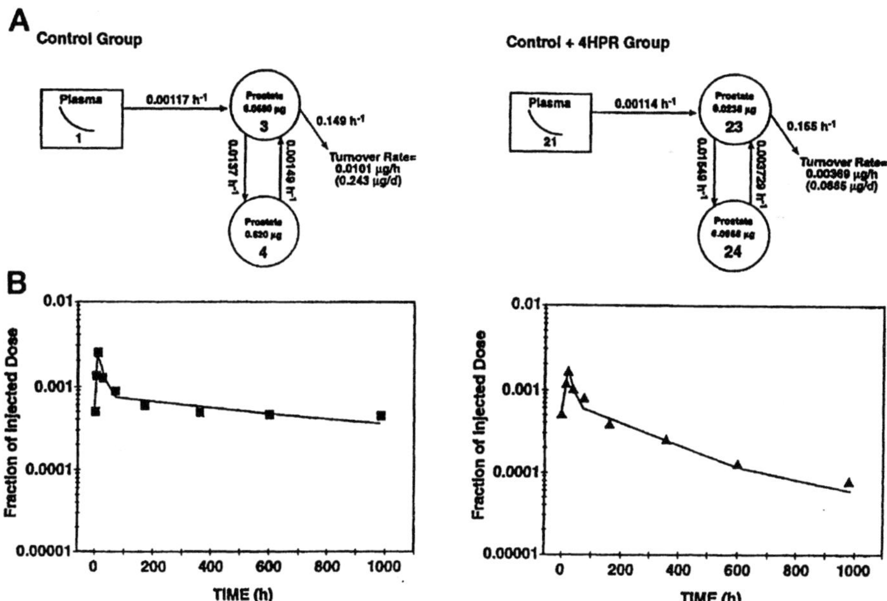

Figure 5. (A) Proposed models of vitamin A kinetics in the prostate gland of control vs. 4-HPR treated rats showing the fractional rate constants (d⁻¹) and flow of vitamin A (μg/d). (B) Tracer response curves from the prostate gland of control (■) vs. 4-HPR treated rats (▲) during the 41 d turnover study. Data are group average values. Reprinted with permission from Lewis and Hochadel (1999).

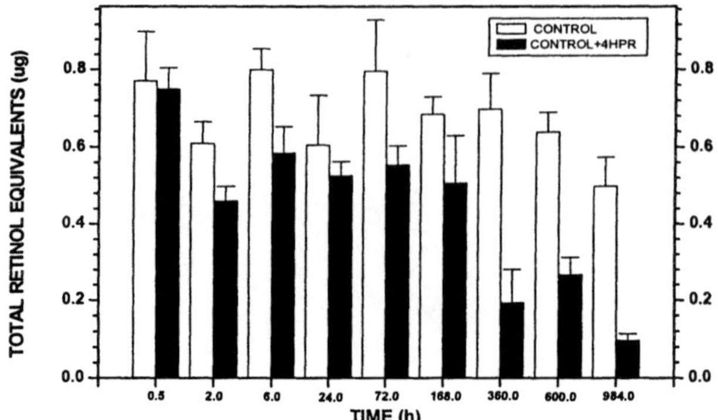

Figure 6. Mass of vitamin A in eyes of control and 4-HPR treated rats during the 41 d turnover study. Data are means ± SD, n = 3-5 rats/group/time.

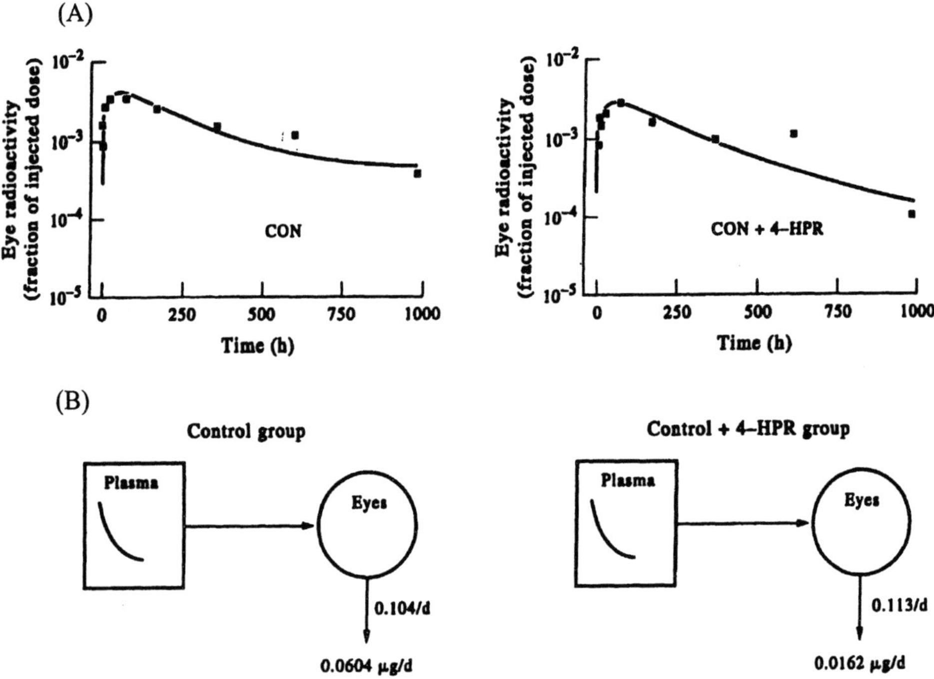

Figure 7. (A) Tracer response curves in eyes of control *vs.* 4-HPR treated rats during the 41-d turnover study. (B) Correspondi[ng] proposed model of vitamin A kinetics in the eyes showing fractional rate constants (d^{-1}) and flow of vitamin A mass out of the ey[e] (μg/d). Reprinted with permission from Lewis et al. (1996).

Besides the observations that have emerged from our work with the eyes, the possible functional consequences of 4-HPR administration on other tissues have not been delineated. We are presently carrying out experiments to identify such consequences. Our findings suggest that there are differential effects of 4-HPR administration on the dynamics of native retinoid metabolism depending on the tissue in question. Our preliminary findings indicate that this also seems to be the case with several other retinoid and/or drug treatments we have examined (unpublished observations).

DEVELOPMENT OF TISSUE CULTURE SYSTEMS TO COMPLEMENT *IN VIVO* RETINOID METABOLISM STUDIES

While recognizing the inherent limitations of any *in vitro* system to fully replicate the complex dynamics of organ and tissue metabolism, we found it necessary to develop simpler, less expensive, and more manageable tissue culture systems to further elucidate details of interest in regard to the mechanisms that give rise to the alterations in native retinoid kinetics observed in our *in vivo* studies. We have developed a number of tissue culture systems designed to mimic *in vivo* metabolism of native retinoids in selected tissues. Here we will focus on our work investigating the actions of the synthetic retinoid 4-HPR in tissue culture systems set up to study retinoid metabolism in the prostate and the eyes.

Choosing an appropriate *in vitro* system to complement and extend *in vivo* findings clearly represents a number of challenges. For our purposes, a tissue culture system had to at least meet certain minimal requirements. An important criteria in choosing a tissue culture system was that there be at least some retinoid stores in the cells so that any retinoid-related metabolic changes that might occur would more closely resemble the metabolism of cells *in vivo*.

To investigate in more detail the role of 4-HPR in the alteration of *in vivo* vitamin A kinetics in the prostate, we used both a normal human prostate cell line (PrEC) and a human adenocarcinoma cell line (LNCaP). We included a vitamin A pretreatment period for both of these cell lines since, in our initial pilot studies, we found that there was little if any detectable

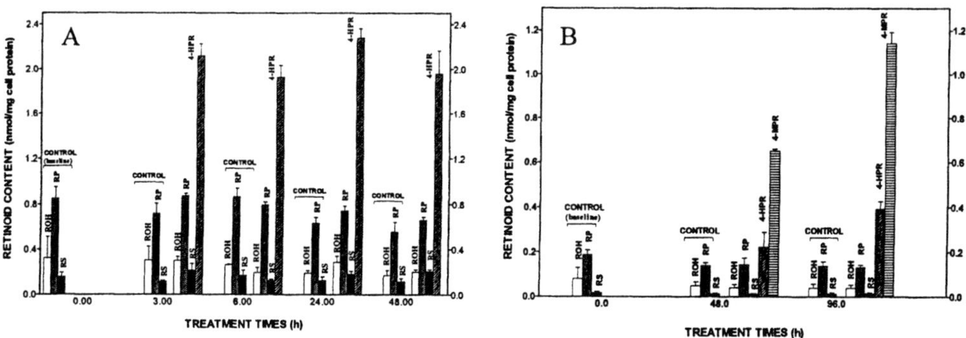

Figure 8. Retinoids identified in PrEC (A) and LNCaP (B) control and 4-HPR treated cells during a 4-d study. Retinoid masses are normalized to measured cell protein. ROH, retinol; RP, retinyl palmitate; RS, retinyl stearate; 4-HPR, *N*-(4-hydroxyphenyl)retinamide; 4-MPR, *N*-(4-methoxyphenyl)retinamide. Reprinted with permission from Lewis and Hochadel (1999).

native retinoid in these cells without preconditioning. We tried different levels of retinol pretreatment and observed retinol uptake and induction of esterification even at the lower levels tested. Changes in the levels of both native retinoids and the synthetic 4-HPR were monitored in the *in vitro* studies. At the time we carried out the *in vivo* studies described earlier, we were not set up to measure tissue levels of 4-HPR and its related metabolites and thus, only the dynamics of native retinoids were monitored.

Measurement of intracellular retinoid content in both PrEC and LNCaP revealed that these cells took up retinol and stored it as such or as retinyl esters, mainly retinyl palmitate and stearate. In addition, we found a number of compounds that we did not specifically characterize, but were able to designate as retinyl esters as determined by photodiode array detection analysis (PDA), which identified these compounds as retinoid in nature based on their maximal absorbance at 325 nm. Intracellular masses of native and synthetic retinoids (i.e., 4-HPR and related metabolites) in PrEC and LNCaP cells are shown in Figure 8. We detected 4-HPR and a number of its metabolites by monitoring these compounds at their wavelength maximum of 362 nm. As shown in Figure 8, neither PrEC or LNCaP cells (preconditioned with retinol prior to treatment with 4-HPR) were significantly different in their stores of retinol, retinyl palmitate, retinyl stearate, or a number of "minor" retinyl esters which were not specifically identified. However, there was substantial uptake and accumulation of 4-HPR by both cell types. In the case of PrEC cells, there was no apparent further conversion of 4-HPR to its normal methoxy metabolite 4-MPR. However, by 24 h, intracellular accumulation of 4-HPR was more than 3 and 7 times higher than the corresponding masses for retinyl palmitate and retinol, respectively. In LNCaP cells, there was substantial conversion to and subsequent accumulation of MPR as well as related metabolites. By 48 h, the total of 4-HPR and its metabolites had accumulated to more than four times the total native retinoid stores (i.e., retinol + retinyl esters). By 96 h, the total 4-HPR stores had risen even further, to the point where they were nearly eight times higher than the total native retinoid stores.

If 4-HPR is taken up by the intact prostate in a similar manner *in vivo* as in both PrEC and LNCaP cells, accumulation of 4-HPR and/or its metabolites might affect metabolism of native retinoids in this tissue, thus helping to account for the observed alterations in vitamin A kinetics in the prostate. We are investigating this possibility in more detail and monitoring possible functional consequences of an accumulation of 4-HPR and/or its metabolites in the prostate.

CURRENT STUDIES

As mentioned earlier, problems with visual function are among the most frequently observed side effects in human trials using 4-HPR as a chemopreventive and/or chemotherapeutic agent. Our *in vivo* findings indicate that administration of 4-HPR in rats is associated with a decrease in native retinoid stores in the eye as well as an alteration in normal retinoid dynamics in this tissue. Similar to the approach we used for the prostate, we conducted *in vitro* experiments using the ARPE-19 human retinal pigment epithelial tissue culture system (kindly provided by Todd Duncan, National Eye Institute). As shown in Figure 9, there was no apparent effect of 4-HPR on the level of native retinoids in the cells; however, we observed a time- and dose dependent uptake of 4-HPR into ARPE-19 cells. There was substantial uptake of 4-HPR between 0 and 240 h, and by the terminal collection point, levels of 4-HPR in cells treated with either 1 or 2.5 μM of 4-HPR were approximately 6 and 17 times higher, respectively, than those of the native retinoid stores in these cells. While we did not measure 4-HPR and/or its metabolites in our *in vivo* studies, we hypothesized that a similarly high accumulation of the synthetic metabolites *in vivo* might contribute to the visual disturbances observed in humans by interfering with the normal functioning of native retinoids that is critical to visual processing. The *in vitro* system did not precisely mimic what we observed *in vivo*. For example, 4-HPR treatment was not associated with a decrease in native retinoid stores in ARPE-19 cells. Although while not in total agreement with our *in vivo* results, we were able to use these data to assist us in the design of a series of new *in vivo* studies to test the hypotheses that emerged from our *in vitro* work (see later). Thus, limitations aside, the two types of studies (*in vitro* and *in vivo*) complemented each other well enough to yield critical information that could be used to formulate rational hypotheses that could be tested in future experiments. The ability to rationally integrate the findings of both our *in vivo* and *in vitro* studies has emerged as a most important component of our methodology.

Coming full circle regarding our use of complementary *in vivo* and *in vitro* systems, we have recently completed a series of *in vivo* studies of retinol turnover in rats fed either a control diet or the control diet containing 4-HPR (782 mg/kg diet). These experiments were designed to test the hypothesis that a decrease in native retinoid stores along with an alteration of native retinoid flux through the eyes combines with an increased accumulation of 4-HPR and related metabolites to account for the disturbances in visual function that are observed in trials using this synthetic retinoid. In agreement with our earlier *in vivo* findings, there was a decrease in total native retinoids in the eyes of the 4-HPR treated group by the end of the experiment (Figure 10A). In agreement with our *in vitro* findings, there was a nearly parallel uptake and subsequent accumulation of 4-HPR and/or its metabolites by the eyes during this same period (Figure 10B). By the end of the 42 d experimental period, more than 90% of the entire pool of native stores of ocular vitamin A had been replaced by synthetic retinoids. Such an accumulation of 4-HPR and its metabolites in the eye, coupled with a parallel decrease in native retinoid stores, could together account for the visual disturbances that have been observed with administration of this retinoid in human trials. We are presently in the process of characterizing the effects of these and other 4-HPR related perturbations in the eyes at the molecular level, focusing in particular on retinoid-related binding proteins, receptors, and enzymes.

SUMMARY

We have used a mathematical/compartmental modeling approach along with a number of rationally designed complementary *in vivo* and *in vitro* systems to investigate the effects of administration of various retinoids and/or drug combinations on normal physiological metabolism of native retinoids. The present paper focuses on our studies of the synthetic retinoid 4-HPR and our use of fairly simple mathematical/compartmental modeling techniques to investigate how this retinoid affects the metabolism of native retinoid overall, as well as in two specific tissues, the prostate and the eyes. We have presented our work with this particular retinoid and these tissues as an example of the type of studies we have been doing and to present some of the information that one can obtain using this approach. In addition, an

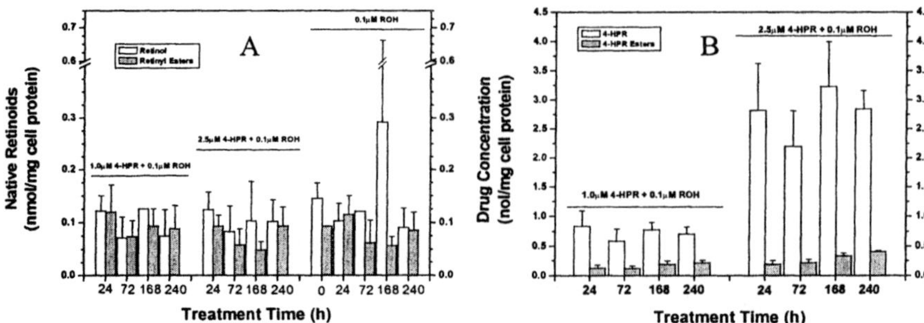

Figure 9. Effects of treatment with 4-HPR on (A) native retinoid content and (B) uptake of 4-HPR by ARPE-19 human retinal pigment epithelium cells. Data are expressed as nmol/mg cell protein. ROH, retinol; 4-HPR, *N*-(4-hydroxyphenyl)retinamide.

important objective of this paper is to highlight the fact that a great deal of critical information can be derived from fairly simple mathematical/compartmental models. When used appropriately, such models provide a powerful tool to direct the design, conduct, and interpretation of experiments. The models we developed for the prostate and the eyes were used as hypotheses to direct our research efforts in both *in vivo* and *in vitro* systems. In the case of the eyes, we were able to elucidate the possible mechanisms involved in one of the most commonly reported complications (i.e., visual function abnormalities) associated with administration of an important chemopreventive and/or chemotherapeutic agent. We are in the process of further expanding our studies with the prostate as well as several other tissues in a similar manner. The immediate clinical relevance and application of our work with the eyes demonstrate the high translational potential of our approach. Without the use of the type of mathematical/compartmental modeling approach we used, which provided the basis for much of this work, we are not aware of any other way that we could have obtained the critical information that we did. We hope that the work presented here demonstrates the usefulness,

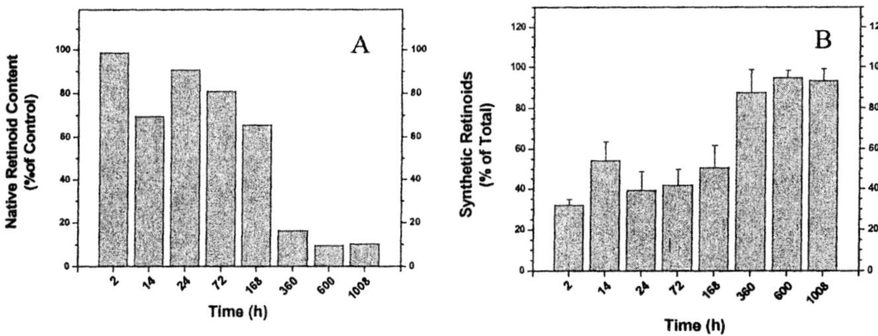

Figure 10. Native retinoid content in eyes of 4-HPR treated rats expressed as (A) percent of control and (B) percent of total retinoids that are synthetic in the eyes of 4-HPR fed animals.

power, and potential clinical applicability of a modeling approach to investigate different retinoid-based treatments as well as a variety of other chemopreventive and/or chemotherapeutic agents.

ACKNOWLEDGMENT

This work presented in this paper is dedicated to the memory of Dr. Loren A. Zech Sr.

CORRESPONDENCE

Please address all correspondence to:
   Kevin C. Lewis
   Drug-Nutrient Interaction Group
   Basic Research Laboratory
   National Cancer Institute
   Building 560, Room 12-27
   Frederick, MD 21702
   lewiskc@mail.ncifcrf.gov

REFERENCES

Berman, M., Beltz, W.F., Greif, P.C., Chabay, R., and Boston, R.C., 1983, *CONSAM User's Guide,* PHS Publication No. 1983-421-132:3279, U.S. Government Printing Office, Washington, DC.

Berman, M., and Weiss, M.F.. 1978, *SAAM Manual; SAAM27 Version,* DHEW Publication No. (NIH) 78-180, U.S. Government Printing Office, Washington, DC.

Cobleigh, M.A., Dowlatshahi, K., Deutsch, T.A., Mehta, R.G., Moon, R.C., Minn, F., Benson, A.B., Rademaker, A.W., Ashenhurst, J.B., Wade, J.L., and Walter, J., 1993, Phase I/II trial of Tamoxifen with or without Fenretinamide, an analog of vitamin A, in women with metastatic breast cancer, *J. Clin. Oncol.* 11:474-477.

Coccheto, D.M., Wargin, W.A., and Crow, J.W., 1980, Pitfalls and valid approaches to pharmacokinetic analysis of mean concentration data following intravenous administration, *J. Pharmacokinet. Biopharm.* 8:539-552.

Costa, A., Malone, W., Perloff, M., Buranelli, F., Campa, T., Dossena, G., Magni, A., Pizzichetta, M., Andreoli, C., Del Vecchio, M., Formelli, F., and Barberi, A., 1989, Tolerability of the synthetic retinoid Fenretinide (HPR), *Europ. J. Cancer Clin. Oncol.* 25: 805-808.

Decensi, A., Torrissi, R., Polizzi, A., Gesi, R., Brezzo, V., Rolando, M. Rondanina, G., Antonietta, M.A., Formelli, F., and Costa, A., 1993, Effects of the synthetic retinoid fenretinide on dark adaptation and the ocular surface, *J. Natl. Cancer Inst.* 86:105-110.

Foster, D.M., and Boston, R.C., 1983, The use of computers in compartmental analysis: The SAAM and CONSAM programs, in: *Compartmental Distribution of Radiotracers,* J.S. Robertson, ed., CRC Press Inc., Boca Raton.

Formelli, F., Carsana, R., Costa, A., Buranelli, F., Campa, T., Dossena, G., Magni, A., and Pizzichetta, M., 1989, Plasma retinol level reduction by the synthetic retinoid fenretinide: a one year follow-up study for breast cancer patients, *Cancer Res.* 49:6149-6152.

Green, M.H., Uhl, L., and Green, J.B., 1985, A multicompartmental model of vitamin A kinetics in rats with marginal vitamin A stores, *J. Lipid Res.* 26:806-818.

Green, M.H., Green, J.B., and Lewis, K.C., 1987, Variation in retinol utilization rate with vitamin A status in the rat, *J. Nutr.* 117:694-703.

Gross, E. G., and Helfgott, M. A., 1992, Retinoids and the eye, *Dermatol.Clin.* 10:521-531.

Hong, W.K. and Itri, L.M., 1994, Retinoids and human cancer, in: *The Retinoids: Biology, Chemistry, and Medicine,*

2nd ed., M.B. Sporn, A.B. Roberts, and D.S. Goodman, eds., Raven Press, New York.

Jacquez, J.A., 1996, Population distributions of parameters, in: *Compartmental Analysis in Biology and Medicine, 3rd ed.,* BioMedware, Ann Arbor.

Kaiser-Kupfer, M.I., Peck, G.L., Caruso, R.C., Jaffe, M.J., DiGiovanna, J., and Gross, E.G., 1986, Abnormal retinal function associated with Fenretinide, a synthetic retinoid, *Arch. Ophthalmol.* 104:69-70.

Keilson, B., Underwood, B.A., and Loerch, J.D., 1979, Effects of retinoic acid on the mobilization of vitamin A from the liver in rats, *J. Nutr.* 109:787-795.

Kingston, T.P., Lowe, N.J., Winston, J., and Heckenlively, J., 1986, Visual and cutaneous toxicity which occurs during N-(4-hydroxyphenyl)retinamide therapy for psoriasis, *Clin. Exp. Dermatol.* 11:624-627.

Lewis, K.C., Green, M.H., and Underwood, B.A., 1981, Vitamin A turnover in rats as influenced by vitamin A status, *J. Nutr.* 111:1135-1144.

Lewis, K.C., Green, M.H., Green, J.B., and Zech, L.A., 1990, Retinol metabolism in rats with low vitamin A status: a compartmental model, *J. Lipid Res.* 31:1535-1548.

Lewis, K.C., Zech, L.A., and Phang, J.M., 1994, Effects of *N*-(4-hydroxyphenyl)retinamide supplementation on vitamin A metabolism, *Cancer Res.* 54:4112-4117.

Lewis, K.C., Zech, L.A., and Phang, J.M., 1996, Effects of chronic administration of *N*-(4-hydroxyphenyl) retinamide (4-HPR) in rats on vitamin A metabolism in the eye, *Europ. J. Canc.* 32:1803-1808.

Lewis, K.C., and Hochadel, J.F, 1999, Retinoid metabolism in the prostate: effects of administration of the synthetic retinoid N-(4-hydroxyphenyl)retinamide, *Cancer Res.* 59:5947-5955.

Lotan, R., 1980, Effects of vitamin A and its analogs (retinoids) on normal and neoplastic cells, *Biochim. Biophys. Acta* 605:33-91.

Modiano, M.R., Dalton, W.S., Lippman, S.M., Joffe, L., Booth, A.R., and Meyskens, F.L., 1990, Phase II study of Fenretinide in advanced breast cancer and melanoma, *Invest. New Drugs* 8:317-319.

Moon, R.C., Mehta, R.C., and Rao, K.V.M., 1994, Retinoids and cancer in experimental animals, in: *The Retinoids: Biology, Chemistry, and Medicine, 2nd ed.,* M.B. Sporn, A.B. Roberts, and D.S. Goodman, eds., Raven Press, New York.

Pienta, K.J., Nguyen, N.N., and Lehr, J.E., 1993, Treatment of prostate cancer in the rat with the synthetic retinoid fenretinide, *Cancer Res.* 53:224-226.

Pollard, M., Luckert, P.H., and Sporn, M.B., 1991, Prevention of primary prostate cancer in Lobund-Wistar rats by N-(4-hydroxyphenyl) retinamide, *Cancer Res.* 51:3610-3611.

Roberts, A.B., and Sporn, M.B., 1984, Cellular biology and biochemistry of the retinoids, *in: The Retinoids: Biology, Chemistry, and Medicine, 2nd ed.,* M.B. Sporn, A.B. Roberts, and D.S. Goodman, eds., Raven Press, New York.

Rotmensz, N., DePalo, G., Formelli, F., Costa, A., Marubini, E., Campa, T., Costa, A., Marubini, E., Campa, T., Crippa, A., Danesini, G.M., Delle Grottaglie, M., DiMauro, M.G., Filiberti, A., Gallazzi, M., Guzzon, A., Magni, A., Malone, W., Mariani, L., Palvarini, M., Perloff, M., Pizzichetta, M., and Veronesi, U., 1991, Long-term tolerability of Fenretinide (4-HPR) in breast cancer patients, *Europ. J. Cancer* 27:1127-1131.

Rescigno, A., and Gurpide, E., 1973, Estimation of average times of residence, recycle, and interconversion of blood-borne compounds using tracer methods, 1973, *J. Clin. Endocrinol. Metab.* 36:263-276.

Rescigno, A., and Segre, G., 1966, *Drug and Tracer Kinetics,* Blaisdell Publishing Co., Waltham.

Shipley, R.A., and Clark, R.E., 1972, *Tracer Methods for In Vivo Kinetics,* Academic Press, New York.

Underwood, B.A., Loerch, J.D., and Lewis, K.C., 1979, Effects of dietary vitamin A deficiency, retinoic acid, and protein quantity and quality on serially obtained plasma and liver levels of vitamin A in rats, *J. Nutr.* 109:796-806.

Welsch, C.W., DeHoog, J.V., and Moon, R.C., 1983, Inhibition of mammary tumorigenesis in nulliparous C3H mice by chronic feeding of the synthetic retinoid, N-(4-hydroxyphenyl) retinamide, *Carcinogenesis* 4:1185-1187.

GLUTAMATE METABOLISM IN PRIMARY CULTURES OF RAT BRAIN ASTROCYTES: RATIONALE AND INITIAL EFFORTS TOWARD DEVELOPING A COMPARTMENTAL MODEL

Mary C. McKenna[*]

INTRODUCTION

During the past 20 years, there has been a tremendous increase in neuroscience research and in the amount of information known about the brain. However, there is a great need to integrate the many bits of knowledge about subcellular processes and molecular mechanisms into the broader context of cellular function and into an overall picture of the dynamic neuronal/glial interactions essential for brain function. The area of brain energy metabolism and neurotransmission has undergone a dramatic resurgence in recent years. It is a field replete with exciting new findings that have overturned long-held 'dogmas' and contributed greatly to our understanding of how closely disturbances in energy metabolism are associated with clinical conditions that involve neurodegeneration. The rapid advances in this field are a result of sophisticated methodologies including ^{13}C, ^{15}N, ^{31}P, and ^{1}H nuclear magnetic resonance spectroscopy (NMR) and GC/MS, *in vivo* microdialysis, implantable electrodes capable of measuring metabolite changes in real time, *in vivo* imaging, and molecular biology techniques. These techniques have enabled researchers to ask more complex questions about mechanisms underlying metabolic alterations in both normal and pathological conditions.

Our studies with primary cultures of brain cells were started with the idea that one could not totally sort out the integrated picture of overall brain metabolism until the intricacies of the regulation of cellular processes in both astrocytes and neurons were better understood. Over many years, our group has obtained considerable information about the transport and oxidation of key metabolic substrates, the subcellular compartmentation of metabolism, regulation of enzymes, and the effects of other

* Mary C. McKenna, Department of Pediatrics, University of Maryland School of Medicine, Baltimore, MD 21201.

compounds on metabolism in astrocytes, neurons, and synaptic terminals (Roeder et al., 1985; Tildon et al., 1985; Zielke et al., 1990; McKenna et al., 1990, 1993; Malik et al., 1993; Tildon et al., 1993). However, the integration of these data was a difficult task.

A number of years ago, I was encouraged by a member of a site-visit team to use mathematical modeling techniques as a means to better understand the complex metabolism in different types of brain cells, as well as the dynamic interactions between astrocytes and neurons in brain. I was enthusiastic about this suggestion in view of the promise of modeling as a powerful and exciting tool for the quantitative analysis of complex systems. However, I had little understanding of the time required for such an undertaking. Modeling of these data was started a number of years ago during my "part-time commuter sabbatical leave" with the late Dr. Loren Zech at the Laboratory of Mathematical Biology, National Cancer Institute, National Institutes of Health.

In this paper, I present and discuss our original conceptual model of glutamate metabolism by astrocytes and demonstrate the application of computer-based compartmental modeling techniques to data published earlier by our group (Zielke et al., 1990) on the metabolism of [U-^{14}C]glutamate by cortical astrocytes from newborn rat brain. It is important to emphasize that the modeling of glutamate metabolism in astrocytes that is presented here is not yet finished. However, the current model is useful for describing glutamate metabolism in astrocytes and for predicting the effect of other metabolites and inhibitors on perturbing glutamate metabolism. Since there are specific considerations and limitations related to modeling studies using cultured cells, part of the discussion deals with those issues.

OVERVIEW OF GLUTAMATE METABOLISM IN THE BRAIN

Glutamate is the major excitatory neurotransmitter released by synaptic terminals in mammalian brain. During neuronal depolarization, glutamate in the synaptic cleft rises from a resting concentration of <10 μM to concentrations between 100 μM and 1 mM. A key function of astrocytes is the rapid uptake and disposal of neurotransmitter glutamate from the synaptic region in order to maintain the low resting concentration of this potentially excitotoxic amino acid that is necessary for continued neuronal function (Hertz, 1979; Hertz and Schousboe, 1988; Shank and Aprison, 1988; Yudkoff et al., 1988; Schousboe et al., 1993). Astrocytes are a specialized type of glial cell with end feet that cover the endothelial cells of the blood-brain barrier and additional processes that envelop glutamatergic synapses. Astrocytes are uniquely suited, both anatomically and metabolically, to maintain the proper extracellular milieu for brain function. Any failure of these cells to remove glutamate from the synaptic cleft can result in a cascade of excitotoxic damage that ultimately results in neurodegeneration.

Much of the glutamate taken up by astrocytes is converted to glutamine and released into the synaptic cleft (Figure 1). Glutamine released by astrocytes is taken up by neurons and used for energy or incorporated into the neurotransmitters glutamate and γ–aminobutyric acid (GABA) (Van den Berg and Garfinkel, 1971; Shank and Aprison, 1988; Schousboe et al., 1997). This continuous, dynamic neuronal/glial pathway of glutamate-glutamine interconversion that is essential for brain function was proposed thirty years ago (Berl and

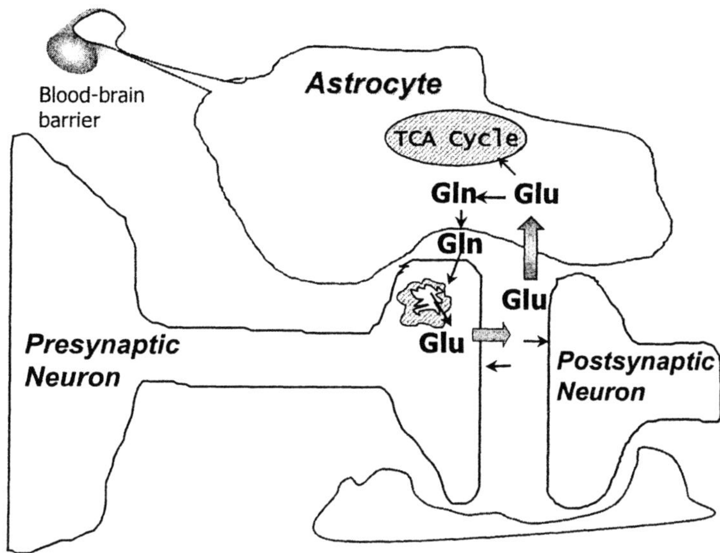

Figure 1. Glutamate metabolism in brain. Glu, glutamate; Gln, glutamine.

Clark 1969; Van den Berg, 1970; Van den Berg and Garfinkel, 1971; Van den Berg et al., 1975). Van den Berg and Garfinkel (1971) used data from their own studies, combined with a variety of findings from other groups, as the basis for the glutamate-glutamine cycle, their model of two compartments of glutamate metabolism in brain. Their model included a 'large' slowly turning-over neuronal pool and a smaller more rapidly turning-over glial (astrocytic) pool of glutamate metabolism. It must be emphasized that this cycle does not operate in a stoichiometric fashion, since it is well established that numerous metabolites are used for glutamate and glutamine formation. It is also known that glutamate from the extracellular milieu is readily used by the astrocytic tricarboxylic acid (TCA) cycle and converted to other products including glutathione and lactate (Yudkoff et al., 1986, 1988; Shank and Aprison, 1988; Schousboe et al., 1988; Yudkoff et al., 1990; Sonnewald et al., 1993a).

Metabolism in astrocytes is very complex. Several groups have demonstrated the existence of subcellular compartmentation, evidence for multiple pools of TCA cycle activity, and evidence for multiple pools of glutamate metabolism in these cells (McKenna et al., 1990; Schousboe et al., 1993; Sonnewald et al., 1993b; McKenna et al., 1996a, b; Waagepetersen, 2000). Understanding the regulation of glutamate metabolism has important clinical relevance, since abnormalities in glutamate metabolism have been implicated in numerous pathological conditions (Perry and Hansen, 1981; Plaitakas et al., 1982; Rothman and Olney, 1986; Harkany et al., 2000). Glutamate metabolism in astrocytes is regulated by a variety of factors including glucocorticoids (Patel et al., 1986; Zielke et al., 1990), cyclic AMP (Browning and Nicklas, 1982; Max et al., 1990; Zielke et al., 1990), ammonia (Yu et al., 1982; Waniewski, 1992), and the concentrations of glutamate (Zielke et al., 1990;

McKenna et al., 1996a), glutamine (Patel et al., 1986; Zielke et al., 1990), and pyruvate (McKenna et al., 1996b) in the extracellular and intracellular milieu.

COMPARTMENTATION IN BRAIN

A number of factors contribute to compartmentation of metabolism in brain. These include the presence of different types of transporters (Assaf et al., 1990; McKenna et al., 1998b; Pellerin et al., 1998; Vannucci et al., 1998), the localization and activity of key enzymes (Norenberg and Martinez-Hernandez, 1979; Yu et al., 1983; Lai and Clark, 1989; McKenna et al., 1995, 2000a), and the transient association and dissociation of key enzymes into complexes or metabolons (Beeckmans and Kanareck, 1981; Fahien et al., 1988, 1989; Beeckmans et al., 1990; McKenna et al., 2000a and references therein) which is mediated, in part, by the intracellular concentration of important metabolites (Malik et al., 1993 and references therein). Although the use of modeling has been extremely important to brain studies because it led to the original formulation of the concept of the glutamate/glutamine cycle between astrocytes and neurons in the brain, the number of studies employing modeling techniques is very small. Indeed, it is significant to note that most of the recent advances in the understanding of cellular and subcellular compartmentation in brain have not come from modeling but rather from groups using other approaches specifically directed at determining more about metabolic compartmentation *per se*.

A number of approaches have been used to obtain information about compartmentation in brain. One type of experiment has been to add unlabeled substrates as competitors in conjunction with studies determining the rates of oxidation or transport and precursor/product relationships of ^{14}C- or ^{13}C-labeled substrates. Studies using stable isotopes of different precursors such as glucose, acetate, glutamate, and glutamine combined with analysis by ^{13}C/^{15}N-NMR or GC/MS have been particularly useful because of the preferential utilization of acetate by glial cells and higher use of glucose in neurons (Bader-Goffer et al., 1990; Sonnewald et al., 1993b; Westergaard et al., 1995a; Waniewski and Martin, 1998). Both subcellular fractionation studies and work with primary cultures of brain cells have provided evidence of mitochondrial heterogeneity and aided in identifying differences in localization, activity, and regulation of key enzymes in the brain (Lai et al., 1977; Lai and Clark, 1989; McKenna et al., 1990, 1995, 2000a, b). Pharmacological tools, including compounds that lead to selective ablation of neurons or glial cells, or selective inhibition of glycolysis *vs.* inhibition of the TCA cycle, have led to further clarification of compartmentation in the brain (Ransom and Fern, 1997).

Many advances in understanding of compartmentation in the brain would not have been possible without the development of methods for preparing and maintaining relatively pure primary cultures of brain cells. Probably the most useful overall approach for identifying compartmentation in the brain has been the use of ^{13}C-NMR spectroscopy and/or GC/MS combined with substrate competition using either primary cultures of astrocytes, neurons, or highly purified preparations of synaptic terminals from discrete areas in the brain (Erecinska et al., 1988; Yudkoff et al., 1988; Schousboe et al., 1993;

Sonnewald et al., 1993a, b; Yudkoff et al., 1994; McKenna et al., 1996b and references therein; Waagepetersen et al., 1998b, 1999).

Since studies that have not used modeling have been so instrumental in identifying cellular and subcellular compartments, one might be tempted to pose this question: is identifying compartments sufficient, or is there much more to be gained by modeling metabolism in brain cells? The primary advantage of compartmental modeling techniques is that, once a model has been developed which fits the data, the model can be used to test assumptions about the data and even to predict the outcome of experiments or treatments (Wastney et al., 1999).

Other Studies Modeling Metabolism in the Brain

Modeling has not been used more widely in studies of brain metabolism for a number of reasons including the fact that data suitable for modeling requires multiple samples over the time of the experiment and sufficient repetitions of the studies. Such experiments are very expensive as well as time consuming to analyze. Most studies that have applied modeling techniques to the brain have used variations on the Van den Berg and Garfinkel (1971) two-compartment model of glutamate metabolism and employed different approaches depending on the goal to be accomplished by modeling. Work by Shulman's group (Petroff et al., 1991; Mason et al., 1995; Shen et al., 1998; Sibson et al., 1998, 2001) has led to the publication of rates of TCA cycle activity, glutamine formation, 'cycling' between glutamate and glutamine, and rates of anaplerosis. The modeling in these studies has led to hypotheses regarding the 'tightness' of the association between various phenomena such as cerebral metabolic rate (measured by glucose uptake) *vs.* the rate of neurotransmitter cycling (Sibson et al., 1998; Magistretti et al., 1999). Other groups have used modeling to identify metabolism that occurs in neuronal *vs.* glial glutamate compartments (Berl and Clark, 1969; Van den Berg and Garfinkel, 1971; Bader-Goffer et al., 1990; Shank et al., 1993). Still other groups have modeled brain metabolism based on the isotopomers formed from ^{13}C- or ^{15}N-labeled substrates such as glutamate, glutamine, and glucose. These studies have provided information about metabolic interrelationships among important substrates, since individual carbon atoms in a molecule can be identified by their NMR spectra and the formation of isotopomers can be studied under various conditions (Cerdan et al., 1990; Badar-Goffer et al., 1992; Kunnecke et al., 1993; Gruetter et al., 1998). None of these other modeling studies have modeled glutamate metabolism in astrocytes like the present study.

The impetus for modeling is often that of "dissecting" the regulation of metabolism with the goal of identifying potential mechanisms related to pathology in clinical conditions. Our goal in modeling glutamate metabolism is both the identification of pathways and, most importantly, the development of a model with sufficient detail that it can be useful in identifying potential mechanisms related to pathology in clinical conditions or in predicting the effect of new drugs on glutamate metabolism in astrocytes.

Specific Considerations Related to Studies Using Cultured Cells

Highly purified primary cultures of brain cells have been invaluable models for revealing many key aspects of brain metabolism. In studies using more intact preparations such as whole brain or even brain slices, important information may be masked since the data obtained generally reflect a composite view of metabolism in many different types of cells. Thus, phenomena such as the selective localization of an enzyme in one type of brain cell [e.g., pyruvate carboxylase and glutamine synthetase are localized almost exclusively in astrocytes (Norenberg and Martinez-Hernandez, 1979; Yu et al., 1983)], evidence that partial TCA cycles are important for providing energy in specific cell types (Yudkoff et al., 1994; McKenna et al., 2000a), and evidence of metabolic trafficking of key substrates between different types of cells (Sonnewald et al., 1993b; Westergaard et al., 1995b) can easily be overlooked. A key advantage of using cultured cells is that they can be studied independently of influences such as blood flow, hormonal input, and interactions with other types of cells. This enables one to study the metabolic capabilities of a specific cell type and determine what factors regulating metabolism are characteristic of that cell type. The information gained from such studies can be particularly useful for designing novel pharmacological approaches to alter or preserve metabolism within a particular type of cell.

However, some cells may develop or act differently *in vitro* than *in vivo* due to the lack of physiological input and/or cell-cell interactions. Therefore, it is important to reconcile and validate the findings from cellular/subcellular work with data from intact animals. A number of *in vivo* studies have validated many findings from cellular studies including the selective localization of key anaplerotic enzymes in astrocytes, clear-cut evidence that metabolic trafficking between astrocytes and neurons occurs *in vivo* (Cerdan et al., 1990; Hassel and Sonnewald, 1995; Bachelard, 1998), and documentation that glutamate and glutamine are oxidized for energy in the brain (Zielke et al., 1999).

PROJECT OBJECTIVES AND METHODOLOGIES

Our objective was to develop a physiologically-based, detailed compartmental model that describes glutamate metabolism in brain astrocytes. The model was developed with the understanding that a well-fitting model can be used to test assumptions about the data and also to predict the outcome of experiments, treatments, and perturbations in metabolism.

Outline of the Modeling Process Used

The modeling efforts essentially followed the steps outlined by Novotny and Caballero (1998), including definition of the objective, collection of system information, development of a conceptual model and diagram of the system, derivation of model equations, determination/estimation of initial conditions and parameter values, incorporation of the model structure and values into an appropriate computer program, fitting the model to the data, and modification of the model structure or specific parameter values to improve fit. The process of "fitting the model to the data" can be extremely involved and necessitates comparing the model predictions to the experimental data and making appropriate adjustments to improve the fit. It is most important that adjusting parameter values or modifying the model structure by adding compartments or altering input and/or output pathways be done in "physiologically meaningful ways" (v. Reinersdorff et al., 1998; Wastney et al., 1999).

Experimental Data

This project used data on glutamate metabolism in astrocytes obtained by Zielke et al. (1990). All cell culture and analytical methods are described in detail in the original paper. To summarize, primary cultures of cortical astrocytes from newborn rat brain were prepared and grown to confluence by standard methods, shaken overnight to remove oligodendroglia, and subcultured after trypsinization. Cells were used for the experiments 4 days after subculturing. Astrocytes were rinsed with sterile phosphate-buffered saline (PBS) and incubated at 37°C with [U-^{14}C]glutamate (200-250 dpm/picomole) in 200 µL minimal essential medium containing 10 µM glutamate, 1 mM glutamine, and 6 mM glucose. After 5-20 min incubation, the medium was removed and aliquots saved for HPLC. Incubations were done in quadriplicate. The astrocytes were rapidly rinsed with PBS, and perchloric acid was added to terminate metabolism and extract intracellular amino acids. An internal standard was added to each sample to determine recovery and facilitate identification of the amino acids separated by HPLC. Prior to analysis, the samples were neutralized and sonicated, and a portion was analyzed for glutamine and acidic amino acids that were separated by ion exchange chromatography on Dowex-1-chloride columns (Zielke et al., 1990). Portions of the medium and astrocyte extracts were analyzed by HPLC (Zielke, 1985) to determine the amount of radioactivity in glutamate and aspartate. The specific activity of each amino acid was determined. Some of the astrocyte cultures were grown under the same conditions in upright 25 cm^2 tissue culture flasks so that the rate of oxidation of 10 µM [U-^{14}C]glutamate could be determined (Zielke et al., 1990).

RESULTS AND MODEL DEVELOPMENT

Development of the Conceptual Model and Diagram

We postulated an initial 7-pool starting model (Figure 2) based on knowledge of glutamate and glutamine metabolism in brain astrocytes (Hertz, 1979; Hertz and Schousboe

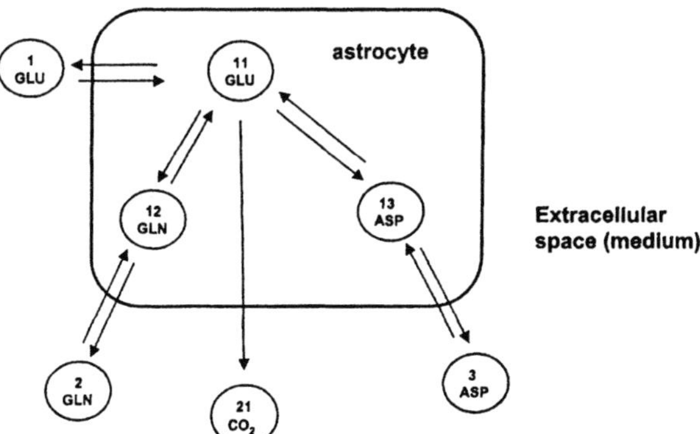

Figure 2. Initial conceptual model of glutamate metabolism in astrocytes. Each circle represents a pool, and the arrows represent flow of metabolites into or out of pools. The solid line represents the astrocyte cell membrane; thus, intracellular pools are shown inside the cell membrane, and pools in the extracellular space are outside the solid line. GLU, glutamate; GLN, glutamine; ASP, aspartate.

1988; Zielke et al., 1990). Due to limitations in the data, it was purposely a simple model that did not include all of the known physiological pathways of glutamate metabolism. This is important since the modeling process used is experimentally based.

Radioactive tracer data were available for glutamate, glutamine, and aspartate in astrocytes and in the medium. Mass data were available for intracellular concentrations of glutamate, glutamine, and aspartate, and for the initial concentrations of added [U-^{14}C]glutamate and unlabeled glutamine in the medium.

In this 7-pool model (Figure 2), glutamate is taken up by astrocytes from the extracellular space. Intracellular glutamate can be converted to glutamine and aspartate which can be converted back to glutamate or released into the extracellular space. Both glutamine and aspartate in the extracellular space can be taken up and metabolized by astrocytes. A portion of the glutamate taken up by astrocytes is irreversibly oxidized to CO_2 (Zielke et al., 1990).

Metabolism of [U-^{14}C]Glutamate

[U-^{14}C]Glutamate was rapidly taken up by cultured astrocytes and converted to glutamine and aspartate in these cells. Radioactivity in glutamate in cells (Figure 3) peaked by 10 min, whereas radioactivity in glutamine (Figure 3) and aspartate (Figure 4) peaked between 10 and 20 min and then gradually declined. Radioactive glutamate in the medium rapidly declined (Figure 5) such that values were 20% of the initial medium radioactivity after 20 min. There was a more gradual increase in radioactive glutamine (Figure 5) and aspartate (Figure 4) in the medium. After two hours, 76% of the added [U-^{14}C]glutamate was recovered in the medium as radioactive glutamine, 2-3% as aspartate, and 7% as glutamate. It was determined in separate oxidation studies that 10% of the glutamate added was oxidized to CO_2. The relatively small proportion of glutamate entering the TCA cycle is consistent with other studies from our lab demonstrating that, when extracellular glutamate concentration is low, glutamate taken up by astrocytes is converted primarily to glutamine. The proportion of glutamate entering the TCA cycle increases considerably in the presence of higher concentrations of extracellular glutamate (McKenna et al., 1996a).

Development of Model Equations

The model equations for the system were derived starting with the sum of the inputs and outputs for each compartment. As an example, the equation for the extracellular glutamate pool would be

$$d(\text{glutamate extracellular})/dt = \text{glutamate returned to the extracellular pool} - \\ \text{glutamate leaving the extracellular pool}$$

The equations for each flow were written with the assumption that the system behaves according to the law of mass action and can therefore be described by first-order linear differential equations. Compartments in the initial model were assigned numbers (Table 1) so that the equations could be written in a format compatible with the Simulation Analysis and Modeling (SAAM) and Conversational SAAM (CONSAM) computer programs (Berman and Weiss, 1978; Berman et al., 1983). Concentration data from the experiments

Table 1. Compartments in the initial conceptual model of glutamate metabolism by astrocytes

Compartment	Tracer model number	Mass model Number	Initial conditions (pmol/dish)
Extracellular			
Glutamate	1	31	1935
Glutamine	2	10	150,000
Aspartate	3	14	0
CO_2	21	27	0
Intracellular			
Glutamate	11	25	3354
Glutamine	12	24	5817
Aspartate	13	7	2462

were used for the initial conditions in the mass model (Table 1). If such data had not been available, the initial conditions could have been estimated from literature values as described by Novotny and Caballero (1998).

In SAAM terminology, L represents the rate coefficient or fractional transfer rate, and F represents the mass of the tracer or tracee (the unlabeled compound being traced). The notation L(1,2) indicates the fractional transfer from compartment 2 to compartment 1 per unit time. Rate equations written for each of the tracer compartments are listed in Table 2.

Modeling Assumptions

Each compartment in the model is assumed to be kinetically homogeneous. This means that all of the molecules in a compartment have the same probability of reversibly or irreversibly leaving the compartment. A fractional standard deviation (FSD) of 0.05 was initially assigned to each data point as a weighting factor for weighted nonlinear regression analysis (Berman et al., 1983). The total radioactive tracer added and the total amounts of glutamate and glutamine in the medium were incorporated into the model as

Table 2. Rate equations for the initial tracer model of glutamate metabolism by astrocytes

$$dF(1)/dt = - L(11,1) * F(1) + L(1,11) * F(11)$$

$$dF(11)/dt = L(11,1) * F(1) - L(1,11) * F(11) - L(12,11) * F(11) + L(11,12) * F(12)$$
$$- L(21,11) * F(11) - L(13,11) * F(11) + L(11,13) * F(13)$$

$$dF(12)/dt = L(12,11) * F(11) - L(11,12) * F(12) + L(12,2) * F(2) - L(2,12) * F(12)$$

$$dF(13)/dt = L(13,11) * F(11) - L(11,13) * F(13) + L(13,3) * F(3) - L(3,13) * F(13)$$

$$dF(2)/dt = L(2,12) * F(12) - L(12,2) * F(2)$$

$$dF(3)/dt = L(3,13) * F(13) - L(13,3) * F(3)$$

initial conditions for the extracellular glutamate and glutamine compartments, respectively, because the medium containing these compounds was added at the beginning of the experiment.

When developing models using cultured cells, one must take into consideration the nutrients in the culture medium and the timing of cell feeding. This is important since compounds in the medium may contribute to intracellular metabolite concentrations, and the composition of many metabolites and nutrients in the medium changes significantly with time. The assumption is often made, although not always correctly, that replacing the medium at the beginning of the experiment with new medium containing the radiolabeled or stable isotopes will not alter the system under study.

A well-fitting compartmental model can be used to determine whether perturbations such as changing the medium are altering the concentration of certain metabolites in the

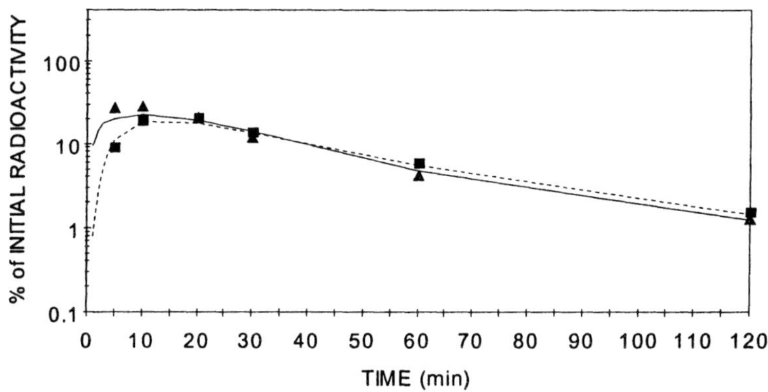

Figure 3. Intracellular glutamate tracer (▲-▲; compartment 11 in Figure 2) and glutamine tracer (■....■; compartment 12) in astrocytes. Symbols are observed data, and lines are model simulations.

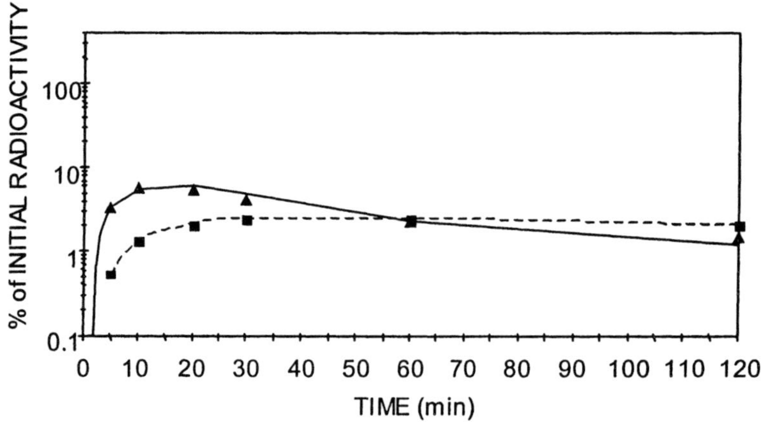

Figure 4. Intracellular aspartate in astrocytes (▲-▲; compartment 13 in Figure 2) and extracellular aspartate in the medium (■...■; compartment 3 in Figure 2). Symbols are observed data, and lines are model simulations.

cells being studied. Although the concentration of some metabolites in the system may be known, the concentrations of other metabolites must be estimated from values in the literature. We were able to use our initial model of glutamate metabolism in astrocytes to determine that steady state conditions were reached within the experimental time-frame (McKenna and Zech, unpublished data), even though the radiotracer was added in new medium containing 1 mM glutamate rather than the 2 mM present in the growth medium (Zielke et al., 1990).

Incorporation of Data into the SAAM Program and Initial Fitting of the Data

The experimental values for the radioactive (tracer) and steady state data (Zielke et al., 1990) and the rate equations describing each reaction in the model were incorporated into a SAAM deck (Berman and Weiss, 1978) in order to develop and test the compartmental model of glutamate and glutamine metabolism in astrocytes. The SAAM (Berman and Weiss, 1978) and CONSAM (Berman et al., 1983) computer programs, run on a Gateway 486/50 computer, were used to fit the model to the experimental data and to compare the predictions of the model with the observed data.

The shape of each data curve was determined initially by using the forcing function feature in SAAM to independently determine the equation that gave an accurate description of the shape of each curve (Foster and Boston, 1983). This process gave some insight into the complexity of glutamate metabolism in astrocytes, since four exponentials were required to describe the curve for the intracellular radioactive glutamate. The two curves formed by combining these exponentials, shown in Figure 6, demonstrate that some of the glutamate turned over very rapidly, and another portion of the intracellular glutamate turned over much more slowly. The equations for the forcing functions were incorporated into the SAAM input file that contained the equations for the fractional transfer rates between compartments. The model was refined numerous times by adjusting it in physiologically meaningful ways to eventually obtain a better visual fit between the simulated values and experimental data. Several changes, described in the

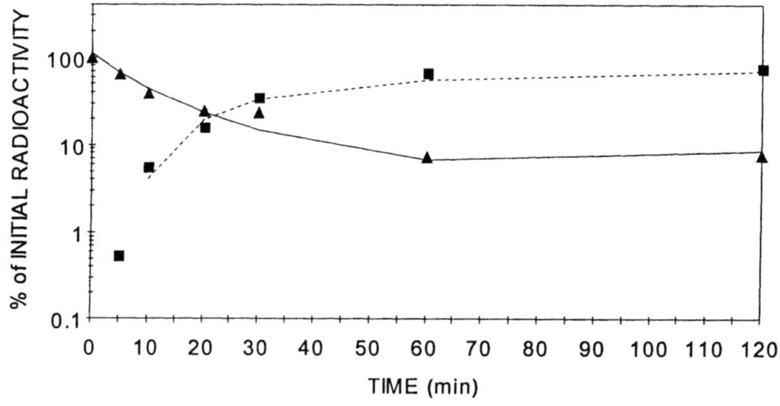

Figure 5. Extracellular glutamate tracer (▲-▲; compartment 1 in Figure 2) and glutamine in the medium (■...■; compartment 2). Symbols are observed data, and lines are model simulations.

next section, were made to the model before the fit was sufficiently close that the SAAM program would permit iteration. Once iteration was possible, the values for the fractional standard deviations were used to evaluate the goodness of fit for each parameter [L(i,j)]. The model was further adjusted to reduce the residual error and obtain a fit with reasonable (e.g., < 0.5) fractional standard deviations for each parameter.

Shown in Figures 4-6 are the experimental values for the tracer and/or metabolites expressed as percent of initial radioactivity. Also shown are the model simulations which were generated by the model. The simulations illustrate how the values predicted by the current model fit the actual experimental data.

Refining the Model

Although the initial model contained 7 compartments, 8 compartments were required to even *begin* to fit the model to the observed data. An early compartmental model for the radioactive tracer is shown in Figure 7. In order for the model to fit the radioactive tracer data more closely, it was necessary to add compartment 43 for intracellular α-ketoglutarate. Even after adding this compartment and adjusting the parameters, the model did not fit the data well: radioactivity increased too quickly in the compartments for glutamine (compartments 12 and 2) and aspartate (compartments 13 and 3), and the radioactivity in compartments 11, 12, and 13 dropped too much at the end. Adding a time delay to the intracellular glutamine compartment (compartment 12) improved the fit for this compartment and for glutamine in the medium (compartment 2). Since the endogenous glutamine concentration in astrocytes is high (Table 1), this time delay would account for enough [^{14}C]glutamine to be formed via glutamine synthetase to label the releasable glutamate stores. However, adding this time delay caused the radioactive glutamate in the medium (compartment 1) to drop too quickly and decreased the amount of intracellular glutamate tracer (compartment 11), since less glutamine was initially converted back to glutamate. To overcome this, it was necessary to add a pool of glutamate (compartment 41) that was taken up from the medium but not rapidly metabolized to either glutamine or α–ketoglutarate. The small amount of glutamate in this compartment stayed in astrocytes and was slowly returned to the medium. It is not known what metabolic fraction of the glutamate in astrocytes this relatively small compartment represents. However, a portion of the glutamate in these cells participates in the malate/aspartate shuttle and also has a role as an amino group donor for the many other transamination reactions in astrocytes (Yudkoff et al., 1986; McKenna et al., 1996a). The highly metabolically active portion of the glutamate pool involved in these processes would be continuously and rapidly transaminated then resynthesized, and it may appear to be a relatively stable compartment in a study design like the one used for this model.

A second time delay was added to the model to prevent the radioactivity from accumulating too rapidly in both intracellular aspartate (compartment 13) and aspartate in the medium (compartment 3). The fit was better when a time delay was added to the α–ketoglutarate compartment (compartment 43) than when it was added to either of the aspartate compartments (Figure 8). The time delay in aspartate formation would account for transport of glutamate into the mitochondria and conversion to α–ketoglutarate, followed by metabolism via several TCA cycle reactions to oxaloacetate. A portion of

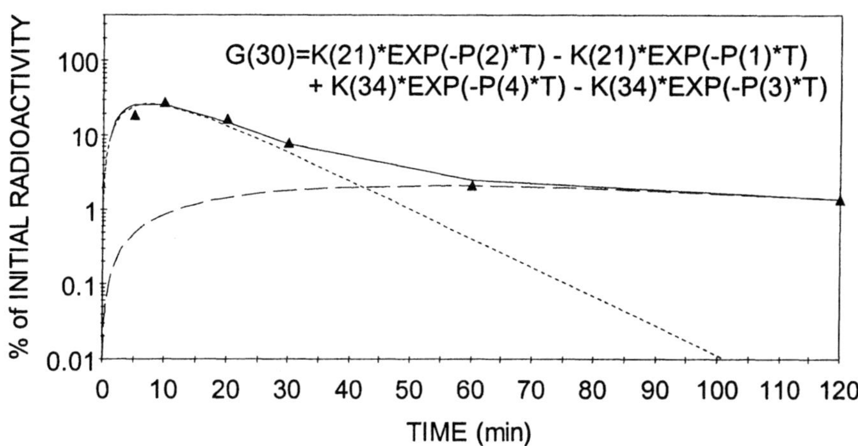

Figure 6. The equation describing the intracellular glutamate compartment, compartment 11 in the tracer model. The lines represent two of the curves contributing to the four-component exponential equation G(30). This equation describes the data points for the intracellular $[^{14}C]$glutamate in astrocytes. In the equation, G is a user-defined function, 30 is the arbitrary name, Ks are intercepts (exponential constants), Ps are slopes (exponential coefficients), and T is time (min). K(21)=35, K(34)=90, P(1)= 0.02, P(2)=0.017, P(3)=0.205, and P(4)=0.09.

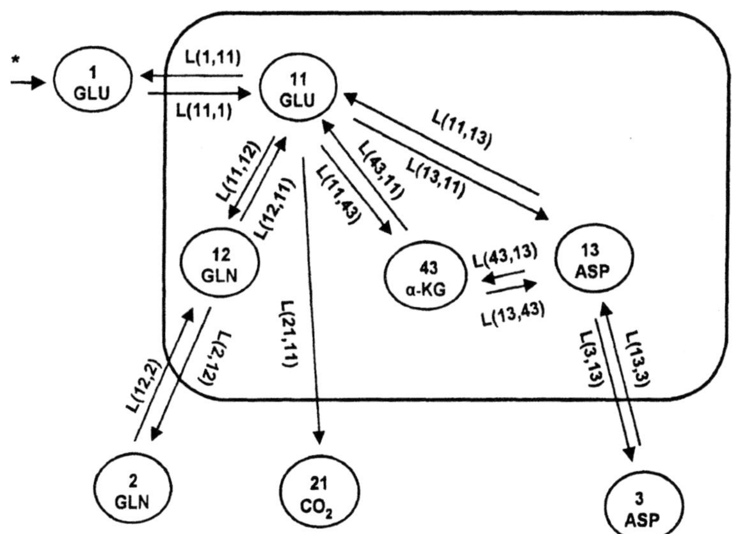

Figure 7. An early version of the tracer model for glutamate metabolism in astrocytes. The asterisk represents the site of tracer input. GLU, glutamate; GLN, glutamine; ASP, aspartate; α-KG, α keto glutarate.

the oxaloacetate formed in the TCA cycle is transaminated to aspartate and participates in the malate/aspartate shuttle, the most important mechanism for transferring reducing equivalents into mitochondria in the brain (McKenna et al., 1995). Adjusting the proportion of α–ketoglutarate that was delayed in compartment 43 (DN 43) led to some improvement in fit at the end of the curves for compartments 1, 11, 12, and 13. However, there was still a tendency for the extracellular glutamate as well as both the intracellular glutamate and glutamine curves to decrease too rapidly between 60 and 120 min. Adjusting the model by adding an additional compartment (compartment 45), which had input from compartment 43 and returned radioactivity to compartment 11, increased the amount of radioactivity in the tail of the three curves. However, when the rates [L(45,43) and L(11,45)] were high enough to make a difference, the input increased the portion of the curve between 60 and 90 min, leading to "bumps" in the tails of the intracellular glutamate and glutamine curves rather than raising the very end of these curves. The addition of a time delay to compartment 45 greatly improved the fit of the curves for intracellular glutamate and glutamine as well as the fit for glutamate in the medium (compartment 1).

After the fit for the calculated data was close to the observed data, iteration was possible. Weighted nonlinear regression analysis in SAAM was used to calculate the model parameters (fractional transfer coefficients). Parameters were further adjusted to decrease the fractional standard deviations for the various components. This led to an improvement in the fit for many of the compartments. The current tracer model is shown in Figure 8.

Fitting the Data for the Parallel Mass Model

Once the model fit the radioactive tracer data, the fit of the parallel model for mass data was addressed. A well-fitting model should adequately describe both mass and tracer data for the system since it is assumed that the behavior of the radioactive tracer and the tracee (i.e., the unlabeled compound that is being traced) are kinetically identical and should have the same rate coefficients (Novotny and Caballero, 1998).

Fitting the mass data was complex. In order to fit the mass data, more mass input into the system was required in addition to glutamate [UF (31)] and glutamine [UF(10)] in the medium. Figure 9 shows a schematic of the model for the mass data. The additional mass input [represented by UF(25) and UF(6)] is from compounds such as glucose, fatty acids, and amino acids in the medium, all of which also contribute carbon atoms to intracellular glutamate, glutamine, and aspartate. It was necessary to divide this additional mass input into two components, one of which is more closely associated with the synthesis of intracellular glutamate [UF(25)] and the other of which is associated more with the biosynthesis of α–ketoglutarate [UF(6)] and other TCA cycle intermediates and less with glutamate biosynthesis.

DISCUSSION

Suitability of the Data Used for Modeling

There are some limitations in the data used for this modeling study that contribute to the difficulty in developing a better-fitting model. These reflect the fact that the study used (Zielke et al., 1990) had not been designed with modeling in mind. The main limitation is the relatively small number of data points obtained during the observation period. It would be easier to model glutamate metabolism if there were more data points, particularly during the first 20 min of incubation when the percent of initial radioactivity was changing rapidly in the intracellular glutamate, glutamine, and aspartate compartments. Having more data points would decrease the standard errors and overcome the major difficulty in fitting the model to the data. It is important to note that the timing of the observations was chosen well, since the data points and curves reflect the complexity of glutamate metabolism. The ideal situation is to design a study specifically for modeling.

Strengths and Limitations of the Current Model of Glutamate Metabolism in Astrocytes

Our modeling results presented here are a promising first step in the development of a more comprehensive model of glutamate/glutamine metabolism in astrocytes. The main advantages of the model are that it is relatively simple, it can be used to predict the effect of alterations in major pathways of glutamate metabolism, and it can be used to help deal with issues related to using cell cultures (e.g., changing media). The model fits most compartments for the radioactive tracer data fairly well (Figures 4-6), and the fractional standard deviations for some of the fractional transfer rates are very low, with the majority being less than 20% [the accepted criterion for a good fit (Clifford and Müller, 1998)]. The model is consistent with data in the literature demonstrating the rapid uptake and conversion

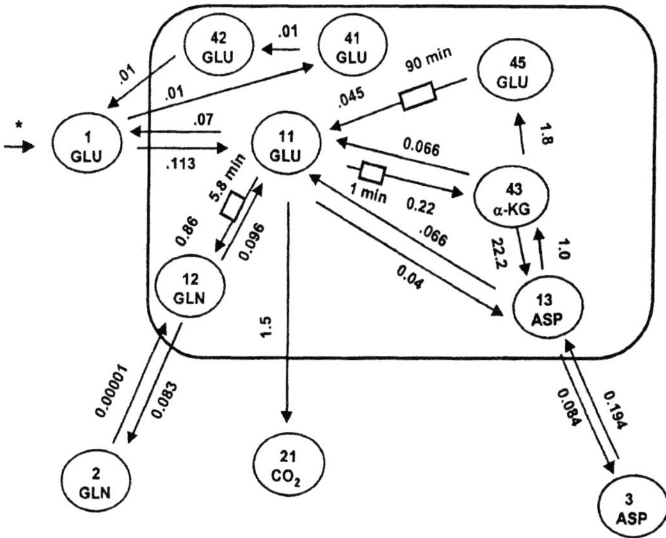

Figure 8. The current version of the tracer model for glutamate metabolism in astrocytes. The asterisk represents the site of tracer input. The numbers over the arrows are the fractional transfer coefficients generated by the model; units are percent of initial radioactivity. The boxes indicate time delays. GLU, glutamate; GLN, glutamine; ASP, aspartate; α-KG, α-ketoglutarate.

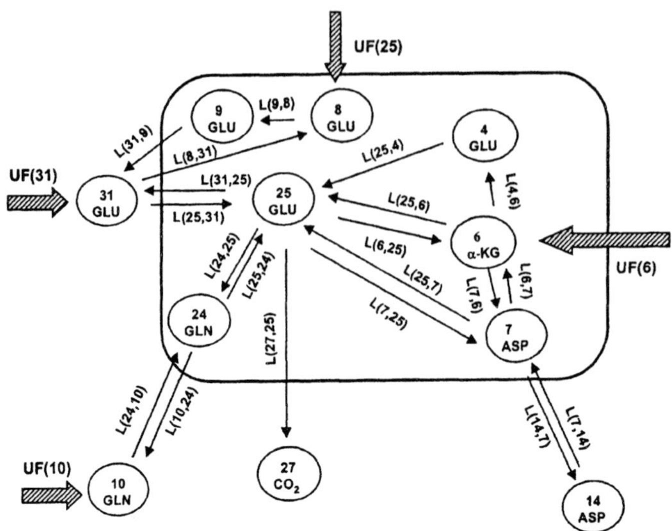

Figure 9. The current mass model for glutamate metabolism in astrocytes. The thick arrows and UF(I)s indicate mass input into the compartments from the medium. The fractional transfer rates are equal to those in the tracer model.

of glutamate to glutamine by astrocytes and the release of glutamine into the medium (Waniewski and Martin, 1986; Hertz and Schousboe, 1988; Zielke et al., 1990). The requirement for multiple glutamate compartments is consistent with current understanding of glutamate metabolism in astrocytes (Schousboe et al., 1993; McKenna et al, 1996b; Waagepetersen, 2000). The requirement for additional input (other than glutamate and glutamine) into the mass model underscores the important contribution of other compounds such as glucose, fatty acids, and amino acids to the intracellular pools of glutamate, glutamine, and aspartate in astrocytes (Van den Berg et al., 1975; Shank et al., 1993; Yudkoff et al., 1997).

The current model for glutamate metabolism in astrocytes has some limitations and weaknesses. The model is oversimplified with respect to some aspects of glutamate metabolism such as the conversion of a portion of glutamate to lactate and incorporation of glutamate into glutathione. In studies using HPLC, metabolism of glutamate via the TCA cycle is evident since aspartate is formed; however, the incorporation of the glutamate carbon skeleton into lactate was detected only relatively recently when ^{13}C-NMR was used to study glutamate metabolism (Sonnewald et al., 1993a).

A few of the fractional transfer coefficients required associated nonlinear functions that alter the metabolic rate with time. The need for such functions suggests that the compartment(s) in question are not in a steady state throughout the experiment. This could be considered a weakness of the model; however, as discussed below, it is to a large extent a characteristic of the system being studied. The fractional transfer coefficient for glutamine re-uptake from the medium [L(12,2); Figure 7] has a relatively high FSD, suggesting that it is not adequately described by the current model. However, this is not surprising since the process of glutamine uptake is mediated by multiple transport mechanisms in astrocytes (Nagaraja and Brookes, 1996).

There is uncertainty in model parameters related to the resynthesis of glutamate from α-ketoglutarate [L(11,43)]. The fact that this resynthesis may occur from multiple simultaneous TCA cycles known to be present in astrocytes (McKenna et al., 1990; Schousboe et al., 1993; McKenna et al., 1996b) suggests that more experiments are needed to be able to more accurately describe this aspect of glutamate metabolism. The parameters related to the formation of α-ketoglutarate from glutamate and aspartate are well described, even though the amount of ^{14}C labeling in α-ketoglutarate is not known.

Requirement for Nonlinear Functions in Order to Fit the Model to the Data

In order to fit the model to the observed data, it was necessary to add nonlinear functions that alter the metabolic rate with time to several of the fractional transfer coefficients [e.g., L(11,12), L(11,13), and L(11,43)]. The need for such functions suggests that the compartment(s) in question are not in a steady state throughout the experiment. Although this could be due to a less than optimal job of fitting the model to the data, there are biochemical explanations of why actual changes in metabolism can occur during the time-frame of the experiment. Nevertheless, a requirement for nonlinearity suggests that some portion of the compound in a particular compartment is turning over at a different rate than the remainder of the compound. Since one of the assumptions in modeling is that all molecules in a compartment are kinetically the same, this would indicate that some of the current compartments may actually be composed of more than one compartment. Additional experiments should help to identify whether some compartments are actually multiple compartments rather than a single one. Support for this concept comes from studies providing evidence of compartmentation of glutamate, aspartate, and pyruvate metabolism in astrocytes (McKenna et al., 1996a; Bakken et al., 1997).

Some of the factors that contribute to these nonlinearities may be difficult to address. For example, it is well established that enzymes are under nonlinear allosteric control by metabolites, and some pathways clearly involve channeling through multienzyme complexes; such processes would dramatically lower the Km of key reactions (Beeckmans and Kanareck, 1981; Fahien et al., 1988, 1989; Beeckmans et al., 1990; McKenna et al., 2000a and references therein). One aspect of glutamate metabolism that would contribute to nonlinearity is the fact that the relative proportion of glutamate converted to glutamine *vs.* entering the TCA cycle is greatly influenced by the concentration of glutamate in the extracellular milieu (McKenna et al., 1996a). Another example is the demonstration that glutamate enters the TCA cycle in astrocyte mitochondria by both glutamate dehydrogenase and transamination (McKenna et al., 1996b; Westergaard et al., 1996) and that the relative proportion of endogenous glutamate entering via transamination is modulated by the availability of amino group acceptors such as pyruvate (McKenna et al., 1996b).

Future Improvements in the Model

One of the advantages of compartmental modeling is that all of the available data about the system must be taken into account when building and identifying the model. In addition, the model developed must fit all of the data (e.g., radioactivity data and mass data), and it must be consistent with other published information about the system being studied. The fact that it has been difficult to get very good fits for all of the parameters suggests that some of the parameters or compartments are not well identified. With additional work, it should be possible to get a better fit and minimize the uncertainties in the model. My goal is to better define the parameters of the model and also to extend the current model by incorporating the information (and compartments) identified through our ongoing ^{13}C-NMR studies. In addition, it will be more useful to other investigators if the mass is expressed on the basis of protein rather than per dish.

The fit of the intracellular glutamate compartment (compartment 11; Figure 7) needs further adjustment to improve the agreement between the model prediction and the tracer data. The fact that the mass model underestimates the amount of glutamate in the intracellular compartment is consistent with the tracer model that underestimates the peak of radioactivity in intracellular glutamate that occurs between 5 and 10 min after the start of the incubation. The experimental data suggest that there was a relatively constant glutamate level in astrocytes. This would necessitate continuous replenishment of the intracellular glutamate pool from other substrates in the medium and turnover of endogenous proteins. Input from glucose, fatty acids, and amino acids in the medium has been added to the mass model. Additional refinements, such as adding amino acid input from protein turnover, may further improve the fit for glutamate and glutamine in the mass model. The mass data are consistent with the model for tracer in that they suggest that some portion of the intracellular glutamate pool is turning over much more slowly than the rest and that there are at least two compartments. The necessity to divide the added mass input into two components is consistent with the requirement for multiple pools of glutamate in the tracer model and also with studies that provide evidence of at least two pools of glutamate in cultured astrocytes (Schousboe et al., 1993; McKenna 1996a, b; Waagepetersen, 2000).

Since it is well established that prolonged exposure of brain cells to excess glutamate can lead to excitotoxic damage to both astrocytes and neurons, it will be particularly important to design studies suitable for modeling that include higher concentrations of glutamate. Any insight into the regulation of the capability of astrocytes to dispose of excess glutamate will have important therapeutic implications. My laboratory is particularly interested in mitochondrial metabolism in astrocytes whereby some of the carbon skeleton from glutamate is metabolized to lactate and released into the medium (Sonnewald et al., 1993a; McKenna et al., 1996a, 1998a, b). This pathway, which becomes more prominent when there is a high extracellular glutamate concentration, may represent a secondary mechanism for detoxification of excess glutamate by astrocytes. In addition, the role of this pathway in disposal of glutamate may become increasingly important when amino group donors are low or the ATP required for glutamine formation is limiting.

The release of lactate by astrocytes will eventually be added to the model, since it is stimulated by neurotransmitter glutamate and it represents an important aspect of neuronal/glial interaction (Prichard et al., 1991; McKenna et al., 1993; Pellerin and Magistretti, 1994; Westergaard et al., 1995b; McKenna et al., 1998b; Gruetter et al., 1998; Magistretti et al., 1999). This lactate, derived primarily from glycolysis, is taken up by synaptic terminals where it is used to meet a portion of the energy requirements and is also incorporated into neurotransmitters (Schurr et al., 1988; McKenna et al., 1993; Ranson and Fern, 1997; McKenna et al., 1998b; Pellerin et al., 1998; Waagepetersen et al., 1998a, b; Waagepetersen, 2000). It will be necessary to use ^{13}C-NMR studies to obtain data to model this aspect of metabolism in astrocytes.

What Insights Have Been Gained Through the Process of Modeling Glutamate Metabolism?

The process of trying to appropriately conceptualize and mathematically describe and determine fractional transfer coefficients for glutamate metabolism in astrocytes has been challenging. Multiple exponentials are required to describe curves to fit the data points for the intracellular glutamate, glutamine, and aspartate compartments. This underscores the complexity of glutamate compartmentation and demonstrates how closely the metabolism of even low concentrations of glutamate are intertwined with astrocyte energy metabolism.

Modeling even the relatively straightforward metabolism of glutamate to glutamine was more complex than anticipated. Glutamate uptake by multiple transporters and regulation of glutamine synthetase by multiple processes contribute to this complexity.

A difficulty with earlier versions of the model was overestimation of the mass of aspartate in compartment 7 because all of the aspartate remaining in astrocytes was connected to the main compartment of glutamate metabolism via α–ketoglutarate metabolism in compartment 43. The addition of compartment 45 to the tracer model, and the parallel addition of compartment 4 to the mass model, resulted in a much better fit for the mass of aspartate in compartment 7. The improved fit of the aspartate compartment with the addition of transfer from aspartate to compartments 45 and 6 is consistent with evidence for more than one compartment of aspartate metabolism in astrocytes, only one of which is tightly linked to glutamate metabolism (Schousboe et al., 1993; Bakken et al., 1997).

It is interesting to note that Van den Berg, Garfinkel, and their collaborators (Van den Berg et al., 1975) identified eight "metabolic spaces" in the small glutamate compartment (glial metabolism—primarily astrocytes) and at least two metabolic spaces within the large glutamate compartment (neuronal TCA cycle pools). These spaces were identified from labeling studies using many different precursors combined with careful studies of the differential distribution of enzyme activities by ultracentrifugation. However, their efforts at modeling these compartments were hindered by their inability to distinguish between metabolic spaces labeled by different precursors that might represent the same pool. They concluded (Van den Berg et al., 1975) that the modeling of these multiple pools was too complex, and they resorted to modeling the glutamate/glutamine cycle in brain using data obtained from essentially one source and in the same species.

Why is More In-depth Modeling of Brain Metabolism so Important?

Modeling has much to offer in terms of increasing our understanding of brain metabolism. Recent studies have shown that the death of neurons in the brain can result from direct damage to neurons (Zeevalk et al., 1998; Heales et al., 1999), damage to astrocytes that compromise the ability of these cells to maintain the proper microenvironment essential for brain functioning (Chan et al., 1990; Dringen and Hamprecht, 1996; McKenna et al., 1998a), and/or by disruption of neuronal/astrocytic metabolic trafficking (Hassel and Sonnewald, 1995; Ransom and Fern 1997; Hassel et al., 1997; Norenberg et al., 1997). Thus, insults to either astrocytes or neurons can ultimately lead to the common end point of neurodegeneration. However, an understanding of the mechanism(s) involved is critical since very different neuroprotective strategies may be required to prevent the damage and neurodegeneration (Ransom and Fern, 1997; Nicklas 1998; McKenna and Edmond, 1998). The insight and more precise understanding of glutamate metabolism in astrocytes and neurons afforded by modeling could be an invaluable tool for helping to identify mechanisms leading to brain damage and for assisting in the development of new treatments for CNS dysfunction.

Alterations in glutamate metabolism are an important feature of many clinical conditions that lead to brain dysfunction and/or degeneration including intractable childhood epilepsy, ischemic brain damage, and stroke. However, certain drugs and protocols that are clinically effective in treating CNS dysfunction act via mechanisms that are not well understood (Daihkin and Yudkoff, 1998; Shank et al., 2000).

Inasmuch as glutamate is the major excitatory neurotransmitter in brain, glutamate metabolism is closely intertwined with many important processes including energy metabolism in both astrocytes and neurons as well as synthesis of the inhibitory neurotransmitter GABA and the antioxidant glutathione. Furthermore, several recent studies provide evidence that metabolic events, including those mediated via metabotrophic glutamate receptors, are even associated with the all important processes of learning and memory (Izquierdo et al., 1997; Szapiro et al., 2000). Thus, there is much to be gained from studies applying modeling to brain metabolism since "the rigors of mathematics provide a framework to indirectly study processes which cannot be observed directly...[and] offer scientists a window with which to view aspects of systems which could otherwise be inaccessible" (Novotny and Caballero, 1998).

ACKNOWLEDGMENTS

This paper is dedicated to the late Loren Zech who introduced me to the art and science of modeling. He taught me to look at a curve "like an engineer" and see multiple exponentials. This modeling could not have been done without his enthusiasm, guidance, sincerity, and infinite patience.

The initial modeling work described in this paper was supported in part by a Pew Foundation Faculty Scholars in Nutrition Fellowship that enabled me to spend time in the laboratory of Dr. Loren Zech at the Laboratory of Mathematical Biology, National Cancer Institute, NIH, to learn how to apply mathematical modeling techniques to my research. I am particularly indebted to my colleague, Dr. H. Ronald Zielke, for giving me access to data from his laboratory, supported in part by National Institutes of Health Program Project Grant HD 16596, that was used as the basis for this model. I thank Peter Grief for his assistance during my sabbatical at NIH, and Dr. David Martin for his helpful suggestions for revising this chapter. Thanks to Drs. Ray Boston, Janet Novotny, and Mike Green for the invitation to present this work on modeling at the 7th Workshop on Mathematical Modeling in Nutrition and the Health Sciences at Penn State, July 29 - August 1, 2000. Special thanks to Joanne Green for her invaluable editorial assistance.

CORRESPONDENCE

Please address all correspondence to:
Mary C. McKenna
University of Maryland School of Medicine
Department of Pediatrics, 10-029 BRB
655 W. Baltimore Street
Baltimore, MD 21201
mmckenna@umaryland.edu

REFERENCES

Assaf, H.M., Ricci, A.J., Whittingham, T.S., LaManna, J.C., Ratcheson, R.A., and Lust, W.D., 1990, Lactate compartmentation in hippocampal slices: evidence for a transporter, *Metab. Brain Dis.* 5:143-154.

Bachelard, H., 1998, Landmarks in the application of ^{13}C-magnetic resonance spectroscopy to studies of neuronal/glial relationships, *Dev. Neurosci.* 20:27-288.

Badar-Goffer, R.S., Bachelard, H.S., and Morris P.G., 1990, Cerebral metabolism of acetate and glucose studied by ^{13}C-NMR spectroscopy: a technique for investigating metabolic compartmentation in the brain, *Biochem. J.* 266:133-139.

Bader-Goffer, R.S., Ben-Yoseph, O., Bachelard, H.S., and Morris, P.G., 1992, Neuronal-glial metabolism under depolarizing conditions. A ^{13}C-n.m.r. study, *Biochem. J.* 282:225-230.

Bakken, I.J., White L.R., Aasly, J., Unsgard, G., and Sonnewald, U., 1997, Lactate formation from [U-^{13}C]aspartate in cultured astrocytes: compartmentation of pyruvate metabolism, *Neurosci. Lett.* 237:117-120.

Beeckmans, S., and Kanarek, L., 1981, Demonstration of a physical interaction between consecutive enzymes of citric acid cycle and of the malate aspartate shuttle. A study involving fumarase, malate dehydrogenase, citrate synthetase, and aspartate aminotransferase, *Europ. J. Biochem.* 117:527-535.

Beeckmans, S., Driessche, E.V., and Kanarek, L., 1990, Clustering of sequential enzymes in the glycolytic pathway and the citric acid cycle, *J. Cell Biochem.* 43:297-306.

Berl, S., and Clark, D.D., 1969, Metabolic compartmentation of glutamate in the CNS, in: *Handbook of Neurochemistry. Chemical and Cellular Architecture, Vol. 1*, A. Lajtha, ed., Plenum Press, New York.

Berman, M., and Weiss, M.F., 1978, *SAAM Manual*, DHEW Publ. #78-180, U.S. Government Printing Office, Washington, DC.

Berman, M., Beltz, W.F., Greif, P.C., Chabay, R., and Boston, R.C., 1983, *CONSAM User's Guide*, PHS Publ. #1983-421, U.S. Government Printing Office, Washington, DC.

Browning, E.T., and Nicklas, W.J., 1982, Induction of glutamine synthetase by dibuturyl cyclic AMP in C-6 glioma cells. *J. Neurochem.* 39:336-341.

Cerdan, S., Kunnecke, B., and Seelig, J., 1990, Cerebral metabolism of [1,2-^{13}C$_2$]acetate as detected by in vivo and in vitro ^{13}C NMR, *J. Biol. Chem.* 265:12916-12926.

Chan, P.H., Chu, L., and Chen, S., 1990, Effects of MK-801 on glutamate-induced swelling of astrocytes in primary culture, *J. Neurosci. Res.* 25:87-93.

Clifford, A.J., and Müller H.-G., 1998, *Mathematical Modeling in Experimental Nutrition*, Plenum Press, New York.

Daihkin, Y., and Yudkoff, M., 1998, Ketone bodies and brain glutamate and GABA metabolism, *Dev. Neurosci.*, 20:358-64.

Dringen, R., and Hamprecht, B., 1996, Glutathione content as an indicator for the presence of metabolic pathways of amino acids in astroglial cultures, *J. Neurochem.* 67:1375-1382.

Erecinska, M., Zaleska, M.M., Nissim, I., Nelson, D., Dagani, F., and Yudkoff, M., 1988, Glucose and synaptosomal glutamate metabolism: studies with [^{15}N]glutamate, *J. Neurochem.* 51:892-902.

Fahien, L.A., Kmiotek, E.H., MacDonald, M.J., Fibrich, B., and Mandic, M., 1988, Regulation of malate dehydrogenase by glutamate, citrate, α-ketoglutarate, and multienzyme interactions, *J. Biol. Chem.* 263:10687-10697.

Fahien, L.A., MacDonald, M.J., Teller, J.K., Fibich, B., and Fahien, C.M., 1989, Kinetic advantages of hetero enzyme complexes with glutamate dehydrogenase and the α-ketoglutarate dehydrogenase complex, *J. Biol. Chem.* 264:12303-12312.

Foster, D.M., and Boston, R.C., 1983, The use of computers in compartmental analysis: the SAAM and CONSAM programs, in: *Compartmental Distribution of Radiotracers,* J.S. Robertson, ed., CRC Press, Boca Raton.

Gruetter, R., Seaquist, E.R., Kim, S., and Ugurbil, K., 1998, Localized in vivo ^{13}C NMR of glutamate metabolism in the human brain. Initial results at 4 Tesla, *Dev. Neurosci.* 20:380-388.

Harkany, T., Abraham, T., Timmerman, W., Laskay, G., Toth, B., Sasvari, M., Konya, C., Sebens, J.B., Korf, J., Nyakas, C., Zarandi, M., Soos, K., Penke, B., and Luiten, P.G., 2000, Beta-amyloid neurotoxicity is mediated by a glutamate-triggered excitotoxic cascade in rat nucleus basalis, *Europ. J. Neurosci.* 12:2735-2745.

Hassel, B., and Sonnewald, S., 1995, Glial formation of pyruvate and lactate from TCA cycle intermediates: implications for inactivation of transmitter amino acids, *J. Neurochem.* 65:2227-2234.

Hassel, B., Bachelard, H., Fonnum, F., Jones, P., and Sonnewald, U., 1997, Trafficking of amino acids between neurons and glia in vivo. Effects of inhibition of glial metabolism by fluoroacetate, *J. Cereb. Blood Flow Metab.* 17:1230-1238.

Heales, S.J., Bolanos, J.P., Stewart, V.C., Brookes, P.S., Land, J.M., and Clark, J.B., 1999, Nitric oxide, mitochondria and neurological disease, *Biochim. Biophys. Acta* 1410:215-228.

Hertz, L., 1979, Functional interactions between neurons and astrocytes. I. Turnover and metabolism of putative amino acid transmitters, *Prog. Neurobiol.* 13:277-323.

Hertz, L., and Schousboe, A., 1988, Metabolism of glutamate and glutamine in neurons and astrocytes in primary cultures, in: *Glutamate and Glutamine in Mammals, Vol. 2*, E. Kvamme, ed., CRC Press, Boca Raton.

Izquierdo, I., Barros, D.M., Izquierdo, L., Mello e Souza, T., Souza, M., and Medina, J.H., 1997, Memory formation: the sequence of biochemical events in the hippocampus and its connection to activity in other brain structures, *Neurobiol. Learn. Mem.* 68:285-316.

Kunnecke, B., Cerdan, S., and Seelig, J., 1993, Cerebral metabolism of [1,2-[13]C_2]glucose and [U-[13]C_4]3-hydroxybutyrate in rat brain as detected by ^{13}C NMR spectroscopy, *NMR Biomed.* 6: 264-277.

Lai, J.C., Walsh, J.M., Dennis, S.C., and Clark, J.B., 1977, Synaptic and non-synaptic mitochondria from rat brain: isolation and characterization, *J. Neurochem.* 28:625-631.

Lai, J.C.K., and Clark, J.B., 1989, Isolation and characterization of synaptic and nonsynaptic mitochondria from the mammalian brain, in: *Neuromethods, Vol. II, Carbohydrates and Energy Metabolism*, A.A. Boulton and G.B. Baker, eds., Humana Press, Clifton.

Magistretti, P.J., Pellerin, L., Rothman, D.L., and Shulman, R.G., 1999, Energy on demand, *Science* 283:496-497.

Malik, P., McKenna, M.C., and Tildon, J.T., 1993, Regulation of malate dehydrogenases from neonatal adolescent and mature rat brain, *Neurochem. Res.* 18:247-257.

Mason, G.F., Gruetter, R., Rothman, D.L., Behar, K.L., Shulman, R.G., and Novotny, E.J., 1995, Simultaneous determination of the rates of the TCA cycle, glucose utilization, alpha-ketoglutarate/glutamate exchange, and glutamine synthesis in human brain by NMR, *J. Cereb. Blood Flow Metab.* 15:12-25.

Max, S.R., Landry, M.E., and Zielke, H.R., 1990, Induction of glutamine synthetase by 8-bromo cyclic AMP in primary cultures of rat astrocytes, *Neurochem. Res.* 15:589-592.

McKenna, M.C., and Edmond, J., 1998, *Energy Metabolism in Brain Function and Neuroprotection. Developmental Neuroscience, Vol. 20*, S. Karger AG, Basel.

McKenna, M.C., Tildon, J.T., Couto, R., Stevenson, J.H., and Caprio, F.J., 1990, The metabolism of malate by cultured rat brain astrocytes, *Neurochem. Res.* 15:1211-1220.

McKenna, M.C., Tildon, J.T., Stevenson, J.H., Boatright, R., and Huang, X., 1993, Regulation of energy metabolism in synaptic terminals and cultured rat brain astrocytes: differences revealed using aminooxyacetate, *Dev. Neurosci.* 15:320-329.

McKenna, M.C., Tildon, J.T., Stevenson, J.H., Huang, X., and Kingwell, K.G., 1995, Regulation of mitochondrial and cytosolic malic enzymes from cultured rat brain astrocytes, *Neurochem. Res.* 20:1491-1501.

McKenna, M.C., Sonnewald, U., Huang, X., Stevenson, J.H., and Zielke, H.R., 1996a, Exogenous glutamate concentration regulates the metabolic fate of glutamate in astrocytes, *J. Neurochem.* 66:386-393.

McKenna, M.C., Tildon, J.T., Stevenson, J.H., and Huang, X., 1996b, New insights into the compartmentation of glutamate and glutamine metabolism in astrocytes, *Dev. Neurosci.*18:380-390.

McKenna, M.C., Sonnewald, U., Huang, X., Stevenson, J.H., Johnsen, S.F., Sande, L.M., and Zielke, H.R., 1998a, α-Ketoisocaproate alters the production of both lactate and aspartate from [U-^{13}C]glutamate in astrocytes: a ^{13}C-NMR study, *J. Neurochem.* 70:1001-1008.

McKenna, M.C., Tildon, J.T., Stevenson, J.H., Hopkins, I.B., Huang, X., and Couto, R., 1998b, Lactate transport by cortical synaptosomes from adult rat brain: characterization of kinetics and inhibitor specificity, *Dev. Neurosci.* 20:300-309.

McKenna M.C., Stevenson J.H., Huang X., and Hopkins I.B., 2000a, Differential distribution of the enzymes glutamate dehydrogenase and aspartate aminotransferase in cortical synaptic mitochondria contributes to metabolic compartmentation in cortical synaptic terminals, *Neurochem. Intl.* 37:229-241.

McKenna M.C., Stevenson J.H., Huang X., Tildon, J.T., Zielke, C.L., and Hopkins I.B., 2000b, Mitochondrial malic enzyme activity is much higher in mitochondria from cortical synaptic terminals compared with mitochondria from primary cultures of cortical neurons or cerebellar granule cells, *Neurochem. Int.* 36:451-459.

Nagaraja, T.N., and Brookes, N., 1996, Glutamine transport in mouse cerebral astrocytes, *J. Neurochem.* 66:1665-1674.

Nicklas, W.J., 1998, Introduction, *Dev. Neurosci.* 20:399-400.

Norenberg, M.D., and Martinez-Hernandez, A., 1979, Fine structural localization of glutamine synthetase in astrocytes of rat brain, *Brain Res.*161:303-310.

Norenberg, M.D., Huo, Z., Neary, J.T., and Roig-Cantesano, A., 1997, The glial glutamate transporter in hyperammonemia and hepatic encephalopathy: relation to energy metabolism and glutamatergic neurotransmission, *Glia* 21:124-133.

Novotny, J.A., and Caballero, B., 1998, Compartmental modeling of human lactation, in: *Mathematical Modeling in Experimental Nutrition*, A.J. Clifford and H.-G. Müller, eds., Plenum Press, New York.

Patel, A.J., Hunt, A., and Faraji-Shadan, F., 1986, Effect of removal of glutamine and addition of dexamethasone on the activities of glutamine synthetase, ornithine decarboxylase and lactate dehydrogenase in primary cultures of forebrain and cerebellar astrocytes, *Dev. Brain Res.* 26:229-238.

Pellerin, L., and Magistretti, P.J., 1994, Glutamate uptake into astrocytes stimulates aerobic glycolysis: a mechanism coupling neuronal activity to glucose utilization, *Proc. Natl. Acad. Sci. USA* 91:10625-10629.

Pellerin, L., Pellegri, G., Martin, J.L., and Magistretti, P.J., 1998, Expression of monocarboxylate transporter mRNAs in mouse brain: support for a distinct role of lactate as an energy substrate for the neonatal versus adult brain, *Proc. Nat. Acad. Sci. USA* 95:3990-3995.

Perry, T.L., and Hansen S., 1981, Amino acid abnormalities in epileptogenic foci, *Neurology* 31:872-876.

Petroff, O.A.C., Burlina, A.P., Black, J., and Prichard, J.W., 1991, Metabolism of [1-^{13}C]glucose in a synaptosomally enriched fraction of rat cerebrum studied by ^{1}H/^{13}C magnetic resonance spectroscopy, *Neurochem. Res.* 16:1245-1251.

Plaitakas, A., Berl, S., and Yaho, M.D., 1982, Abnormal glutamate metabolism in an adult-onset degenerative neurological disease, *Science* 216:193-196.

Prichard, J., Rothman, D., Novotny, E., Petroff, O., Kuwabara, T., Avison, M., Howseman, A., Hanstock, C., and Shulman, R., 1991, Lactate rise detected by ^{1}H NMR in human visual cortex during physiologic stimulation, *Proc. Natl. Acad. Sci. USA* 88:5829-5831.

Ranson, B.R., and Fern, R., 1997, Does astrocytic glycogen benefit axon function and survival in CNS white matter during glucose deprivation? *Glia* 21:134-141.

Roeder, L.M., Williams, I.B., and Tildon, J.T., 1985, Glucose transport in astrocytes: regulation by thyroid hormone, *J. Neurochem.* 45:1653-1657.

Rothman, S.M., and Olney, J.W., 1986, Glutamate and the pathophysiology of hypoxic-ischemic brain damage, *Ann. Neurol.* 19:105-111.

Schousboe, A., Drejer, J., and Hertz, L., 1988, Uptake and release of glutamate and glutamine in neurons and astrocytes in primary culture, in: *Glutamine and Glutamate in Mammals, Vol. 2*, E. Kvamme, ed., CRC Press, Boca Raton.

Schousboe, A., Westergaard, N., Sonnewald, U., Petersen, S.B., Huang, R., Peng, L., and Hertz, L., 1993, Glutamate and glutamine metabolism and compartmentation in astrocytes, *Develop. Neurosci.* 15:359-366.

Schousboe, A., Westergaard, N., Waagepetersen, H., Larsson, O.M., Bakken, I.J., and Sonnewald, U., 1997, Trafficking between glia and neurons of TCA cycle intermediates and related metabolites, *Glia* 21:99-105.

Schurr, A., West, C.A., and Rigor, B.M., 1988, Lactate-supported synaptic function in the rat hippocampal slice preparation, *Science* 240:1326-1328.

Shank, R.P., and Aprison, M.H., 1988, Glutamate as a neurotransmitter, in: *Glutamine and Glutamate in Mammals, Vol. 2,* E. Kvamme, ed., CRC Press, Boca Raton.

Shank R.P., Leo G.C., and Zielke, H.R., 1993, Cerebral metabolic compartmentation as revealed by nuclear magnetic resonance analysis of D-[1-^{13}C]glucose metabolism, *J. Neurochem.* 61:315-323.

Shank, R.P., Gardocki, J.F., Streeter, A.J., and Maryanoff, B.E., 2000, An overview of the preclinical aspects of topiramate: pharmacology, pharmacokinetics, and mechanism of action, *Epilepsia* 41:S3-9.

Shen, J., Sibson, N.R., Cline, G., Behar, K.L., Rothman, D.L., and Shulman, R.G., 1998, ^{15}N-NMR spectroscopy studies of ammonia transport and glutamine synthesis in the hyperammonemic rat brain, *Neuroscience* 20:434-443.

Sibson, N.R., Dhankhar, A., Mason, G.F., Rothman, D.L., Behar, K.L., and Shulman, R.G., 1998, Stoichiometric coupling of brain glucose metabolism and glutamatergic neuronal activity, *Proc. Natl. Acad. Sci. USA* 95:316-321.

Sibson, N.R., Mason, G.F., Shen, J., Cline, G.W., Herskovits, A.Z., Wall, J.E.M., Behar, K.L., Rothman, D.L., and Shulman, R.G., 2001, *In vivo* ^{13}C NMR measurement of neurotransmitter glutamate cycling, anaplerosis and TCA cycle flux in rat brain during [2-^{13}C]glucose infusion, *J. Neurochem.* 76:975-989.

Sonnewald, U., Westergaard, N., Petersen, S.B., Unsgard, G., and Schousboe, A., 1993a, Metabolism of [U-^{13}C]glutamate in astrocytes studied by ^{13}C NMR spectroscopy: incorporation of more label into lactate than into glutamine demonstrates the importance of the tricarboxylic acid cycle, *J. Neurochem.* 61:1179-1182.

Sonnewald, U., Westergaard, N., Schousboe, A., Svendsen, J.A., Unsgard, G., and Petersen, S.B., 1993b, Direct demonstration by [^{13}C]NMR spectroscopy that glutamine from astrocytes is a precursor for GABA synthesis in neurons, *Neurochem. Intl.* 22:19-29.

Szapiro, G., Izquierdo, L.A., Alonso, M., Barros, D., Paratcha, G., Ardenghi, P., Pereira, P., Medina, J.H., and Izquierdo, I., 2000, Participation of hippocampal metabotropic glutamate receptors, protein kinase A and mitogen-activated protein kinases in memory retrieval, *Neuroscience* 99:1-5.

Tildon, J.T., Roeder, L.M., and Stevenson, J.H., 1985, Substrate oxidation by isolated rat brain mitochondria and synaptosomes, *J. Neurosci. Res.* 14:207-215.

Tildon, J.T., McKenna, M.C., Stevenson, J.H., and Couto, R., 1993, Transport of L-lactate by cultured rat brain astrocytes, *Neurochem. Res.*18:177-184.

Van den Berg, C.J., 1970, Compartmentation of glutamate metabolism in the developing brain: experiments with labeled glucose, acetate, phenylalanine, tyrosine and proline, *J. Neurochem.* 17:973-983.

Van den Berg, C.J., and Garfinkel, D., 1971, A simulation study of brain compartments. Metabolism of glutamate and related substances in mouse brain, *Biochem. J.* 123:211-218.

Van den Berg, C.J., Matheson, D.F., Ronda, G., Reijnierse, G.L.A., Blokhuis, G.G.D., Kroon, M.C., Clarke, D.D., and Garfinkel, D., 1975, A model of glutamate metabolism in brain: a biochemical analysis of a heterogeneous structure, in: *Metabolic Compartmentation and Neurotransmission—Relation to Brain Function,* S. Berl, D.D. Clarke, and S. Schneider, eds., Plenum Press, New York.

Vannucci, S.J., Clark, R.R., Koehler-Stec, E., Li, K., Smith, C., Davies, P., Maher, F., and Simpson, I.A., 1998, Glucose transporter expression in brain: relationship to cerebral glucose utilization, *Dev. Neurosci.* 20:369-379.

v. Reinersdorff, D., Green, M.H., and Green, J.B., 1998, Development of a compartmental model describing the dynamics of vitamin A metabolism in men, in: *Mathematical Modeling in Experimental Nutrition,* A.J. Clifford and Müller, H.-G., eds., Plenum Press, New York.

Waagepetersen, H.S., 2000, *Compartmentation of Energy and Amino Acid Metabolism in Neurons and Astrocytes: Implications for Glutamate and GABA Biosynthesis,* Ph.D. Thesis, Department of Pharmacology, Royal Danish School of Pharmacy, Copenhagen.

Waagepetersen, H., Bakken, I.J., Larsson, O.M., Sonnewald, U., and Schousboe, A., 1998a, Metabolism of lactate in cultured GABAergic neurons studied by ^{13}C NMR spectroscopy, *J. Cereb. Blood Flow. Metab.* 18:109-117.

Waagepetersen, H., Bakken, I.J., Larsson, O.M., Sonnewald, U., and Schousboe, A., 1998b, Comparison of lactate and glucose metabolism in cultured neocortical neurons using ^{13}C NMR spectroscopy, *Dev. Neurosci.* 20:310-321.

Waagepetersen, H., Sonnewald, U., Larsson, O.M., and Schousboe, A., 1999, Synthesis of vesicular GABA from glutamine involves TCA cycle metabolism in neocortical neurons, *J. Neurosci. Res.* 57:342-349.

Waniewski, R.A., 1992, Physiological levels of ammonia regulate glutamine synthesis from extracellular glutamate in astrocyte cultures, *J. Neurochem.* 58:167-174.

Waniewski, R.A., and Martin, D.L., 1986, Exogenous glutamate is metabolized to glutamine and exported by rat primary astrocyte cultures, *J. Neurochem.* 47:304-313.

Waniewski, R.A., and Martin, D.L., 1998, Preferential utilization of acetate by astrocytes is attributable to transport, *J. Neurosci.* 18:5225-5233.

Wastney, M.E., Patterson, B.H., Linares, O.A., Greif, P.C., and Boston, R.C., 1999, *Investigating Biological Systems Using Modeling: Strategies and Software*, Academic Press, San Diego.

Westergaard, N., Sonnewald, U., Petersen, S.B., and Schousboe, A., 1995a, Glutamate and glutamine metabolism in cultured GABAergic neurons studied by ^{13}C NMR spectroscopy may indicate compartmentation and mitochondrial heterogeneity, *Neurosci. Lett.* 185:24-28.

Westergaard, N., Sonnewald, U., and Schousboe A., 1995b, Metabolic trafficking between neurons and astrocytes: the glutamate/glutamine cycle revisited, *Dev. Neurosci.* 17:203-211.

Yu, A.C.H., Schousboe, A., and Hertz, L., 1982, Metabolic fate of [^{14}C]-labelled glutamate in astrocytes, *J. Neurochem.* 39:964-966.

Yu, A.C.H., Drejer, J., Hertz, L., and Schousboe, A., 1983, Pyruvate carboxylase activity in primary cultures of astrocytes and neurons, *J. Neurochem.* 41:1484-1487.

Yudkoff, M., Nissim, I., Hummeler, K., Medow, M., and Pleasure, D., 1986, Utilization of [^{15}N]glutamate by cultured astrocytes, *Biochem. J.* 234:185-192.

Yudkoff, M., Nissim, I., and Pleasure, D., 1988, Astrocyte metabolism of ^{15}N and ^{13}C glutamine: implications for the glutamine-glutamate cycle, *J. Neurochem.* 51:843-850.

Yudkoff, M., Pleasure, D., Cregar, L., Lin, Z.-P., Nissim I., Stern, J., and Nissim I., 1990, Glutathione turnover in cultured astrocytes: studies with [^{15}N]glutamate, *J. Neurochem.* 55:137-145.

Yudkoff, M., Nelson, D., Daikhin, Y., and Erecinska, E., 1994, Tricarboxylic acid cycle in rat brain synaptosomes. Fluxes and interactions with aspartate aminotransferase and malate/aspartate shuttle, *J. Biol. Chem.* 269:27414-27420.

Yudkoff, M., Daikhin, Y., Nissim, I., Grunstein, R., and Nissim, I., 1997, Effect of ketone bodies on astrocyte amino acid metabolism, *J. Neurochem.* 69:682-692.

Zeevalk, G.D., Bernard, L.P., Sinha, C., Ehrhart, J., and Nicklas, W.J., 1998, Excitotoxicity and oxidative stress during inhibition of energy metabolism, *Dev. Neurosci.* 20:444-453.

Zielke, H.R., 1985, Determination of amino acids in brain by high-pressure liquid chromatography with isocratic elution and electrochemical detection, *J. Chromat.* 349:320-324.

Zielke, H.R., Tildon, J.T., Landry, M.E., and Max, S.R., 1990, Effect of 8-bromo-cAMP and dexamethasone on glutamate metabolism in rat astrocytes, *Neurochem. Res.* 15:1115-1122.

Zielke, H.R., Collins, R.M. Jr., Baab, P.J., Huang, Y., Zielke, C.L., and Tildon, J.T., 1999, Compartmentation of [^{14}C]glutamate and [^{14}C]glutamine oxidative metabolism in the rat hippocampus as determined by microdialysis, *J. Neurochem.* 71:1315-1320.

PART VIII
STATE-OF-THE-ART MODELING SOFTWARE

WinSAAM: APPLICATION AND EXPLANATION OF USE

Janet A. Novotny, Peter Greif, and Raymond C. Boston[*]

INTRODUCTION

The WinSAAM modeling software is an integrated package of mathematical, simulation, and graphical tools for analysis of biokinetic data. WinSAAM is the Windows version of SAAM, the Simulation, Analysis, and Modeling computer program. SAAM, a collection of scientific subroutines, was created in the 1950s by Dr. Mones Berman to meet the mathematical needs of his research program. Over the decades, SAAM has been continually developed and enhanced (Berman and Weiss, 1978; Boston et al., 1981), and today, WinSAAM is an elegant and powerful tool for mathematical and compartmental modeling.

OVERVIEW OF MODELING WITH WinSAAM

Among WinSAAM's capabilities are two very basic functions for modeling: solving equations (which describe a model) and performing least squares fitting of model solutions to data for parameter estimation. WinSAAM also provides other facilities related to modeling, including graphical displays of results, statistical measures of model fit, and many other features which are important for a thorough modeling investigation.

The control center for modeling with WinSAAM is called the terminal window. Commands are entered into the terminal window, and other modeling zones (windows containing different tools) are accessed through the terminal window. A window-contained zone called the *text editor* is a specially refined editor which allows easy creation and modification of model input files. The *charting system* is a flexible plotting environment

[*] Janet A. Novotny, USDA, Beltsville Human Nutrition Research Center, Beltsville, MD 20705. Peter Greif, Laboratory of Experimental and Computational Biology, National Cancer Institute, Bethesda, MD 20892. Raymond C. Boston, University of Pennsylvania, Clinical Studies Department, New Bolton Center, Kennett Square, PA 19348.

for viewing results. The *spreadsheet manager* is a data organization center. The *logger* allows the modeler to retain a detailed modeling history. The *batch window* functions to collect and save processing information. The terminal window and the other modeling zones are displayed in Figure 1.

Upon opening the WinSAAM program, the terminal window is immediately accessed. From the terminal window, one may open an existing model, access other windows (or modeling zones) to facilitate various aspects of modeling, perform standard operations on files (*new*, *open*, *save*, etc.), perform compiling and solving procedures, and access WinSAAM's online help files. Through the terminal window, one also enters WinSAAM commands. The most common commands [*deck* (which compiles), *solve*, *iterate*, and *SAAM* (which runs the model in batch mode)] are available through a pull-down menu, while other commands such as *plot* (which creates a graph of results) are entered only via the command line. Through the pull-down menus and through commands entered at the terminal window prompt, the modeler interacts with the WinSAAM program to control the modeling process.

ENTERING A MODEL AND DATA

Problems are defined in WinSAAM by entering all model components (model structure, parameter specifications, mathematical controls, and data) into one input file, commonly referred to as the SAAM deck. The deck consists of several sections which are separated by header lines. Each section contains a list of statements containing information about the model. These statements describe the model structure, the initial values of the

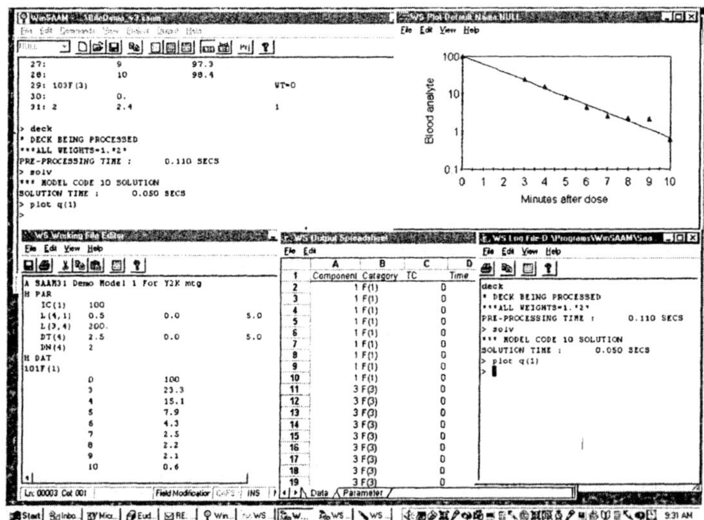

Figure 1. Windows for modeling utilities in the WinSAAM program: Terminal Window (upper left), Plot Window (upper right), File Editor (lower left), Output Spreadsheet (lower middle), and Logging Window (lower right).

model pools, initial estimates of transfer parameters, index numbers for accessing stored data, and similar information.

Because many problems in physiological kinetics are governed by first order linear differential equations, WinSAAM has been written to make entry of such problems very easy. A model's structure is defined by listing transfer coefficients characterizing the model in the section of the input file entitled H PAR (for *parameters*). The listing of transfer parameters defines all compartment connections in the model as well as compartments comprising the system. Transfer parameters are given the syntax L(i,j), where i is the index number of the recipient compartment and j is the index number of the donor compartment. The H PAR section also includes estimates of parameter values and minimum and maximum values for adjustable parameters.

Experimental data are entered directly into the file that defines the model structure. In a section entitled H DAT, data values are associated with time points, data weighting is specified, and indexing for data retrieval is defined. Data may be entered into the input file in several ways through the terminal window, the text input window, or the spreadsheet window. Entering data through the terminal window is achieved by special tabulation facilities. Data may be entered easily in the text window using pre-configured tabulation, by inserting text from tab-delimited files, or by pasting tab-delimited data, such as that copied from a spreadsheet. Using standard spreadsheet data entry techniques, data may also be entered through the spreadsheet window. The versatility by which data may be entered makes data entry easy for many data formats.

An example model and input file (or "deck") is shown in Figure 2. In this example taken from the MLAB User's Guide (1997), a dose of a radioactive molecule is injected into the blood stream of a rat. From the blood stream, the radioactive molecule appears in the rat's bile. Figure 2A shows a blood compartment (compartment 1) into which the radioactivity is injected and a compartment for bile (compartment 3). Plotting of the data suggests that a delay exists before the radioactivity enters the bile; therefore a delay compartment separates compartments 1 and 3. The WinSAAM input file representing this model is shown in Figure 2B. In the first section, entitled H PAR, the initial condition of the radioisotope dose in the blood [IC(1)] is set to 100%, and the initial values of rate parameters are estimated with lower and upper limits. In the section entitled H DAT, the observed radioactivity over time is entered for compartment 1 [blood, designated F(1)] and compartment 3 [bile, designated F(3)]. As seen from this example, WinSAAM allows entry of a first order linear model with very little typing by the modeler. Simply listing the rate constants informs WinSAAM of the existence of model compartments, compartmental connectivity, and equations defining transfer of material between compartments.

WinSAAM easily handles more complex models and models which are not governed by first order kinetics, as seen in the example in Figure 3. Figure 3 shows the classic model of the rumen by France, Thornley, and Beever (1982). In Figure 3A, the model diagram is shown with a description of parameters. One can see that this model encompasses many parameters and processes. Figure 3B shows this model in WinSAAM syntax. WinSAAM handles the equations easily, and one can see that the equation section of the model is fairly brief.

COMPILING, SOLVING, AND FITTING

Compiling, solving, and fitting are achieved by entering WinSAAM commands through the terminal window. These commands may be entered at the terminal window prompt, and common commands may also be accessed through the terminal window command menu list. The *deck* command compiles the input file and makes it available for solution through the *solv* (solve) command. Solutions for specific compartments may be retrieved if their index number has been included as a category item on the component definition line (as shown in the example input file for the rumen model in Figure 3).

Fitting of the model prediction to experimental data is performed through the *iter* (iterate) command. When this command is invoked, the WinSAAM program iteratively adjusts the model parameters to reduce the sum of squares between the experimental data and the model prediction.

MODEL ASSESSMENT

The modeler monitors the fitting procedure through messages received in the terminal window. Following iteration, available output includes the percent reduction in the sum of squares, identification of the parameter most altered by the fitting process, and the percent change in the parameter most affected by the iteration (Figure 4). Comparison of the model solution to experimental data can be achieved in several ways. The command *prin* (print) causes calculated values, observed values, and their ratios to be listed with their time values in the terminal window (Figure 5). The command *plot* produces a graphics window to appear with the specified compartment values as a

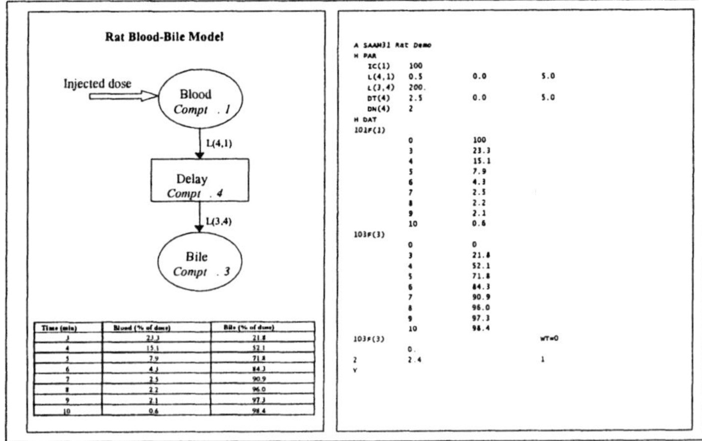

Figure 2. An example model diagram (left) and corresponding WinSAAM input file (right). The model represents an experiment in which a radioactive compound is injected into the blood stream of a rat, and the appearance of radioactivity is subsequently measured in bile.

3A

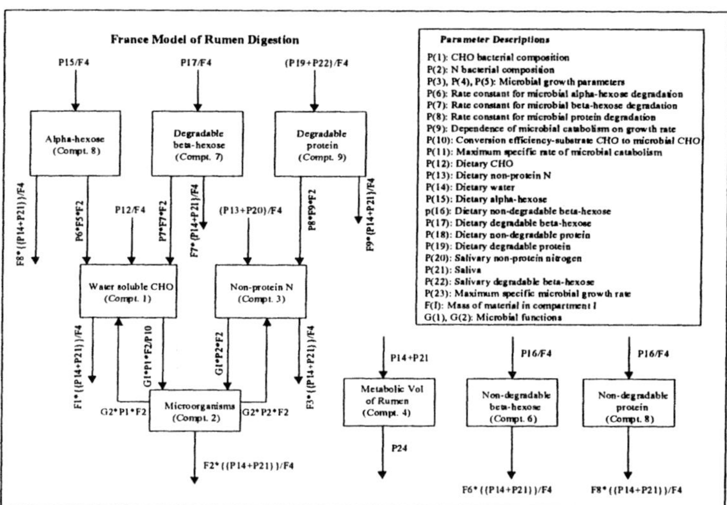

3B

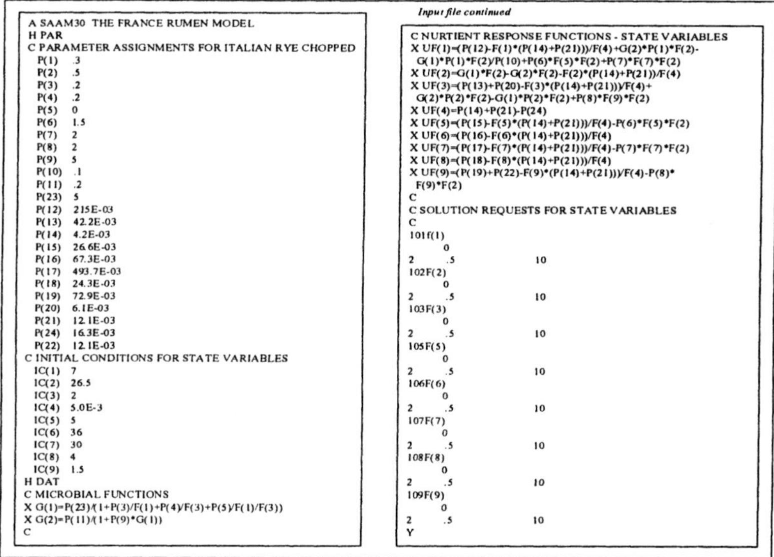

```
A SAAM30  THE FRANCE RUMEN MODEL
H PAR
C PARAMETER ASSIGNMENTS FOR ITALIAN RYE CHOPPED
   P(1)    .3
   P(2)    .5
   P(3)    .2
   P(4)    .2
   P(5)    0
   P(6)    1.5
   P(7)    2
   P(8)    2
   P(9)    5
   P(10)   .1
   P(11)   .2
   P(23)   5
   P(12)   215E-03
   P(13)   42.2E-03
   P(14)   4.2E-03
   P(15)   26.6E-03
   P(16)   67.3E-03
   P(17)   493.7E-03
   P(18)   24.3E-03
   P(19)   72.9E-03
   P(20)   6.1E-03
   P(21)   12.1E-03
   P(24)   16.3E-03
   P(22)   12.1E-03
C INITIAL CONDITIONS FOR STATE VARIABLES
   IC(1)   7
   IC(2)   26.5
   IC(3)   2
   IC(4)   5.0E-3
   IC(5)   5
   IC(6)   36
   IC(7)   30
   IC(8)   4
   IC(9)   1.5
H DAT
C MICROBIAL FUNCTIONS
X G(1)=P(23)/(1+P(3)/F(1)+P(4)/F(3)+P(5)/F(1)/F(3))
X G(2)=P(11)/(1+P(9)*G(1))
C

Input file continued

C NURTIENT RESPONSE FUNCTIONS - STATE VARIABLES
X UF(1)=(P(12)-F(1)*(P(14)+P(21)))/F(4)+G(2)*P(1)*F(2)-
  G(1)*P(1)*F(2)/P(10)+P(6)*F(5)*F(2)+P(7)*F(7)*F(2)
X UF(2)=G(1)*F(2)-G(2)*F(2)-F(2)*(P(14)+P(21))/F(4)
X UF(3)=(P(13)+P(20)-F(3)*(P(14)+P(21)))/F(4)+
  G(2)*P(2)*F(2)-G(1)*P(2)*F(2)+P(8)*F(9)*F(2)
X UF(4)=P(14)+P(21)-P(24)
X UF(5)=(P(15)-F(5)*(P(14)+P(21)))/F(4)-P(6)*F(5)*F(2)
X UF(6)=(P(16)-F(6)*(P(14)+P(21)))/F(4)
X UF(7)=(P(17)-F(7)*(P(14)+P(21)))/F(4)-P(7)*F(7)*F(2)
X UF(8)=(P(18)-F(8)*(P(14)+P(21)))/F(4)
X UF(9)=(P(19)+P(22)-F(9)*(P(14)+P(21)))/F(4)-P(8)*
  F(9)*F(2)
C
C SOLUTION REQUESTS FOR STATE VARIABLES
C
101f(1)
        0
2       .5              10
102F(2)
        0
2       .5              10
103F(3)
        0
2       .5              10
105F(5)
        0
2       .5              10
106F(6)
        0
2       .5              10
107F(7)
        0
2       .5              10
108F(8)
        0
2       .5              10
109F(9)
        0
2       .5              10
Y
```

Figure 3. (A) A model diagram and parameter list for the France model of rumen digestion and (B) the corresponding WinSAAM input file.

function of time (Figure 6). A nice graphics tool is included in the WinSAAM software to facilitate a wide variety of adjustments to the graphical display. Finally, the spreadsheet window also stores solution values.

Other aspects of the modeling process can be monitored easily in WinSAAM. The modeler may issue the *part* command to view the partials matrix for assessment of sensitivity. Identifiability of parameters is readily available by issuing the command *fsd(i)*, which returns fractional standard deviation values for adjustable parameters. This list includes only a few of the tools in WinSAAM by which models may be assessed.

PROJECT MANAGEMENT

A fairly new feature of the WinSAAM package is the Project Manager. This tool allows the modeler to investigate the distribution of parameter values associated with a population of individuals, rather than just the individuals themselves. This application, which is embedded in the WinSAAM software, is called the Extended Multiple Studies Analysis (EMSA) tool. With EMSA, parameter estimates for individuals are first obtained. The optimized parameter estimates and the data for individuals are then concatenated and processed with EMSA. The individual data are reanalyzed to yield improved parameter and variance-covariance estimates. The analysis continues iteratively until the conditions are satisfied for a convergent maximum likelihood solution, yielding final estimates for the population mean and the variance-covariance matrix.

EMSA is an important tool in modeling because pooling of parameter estimates to calculate simple means and standard deviations does not provide an accurate description of the population parameters. The EMSA tool is invoked through the Project pull-down menu. A box appears, as shown in Figure 7, which allows the modeler to attach files to or detach files from the project analysis, edit the population project, or run the population analysis. The output data from the population analysis are available via a batch output window or via the WinSAAM spreadsheet tool, as shown in Figure 8.

CONCLUSION

WinSAAM's many strengths contribute to the ease and power with which compartmental modeling studies may be performed. WinSAAM allows complex models to be entered and explored with ease. The absence of a need to explicitly specify differential equations for a first-order model makes entering most linear problems quick and easy. The ability to support a set of highly refined modeling constructs, such as delays, allows WinSAAM to meet the needs of very complex modeling projects. The support of commands entered via command line in the terminal window or via menus in the Windows graphical user interface meet the variety of style preferences that modelers may have in interacting with a computer. And WinSAAM is continually evolving to expand its capabilities.

To learn more about WinSAAM, the reader is encouraged to refer to the book *Investigating Biological Systems Using Modeling: Strategies and Software* (Wastney et al., 1999). This text describes capabilities of WinSAAM as well as basic modeling issues in detail. Further, the reader is encouraged to download a free copy of the WinSAAM software from http://www-saam.nci.nih.gov. WinSAAM is a powerful, elegant, and exciting tool which is a wonderful enhancement to modeling investigations.

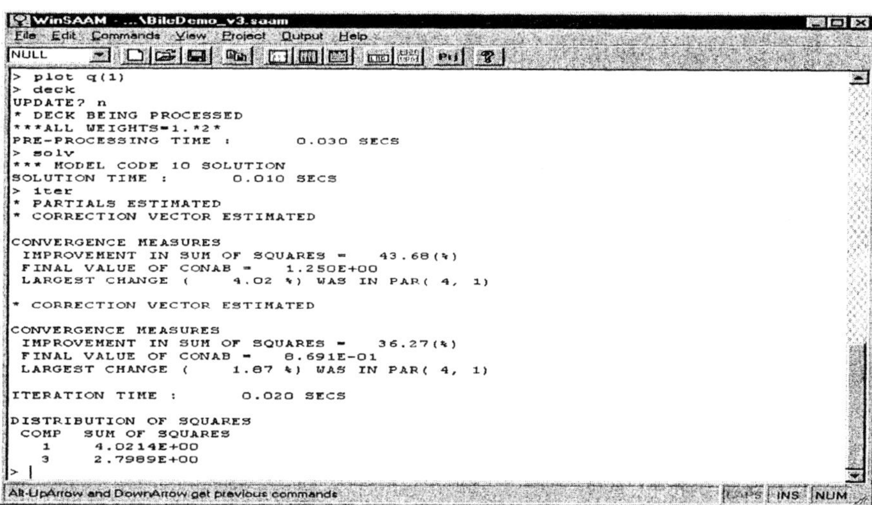

Figure 4. A WinSAAM terminal window displaying program messages following an *iterate* command. Output includes percent reduction in the sum of squares, identification of the parameter most affected by the fitting process, and the percent change in the parameter most affected by the iteration.

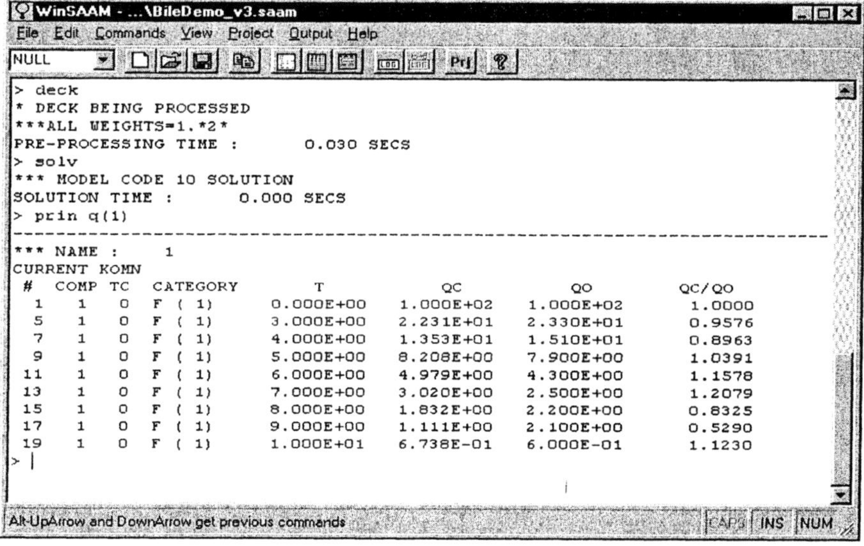

Figure 5. An example of output generated in the WinSAAM terminal window by the *prin* (print) command. The first several columns identify the observation number; T is time, QC is the calculated values, QO is observed values, and QC/QO is the ratio of calculated to observed values.

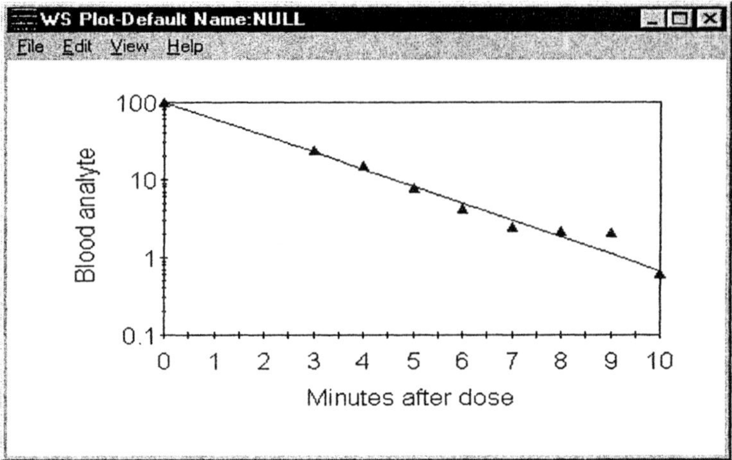

Figure 6. A plot of observed and calculated data as a function of time and generated by WinSAAM for the rat bile model shown in Figure 2.

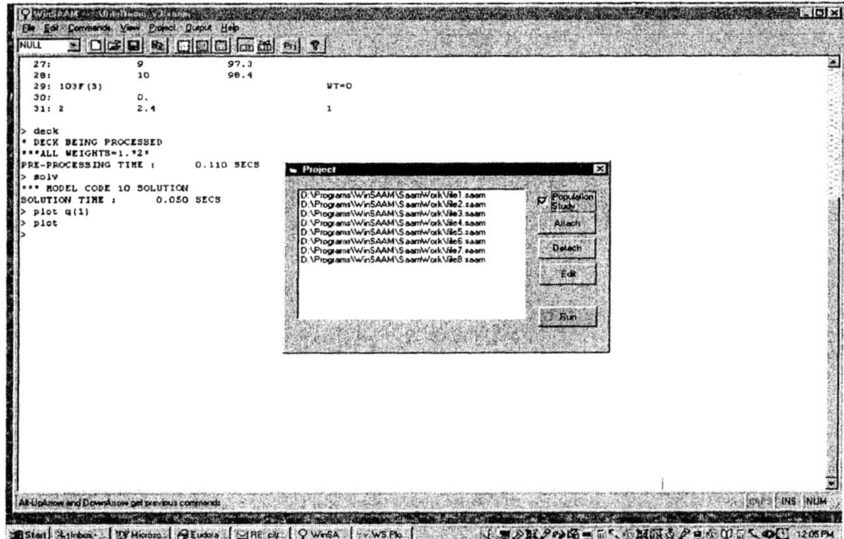

Figure 7. The WinSAAM dialogue box for entering population studies. Files for population analysis are added by clicking on the *attach* button, and the population analysis is performed by clicking the *run* button.

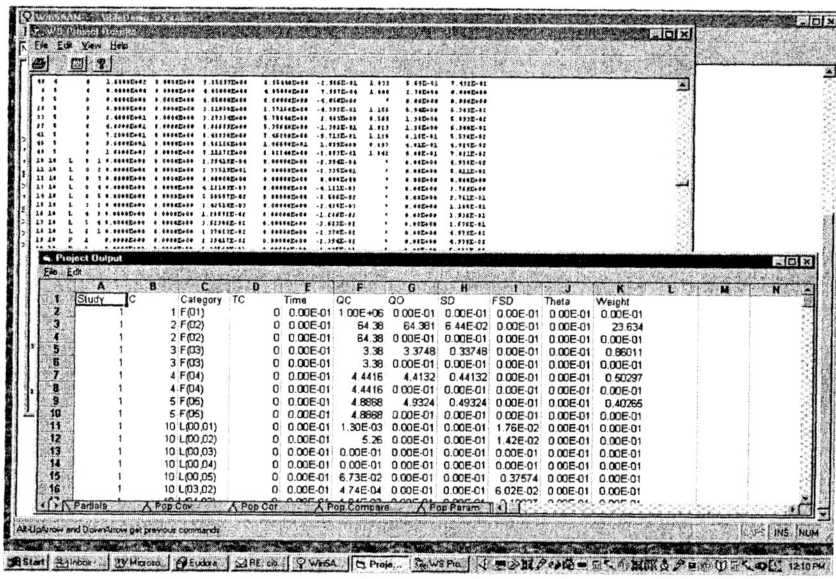

Figure 8. An example of output from a population modeling study performed using WinSAAM and displayed using the WinSAAM spreadsheet tool.

CORRESPONDENCE

Please address all correspondence to:
Janet A. Novotny
USDA, BHNRC, DHPL
Beltsville Human Nutrition Research Center
Building 308, Room 201 BARC-East
Beltsville, MD 20705
novotny@bhnrc.arsusda.gov

REFERENCES

Berman, M., and Weiss, M.F., 1978, *SAAM Manual,* DHEW Publ. (NIH) 78-180, U.S. Government Printing Office, Washington, DC.

Boston, R.C., Greif, P.C., and Berman, M., 1981, Conversational SAAM-an interactive program for the kinetic analysis of biological systems, *Comput. Prog. Biomed.* 13:111-119.

France, J., Thornley, J.H.M., and Beever, D.E., 1982, A mathematical model of the rumen, *J. Agric. Sci. Camb.* 99:343-353.

MLAB User's Guide, p. 316, 1997, Civilized Software, Inc., Bethesda.

Wastney, M.E., Patterson, B.H., Linares, O.A., Grief, P.C., and Boston, R.C., 1999, *Investigating Biological Systems Using Modeling: Strategies and Software,* Academic Press, New York.

STATA: A STATISTICAL ANALYSIS SYSTEM FOR EXAMINING BIOMEDICAL DATA

Ray C. Boston and Anne E. Sumner[*]

INTRODUCTION

STATA is a statistical analysis program developed by STATA Corporation, College Station, TX. In addition to supporting a very extensive array of statistical tools, STATA incorporates a critical complement of data management and graphics facilities as well. As demonstrated in another article in this volume (Boston et al., 2002), STATA fits very well into the kinetic analysis suite of tools. In that paper, we show how some preliminary analyses of intravenous tolerance test (IVGTT) data leading to estimates of key parameters and their errors in the minimal model (see Bergman and Bowden, 1981) could easily be performed with STATA. We also demonstrate the transfer of some WinSAAM Population Management Results from WinSAAM to STATA for further appraisal as well as for graphical display. Here we will present aspects of the organization, operation, use, and availability of STATA.

USING STATA – AN INTRODUCTION

STATA, like WinSAAM, is a command-line interface (CLI)-driven system. This implies that STATA users interact with its 'processing machinery' by typing commands in the command window. The command is executed as soon as the 'enter' or 'carriage return' button is pressed.

Before we give the full syntax of STATA commands, let's look at several commands and their responses. The command

---

[*] Ray C. Boston, Department of Clinical Studies, School of Veterinary Medicine, University of Pennsylvania, Kennett Square, PA 19348. Anne E. Sumner, Diabetes Branch, National Institute of Diabetes and Digestive and Kidney Disease, National Institutes of Health, Bethesda, MD 20892.

```
use "STATA Demo_A-001
```

causes the saved STATA dataset, 'STATA Demo_A-001.dta' to be read into memory, providing access for the user to the variables and observations it contains. First, let us ask STATA to describe (de) the data set. We see that it comprises 19 observations on three variables.

```
. de

Contains data from STATA Demo_A-001.dta
  obs:              19
  vars:              3                             10 Jun 2000 11:49
  size:            266 (99.9% of memory free)
-------------------------------------------------------------------
1. time       int    %8.0g
2. glc         float  %9.0g
3. ins         float  %9.0g
-------------------------------------------------------------------
Sorted by:
```

Next let's summarize (su) the dataset.

```
. su

Variable|  Obs       Mean       Std. Dev.       Min          Max
--------+----------------------------------------------------------
   time|   19    38.31579      50.51188         0            180
    glc |   19   144.6579       54.7748         58           234
    ins |   19   203.8579      144.9388         14           452
```

Note that, in the two commands used so far, we've simply typed 'de' for describe and 'su' for summarize. STATA permits abbreviation of non-destructive commands. Indeed, any non-destructive command can be shortened to its simplest unique character pattern.

Now let's plot the data. In this dataset, 'time' denotes time, 'glc' denotes plasma glucose concentration ($mg \cdot L^{-1}$), and 'ins' denotes plasma insulin concentration ($mU \cdot L^{-1}$).

```
. gra glc ins time, c(ll) s(po) xlabel ylabel rlabel res
```

In Figure 1, we see that this command has produced separately-scaled profiles of glucose level and insulin level *vs.* time. A feature of STATA's graphics is its very flexible approach, in that any useful data display seems to be producible. For example, we may wish to see a hysteresis plot of glucose concentration against insulin concentration (Figure 2). Derendorf and Hochhaus (1995) suggest that clockwise patterns of such plots may be associated with pharmacodynamic (PD) tolerance.

```
. gra glc ins, c(l) s([time]) ylabel xlabel(0 100 200 300 400 500)
            border
```

In Figure 2, we used the value of the plot parameter 'time' as the symbol at each plot point.

Two recurring operations needed for the investigation of the glucose-insulin system are differentiation and integration. Specifically, the time derivative of glucose is used, in part, to explore its net disposal, and the time integral of insulin is used to examine insulin

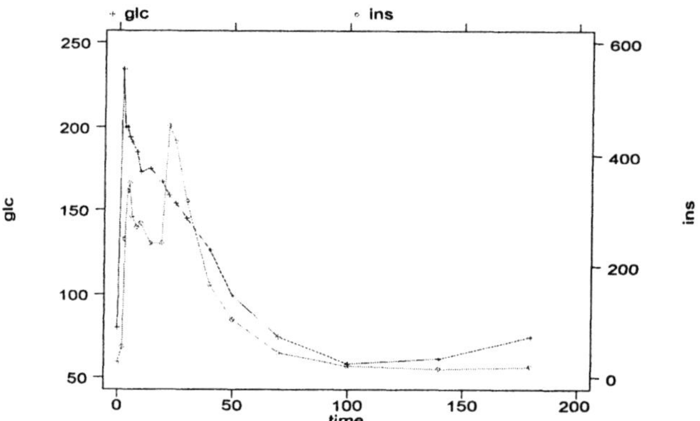

Figure 1. A plot of insulin (o) and glucose profiles (+) following an IVGTT. Note the rescaling that exposes the structural details of each response as well as the left and right axis labeling for glucose (mg•dL^{-1}) and insulin (mU•L^{-1}), respectively.

response. The integral of the insulin profile, above basal, with time for the first 10 minutes is in fact referred to as the acute insulin response to glucose. We will now generate the appropriate variables.

```
. dydx glc time, gen(gp)
. gen iab=ins-ins[1]
. integ iab time, gen(iins)
```

In Figure 3, we present a time profile of the area under the insulin profile above basal.

It would be helpful in the glucose response analysis to know when the glucose level is falling at its fastest rate and also when the fractional disposal rate ('gd') is greatest.

```
. quietly su gp, detail
```

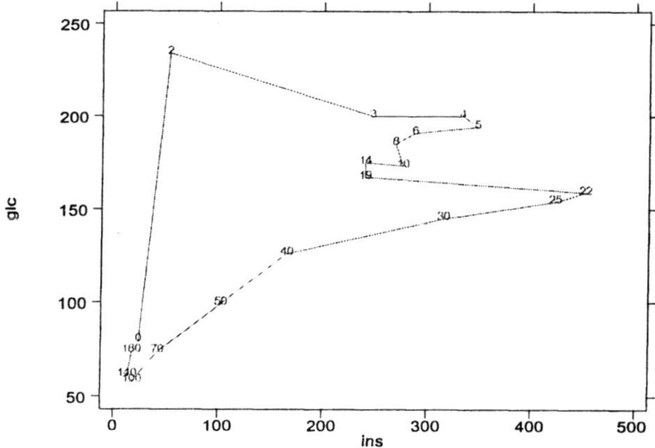

Figure 2. A hysteresis plot of glucose *vs.* insulin. The hysteresis parameter is time, and the numbers lying along the hysteresis coincide with time after the IVGTT. Other units are the same as in Figure 1.

```
. scalar def gpmax=r(min)
. li time if gp==gpmax

          time
3.          3
```

For the fractional disposal rate, we proceed as follows:

```
. gen gd=gp/glc
. quietly su gd, det
. scalar gdmax=r(min)
. li time if gd==gdmax

          time
3.          3
```

We see that 'gp' and 'gd' both fall fastest at the same time point for this observation set. To generate a STATA variable containing this value, we would enter

```
. gen tmax=time if gd==gdmax

. sort tmax

. replace tmax=tmax[1] if tmax==.
```

Of course, all of our tmax values are the same since there is only one time point coinciding with the highest fractional glucose disposal rate.

It would be helpful to see plots of 'gp' and 'gd' *vs.* time (Figure 4).

```
. gra gd gp time if time>1, c(ll) s(OT) xlabel ylabel sort res rlabel
```

It may also be interesting to see a plot of gd *vs.* integrated insulin, with a horizontal line coinciding with a fractional disposal rate of 0.024 min^{-1} (Figure 5). This is the value estimated by Ader et al. (1985) for S_G (the glucose-mediated glucose disposal; i.e., the

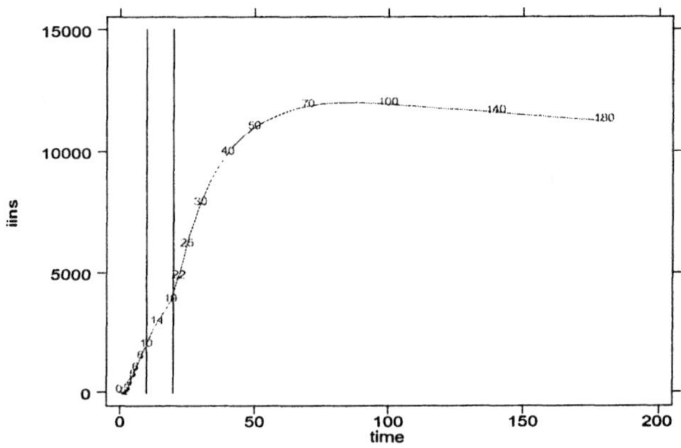

Figure 3. A plot of the area under insulin above basal (mU• L^{-1}•min) *vs.* time following the start of the IVGTT. The two vertical lines coincide with the 10 and 20 minute time points. The area under the plasma insulin above basal at the 10 minute point is known as the 'acute insulin response to glucose.'

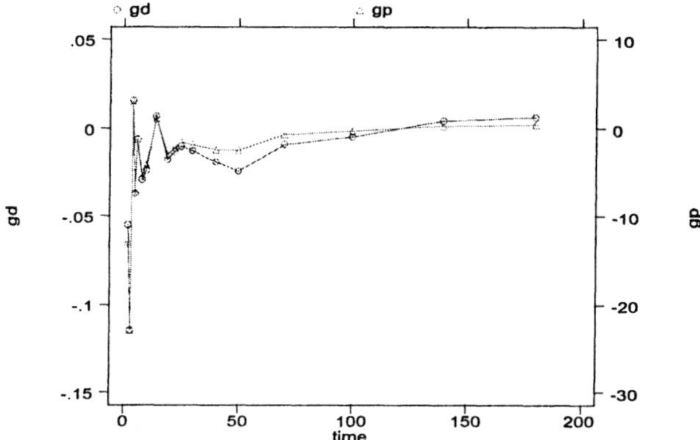

Figure 4. A plot of the rate (gp, mg dL^{-1} min^{-1}) (Δ) and fractional rate (gd, min^{-1}) (o) of glucose disposal during the IVGTT. Note that we have rescaled each plot to expose details appropriately.

average value of gd during the glucose disposal phase when glucose *per se* dominates disposal).

```
. gra gd iins if time>2, c(l) sort s([ti]) yline(-.024) ylabel xlabel
```

As another example of the use of STATA to assist with the management and investigation of biokinetic data, we will examine some antibacterial drug kinetics. The population dataset can be described as follows:

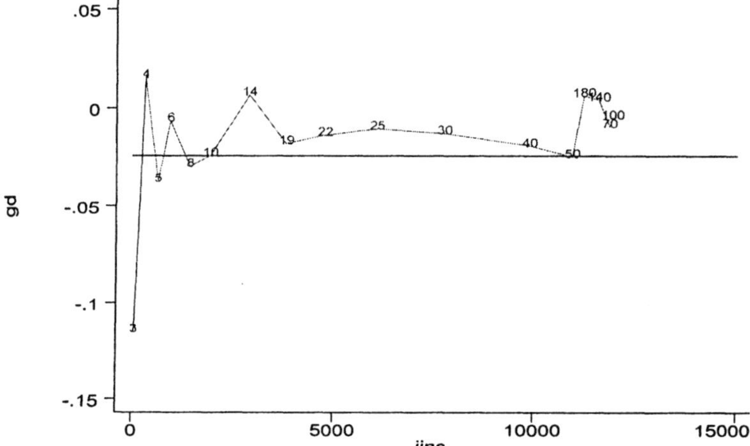

Figure 5. A plot of the fractional rate of glucose disposal (min^{-1}) against the area under insulin above basal (mU l^{-1} .min). The horizontal line represents a glucose disposal rate of 2.4% min^{-1}.

```
. de

Contains data from C:\Talks\ Chapter\Mid Dose population Cefotax.dta
  obs:        25
  vars:        5                                          12 Jun 2000 15:36
  size:       875 (99.8% of memory free)
-------------------------------------------------------------------------------
    1. Paramete      str8     %9s                Parameter
    2. Study         byte     %8.0g              Study
    3. Studynam      str14    %14s               Study Name
    4. Studyval      float    %9.0g              Study Value
    5. Popmean       float    %9.0g              Pop. Mean
-------------------------------------------------------------------------------
Sorted by:
```

We can explore the pattern of variation of the various parameters by replicate as follows:

```
. encode par, gen(epar)
. byvar epar
-> epar==K(01)
    Variable |     Obs        Mean     Std. Dev.        Min         Max
-------------+---------------------------------------------------------------
    studyval |       5    .0000103      2.37e-06    7.86e-06    .0000138
-> epar==K(03)

    Variable |     Obs        Mean     Std. Dev.        Min         Max
-------------+---------------------------------------------------------------
    studyval |       5    .0000189      6.01e-06    .0000117    .0000265
-> epar==L(00,01)

    Variable |     Obs        Mean     Std. Dev.        Min         Max
-------------+---------------------------------------------------------------
    studyval |       5     .03766      .0048258       .0345        .046
-> epar==L(01,03)

    Variable |     Obs        Mean     Std. Dev.        Min         Max
-------------+---------------------------------------------------------------
    studyval |       5    .013364      .0068843      .00651       .0233
-> epar==L(03,01)
    Variable |     Obs        Mean     Std. Dev.        Min         Max
-------------+---------------------------------------------------------------
    studyval |       5    .012646      .0046552      .00583       .0172
```

We can clearly sort the observations and explore the regularity of the position of each subject (represented by the variable 'studyna' here). If a specific subject remains in the same extreme sorted position for each parameter, we would be concerned.

```
. sort par studyval

. li par studyna studyval
```

```
          Paramete          Studynam    Studyval
   1.       K(01)             B_MID      7.86e-06
   2.       K(01)             M_MID      8.34e-06
   3.       K(01)             D_MID      .0000104
   4.       K(01)             C_MID      .000011
   5.       K(01)             W_MID      .0000138
   6.       K(03)             B_MID      .0000117
   7.       K(03)             M_MID      .0000154
   8.       K(03)             D_MID      .0000174
   9.       K(03)             C_MID      .0000234
  10.       K(03)             W_MID      .0000265
```

```
11.  L(00,01)          D_MID    .0345
12.  L(00,01)          W_MID    .0346
13.  L(00,01)          C_MID    .0356
14.  L(00,01)          B_MID    .0376
15.  L(00,01)          M_MID    .046
16.  L(01,03)          D_MID    .00651
17.  L(01,03)          W_MID    .00721
18.  L(01,03)          B_MID    .0141
19.  L(01,03)          C_MID    .0157
20.  L(01,03)          M_MID    .0233
21.  L(03,01)          W_MID    .00583
22.  L(03,01)          D_MID    .0102
23.  L(03,01)          C_MID    .0139
24.  L(03,01)          M_MID    .0161
25.  L(03,01)          B_MID    .0172
```

We can explore correlations among the kinetic parameters after making the data into wide form using the reshape function as follows:

```
. restore
. gen v=studyval
. encode param, gen(pp) label(name)
. keep pp v study
. reshape wide v, i(study) j(pp)

. de
Contains data
  obs:                5
  vars:               6
  size:             125 (94.6% of memory free)
-------------------------------------------------------------------------
  1. study     byte    %8.0g                   Study
  2. v1        float   %9.0g                   1 v
  3. v2        float   %9.0g                   2 v
  4. v3        float   %9.0g                   3 v
  5. v4        float   %9.0g                   4 v
  6. v5        float   %9.0g                   5 v
-------------------------------------------------------------------------
Sorted by:  study
. pwcorr v*, star(.05)


             |      v1       v2       v3       v4       v5
--------+------------------------------------------------------
      v1 |   1.0000
      v2 |   0.9471*   1.0000
      v3 |  -0.6256   -0.4876    1.0000
      v4 |  -0.6403   -0.3993    0.8917*   1.0000
      v5 |  -0.9299*  -0.7810    0.6069    0.7747   1.0000
```

Here v1 denotes K(1), v2 denotes K(2), v3 denotes L(0,1), v4 denotes L(1,3), and v5 denotes L(3,1), where the Ks and Ls denote parameters of the kinetic model. We have asked STATA to insert an asterisk in this table beside any pairwise correlation estimate that is significantly different from zero at the 5% level.

THE STATA PROCESSING MACHINERY

STATA is extremely fast, conservative of computer space, and conservative of user input (that is, we often get a lot accomplished for relatively little input). These advantages derive, at least partially, from the ingenious file-structured nature of STATA. STATA may be thought of as two distinct entities: a processing kernel which manages all the frequently used STATA commands and features plus a vast array of 'ado' ('automated do') files which manage the remainder of STATA's functions.

The 'ado' files are just one of the file structures supported by STATA. Users can write their own 'do' files (and indeed, this is actually the only appropriate way to maintain a record of one's work), they can save data sets in a variety of ways in addition to STATA's dataset file format, and they can save plots as well in graphic file format (gph format).

In addition to assignment and replacement operations, STATA supports a powerful array of flow control services. At any time in the command line, for example, we can initiate looped analysis using either the 'while' or the 'for' flow controls. Furthermore, in our 'do' and 'ado' files, we can use selection (if), as needed, for flow control.

The full syntax of the STATA commands is as follows:

```
[by varlist:] command [varlist]
              [if] [in] [weights] [expression], [options]
```

Here items in the square brackets ([ ]) are optional for some of the commands and for the program when it is in an appropriate state. It is important to note that the comma (,) separates the command and its arguments from the command options in the command line. Because the STATA language is continually being expanded, the syntax is apt to be extended at any point in time.

Let's explore some additional examples which illustrate this: first we'll summarize plasma glucose for time points on either side of t = 20 minutes (i.e., for time < 20 and also for time > 20).

```
. gen x=time<20
. sort x
. by x: su glc

-> x=         0
Variable|   Obs      Mean        Std. Dev.      Min       Max
--------+-----------------------------------------------------
   glc  |    9      105.5556     41.01558       58        159

-> x=         1
Variable |  Obs      Mean        Std. Dev.      Min       Max
---------+-----------------------------------------------------
   glc   |  10      179.85       40.00142       79.5      234
```

Occasionally, we run into trouble with the 'by' sub-group selector, in that it may not function with a particular command. This doesn't mean that something is in error, but rather it implies that we've hit a functional limitation of 'by' as implemented by STATA. Not to worry! To circumvent this (and remember that we said that STATA was continually being enhanced by the very capable user community), we can use Royston's 'byvar' selector. Royston, a British biostatistician has, over the years, made available some very ingenious and helpful tools to facilitate the use of STATA. The following command, which robustly regresses the natural log of plasma glucose on time for the two time zones just introduced, does not work with 'by' but functions beautifully with 'byvar.'

```
. gen lgc=ln(glc)
. byvar x, tab return b(time) se(time) e(r2 F) pause: rreg lgc time

x          |    e(r2)         e(F)          _b[time]        _se[time]
-----------+-----------------------------------------------------------
0          |    .81107857     25.759235     -.00954594       .00188084
1          |    .93478607     100.33903     -.01148366       .00114642
```

You will find that, with graph commands (see above), you often need to use many options to generate the plot you desire. Glance back at our introductory demonstrations.

The following pointers about STATA will assist you if you're considering taking it on board.

- All commands are entered in lower case.
- STATA is case sensitive.
- Variables may be abbreviated in the same fashion as non-destructive commands.
- You may use uppercase for variable names but, once chosen, the capitalized part needs to be entered as such if it is essential.
- The command line terminator is, by default, a carriage return, and the command/ options separator is a comma.
- You may use the PgUp and PgDn keys to retrieve and re-execute past commands.
- Past commands can be edited in the usual way in the command window.
- Variables for use in commands can be retrieved from the variables windows.
- If your job is too large for the current memory assignment, you can increase the allocation, but you need to appreciate that abusing memory can seriously degrade performance.
- If you need to be able to reproduce your analysis of data for some reason or another, employ a 'do' file.

RELEASE VERSIONS AND UPDATES OF STATA

New STATA releases become available for purchase from STATA approximately every three years. However, within a release, updates are always available on the Internet, free of cost, and it pays to keep abreast of developments there. Whereas the new STATA releases and updates come from STATA Corp., the STATA user community makes very smart 'ados' (see 'byvar' above) available about every second month for general improvements to STATA within a release. Furthermore, these user updates can be acquired over the Internet or simply explored in the STB (STATA Technical Bulletin) which also comes out about every second month.

STATA has an array of very efficient tools which assist with maintenance of the product. The downloads of updates to either the kernel or the 'ados' always run smoothly. It is possible also to download the Technical Bulletin 'ados,' either so that they are immediately available, as though a basic part of STATA, or to download them and simply extract features of special interest to you. We elect to use the latter updating procedure.

USING STATA – AN ADVANCED DEMONSTRATION

For our advanced demonstration of STATA, aimed at highlighting its utility to kinetic modeling, we take as our topic 'finding good initial guesses for the glucose effectiveness parameter *apropos* of an IVGTT.' Furthermore, we will demonstrate how an exploration of any bias of that estimate might lead to an automated means for improving such an initial estimate, possibly even providing an additional index of insulin action.

Here is the problem. In the early stages of glucose response following a glucose injection into plasma, glucose is primarily responsible for its own removal. Indeed, once glucose is uniformly distributed in plasma, say about 10 minutes after injection, the fractional rate of glucose disposal is considered a good approximation to S_G (glucose-mediated glucose disposal). What we would like to do here is to see just how close this number is to S_G and what other measured features associated with the IVGTT may help explain any bias between S_G and the mentioned slope (let's call the latter $S_{G,est}$).

Here is the approach to the problem using STATA. The data for this analysis can be visualized in two forms, one long and one wide. The long form is the concatenated glucose and insulin time profiles for each subject, stacked end to end. The actual variables in this dataset are subject (actually subject_id), time, glucose, insulin, and free fatty acid (ffa). Note that the subject variable has repeated values along an entire subject entry. The wide form is the demographic/clinical dataset in which important measurements and determinations are stored for each subject.

There are three analysis steps. First, the long dataset is analyzed for a 'first time' and the $S_{G,est}$ values, as well as G0 (the level of glucose in plasma when its concentration is uniform throughout plasma) and the uncertainties of each of these, are estimated and stored with the subject IDs in a STATA dataset file. Second, the long dataset is examined a second time to produce other subject level determinations. This second set of determinations has the property that they rely on arithmetic steps, as opposed to statistical steps (c.f. step 1). The determinations here are summarized in Table 1. These results are saved in a STATA dataset file. Finally, using sequences of sorted, keyed ('subject') merges, the three datasets are merged as one for analysis. In Figure 6 A-C, we present the STATA instruction sequence needed for this analysis.

One of the determinations in the third dataset is the S_G values produced by the minimal model analysis of IVGTT data using Bergman's MINMOD program. We will use this as a basis for isolating the bias of our $S_{G,est}$ values from the STATA analysis. If we robustly regress our $S_{G,est}$ on MINMOD S_G (respectively St_sg and sg in the STATA dataset), we obtain

Table 1. STATA determinations from IVGTT data in which insulin, glucose, and free fatty acids profiles are monitored

Parameter	Measurement Implicated	Meaning
AIRG	Insulin	Acute Insulin Response to Glucose
Ib	Insulin	Basal Insulin
Gb	Glucose	Basal Glucose
Fb	FFA	Basal FFA
Imax10	Insulin	Phase I Insulin peak
Imax25	Insulin	Phase II Insulin peak
Mit1	Insulin	Time of Phase I Peak
Mit2	Insulin	Time of Phase II Peak
Gmax	Glucose	Peak Glucose
Tgmax	Glucose	Time of glucose peak
Gmin	Glucose	NADIR of glucose
Tgmin	Glucose	Time of glucose NADIR
Tcross	Glucose	Time glucose crosses Gb
Gpmin	Glucose	Maximum glucose disposal rate
Tgpmin	Glucose	Time of maximum glucose disposal rate
Fpmin	FFA	Maximum FFA disposal rate
Tfpmin	FFA	Time of maximum FFA disposal rate
Gdmin	Glucose	Maximum fractional glucose disposal rate
Fdmin	FFA	Maximum fractional FFA disposal rate

```
. rreg St_sg sg
Robust regression estimates                      Number of obs =       70
                                                 F(  1,  68)   =   67.88
                                                 Prob > F      = 0.0000

--------------------------------------------------------------------------
St_sg |  Coef.     Std. Err.    t      P>|t|     [95% Conf. Interval]
----- +-------------------------------------------------------------------
   sg | .8499373   .1031587   8.239    0.000    .6440872      1.055787
_cons | .0045228   .0020468   2.210    0.030    .0004386      .0086071
--------------------------------------------------------------------------
```

This suggests that our STATA estimates of S_G may be initially 0.005 min^{-1} larger than the MINMOD S_{GS} but, as the MINMOD predicted value increases, the STATA value increases at only 85% of the rate. The values actually are in agreement at 0.03 min^{-1}. From this point on, the MINMOD S_G values exceed the STATA S_G values. This is not too bad since the estimated population value for S_G given by Ader et al. (1985) is ~0.024 min^{-1} (see above).

How can we examine this departure further? We could attempt to explain the difference in terms of the aspects of glucose present in the MINMOD analysis but omitted from the STATA analysis; specifically, insulin action. Let's see if the acute insulin response to glucose (AIRg) is a culprit. First, we form the difference between S_G and S_{Gest} and then regress this difference on AIRg robustly.

```
clear
set mem 32m
use "all.dta"
postfile sg_g0 subint sg sesg g0 seg0 using sg_g0, replace
set more off
encode subj, gen(esub)
sort subj
dydx glc ti if ti<=30, gen(gp) by(subj)
* sort subj time
* quietly by subject: gen gd=gp/(glc-glc[1]) if time>0
gen gd=gp/glc
sort subj
su esub
local s=r(max)
gen z=.
quietly {
#delimit ;
for num 1/`s':
    preserve \
    drop if esub~=X \
    rreg gd if time>10 & time<=25, nolog \
    scalar define sg=_b[_cons] \
    scalar sesg=_se[_cons] \
    replace z=ln(glc)-sg*time \
    rreg z if time>10 & time<25, nolog \
    scalar g0=exp(_b[_cons]) \
    scalar seg0=g0*_se[_cons] \
    scalar sint=real(substr(subj[1], length(subj[1])-2,3)) \
    post sg_g0 sint sg sesg g0 seg0 \
    restore
    ;
# delimit cr
  }
postclose sg_g0

use sg_g0, clear
gen str6 subject="RF-0"+string(subint) if subint>9
replace subject="RF-00"+string(subint) if subint<=9
replace subject="RF-"+string(subint) if subint>99
drop subint
sort subject
save sg_g0, replace
use sg_g0, clear
li
su
use "all.dta", clear
sort subject
merge subject using sg_g0
sort subject
gen St_sg=-sg
gen St_fsdsg=-100*sesg/sg
gen St_g0=g0
gen St_fsdg0=100*seg0/g0
gen str6 Subject=subject
encode subject, gen(esub)
quietly byvar esub, generate b(time) se(time) return:  reg glc time if time>=23 &
time<=40
rename Btime_ St_sg40
replace St_sg40=-St_sg40
rename Stime_ St_se40
keep St_sg St_fsdsg St_g0 St_fsdg0  St_sg40 St_se40 Subject
collapse St_sg St_fsdsg St_g0 St_fsdg0 St_sg40 St_se40, by(Subject)
rename Subject subject
sort subject
save sg_g0, replace
```

```
use "all.dta", clear
quietly egen airg=sum(ins) if time<=10, by(subj)
sort subject time
quietly by subj: replace airg=airg[_n-1] if airg==.
quietly by subj: gen ib=ins[1]
qby subj: replace airg=airg-10*ib
quietly by subj: gen gb=glc[1]
quietly egen imax10=max(ins) if time<=10, by(subj)
quietly egen imax25=max(ins) if time>20, by(subj)
sort subj imax10
quietly by subj: replace imax10=imax10[1] if _n~=1
sort subj imax25
quietly by subj: replace imax25=imax25[1] if _n~=1
sort subj time
quietly by subj: gen mit1=time if ins==imax10
quietly by subj: gen mit2=time if ins==imax25
sort subj mit1
quietly by subj: replace mit1=mit1[1] if mit1==.
sort subj mit2
quietly by subj: replace mit2=mit2[1] if mit2==.
quietly egen gmax=max(glc), by(subj)
sort subj
quietly by subj: gen tgmax=time if glc==gmax
sort subj tgmax
quietly by subj: replace tgmax=tgmax[1] if _n~=1
quietly egen gmin=min(glc), by(subj)
sort subj
quietly by subj: gen tgmin=time if glc==gmin
sort subj tgmin
quietly by subj: replace tgmin=tgmin[1] if _n~=1
sort subj time
quietly by subj: gen tcross= time if glc<gb & time>5
sort subj tcross
quietly by subj: replace tcross=tcross[1] if _n~=1
dydx glc time, gen(gp) by(subject)
sort subj
qby subj: gen gd=gp/glc
quietly egen gpmin=min(gp) if time>10, by(subj)
quietly egen gdmin=min(gd) if time>10, by(subj)
sort subj time
quietly by subj: replace gpmin=gpmin[_N] if gpmin==.
qby subj: replace gdmin=gdmin[_N] if gdmin==.
sort subj
quietly by subj: gen tgpmin=time if gp==gpmin & gp~=.
sort subj tgpmin
quietly by subj:replace tgpmin=tgpmin[1] if _n~=1
dydx ffa time, gen(fp) by(subject)
sort subj
qby subj: gen fd=fp/ffa
quietly egen fpmin=min(fp) if time>10, by(subj)
quietly egen fdmin=min(fd) if time>10, by(subj)
sort subj time
quietly by subj: replace fpmin=fpmin[_N] if fpmin==.
qby subj: replace fdmin=fdmin[_N] if fdmin==.
sort subj
quietly by subj: gen tfpmin=time if fp==fpmin & fp~=.
sort subj tfpmin
quietly by subj:replace tfpmin=tfpmin[1] if tfpmin[1]~=.
sort sub time
quietly by subj: gen fb=ffa[1] if ffa[1]~=.
#delimit ;
collapse airg ib gb imax10 imax25
         mit1 mit2 gmax tgmax gmin tgmin
         tcross tgpmin gpmin fpmin tfpmin fb
         gdmin fdmin, by(subject);
#delimit cr

*drop _
su
* pwcorr airg - fb, sig
* swilk airg - fb
* for X in var airg-fb : for Y in var X-fb : spearman X Y
save "sundry values", replace
```

```
set more off
log using "merge analyser", replace
quietly do "sg estimation
use "flow sheet May 29 2000", clear
sort subject
save "flow sheet May 29 2000", replace
quietly do "sundry values"
use "sundry values", clear
for var gpmin fpmin fdmin gdmin: replace X=-X
sort subject
merge subject using "flow sheet May 29 2000"
tab _merge
drop _merge
more
sort subject
merge subject using sg_g0
sort subject
save "merge all 1", replace
```

Figure 6. (A) Phase 1 of the use of STATA for the analysis of a population of IVGTT studies: estimate S_G (glucose effectiveness) and G0 (level of glucose in plasma following the IVGTT at the point when glucose is fully mixed in the plasma space). (B) Phase 2 of the use of STATA for the analysis of a population of IVGTT studies: determine the properties of the IVGTT response. (C) Phase 3 of the use of STATA for the analysis of a population of IVGTT studies: merge S_G/G0 estimations, IVGTT determinations (see Table 1), and the Subject Demographics.

```
. gen sgdiff=St_sg-sg
. rreg sgdiff airg
Robust regression estimates                    Number of obs =      70
                                               F(  1,  68)    =    8.08
                                               Prob > F       =   0.005
  sgdiff | Coef.      Std. Err.      t      P>|t|   [95% Conf. Interval]
---------+-------------------------------------------------------------
    airg | 3.90e-06   1.37e-06    2.842    0.006   1.16e-06    6.64e-06
   _cons | -.0009796  .0014568   -0.672    0.504  -.0038867   .0019275
-----------------------------------------------------------------------
```

There is some evidence of an association but not a very strong one. Let's take a closer look at the distribution of 'sgdiff'(i.e., is it normally distributed or not?).

```
. swilk sgdiff

                       Shapiro-Wilk W test for normal data
  Variable |    Obs       W        V        z       Pr > z
-----------+-------------------------------------------------
    sgdiff |     70    0.98028   1.214    0.421   0.33687
```

It passes the Shapiro Wilks normality test. How about the t-test for a zero mean?

```
. ttest sgdiff=0

One-sample t test
------------------------------------------------------------------------------
  Variable |  Obs    Mean    Std. Err.   Std. Dev.    [95% Conf. nterval]
-----------+------------------------------------------------------------------
    sgdiff |  70  .0022693  .0009198     .0076952     .0004345   .0041042
------------------------------------------------------------------------------
Degrees of freedom: 69

Ho: mean(sgdiff) = 0
```

```
Ha: mean < 0            Ha: mean ~= 0           Ha: mean > 0
t =    2.4673           t =     2.4673          t =   2.4673
P < t =    0.9920       P > |t| =    0.0161     P > t =   0.0080
```

Evidently, there is divergence of 'sgdiff' from zero. Our original dataset in fact comprised 4 distinct populations, males and females, and impaired and non-impaired glucose tolerant (IGT) subjects. Let's split up the dataset appropriately, and redo our t-test (note that the output has been attenuated).

```
. egen group=group(sex igt)
. byvar group: ttest sgdiff=0

-> group==1
----------------------------------------------------------------------
Variable | Obs   Mean    Std. Err.    Std. Dev.    [95% Conf. Interval]
---------+------------------------------------------------------------
  sgdiff |  30  .0023777  .0015145     .008295     -.0007197   .0054751
----------------------------------------------------------------------
      H: mean < 0            Ha: mean = 0            Ha: mean > 0
      t = 1.5700             t = 1.5700              t = 1.5700
      P < t = 0.9364         P > |t| = 0.1273        P > t = 0.0636
-> group==2
----------------------------------------------------------------------
Variable | Obs   Mean    Std. Err.  Std. Dev.    [95% Conf. Interval]
---------+------------------------------------------------------------
  sgdiff |  11  .0014473  .0013882   .0046041     -.0016458   .0045403
----------------------------------------------------------------------
      Ha: mean < 0          Ha: mean ~= 0          Ha: mean > 0
      t = 1.0426            t = 1.0426             t = 1.0426
      P < t = 0.8391        P > |t| = 0.3217       P > t = 0.1609
-> group==3
----------------------------------------------------------------------
Variable |  Obs   Mean    Std. Err.  Std. Dev.    [95% Conf. Interval]
---------+------------------------------------------------------------
  sgdiff |  25  .0027458  .0016833   .0084167     -.0007284   .0062201
----------------------------------------------------------------------
      Ha: mean < 0          Ha: mean ~= 0          Ha: mean > 0
      t = 1.6312            t = 1.6312             t =   1.6312
      P < t = 0.9420        P > |t| = 0.1159       P > t = 0.0580
-> group==4
----------------------------------------------------------------------
Variable | Obs  Mean    Std. Err.    Std. Dev.    [95% Conf. Interval]
---------+------------------------------------------------------------
  sgdiff |  4  .0007394  .0036002     .0072004     -.010718   .0121969
----------------------------------------------------------------------
      Ha: mean < 0              Ha: mean ~= 0            Ha: mean > 0
      t =   0.2054              t =   0.2054             t =   0.2054
      P < t = 0.5748           P > |t| = 0.8504         P > t = 0.4252
```

The differences have now, as we suspected, vanished. We presume that, on a group basis, our STATA S_G estimates agree well with Bergman's MINMOD values.

Of course, we don't actually know from the above how our encoding of the group variable relates to our subject groups. To expose this, we use the 'tabdisp' command.

```
. tabdisp sex igt, c(group)
----------+-----------
          |    igtt
      sex |  N      Y
----------+-----------
        F |  1      2
        M |  3      4
----------+-----------
```

Thus, group 1 coincides with normal females, group 2 with IGT females, group 3 with normal males, and group 4 with IGT males.

ACQUIRING STATA

STATA is acquired either directly from STATA Corporation or from agencies licensed to sell the software. There are very competitive pricing arrangements for academic institutions, for which the current price is ~$100 for the software and $150 for the documentation. By purchasing STATA, you acquire a perpetual license. Updates to the version of STATA you purchase are freely available over the Internet.

For information on purchasing STATA software, contact:

STATA Corporation
4905 Lakeway Drive
College Station, TX 77845
800-STATA-PC
979-696-4600
979-696-4601 (fax)
stata@stata.com
http://www.stata.com

SUPPORT FOR STATA

STATA is not a trivial software tool. Being command-line driven means that you must be familiar with the format of a number of commands to advance an analysis efficiently. On the bright side, though, the commands are mnemonic, and they can be simplified (abbreviated). Furthermore, each command contains a sensible array of options which substantially enhances its power.

There are a number of lines of support enabling you to obtain help with your use of STATA. You can subscribe to a STATA list server and post questions to the STATA community. You can also enroll in STATA's excellent Internet courses and be tutored through a series of compact study programs. For urgent concerns, you can email or phone STATA's help desk. Author RC Boston, a STATA instructor in U. Penn's Biostatistics Department, will be happy to assist readers in overcoming problems they may run into.

In closing, we reiterate that STATA is a very powerful and elegant statistical suite that has been beautifully crafted by a first rate company.

CORRESPONDENCE

Please address all correspondence to:
Ray Boston
University of Pennsylvania
School of Veterinary Medicine
Department of Clinical Studies
New Bolton Center
382 West Street Road
Kennett Square, PA 19348
boston@cahp2.nbc.upenn.edu

REFERENCES

Ader, M., Pacini, G., and Bergman, R.N., 1985, Importance of glucose *per se* to intravenous glucose tolerance: comparison of the minimal model prediction with direct measurement, *Diabetes* 34:1092-1103.

Bergman, R.N., and Bowden, C.R., 1981, The minimal model approach to quantification of factors controlling glucose disposal in man, in: *Carbohydrate Metabolism,* C. Cobelli and R.N. Bergman, eds., Wiley, New York.

Boston, R., Stefanovski, D., Moate, P., Linares O., and Grief, P., 2002. Cornerstones to shape Modeling in the 21st Century: Introducing the AKA-Glucose Project, in: *Mathematical Modeling in Nutrition and Health Sciences*, J.A. Novotny, M.H. Green, and R. Boston, eds., Kluwer Academic Publishing, New York.

Derendorf, H., and Hochhaus, G., 1995, Handbook of Pharmacokinetic/ Pharmacodynamic Correlation, CRC Press, Washington, DC.

USING ADVANCED CONTINUOUS SIMULATION LANGUAGE (ACSL) TO SIMULATE, SOLVE, AND FIT MATHEMATICAL MODELS IN NUTRITION

Heidi A. Johnson[*]

INTRODUCTION

'Advanced Continuous Simulation Language' (ACSL; AEgis Technologies, Huntsville, AL) is a Fortran-based language that can be used to build, run, and evaluate biological models as well as to fit and optimize parameters. In this paper, two models will used to illustrate the functionality of ACSL: a blood/bile model presented in the MLAB User's Guide (1997) and a rumen model developed by France et al. (1982). These two problems will be used to demonstrate building a model, defining experimental observations, simulating and solving models, and evaluating results using ACSL.

MODEL OBJECTIVES

Model objectives for the blood/bile model were as follows: given time course data on the disappearance of a radiolabeled drug from blood and appearance of label in bile, predict the time delay for appearance of the drug in bile and the fractional rate of elimination of the drug from blood. For the rumen model, model objectives were: given feed inputs, predict rumen outflow of β-hexose (Hb), α-hexose (Ha), water-soluble carbohydrate (Sc), true protein (Pr), non-protein nitrogen (NPN), and microbial mass (M).

MODEL DESCRIPTIONS

Annotated model code (csl file and cmd file) for the two models is listed in Appendices 1 (blood/bile model) and 2 (rumen model). The blood/bile model was created using

---

[*] Heidi A. Johnson, Department of Animal Science, University of California, Davis, CA 95616.

the graphic interface tool, Graphic Modeler, to generate computer code. The rumen model was built directly from code.

Blood/Bile Model

A radiolabeled drug injected into the blood of a rat passes from the blood stream into bile. As shown in Figure 1, there is a delay associated with the drug's appearance in bile. Boxes or symbols represent functions such as integration, time delay, and gain [K * state variable as in the FRE (fractional rate of elimination) and FRA (fractional rate of appearance) boxes], and wires or lines represent state variables. Therefore, this diagram represents the transformations of model variables (wires) as they pass through different mathematical functions (the boxes/symbols).

Rumen Model

The rumen model diagram (Figure 2) represents state variables as boxes and ar-rows/lines as mathematical functions. Seven dietary nutrients (prefix D), plus drinking (Dv) and salivary (Sv) functions, are inputs into the model. Products of nutrient breakdown from α-hexose (Ha) and degradable β-hexose (Hbr), signified by the prefix U, contribute to soluble carbohydrates (Sc). Breakdown of rumen degradable protein (Urdp) contributes to non-protein nitrogen (NPN). Microbial production or mass (M) is a function of the utilization of Sc and NPN. Microbial death and breakdown recycles Sc

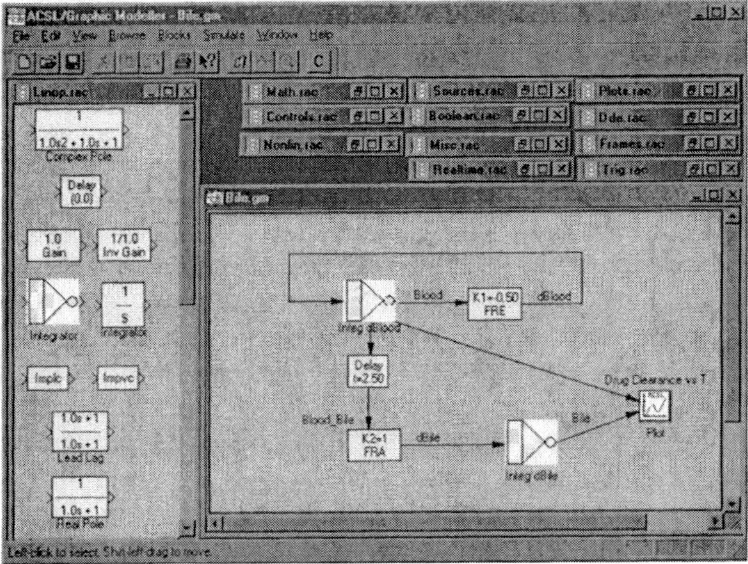

Figure 1. ACSL Graphic Modeler. The blood/bile model diagram is shown with one of the operations windows (linear operators or Lineop) open. Diagrams/models are built by clicking and dragging an operator to the model window and then connecting operations with lines (wires).

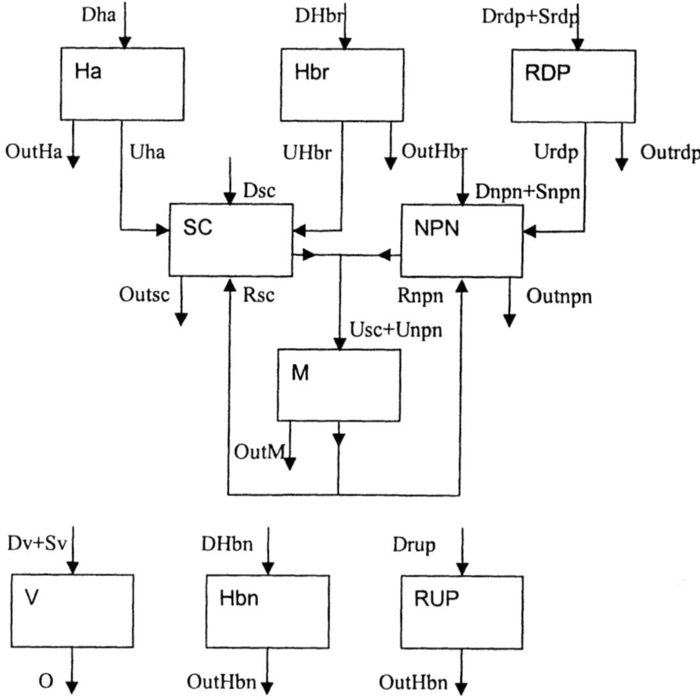

Figure 2. Diagram of the rumen model [modified from Figure 3 of France et al., (1982)]. See text for definition of parameters.

and NPN as RSc and Rnpn. Non-rumen degradable nutrients, non-rumen degradable β-hexose (Hbn) and non-rumen degradable protein (RUP) pass through the rumen to the abomasum. Flow of nutrients out of the rumen to the abomasum is represented by the prefix Out; flow is dependent on drinking, salivation, metabolic volume of the rumen (V), and nutrient utilization (microbial mass M).

OVERVIEW OF ACSL

ACSL consists of three programs that work together for simulations, fitting of model parameters, optimization, sensitivity analyses, and (data) matrix manipulation. ACSL Builder or Sim (Figure 3) compiles Fortran-based code and runs model simulations. ACSL Graphic Modeler (Figure 1) contains preprogrammed operations so that, through building a diagram, model code can be generated. Then, Graphic Modeler will call ACSL Builder to run simulations. ACSL Math (Figure 4) manipulates data matrices and contains the ACSL Optimize functions (parameter fitting, sensitivity analyses, and model optimization); it will run scripts or macros from m files. DOS and ACSL Builder

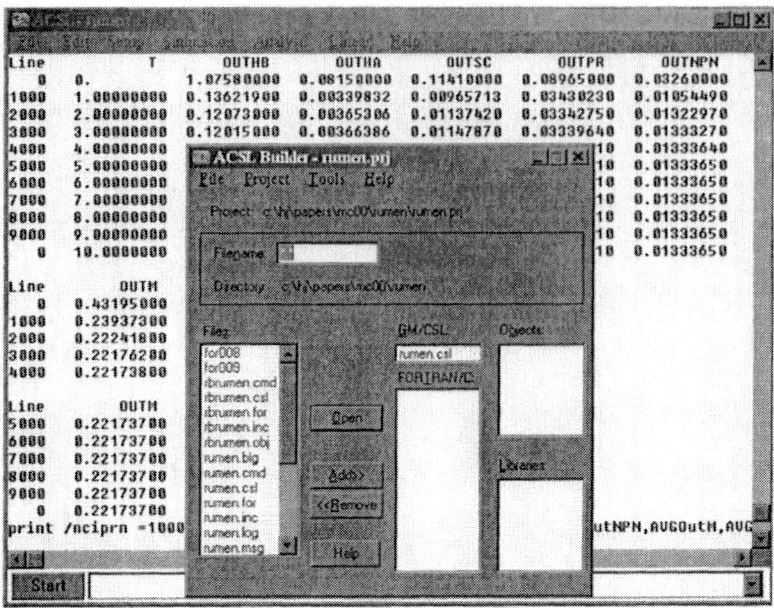

Figure 3. ACSL Builder. The window in the foreground shows the initial settings and files, and the window in the background is the runtime window. The runtime window shows results of the simulation; temporary changes to parameter values can be made here. Output from the rumen model is shown.

(runtime) commands can also be run within the Math output window. In addition, Math is Matlab compatible.

GETTING OBSERVATIONS INTO THE ACSL SOFTWARE

Getting data into ACSL for producing observed and predicted plots or for fitting parameters is fairly easy, but the file must be set up correctly. Time must be listed in the first column. The file should only contain data with no column headings. Therefore, the user must remember the order of data columns when fitting parameters. Text files (ASCII) are the easiest format to produce (from Excel, etc.) and use. All models must be compiled in double precision format. In addition, if more decimal places are required to define the data, the command 'set @format=long' must be entered first.

Data are brought into ACSL Math by the command 'load newfilename @file=oldfilename @format=ascii.' Alternatively, data may be brought in using an edit/paste command or the import wizard. Both methods (edit/paste and import wizard) create a spreadsheet within Math. The data are essentially a matrix within Math that can be manipulated by removing columns and/or rows, inverting rows and columns, or truncating rows or columns, thereby creating subsets of the original data file. Math is compatible with and uses syntax that is very similar to Matlab.

Data for plotting can be specified in the command file to automatically print observed and predicted plots during simulations (with ACSL Builder). Observed and predicted plots can also be automatically generated during data fitting by checking the 'cmpplain.m' box on the 'Report' tab (Figure 4).

DEFINING THE ERROR STRUCTURE OF THE OBSERVATIONS

Data errors are assumed to be normally distributed and independent. The error model used for fitting parameters within ACSL Optimize (Math) is defined with two settings: absolute or relative heteroscedasticity (weighting) parameter and whether the weighting parameter should be fit by the optimizer or set by the user (Figure 5). 'Absolute' or 'relative' is a sliding scale indicating the setting (from 0 to 2) for the weighting parameter when the user sets it. If 'absolute' is chosen (weighting parameter is 0), scatter (variability) is constant. If 'relative' is chosen (weighting parameter is 2), variability changes. The weighting parameter can be fit by the optimizer along with the parameters ('Fit' is checked) or set by the user ('Set' is checked).

In order to fit the blood/bile data to the model and predict the fractional rate of elimination of the drug from blood and delay into bile, the heteroscedasticity parameter was initially set to 0. Errors were assumed to be independent of predicted values (or standard deviations were assumed to be the same for each data point) and ACSL was allowed to 'fit' the heteroscedasticity parameter. In the final solution (Figure 6), the weighting parameter for blood was 0.1198 and for bile was 2.

ENTERING A MODEL: SPECIAL PREFABRICATED MODELING CONSTRUCTS

The Graphic Modeler contains several 'racks' of modeling functions which can be connected by wires (representing model variables) to construct a model. In Figure 1, the 'linear operators' rack is shown. Other built functions (or racks) are plots, dynamic data exchange (DDE), sources, boolean, math, control operators, frame operators, nonlinear functions, real time functions, trigonometry functions, and miscellaneous. An operator is dragged from the operators window and dropped in the model diagram window. Operators are then connected with lines in order to construct the model. Model code (csl file) can be generated from the diagram, or the model can be run from within the Graphic Modeler. Models can also be combined (in hierarchies) and diagrams from files can be brought in to create an easy-to-use interface with Graphic Modeler.

ASSOCIATING MODEL SOLUTIONS WITH OBSERVATIONS

Simulation output from ACSL Builder is printed immediately into the runtime window. It can be cut and pasted into any text editor or spreadsheet, or the log file can be edited down to the model solution. Observed and simulated data can be plotted by using a data command in the cmd (command file) which contains the observed data. Simulated data alone can be plotted at the end of the simulation run by using the 'analysis' menu from the output window. Running simulations from within the Graphic Modeler produces similar

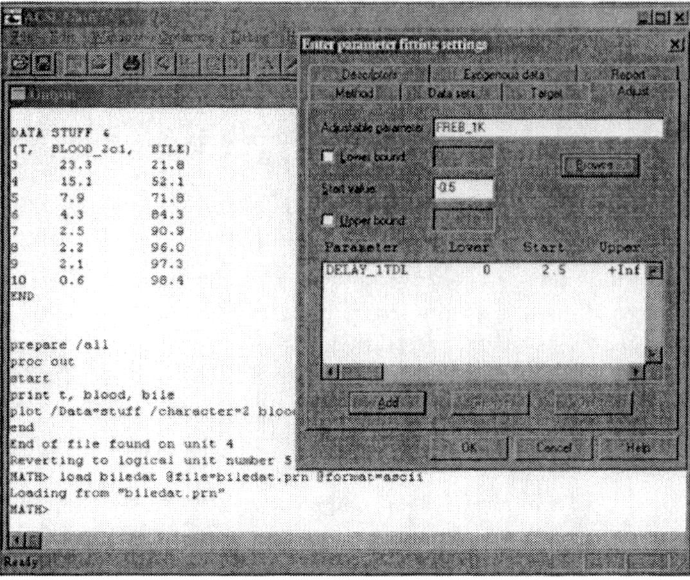

Figure 4. ACSL Math/Optimize. The window for specifying initial settings to estimate model parameters is shown in the 'Enter parameter fitting settings' window. The output window is where results from estimating parameters, optimizing, sensitivity analysis, or data matrix manipulation are shown. Initial settings for the blood/bile model are shown.

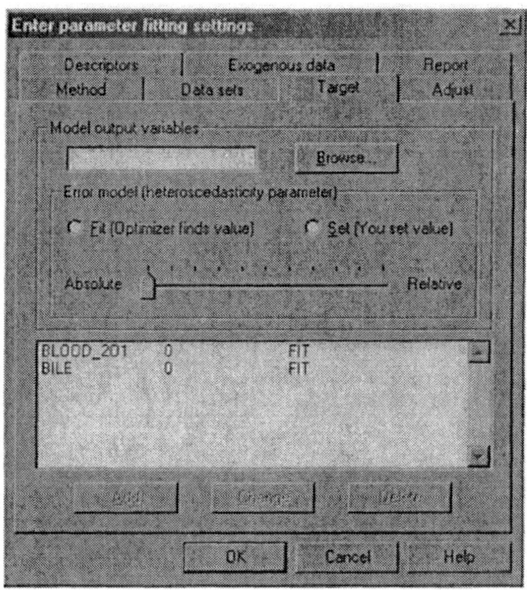

Figure 5. Specifying the variables to be fit, their order in the data, and the error model for each within ACSL Math/Optimize.

output to ACSL Builder (since Builder is running in the background). Dragging and dropping a plot function into the model diagram specifies plotting. A plot is produced after a simulation run by double clicking on the 'plot' icon (Figure 1). Solutions to parameter fitting from ACSL Math/Optimize are output immediately into the output window where they can be cut and pasted into an editor/spreadsheet. Alternatively, a diary file can be specified and all subsequent runs will be saved in the diary file. Observed and predicted data can automatically be plotted as mentioned before. Figure 6 shows a solution to parameter fitting from the blood/bile model.

COMPILING, SOLVING, AND FITTING THE MODEL

Compiling and simulation runs are done in ACSL Builder or in the Graphic Modeler. Eight integration algorithms are available [Adams-Moulton, Gear, Runge-Kutta (Euler), Runge-Kutta second order, Runge-Kutta fourth order, Runge-Kutta-Fehlberg second order, Runge-Kutta-Fehlberg fifth order, and differential algebraic system solver]. These are specified in the model code or csl file. For parameter fitting, Builder must be shut down, and the model is loaded into ACSL Math/Optimize. Three optimizers are available: test (not an algorithm), Nelder-Mead Simplex, and Generalized Reduced Gradient. The 'test' fit option will give results on initial settings only. Even if the solution does not converge, the current solution is shown. This is useful for determining why the model was unable to fit the data. All of the ACSL programs provide fairly rapid solutions for most models. Depending on processor speed, most programs run in less than 30 seconds.

ASSESSING THE GOODNESS OF MODEL FIT

Model fit is assessed by examining plots (Figures 7, 8, and 9) and analyzing the output from 'Math/Optimize' (Figure 6). As the log likelihood function gets bigger, the better the fit. Small standard deviations of parameter predictions indicate uniqueness of the solution. Standard deviations are the diagonal of the variance/covariance matrix at the optimum. Observed and predicted values, percent errors of predictions, and standardized residuals for each data point are reported, as well as is more detailed information on the log likelihood function and variance/covariance matrices. Preprogrammed plots are also available as described in the next section.

PRODUCING STANDARD PLOTS: MANIPULATION OF GRAPHICS

In running simulations (ACSL Builder, Graphic Modeler), plots are produced by an explicit 'plot' statement in the runtime window or cmd file, or by using the 'analysis, plot' menu in the runtime window. In fitting parameters (ACSL Math/Optimize), two dimensional standard plots of individual parameter observed and predicted values *vs.* time (cmpplain.m), log plots of individual parameter observed and predicted values *vs.* time (cmplog.m), all observed and predicted parameters (cmpjoint.m; Figure 7), and standardized residuals (Figure 8) can be produced easily. Three-dimensional plots of the dependence of the log likelihood function on the parameters (llfplot.m and llfplt.m; Figure 9) can also be produced.

```
                DESCRIPTION        PARAMETER ESTIMATES   STANDARD
                -----------        --------------------
                DEVIATION
                                   INITIAL        FINAL
LOG LIKELIHOOD FUNCTION             -49.417      -27.338
              DELAY_1TDL                 2.5        1.247    0.026529
                FREB_1K            -0.48406      -0.49637    0.005414
Variable: BLOOD_2Ol
                BLOOD_2Ol         BLOOD_2Ol       PERCENT   STANDARDIZED
       T         OBSERVED         PREDICTED         ERROR       RESIDUAL
       0              100               100             0              0
       3             23.3           22.5572       3.18782       0.957535
       4             15.1           13.7314       9.06363        1.81759
       5              7.9           8.35879      -5.80741      -0.627678
       6              4.3           5.08829      -18.3323       -1.11103
       7              2.5           3.09742      -23.8969      -0.867425
       8              2.2           1.88551        14.295         0.4704
       9              2.1           1.14778       45.3439        1.46728
      10              0.6          0.698693      -16.4488      -0.156665
Variable: BILE
                    BILE              BILE       PERCENT   STANDARDIZED
       T         OBSERVED         PREDICTED         ERROR       RESIDUAL
       0                0                 0             0              0
       3             21.8           22.8234      -4.69433       -1.15261
       4             52.1           48.0044       7.86102        2.19315
       5             71.8           70.1717       2.26779       0.596484
       6             84.3           84.7095     -0.485808      -0.124278
       7             90.9           92.5363      -1.80012      -0.454554
       8               96           97.8541      -1.93135      -0.487065
       9             97.3           100.923      -3.72336      -0.922768
      10             98.4            102.77      -4.44137       -1.09315
STATISTICAL SUMMARY
                MAXIMIZED          WT RESID      WEIGHTED     PERCENTAGE
                      LOG           SUM OF      RESIDUAL      VARIATION     WEIGHTING
                LIKELIHOOD         SQUARES           SUM      EXPLAINED     PARAMETER
      ----------  -------------  -------------  -------------  -------------  -------------
BLOOD_2Ol            -9.741       3.728e+000    1.255e+000          99.94         0.1198
     BILE            -17.6        1.362e-002   -5.620e-002          99.83              2
  OVERALL            -27.34       3.742e+000    1.199e+000          99.94

CORRELATION MATRIX
                DELAY_1TDL           FREB_1K
DELAY_1TDL               1
FREB_1K            0.58494                 1

VARIANCE-COVARIANCE MATRIX
                DELAY_1TDL           FREB_1K
DELAY_1TDL      0.00070376
FREB_1K         8.4011e-005       2.9311e-005

Optimization Method:  Generalized Reduced Gradient
INFORM = 0
GRG2 concluded successfully.
Time elapsed:                      2.75 seconds
Number of function evaluations:    163
Time limit has been reached.
```

Figure 6. Solution to parameter fitting for the blood/bile model with associated statistics of the fit.

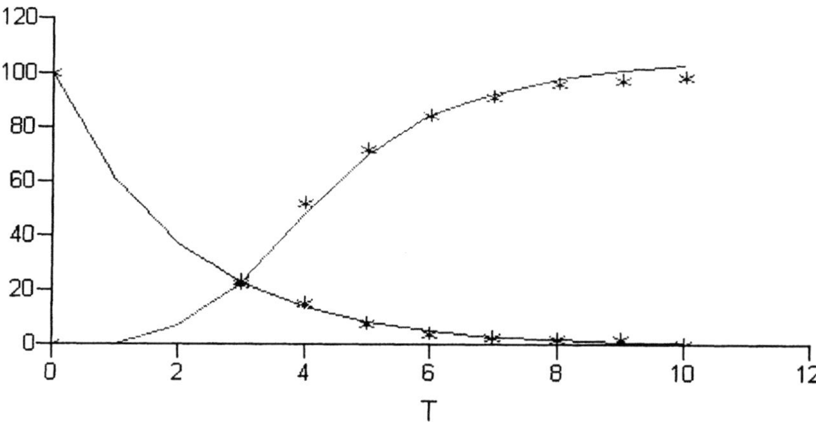

Figure 7. Predicted (line) and observed (*) values for blood (decreasing with time) and bile (increasing with time).

PROCESS LOGGING, PROCESSING MACROS, AND BLOCK PROCESSING

All processes appear in real time windows as they occur. Log files are also created automatically and can be edited. Procedures can be created in the cmd file to run and store sequential simulation data. Macros can be created (m files) to run specific plots, manipulate data, or run specific, routine processes within ACSL Math. If the user is familiar with Matlab, the macro language is fairly easy to use. Otherwise, it is complex enough to take some time to learn.

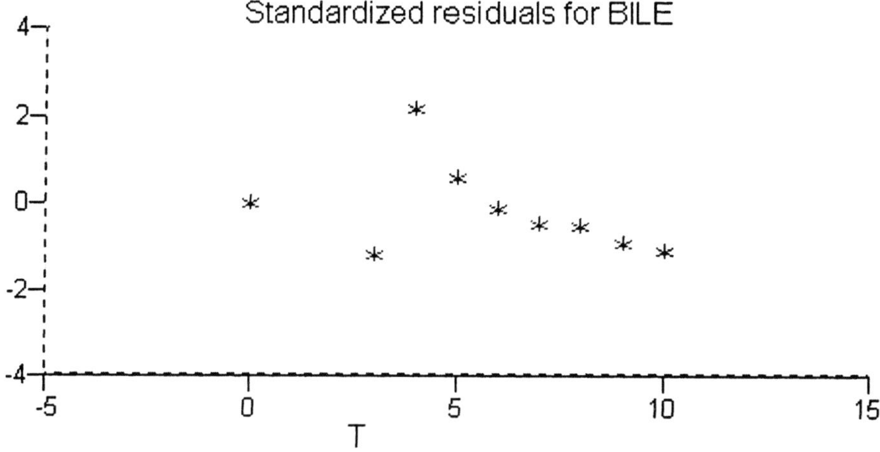

Figure 8. Plots of standardized residuals for blood/bile model.

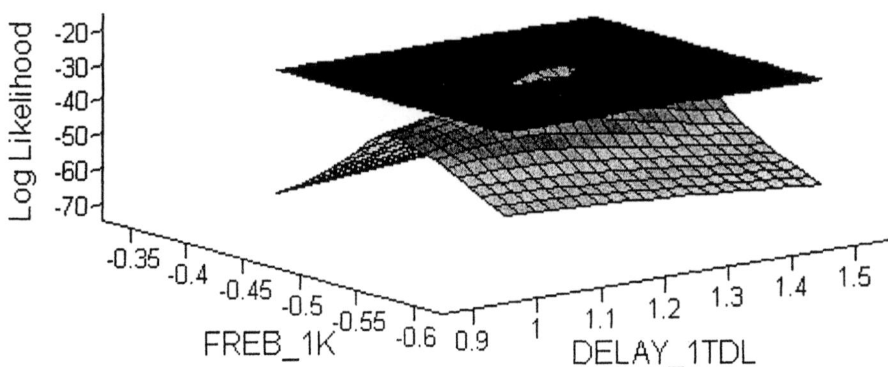

Figure 9. Dependence of log likelihood function on fitted parameters FRE (FREB_1K) and time delay (Delay_1tdl).

BLOOD/BILE MODEL SOLUTION

Final estimates of FRE and delay for the blood/bile model were –0.49637 /min and 1.247 min, respectively (Figure 6). Certainty of the estimates can be assessed by standard deviation estimates (0.005414 for FRE and 0.026529 for delay). Large standard deviations (50% of the parameter estimate) indicate that either there is too much variability in the data, the model is over-specified, or the parameters are coupled (correlated). Standard deviations (SD) are the square root of the diagonal of the variance/covariance matrix (inverse of Hessian matrix).

Observed (obs) and predicted (pred) values for each data point are listed in Figure 6 with percent error and standardized residuals. Percent error is calculated as 100 * (obs - pred) / obs. Standardized residual is computed as (obs - pred) / SD. Standardized residuals are used to identify bias in model solutions. Standardized residuals should be distributed about 0 in a random fashion. In the blood/bile model, percent errors were very small for bile predictions and somewhat larger for blood. The 9 min blood value (2.1 /min) has a large percent error (45.3%) and standard residual (1.47) indicating that it may be an outlier.

In the statistical summary (Figure 6), the maximum log likelihood shows the contri-bution of each parameter to the log likelihood function. If the parameters are independent, the maximum log likelihood for each parameter is additive. In the model, 'blood' and 'bile' are almost additive, and 'blood' accounts for a greater amount of the likelihood. 'Weighted residual sums of squares' and 'weighted residual sums' reflect the weighting parameter set by the user or fit by the computer. If the weighting parameter is 0, there is no weighting on the residual sums of squares or the residual sums. Both the

Table 1. Comparison of rumen model steady state nutrient outflows with rumen model in France et al. (1982) for chopped Italian rye

Parameter (kg/d)	France solution	ACSL Builder solution
OutHbr	0.1201	0.1201
OutHa	0.0037	0.003664
Outsc	0.0115	0.01148
Outrdp	0.0334	0.03340
Outnpn	0.0133	0.01334
OutM	0.2217	0.2217

residual sums of squares and the residual sums can be used to determine the relative importance of each fitted parameter to the model. The blood parameter accounts for a higher amount of the sums of squares and residual sums relative to overall.

The correlation matrix (Figure 6) can reflect an over-specified model or show if parameters are coupled (i.e., if colinearities exist between fitted parameters). The delay and FRE parameters are fairly highly related with a correlation of 0.585. This is expected since the time of delay influences the amount of drug appearing (and when it appears) in bile. High correlations are usually undesirable because they indicate parameters are not independent and that the model may be over-parameterized. The covariance in the variance/covariance matrix also indicates how closely the parameter estimates are related, similar to the correlation matrix. The diagonal of the variance/covariance matrix is the variance of the parameter estimates. Figures 7, 8, and 9 show some of the plots generated by ACSL Optimize for the blood/bile model parameter estimation.

RUMEN MODEL SOLUTION

For the rumen model, steady state solutions were derived using ACSL Builder. Table 1 shows the solutions from ACSL Builder compared to the solutions from France et al. (1982) for chopped Italian ryegrass fed continuously. The solutions matched exactly.

The rumen model reached steady state within 2 minutes. The plot of the approach of rumen outflows to steady state is shown in Figure 10. The plot was produced from 'analysis, plot' within the ACSL Builder runtime window.

Also incorporated into the rumen model code (csl file; Appendix 2) is an array of feed input values (Italian rye chopped and pelleted, and timothy chopped and pelleted) and pulsing functions to mimic feeding, drinking, and salivation. The feed array is defined by the 'dimension' and 'parameter' commands. Different feeds can be specified by changing the feed column number (via parameter fd). The pulsing functions are coded explicitly by creating a timer that executes the pulse code at specific intervals through if/then/else statements. The advantages of creating a model directly through code are that the code is much more flexible, concise, and explicit. In contrast, the Graphic Modeler generates much more code, and it is more difficult to identify how the code functions. For example, the blood/bile model is a much simpler model; however, the code is as long as the code for the rumen model.

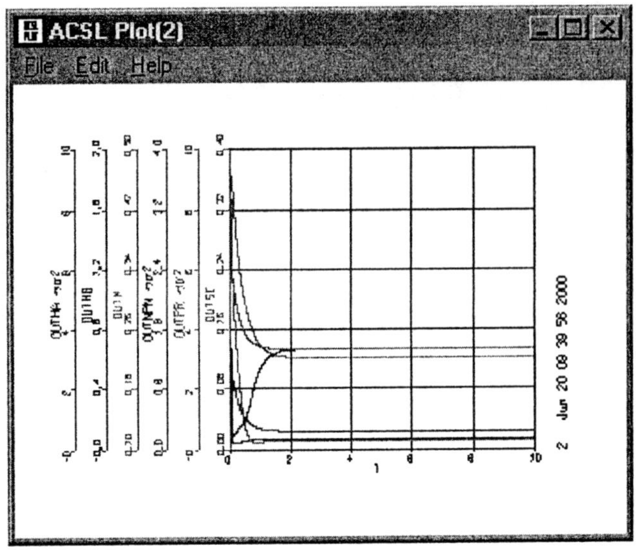

Figure 10.   Approach of the rumen model parameters to steady state with ACSL Builder.

SUMMARY

The ACSL programs (AEgis Technologies, 2000a, b) are fairly easy to use and provide a good combination of canned programming with the flexibility to do more, if one is willing to learn the language (for Fortran and m files). The manuals are fairly easy to understand but are not detailed enough. The manuals are geared toward operating the software. For the Optimize software, the best references are the Simusolv manuals (Steiner et al., 1990). ACSL also provides an interface (Open API or ACSL Server) which can be purchased separately. The interface allows compiled models to be distributed and run as independent programs with Visual Basic or C/C++. For models built with the Graphic Modeler, the interface is shareware.

For more information on purchasing the software, contact

AEgis Technologies
6703 Odyssey Dr., Suite 103
Huntsville, AL 35806
Phone: (256) 922-0802; FAX (256) 883-5516
http://www.acslsim.com/

CORRESPONDENCE

Please address all correspondence to:
Heidi Johnson
University of California
Department of Animal Science
One Shields Avenue
Davis, CA 95616
HAJohnson@UCDavis.edu

REFERENCES

ACSL Model Reference Manual v. 11.7, 2000a, AEgis Technologies, Huntsville, AL.

ACSL Optimize, 2000b, AEgis Technologies, Huntsville, AL.

France, J., Thornley, J.H.M., and Beever, D.E., 1982, A mathematical model of the rumen, *J. Agric. Sci. Camb.* 99:343-353.

Steiner, E.C., Rey, T.D., and McCroskey, P.S, 1990, *Simusolv: Modeling and Simulation Software, volume 2,* Simusolv Reference Guide, Dow Chemical Co., Midland.

MLAB User's Guide, p. 316, 1997, Civilized Software, Inc., Bethesda.

APPENDIX 1. THE BLOOD/BILE (GENERATED) MODEL CODE (Bile.csl)

```
PROGRAM !C:\HJ\papers\MC00\Bile\Bile.gm

INITIAL                            $!Initial values for integration and constants
                                             $!are set
equivalence (Bile_1o1, Bile)       $!names represent same variables
END                                          $!end of INITIAL section

DYNAMIC                            $!iterative section
CINTERVAL  Cint =  1.0                       $!communication interval (output every min)
CONSTANT TimeLimit=10.0
VARIABLE T =0.0

DERIVATIVE DEFAULT          $!section integrated continuously
MINTERVAL DEFAULTMint = 1.0e-10              $!minimum iteration interval
MAXTERVAL DEFAULTMaxt = 1.0e10               $!maximum iteration interval
NSTEPS DEFAULTNstp = 100                     $!number of steps/interval
ALGORITHM DEFAULTIalg = 5                    $!integration algorithm = Runge-Kutta 4
TERMT (zzfrfl .and. T .ge. TimeLimit, 'Time limit has been reached.')
                                             $! Terminate when timelimit (10 min) is reached
!-----------------------------------------------------------------------------------
------! ACSL Block: FREB_1    $!fractional rate of elimination of drug from blood
! Type: Gain
! Author: J. S. Gauthier
! Version: 4.1
! Last Change: 13:28:02 06/15/00
! Description:
!        Fractional Rate of Elimination of Drug from Blood (%)/min
! Input Ports:
!        Port Name: Blood
!        Connected Variable: i1
! Output Ports:
!        Port Name: dBlood
!        Connected Variable: o1
!-----------------------------------------------------------------------------------
-------
equivalence (Blood_2o1, Blood)
CONSTANT FREB_1k=-0.5 ! (/min)               $!initial value FRE
!----- gain                                  $!rac operation used
FREB_1o1  = FREB_1k  * Blood_2o1             $!1o1 is change in drug amt output, 1k
                                             is          $!elmination constant, 2o1
                                             is blood amt

!-----------------------------------------------------------------------------------
-------
! ACSL Block: Blood_2
! Type: ScalarIntegrator2
! Author: J. S. Gauthier
! Version: 4.1
```

```
! Last Change: 12:58:54 06/15/00
! Description:
!          Integrate change in blood level of drug(dblood)to get amount in blood
!          Integrate a state value from a numerically integrated derivative.
!          Different bitmap to normal scalar integrator.
! Input Ports:
!          Port Name: dBlood
!          Connected Variable: i1
! Output Ports:
!          Port Name: Blood
!          Connected Variable: o1
!          Port Name: Blood_Bile
!          Connected Variable: o1
!          Port Name: Plot
!          Connected Variable: o1
!-------------------------------------------------------------------------------
-------
equivalence (FREB_1o1, dBlood)
!! initially 100% of drug is in blood
CONSTANT Blood_2ic=100 ! (%)                          $!initial amt of drug in blood
!-----integrator                                      $!integration of change in amt drug
blood =
Blood_2o1  = integ (FREB_1o1 , Blood_2ic )                     $!amt in blood at t
!-------------------------------------------------------------------------------
-------
! ACSL Block: FRAB_1                               $!fractional rate of appearance of drug
in bile
! Type: Gain
! Author: J. S. Gauthier
! Version: 4.1
! Last Change: 13:10:07 06/15/00
! Description:
!          Fractional Rate of Appearance of Drug in Bile (%/min)
!          K2t is basically meaningless since it equals 1 and the transport
!          of drug from blood to bile to controlled by the time of delay
! Input Ports:
!          Port Name: Blood_Bile
!          Connected Variable: i1
! Output Ports:
!          Port Name: dBile
!          Connected Variable: o1
!-------------------------------------------------------------------------------
-------
!! %/min drug from blood through delay (liver) into Bile
CONSTANT FRAB_1k=1 ! (/min)
!----- gain
FRAB_1o1  = FRAB_1k  * Delay_1o1 $!1o1 is change in drug amt output, 1k is
!appearance constant, 2o1 is bile amt, delay is how long takes drug to appear
!in bile
!-------------------------------------------------------------------------------
-------
! ACSL Block: Bile_1
! Type: ScalarIntegrator2
! Author: J. S. Gauthier
! Version: 4.1
! Last Change: 12:58:35 06/15/00
! Description:
!          Integrate a state value from a numerically integrated derivative.
! Input Ports:
!          Port Name: dBile
!          Connected Variable: i1
! Output Ports:
!          Port Name: Bile
!          Connected Variable: o1
!-------------------------------------------------------------------------------
-------
equivalence (FRAB_1o1, dBile)
CONSTANT Bile_1ic=0.0 ! (%)
!-----integrator                          $!integration of change in FRA=amt in bile at t
Bile_1o1 = integ (FRAB_1o1 , Bile_1ic )
!-------------------------------------------------------------------------------
-------
! ACSL Block: Delay_1
! Type: Delay
! Author: J. S. Gauthier
! Version: 4.1
! Last Change: 13:59:02 06/15/00
! Description:
!          transport delay of drug from blood into bile
! Input Ports:
!          Port Name: Blood
!          Connected Variable: i1
! Output Ports:
!          Port Name: Blood_delay
!          Connected Variable: o1
!-------------------------------------------------------------------------------
-------
CONSTANT Delay_1tdl=2.5
!! tdl is time of delay (min), the fitted variable
CONSTANT Delay_1t1=1.0
!! t1 I'm not sure what it is and it doesn't seem to affect solutions
CONSTANT Delay_1delmin=2.5
```

```
!! delmin is buffer of time to record transfer values from blood into bile.
!! It WILL change the answer but not the actual delay time.
CONSTANT Delay_1ic=0.0
!! ic is intial value of transport of drug from blood to bile BEFORE delay
!-----transport delay
integer Delay_1nmx
parameter (Delay_1nmx  = 100)
delay (Delay_1o1=Blood_2o1, Delay_1ic, Delay_1td1, Delay_1nmx, Delay_1delmin)

END                            $! end of DERIVATIVE
END                            $! end of DYNAMIC
END                            $! end of PROGRAM
```


BLOOD/BILE COMMAND FILE (Bile.cmd)

```
DATA STUFF &                   $!listing of obs data so can plot obs and pred
(T,    BLOOD_2o1,      BILE)
3      23.3                    21.8
4      15.1                    52.1
5      7.9                     71.8
6      4.3                     84.3
7      2.5                     90.9
8      2.2                     96.0
9      2.1                     97.3
10     0.6                     98.4
END
prepare /all                   $!store variables for output and plotting
proc out                       $!procedure which can be called within builder runtime
start                          $!begin simulation
print t, blood, bile
plot /Data=stuff /character=2 blood_2o1,bile /xhi=10 /hi=110
end                            $!of procedure out
```


APPENDIX 2. THE RUMEN MODEL (rumen.csl)

```
PROGRAM rumen
!France, Thornley & Beever
!J AG Sci Camb (1982)99:343-353
!**********************************
!              Def'ns
!**********************************
!O = Rate Outflow from rumen to omasum (kg/d)
!Uj = Utilization rate of material j(kg/d)
!Rj = Rate release of C & N from M Catabolism (kg/d)
!DRIVING VARIABLES
!Dj = Dietary input of j(kg or m3/d)
!Sj = Salivary input of j (kg or m3/d)
!STATE VARIABLES (j)
!v = volume of rumen (m3)
!Hbn = Beta Hexose not degraded in rumen (kg CHO/m3)
!Hbr = Beta Hexose degraded in rumen (kg CHO/m3)
!Ha = Alpha Hexose (kg CHO/m3)
!Sc = Water Sol CHO (kg CHO/m3)
!Npn = Non protein N (kg CHO/m3)
!Rdp = Rumen Degradable Protein (kg CHO/m3)
!Rup = Rumen Undegradable Protein (kg CHO/m3)
!M = micro-organisms
!PARAMETERS
!fj = Microbial composition (kg CHO(NPN)/kg M DM)
!kj = Parameters of microbial growth eqn
!kjM = Rate constants for microbial degradation of
!      Ha, Hb and Protein (m3/kg M DM)
!kg = Constant determining dependence of microbial
!      catabolism on microbial growth rate (d)
!Ym = Conversion eff of substrate CHO into M CHO
!      (kg M CHO/kg CHO)
!MC = Maximum rate of M catabolism (/d)
!Mg = Maximum M growth rate (/d)
!MGrow = Specific growth rate of of microbed dep on SC and NPN (/d)
!PoPMGrow = Growth rate of microbial population (kg/d)
!W = Washout or dilution rate (/d)

INITIAL
!**********************************
!   Initial Values (i)
!**********************************
!STATE VARIABLES
CONSTANT iv = 0.005   $!m3
CONSTANT iRUP = 4.0   $!kg/m3
CONSTANT iHbr = 30.0  $!kg/m3
CONSTANT iSC = 7.0    $!kg/m3
CONSTANT iM = 26.5    $!kg/m3
CONSTANT iHbn = 36.0  $!kg/m3
CONSTANT iHa = 5.0    $!kg/m3
CONSTANT iRDP = 1.5   $!kg/m3
CONSTANT iNPN = 2.0   $!kg/m3
```

```
!For Pulse Feeding, Salavation, Drinking
CONSTANT intF=2
CONSTANT xD=1
CONSTANT intD=4
CONSTANT DurF=0.1 $!0.1 for chop, 0.03 for pellet
CONSTANT Dur=0.1
CONSTANT PULSEON=0.0 $!=1.0 if pulse is on, else=0.0

PARAMETER (COL=4)                  $!setup of 4 column array of feeds
INTEGER Fd
Fd = 1                                   $!fd=1-4 to specify different feeds
DIMENSION O(COL)                   $!defines row for 4 poss values
DIMENSION Sv(COL)
DIMENSION Snpn(COL)
DIMENSION Srdp(COL)
DIMENSION Dv(COL)
DIMENSION DHbn(COL)
DIMENSION Drup(COL)
DIMENSION DHa(COL)
DIMENSION DHbr(COL)
DIMENSION Drdp(COL)
DIMENSION Dsc(COL)
DIMENSION Dnpn(COL)

!INITIAL FEED VALUE ARRAYS FOR Dj,Sj,O (Diet inputs)
!               ----It Rye----   ---Timothy----
!               Chop    Pellet   Chop    Pellet
CONSTANT O    = 0.0163, 0.0163, 0.0176, 0.0176
CONSTANT Sv   = 0.0121, 0.0122, 0.0134, 0.0135
CONSTANT Snpn = 0.0061, 0.0061, 0.0067, 0.0067
CONSTANT Srdp = 0.0121, 0.0122, 0.0134, 0.0135
CONSTANT Dv   = 0.0042, 0.0041, 0.0042, 0.0041
CONSTANT DHbn = 0.0673, 0.0971, 0.0673, 0.1770
CONSTANT Drup = 0.0243, 0.0648, 0.0308, 0.0614
CONSTANT DHa  = 0.0266, 0.0409, 0.0586, 0.0551
CONSTANT DHbr = 0.4937, 0.4742, 0.5334, 0.5310
CONSTANT Drdp = 0.0729, 0.0278, 0.0719, 0.0330
CONSTANT Dsc  = 0.2150, 0.2261, 0.0422, 0.0686
CONSTANT Dnpn = 0.0422, 0.0462, 0.0567, 0.0450
!*********************************
!        Parameter Values
!*********************************
CONSTANT kHaM = 1.5      $!m3/kg M/d
CONSTANT kHbrM = 2.0     $!m3/kg M/d
CONSTANT kRdpM = 2.0     $!m3/kg M/d
CONSTANT Mg = 5.0        $! /d
CONSTANT ksc = 0.2
CONSTANT knpn = 0.2
CONSTANT kscnpn = 0.0
CONSTANT Mc = 0.2        $! /d
CONSTANT kg = 5.0
CONSTANT fsc = 0.3
CONSTANT fnpn = 0.5
CONSTANT Ym = 0.1

      ALGORITHM IALG = 5
      NSTEPS    NSTP = 1
      MAXTERVAL MAXT = 0.0025 $!0.0025 for contin,0.001 for pulse
      MINTERVAL MINT = 1D-09
      CINTERVAL CINT = 0.001
      CONSTANT  TSTOP = 10

END $!of INITIAL

DYNAMIC
!***************************
!      PULSE ON/OFF
!***************************
If (PULSEON.LE.0.1) THEN        $!turns on pulse
    PFeed = 1.0
    PSalv = 1.0
    PDrink = 1.0
    GO TO 60
END IF

timer = ABS(INT(t)-t)                   $! Converts t into whole number for timer
!***************************
!      PFeed PULSE
!***************************
!PFeed Pulse function pulses feeding at 2x/d
!for 0.1/d,beginning at 0.4d and 0.7 for chop
!for 0.03/d for pellet
!xF=amt nutrient,intF=times feed/d,Dur=duration of feeding
If (timer.GE.0.4 .AND. timer.LT.0.5 &
.OR. timer.GE.0.7 .AND. timer.LT.0.8) THEN
    PFeed = 1/(intF*DurF)
ELSE
    PFeed = 0
END IF
!***************************
!      PSalv PULSE
```

```
!***************************
!PSalv Pulse function pulses Rumen Salivary flow
!Feed t = rumen t,PSalv 2x during both
!Feed+Rumen t=0.2
If (timer.GE.0.4 .AND. timer.LT.0.6 &
.OR. timer.GE.0.7 .AND. timer.LT.0.9) THEN
   PSalv = 2/(1+(intF*2*Dur))
ELSE
   PSalv = 1/(1+(intF*2*Dur))
END IF
!***************************
!       PDrink PULSE
!***************************
!PDrink Pulse function pulses drinking at 4x/d
!Chop feeds=for 0.01/d,beginning at 0.01d,0.31d,0.51d,0.81d
!Pellet feeds=for 0.01/d,beginning at 0.24d,0.44d,0.74d,0.94d
!xD=amt drink,intD=times drink/d
Procedural (PDrink=timer,intD,Dur) $!FOR CHOP
     If (Fd.EQ.1 .OR. Fd.EQ.3) GO TO 10
     GO TO 30
10.. If (timer.GT.0.01 .AND. timer.LT.0.02 &
    .OR. timer.GT.0.31 .AND. timer.LT.0.32 &
    .OR. timer.GT.0.51 .AND. timer.LT.0.52 &
    .OR. timer.GT.0.81 .AND. timer.LT.0.82) &
    GO TO 20
    PDrink = 0
    GO TO 30
20.. PDrink = 1/(intD*(Dur/10))
    GO TO 30
30.. CONTINUE
END $!of Procedural

Procedural (PDrink=timer,intD,Dur) $! FOR PELLET
     If (Fd.EQ.2 .OR. Fd.EQ.4) GO TO 40
     GO TO 60
40.. If (timer.GT.0.24 .AND. timer.LT.0.25 &
    .OR. timer.GT.0.44 .AND. timer.LT.0.45 &
    .OR. timer.GT.0.74 .AND. timer.LT.0.75 &
    .OR. timer.GT.0.94 .AND. timer.LT.0.95) &
    GO TO 50
    PDrink = 0
    GO TO 60
50.. PDrink = 1/(intD*(Dur/10))
    GO TO 60
60.. CONTINUE
END $!of Procedural

DERIVATIVE
!************************************
!         Rumen H2O Volume
!************************************
dvdt = Dv(Fd)*PDrink + Sv(Fd)*PSalv - O(Fd)
W = (O(Fd))/V
V = INTEG(dvdt,iv)
!************************************
!    Rumen Dietary State Vars
!************************************
!Equations in the form
!d(substrate pool)dt = Inflow-Outflow+Synthesis-Utilization

dHadt = ((DHa(Fd)*PFeed - (Ha*Dv(Fd)*PDrink) &
        - (Ha*Sv(Fd)*PSalv))/V) - (kHaM*Ha*M)
UHa = V*kHaM*Ha*M
Ha = INTEG(dHadt,iHa)
OutHa = Ha*O(Fd)
AVGOutHa = INTEG(OutHa,0.0)/TSTOP

dHbrdt = (((DHbr(Fd)*PFeed) - (Hbr*Dv(Fd)*PDrink) &
        - (Hbr*Sv(Fd)*PSalv))/V) - (kHbrM*Hbr*M)
UHbr = V*kHbrM*Hbr*M
Hbr = INTEG(dHbrdt,iHbr)

dHbndt = ((DHbn(Fd)*PFeed) - (Hbn*Dv(Fd)*PDrink) &
        - (Hbn*Sv(Fd)*PSalv))/V
Hbn = INTEG(dHbndt,iHbn)
OutHb = (Hbn+Hbr)*O(Fd)
AVGOutHb = INTEG(OutHb,0.0)/TSTOP

dRUPdt = (Drup(Fd)*PFeed - (RUP*Dv(Fd)*PDrink) &
        - (RUP*Sv(Fd)*PSalv))/V
RUP = INTEG(dRUPdt,iRUP)

dRDPdt = ((Drdp(Fd)*PFeed + Srdp(Fd)*PSalv &
        - (RDP*Dv(Fd)*PDrink) - (RDP*Sv(Fd)*PSalv))/V) &
        - (kRdpM*RDP*M)
Urdp = V*kRdpM*RDP*M
RDP = INTEG(dRDPdt,iRDP)
OutPr = (RUP+RDP)*O(Fd)
AVGOutPr = INTEG(OutPr,0.0)/TSTOP

dSCdt = ((Dsc(Fd)*PFeed - (SC*Dv(Fd)*PDrink) &
        - (SC*Sv(Fd)*PSalv))/V) + (MCatab*fSC*M) &
```

```
            - ((MGrow*fsc*M)/Ym) + (kHaM*Ha*M) + (kHbrM*Hbr*M)
SC = INTEG(dSCdt,iSC)
OutSC = SC*O(Fd)
AVGOutSC = INTEG(OutSC,0.0)/TSTOP

dNPNdt = ((Dnpn(Fd)*PFeed + Snpn(Fd)*PSalv - (NPN*Dv(Fd)*PDrink) &
           - (NPN*Sv(Fd)*PSalv))/V) + (MCatab*fnpn*M) &
           - (MGrow*fnpn*M) + (KRdpM*RDP*M)
NPN = INTEG(dNPNdt,iNPN)
OutNPN = NPN*O(Fd)
AVGOutNPN = INTEG(OutNPN,0.0)/TSTOP
!**********************************
!        AVG Diet Inputs
!**********************************
AVGDnpn = INTEG(Dnpn(Fd),0.0)/TSTOP
AVGDrup = INTEG(Drup(Fd),0.0)/TSTOP
AVGDrdp = INTEG(Drdp(Fd),0.0)/TSTOP
AVGDHa = INTEG(DHa(Fd),0.0)/TSTOP
AVGDHbr = INTEG(DHbr(Fd),0.0)/TSTOP
AVGDHbn = INTEG(DHbn(Fd),0.0)/TSTOP
AVGDsc = INTEG(Dsc(Fd),0.0)/TSTOP
AVGSnpn = INTEG(Snpn(Fd),0.0)/TSTOP
AVGSrdp = INTEG(Srdp(Fd),0.0)/TSTOP
AVGRespR = INTEG(RespR,0.0)/TSTOP
!**********************************
!        Microbial Metabolism
!**********************************
MGrow = Mg/(1 + (ksc/SC) + (kNpn/NPN) + (kscnpn/SC*NPN))
PoPMGrow = Mgrow*V*M

MCatab = Mc/(1 + (kg*MGrow))
Mdeath = MCatab*V*M

dMdt = (MGrow*M) - (MCatab*M) - (M*(Dv(Fd)*PDrink+Sv(Fd)*PSalv))/V
M = INTEG(dMdt,iM)
OutM = M*O(Fd)
AVGOutM = INTEG(OutM,0.0)/TSTOP

Rsc = fsc*Mdeath
Rnpn = fnpn*Mdeath

Usc = (fsc*PoPMGrow)/Ym
Unpn = fnpn*PoPMGrow

RespR = ((1-Ym)/Ym)*fsc*V*Mgrow*M

      TERMT( T .GE. TSTOP )

END   $!of Derivative
END   $!of Dynamic
TERMINAL
END   $! of Terminal
END   $! of Program
```

RUMEN COMMAND FILE (Rumen.cmd)

```
!File:rumen.cmd
PROCEDURE MTH
SET WEDITG=.FALSE.
SET HVDPRN=.F.
FILE/PRNFILE='NUL'
END

PROC OUT
PREPARE /all
PREPARE O,SV,SRDP,DV,DNPN,DRUP,DRDP,SNPN,DHBN,DHBR,DHA,DSC
set ndbug=1
s pcwprn=120
START
print /nciprn=1000 t,V,RUP,Hbr,SC,M,Hbn,Ha,RDP,NPN
print /nciprn=10 t,timer,PFeed,PSalv,PDrink,O,Sv,Dv
print /nciprn=1000 t,OutHb,OutHa,OutSC,OutPr,OutNPN,OutM,RespR
END
```

SOLVING AND FITTING FRANCE'S RUMEN MODEL WITH MLAB

Gary Knott and Daniel Kerner[*]

INTRODUCTION

Ruminants (cattle, sheep, etc.) "ferment" their ingested food in their "first stomach," the rumen. There, food is partially digested due to mastication and microbial action. As a result of this "pre-processing," the animal is able to extract more nutrient substances from its food than would otherwise be the case.

A mathematical description for the action of the rumen was published in "A Mathematical Model of the Rumen" by France, Thornley, and Beever (1982). This model treats the rumen as a fermentation tank into which flow water and feed (which consists of carbohydrates and proteins "bound" within plant cells). A microbial mass in the rumen then enzymatically "converts" these inputs into various amounts of unencapsulated proteins, sugars, and other carbohydrates which are thereby made available for absorption and utilization by the animal. The microbial mass also utilizes some of the ingested materials to support its growth and enzymatic action. France's model consists of nine differential equations that describe the input, processing, and output of the various materials of interest. These differential equations involve twelve parameters and various other input and output rate functions, all of which must be somehow assigned values in order to establish a definite model.

France et al. originally solved their model differential equations using the CSMP software (Continuous System Modeling Program), employing the first-order Euler method. By manually solving the equations many times with different parameter values, they essentially curve-fit their model to data by choosing parameter values by trial and error. Here, we will demonstrate the use of the MLAB Mathematical and Statistical Modeling System (Civilized Software, Inc., www.civilized.com) in both solving and fitting France's model.

---

[*] Gary Knott and Daniel Kerner, Civilized Software Inc., Silver Spring, MD 20906.

APPLYING MLAB TO THE RUMEN MODEL OF FRANCE ET AL.

France et al. modeled a sheep eating either chopped or pelleted Italian ryegrass, or timothy. They also studied both continuous input (obviously an approximation to reality) and "pulsed" input where feeding occured intermittently. Here we will use only chopped Italian ryegrass as an example.

The inputs consist of (1) water from drinking and from saliva, (2) water-soluble carbohydrates (sugars and short-chain polymers), (3) α-hexose (pectin and α-linked polymers), (4) rumen degradable β-hexose (cellulose and lignin), (5) non-rumen degradable β-hexose, (6) rumen degradable protein (including nucleic acids), (7) non-rumen degradable protein, (8) non-protein nitrogen (amino acids, urea, ammonia, etc.), and (9) microbial mass. Inputs 2-8 are contributed by the Italian ryegrass feed. The outputs consist of these same materials after extraction from the feed minus the amounts utilized and/or synthesized by the microbial content of the rumen.

The fermentation tank rumen model requires a differential equation that defines the volume of water in the rumen (measured in meters^3, where 1 meter^3 = 10^3 liters). This volume changes as water is taken in and out mixed with various nutrients. The nutrient materials are measured as concentrations (kg/m^3), and these concentrations change, both as the amount of materials change and as the amount of water present changes. There is also a differential equation defining the growth and destruction of the microbial population. Thus, altogether, each material listed above has an associated mass-balance differential equation that together define the following functions of time:

$V(t)$ = volume of water from drinking and from saliva (m^3)
$Bn(t)$ = non-rumen degradable β-hexose (kg/m^3)
$Br(t)$ = rumen degradable β-hexose (kg/m^3)
$A(t)$ = α-hexose (kg/m^3)
$C(t)$ = water-soluble carbohydrates (kg/m^3)
$N(t)$ = non-protein nitrogen (kg/m^3)
$Pr(t)$ = rumen degradable protein (kg/m^3)
$Pn(t)$ = non-rumen degradable protein (kg/m^3)
$M(t)$ = microbial population concentration (kg/m^3)

The derivation of these equations is well presented in France et al. (1982).

The MLAB 'do-file' that defines France's rumen model, solves the differential equations, and graphs them for comparison to France's results and to the data given in France's paper is presented next.

```
/*      file: rumen.do                                3/26/00 */
/*      This file sets up and solves the compartmental model of the
        rumen described in J.France, et.al, J.Agric.Sci.Camb. 99
        (1982) 343-353.

The state functions are:
        V(t) = rumen water volume (m^3) (10^6 cm^3 = 1 m^3 = 10^3 liters)
        Bn(t) = non-rumen degradable beta hexose (kg/m^3)
        Br(t) = rumen degradable beta hexose (kg/m^3)
        A(t) = alpha hexose (kg/m^3)
        C(t) = water soluble carbohydrates (kg/m^3)
```

```
              N(t) = non-protein nitrogen (kg/m^3)
              Pr(t) = rumen degradable protein (kg/m^3)
              Pn(t) = non-rumen degradable protein (kg/m^3)
              M(t) = microbial population concentration (kg/m^3)

The values of the constants and input/output rates are taken from France et al.
for sheep on a continuous diet of chopped Italian ryegrass.
*/
reset
echodo = 3

/* define first order differential equations for state functions */
fct A't(t) = (DA(t)-A(t)*Dv(t)-A(t)*Sv(t))/V(t)-kam*A(t)*M(t)
fct Br't(t) = (DBr(t)-Br(t)*Dv(t)-Br(t)*Sv(t))/V(t)-kbrm*Br(t)*M(t)
fct Bn't(t) = (DBn(t)-Bn(t)*Dv(t)-Bn(t)*Sv(t))/V(t)
fct Pn't(t) = (DPn(t)-Pn(t)*Dv(t)-Pn(t)*Sv(t))/V(t)
fct Pr't(t) = (DPr(t)+SPr(t)-Pr(t)*Dv(t)-Pr(t)*Sv(t))/V(t)-kprm*Pr(t)*M(t)
fct C't(t) = (DC(t)-C(t)*Dv(t)-C(t)*Sv(t))/V(t)+(L(t)-mu(t)/Ym)*fC*M(t)+ \
               kam*A(t)*M(t)+kbrm*Br(t)*M(t)
fct N't(t) = (DN(t)+SN(t)-N(t)*Dv(t)-N(t)*Sv(t))/V(t)+(L(t)-mu(t))*fN*M(t)+ \
               +kprm*Pr(t)*M(t)

/* ODE for the volume of the rumen */
fct V't(t) = Dv(t)+Sv(t)-vr(t)

/* ODE for the microbial mass */
fct M't(t) = mu(t)*M(t)-L(t)*M(t)-M(t)*(Dv(t)+Sv(t))/V(t)

/* model for the specific daily growth rate of microbes */
fct mu(t) = mum/(1+(kc/C(t))+(kn/N(t))+(kcn/(C(t)*N(t))))

/* model for the daily catabolism(destruction) rate for microbes */
fct L(t) = Lm/(1+kmu*mu(t))

/* initial conditions, all in units of kg/m-=^3, except V is in m^3 units. */
init V(0) = .005
init Bn(0) = 36
init Pn(0) = 4
init A(0) = 5
init Br(0) = 30
init Pr(0) = 1.5
init C(0) = 7
init N(0) = 2
init M(0) = 26.5

/* constants */
kam = 1.5 /* m^3/kg M per day */
kbrm = 2.0 /* m^3/kg M per day */
kprm = 2.0 /* m^3/kg M per day */
mum = 5.0 /* per day */
kn = .2
kc = .2
kcn = 0.
Lm = .2 /* per day */
kmu = 5.0 /* per day */
fC = .3
fN = .5
YM = .1

/* input functions for continuous diet model */
fct DBn(t) = .0673 /* kg/day */
fct DPn(t) = .0243
fct DA(t) = .0266
fct DBr(t) = .4937
fct DPr(t) = .0729
fct DC(t) = .215
fct DN(t) = .0422
fct SN(t) = .0061
```

```
fct SPr(t) = .0121

/*   input/output flow rate functions. all constant, in m^3/day units
     Note the input and output balances. */
fct vr(t) = .0163 /* total water output from rumen */
fct Dv(t) = .0042 /* water input from drinking */
fct Sv(t) = .0121 /* water input from saliva */

/* Integrate from 0 to tmax days, reporting at np points in the interval. 3 days
is sufficient time to reach equilibrium (within 1 percent or so of the stationary
point.) */
tmaLx=3; np=400
t = 0:tmax!np
Res = points(V,Bn,Pn,A,Br,Pr,C,N,M,t);

/* Res col 1 = t, Res col 2:10 = [V,Bn,Pn,A,Br,Pr,C,N,MI values */

F = (Res row np)' /* F = steady-state values to be used later.*/

/*   Multiply concentrations by outflow rate value vr(0) to obtain
output per day amounts in kilograms */

Res col (3:10) = (Res col (3:10)) * vr(0)

/*   data from J.France paper for observed outputs from chopped Italian ryegrass
in kg/m^3 per day units. To be used for graphics output. */

ddb row 1 = list(tmax-.6,.1121)'
dda row 1 = list(tmax-.6,.0027)'
ddc row 1 = list(tmax-.6,.005)'
ddp row 1 = list(tmax-.6,.0365)'
ddm row 1 = list(tmax-.6,.1756)'
ddn row 1 = list(tmax-.6,.0495)'

draw Res col (1,5) lt dashed                    /* A = Alpha hexose */
title "A" at (3.1,Res[np,5]) world
draw dda pt circle ptsize .025 ffract
draw dda pt "A" ptsize .017 ffract

draw Res col (1,8) lt ldash                     /* C = carbohydrate */
title "C" at (3.1,Res[np,8]) world, place (left,top)
draw ddc pt circle ptsize .025 ffract
draw ddc pt "C" ptsize .017 ffract

/* Res col 7 = digestable protein + non-digestable protein */
Res col 7 = (Res col 7)+(Res col 4)
draw Res col (1,7) lt solid                     /* Pr+Pn = total protein */
title "P" at (3.1,Res[np,7]) world
draw ddp pt circle ptsize .025 ffract
draw ddp pt ''P" ptsize .017 ffract

draw Res col (1,9) lt ddash                     /* N = non-protein nitrogen */
title "N" at (3.1,Res[np,9]) world, place (left,bottom)
draw ddn pt circle ptsize .025 ffract
draw ddn pt "N" ptsize .017 ffract

window 0 to tmax, 0 to .05
bottom title "days"
left title "kg/day"
frame 0 to 1, .5 to 1
w1 = w

draw Res col (1,10) lt solid                    /*  M = microbial concentration */
title "M" at (3.1,Res[np,10]) world

draw ddm pt circle ptsize .025 ffract
draw ddm pt "M" ptsize .017 ffract
```

```
/* Res col 6 = digestable and non-digestable beta hexose */
Res col 6 = (Res col 6)+(Res col 3)
draw Res col (1,6) lt dashed                    /* Br+Bn = total Beta hexose */
title "B" at (3.1,Res[np,6]) world
draw ddb pt circle ptsize .025 ffract
draw ddb pt "B" ptsize .017 ffract

bottom title "days"
left title "kg/day"
frame 0 to 1, 0 to .5
window 0 to tmax, 0 to 1.1

view
unview
delete w,w1
```

The graphs produced by this MLAB do-file are shown in Figure 1. Note that the observed steady state output values reported in France et al. are drawn as single labeled points, with the labels enclosed in circles. Except for non-protein nitrogen N, France's parameter values cause the model to fit the data reasonably well.

Now we introduce various step functions for the inputs in order to model intermittent feeding and drinking following the regimes described by France et al. The MLAB statements shown next redefine the model and solve it for such 'pulsed' inputs. These statements may be provided as a continuation of the do-file given earlier. The results for pulsed feeding and drinking are shown in Figure 2.

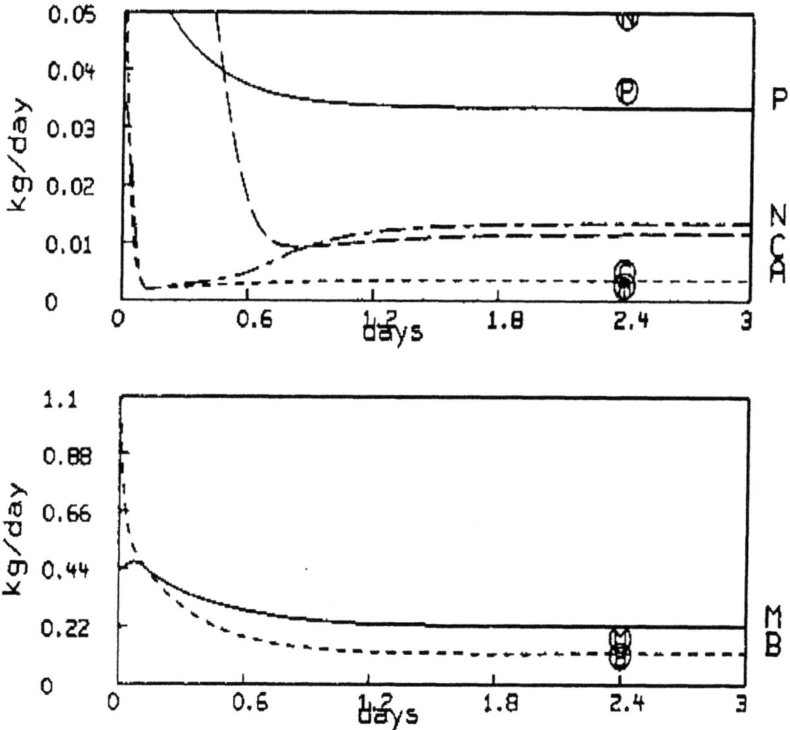

Figure 1. MLAB replication of the results of France et al. (1982).

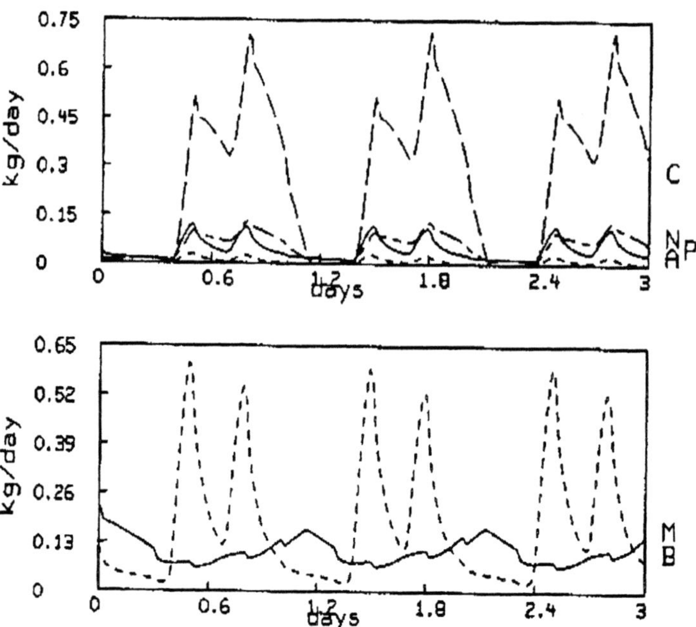

Figure 2. MLAB results for pulsed feeding and drinking.

```
/* use the steady state values computed earlier above for the initial conditions
*/

/* initial conditions, all in units of kg/m^3 */

init V(0) = F[2]
init Bn(0) = F[3]
init Pn(0) = F[4]
init A(0) = F[5]
init Br(0) = F[6]
init Pr(0) = F[7]
init C(0) = F[8]
init N(0) = F[9]
init M(0) = F[10]

/* h(t) is a smoothly transitioning sigmoidal curve representing a smoothed step
function. Using h in preference to a pure step-function avoids integration warn-
ings due to discontinuities. */
fct h(t,b) = .5*(1+sign(t)*(1-exp(-b*abs(t))))

/* pf(t) is the feeding pulse function */
fct pf(t) = sum(i,1,nf,h(mod(t,1)-ts[i],100)-h(mod(t,1)-ts[i]-df,100))/(nf*df)
nf = 2
df = .1
ts[1] = .4; ts[2] = .7

/* ps(t) is the salivary flow pulse function */
fct ps(t) = (1+sum(i,1,nf,h(mod(t,1)-ts[i],100)-h(mod(t,1)-ts[i]-2*df,100)))\
            /(1+2*nf*df)
```

```
/* pd(t) is the drinking pulse function */
fct pd(t) = sum(i,1,nd,h(mod(t,1)-tr[i],200)-h(mod(t,1)-tr[i]-dd,200))/(nd*dd)
nd = 4
dd = .01
tr[1] = .01; tr[2] = .31; tr[3] = .51; tr[4] = .81

/* redefine the input functions to incorporate pulse variation */
fct DBn(t) = .0673*pf(t)
fct DPn(t) = .0243*pf(t)
fct DA(t) = .0266*pf(t)
fct DBr(t) = .4937*pf(t)
fct DPr(t) = .0729*pf(t)
fct DC(t) = .215*pf(t)
fct DN(t) = .0422*pf(t)

fct SN(t) = .0061*ps(t)
fct SPr(t) = .0121*ps(t)

/* redefine fluid input flow rate functions */
fct Dv(t) = .0042*pd(t)
fct Sv(t) = .0121*ps(t)

/* integrate from 0 to tmax days reporting at np points in the interval */
jacsw = 0
method = gear2
disastersw = -2
Res = points(V,Bn,Pn,A,Br,Pr,C,N,M,t);
/* Res col 1 = t, Res col 2:10 = EV,Bn,Pn,A,Br,Pr,C,N,M] values */

/* multiply concentrations by outflow rate value to obtain output
   per day amounts in kilograms */
Res col (3:10) = (Res col (3:10)) * vr(0)

draw Res col (1,5) lt dashed                   /* A = Alpha hexose */
title "A" at (3.1,Res[np,5]) world

draw Res col (1,8) lt ldash                    /* C = carbohydrate */
title "C" at (3.1,Res[np,81]) world, place (left,top)

/* Res col 7 = digestable protein + non-digestable protein
Res col 7 = (Res col 7)+(Res col 4)
draw Res col (1,7) lt solid                    /* Pr+Pn = total protein */
title "P" at (3.1,Res[np,7]) world

draw Res col (1,9) lt ddash                    /* N = non-protein nitrogen */
title "N" at (3.1,Res[np,9]) world, place (left,bottom)

bottom title "days"
left title "kg/day"
frame 0 to 1, .5 to 1
w1 = w

draw Res col (1,10) lt solid                   /* M = microbial concentration */
title "M" at (3.1,Res[np,10]) world

/* Res col 6       = digestable and non-digestable beta hexose */
Res col 6 = (Res col 6)+(Res col 3)
draw Res col       (1,6) lt dashed             /* Br+Bn = total Beta hexose */
title "B" at       (3.1,Res[np,6]) world

bottom title       "days"
left title "kg/day"
frame 0 to 1, 0 to .5
view
unview
```

Finally, we used the MLAB system to try to adjust some of the parameters in France's model to more closely fit the observed steady state data reported by France et al. This fitting was done with a do-file that defined the model, adjusted the parameters, and graphed the resulting fit. The output log-file from this computation is shown next; it incorporates the numerical fitting results.

```
MLAB Mathematical Modeling System, Revision: January 12, 1999
Copyright: Civilized Software, Inc. (301)962-3711, email: csi@civilized.com
Web-site: WWW.CIVILIZED.COM

Thu Apr 13 18:00:10 2000
'* ' is the command prompt

* casesw = 1
*
* /* define first order differential equations for state variables */
* fct A't(t) = (DA(t)-A(t)*Dv(t)-A(t)*Sv(t))/V(t)-kam*A(t)*M(t)
* fct Br't(t) = (DBr(t)-Br(t)*Dv(t)-Br(t)*Sv(t))/V(t)-kbrm*Br(t)*M(t)
* fct Bn't(t) = (DBn(t)-Bn(t)*Dv(t)-Bn(t)*Sv(t))/V(t)
* fct Pn't(t) = (DPn(t)-Pn(t)*Dv(t)-Pn(t)*Sv(t))/V(t)
* fct Pr't(t) = (DPr(t)+SPr(t)-Pr(t)*Dv(t)-Pr(t)*Sv(t))/V(t)-kprm*Pr(t)*M(t)
* fct C't(t) = (DC(t)-C(t)*Dv(t)-C(t)*Sv(t))/V(t)+(L(t)-mu(t)/Ym)*fC*M(t)+ \
: kam*A(t)*M(t)+kbrm*Br(t)*M(t)
* fct N't(t) = (DN(t)+SN(t)-N(t)*Dv(t)-N(t)*Sv(t))/V(t)+(L(t)-mu(t))*fN*M(t)+ \
: +kprm*Pr(t)*M(t)
*
* /* model for the volume of the rumen */
* fct V't(t) = Dv(t)+Sv(t)-vr(t)
*
* /* model for the microbial mass */
* fct M't(t) = mu(t)*M(t)-L(t)*M(t)-M(t)*(Dv(t)+Sv(t))/V(t)
*
* /* model for the specific growth rate of microbes per day */
* fct mu(t) = mum/(1+(kc/C(t))+(kn/N(t))+(kcn/(C(t)*N(t))))
*
* /* model for catabolism rate of microbes per day */
* fct L(t) = Lm/(1+kmu*mu(t))
*
* /* initial conditions, all in units of kg/m^3
* init V(0) = .005
* init Bn(0) = 36
* init Pn(0) = 4
* init AM = 5
* init Br(0) = 30
* init Pr(0) = 1.5
* init C(0) = 7
* init N(0) = 2
* init M(0) = 26.5
*
* /* constants */
* kam = 1.5 /* m^3/kg M per day */
* kbrm = 2.0 /* m^3/kg M per day */
* kprm = 2.0 /* m^3/kg M per day */
* mum = 5.0 /* per day */
* kn = .2
* kc = .2
* kcn = 0.
* Lm = .2 /* per day */
* kmu = 5.0 /* per day */
* fC = .3
* fN = .5
* YM = .1
*
* /* input functions for continuous diet model */
* fct DBn(t) = .0673 /* kg/day */
```

```
*
* fct DPn(t) = .0243
* fct DA(t) = .0266
* fct DBr(t) = .4937
* fct DPr(t) = .0729
* fct DC(t) = .215
* fct DN(t) = .0422
* fct SN(t) = .0061
* fct SPr(t) = .0121
*
* /* output flow rate function */
* fct vr(t) = .0163
* fct Dv(t) = .0042 /* m^3/day
* fct Sv(t) = .0121
*
*/*     try fitting the functions tb, tp, M, N, C, and A to improve
        the match to the experimental data */
* fct tb(t) = Bn(t)+Br(t)
* fct tp(t) = Pr(t)+Pn(t)
*
* /* data from J.France paper for chopped Italian ryegrass (converted
* to concentration units kg/m^3 for day 5). */
* ddb row 1 = list(5,.1121/vr(5))'
* dda row I = list(5,.0027/vr(5))'
* ddc row 1 = list(5,.005/vr(5))'
* ddP row 1 = list(5,.0365/vr(5))'
* ddm row 1 = list(5,.1756/vr(5))'
* ddn row I = list(5,.0495/vr(5))'
*
* jacsw = 0
* lsqrpt = 15
* constraints q = {0 < kam, kam < 5,\
: 0 < kbrm, kbrm < 5,\
: 0 < kprm, kprm. < 5,\
: 0 < mum, mum < 10, \
: 0.05 < kn, kn < 1,\
: 0 < kc, kc < 1,\
: 0.002 < kcn, kcn < .5,\
: 0 < Lm, Lm < 1,\
: 0 < kmu, kmu < 10,\
: 0 < YM, YM < .5}
* fit (kam,kbrm,kprm,milm,kn,kc,kcn,Lm,kmu,YM), tb to ddb, tp to ddp, \
: M to ddm, N to ddn, C to ddc, A to dda, constraints q
final parameter values
```

value	error	dependency	parameter
2.734700822	12.17292675	0.088769606	kam
3.09562705	1.234762472	0.636473194	kbrm
4.943072645	8.39997299	0.083733626	kprm
8.437565766	186030.7204	0.999999999	mum
0.727738472	113630.9435	0.999999999	kn
0.036442671	339101.8347	1	kc
0.469382511	877671.2675	1	kcn
1	2406.006822	0.99999999	Lm
0.003719028	1407.826988	0.999999998	kmu
0.096196936	157.4025003	0.999999999	YM

```
3 iterations
CONVERGED

best weighted sum of squares = 4.390074e-01
weighted root mean square error = 6.625764e-01
weighted deviation fraction = 2.651032e-02
lagrange multiplier[16] = 6.158843348e-07
*
* t = 0:6!160; np=160
* Res = points(V,Bn,Pn,A,Br,Pr,C,N,M,t);
*
* /* multiply concentrations by outflow rate value to obtain output
```

```
* per day amounts in kilograms */
* Res col (3:10) = (Res col (3:10)) * vr(0)

* /*    data from J.France paper for observed outputs from chopped
*        Italian ryegrass in kg/m^3 per day units. Used to fit above.
*        Also to be used for graphics output. */
* ddb row 1 = list(5,.1121)'
* dda row 1 = list(5,.0027)'
* ddc row 1 = list(5,.005)'
* ddp row 1 = list(5,.0365)'
* ddm row 1 = list(5,.1756)'
* ddn row 1 = list(5,.0495)'

* draw Res col (1,5) lt dashed                   /* A = Alpha hexose */
* title "A" at (6.2,Res[np,5]) world
* draw dda pt circle ptsize .025 ffract
* draw dda pt "A" ptsize .017 ffract

* draw Res col (1,8) lt ldash                    /* C = carbohydrate */
* title "C" at (6.2,Res[np,8]) world, place (left,top)
* draw ddc pt circle ptsize .025 ffract
* draw ddc pt "C" ptsize .017 ffract

* /* Res col 7 = digestable protein + non-digestable protein */
* Res col 7 = (Res col 7)+(Res col 4)
* draw Res col (1,7) lt solid                    /* Pr+Pn = total protein */
* title "P" at (6.2,Res[np,7]) world
* draw ddp pt circle ptsize .025 ffract
* draw ddp pt "P" ptsize .017 ffract


* draw Res col (1,9) lt ddash                    /* N = non-protein nitrogen */
* title "N" at (6.2,Res[np,9]) world, place (left,bottom)
* draw ddn pt circle ptsize .025 ffract
* draw ddn pt "N" ptsize .017 ffract

* window 0 to 6, 0 to .05
* bottom title "days"
* left title "kg/day"
* window 0 to 6, 0 to .1
* frame 0 to 1, .5 to I
* w1 = w

* draw Res col (1,10) lt solid                   /* M = microbial concentration */
* title "M" at (6.2,Res[np,10]) world
* draw ddm pt circle ptsize .025 ffract
* draw ddm pt "M" ptsize .017 ffract

* /* Res col 6 = digestable and non-digestable beta hexose */
* Res col 6 = (Res col 6)+(Res col 3)
* draw Res col (1,6) lt dashed                   /* Br+Bn total Beta hexose */
* title "B" at (6.2,Res[np,6]) world
* draw ddb pt circle ptsize .025 ffract
* draw ddb pt "B" ptsize .017 ffract

* bottom title "days"
* left title "kg/day"
* frame 0 to 1, 0 to .5
* window 0 to 6, 0 to .5
* view
* unview
*
*/* end of file */
*
*  exit
end of MLAB.LOG
```

Figure 3 presents the graphical output produced by the graphics commands in the log-file given previously showing the curve-fit results. All the steady state data points given in France et al. are fit reasonably well, and these parameters may be preferred over those proposed in France et al. By adjusting other parameters, it is possible that we could improve the fit even more.

CONCLUSION

Application of the MLAB Mathematical and Statistical Modeling System software to France's classic mathematical description of rumen action (France et al., 1982) results in predictions that greatly improve the original trial and error choice of parameter values. Here we have shown how MLAB provides a good fit to the steady state data reported in the original paper as well as a simulation of pulsed eating and drinking.

For more information on MLAB Mathematical and Statistical Modeling System software, contact Civilized Software, Inc. (301-962-3711; email: csi@civilized.com; URL: http://www.civilized.com).

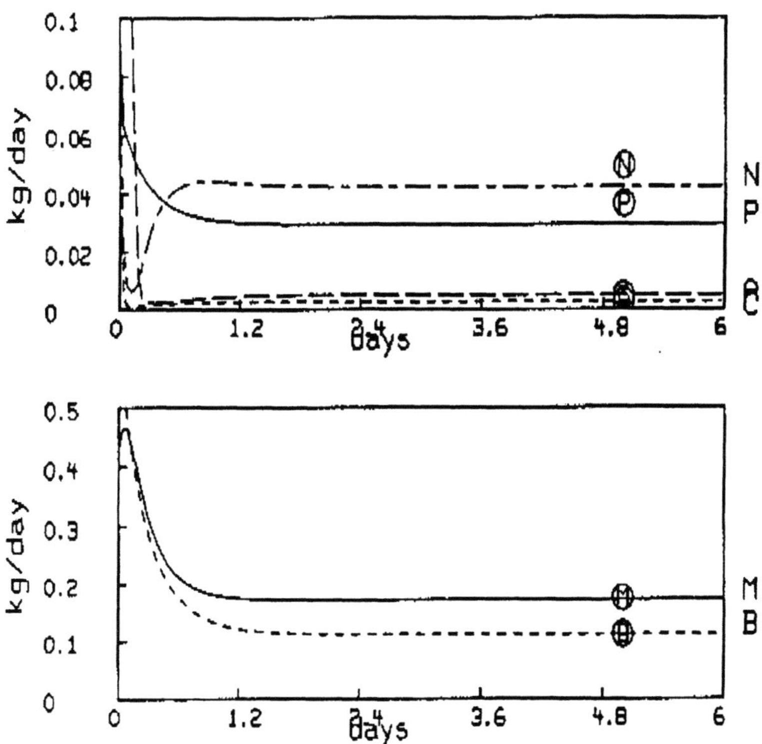

Figure 3. Curve fitting results from MLAB.

CORRESPONDENCE

Please address all correspondence to:
Gary Knott
Civilized Software Inc.
12109 Heritage Park Circle
Silver Spring, MD 20906
Knott@civilized.com
URL: www.civilized.com

REFERENCES

France, J., Thornley, J.H.M., and Beever, D.E., 1982, A mathematical model of the rumen, *J. Agric. Sci. Camb.* 99:343-353.
MLAB Reference Manual, 1998, Civilized Software, Silver Spring.

STELLA® RESEARCH SOFTWARE FOR TEACHING CONCEPTS OF NUTRIENT DYNAMICS IN THE UNDERGRADUATE CLASSROOM

James L. Hargrove[*]

INTRODUCTION

STELLA® Research software[†] is widely used in secondary schools and colleges. It was designed to teach elements of systems thinking that include concepts of dynamic change and feedback. Modeling such responses requires the use of built-in functions that may be algebraic, logical, probabilistic, or trigonometric. Unlike programs developed by the professional modeling community, STELLA software does not perform statistical analysis or derive parameters that provide the best fit between data and a proposed model. However, its advantages as a presentation tool include ease of use, ability to design and share kinetic models for special purposes, and availability of tools that allow for importing sounds, images, and video files. STELLA software is most appropriate for students who have not been trained in mathematics but who should understand the implications of models that deal with growth, accumulation of fat- *vs.* water-soluble vitamins, feedback controls, tissue effects, and issues related to metabolic regulation. The software is ideal for introducing the concept of hypothesis testing, and it can facilitate learning about complex processes such as transport across the placenta or blood-brain barrier. Whereas traditional educational methods lack a dynamic aspect, modeling programs allow the temporal dimension of cause and consequence to be explored. Because foods are widely considered to be medicinal, it would be useful for students in nutrition to learn about the plateau principle, models of effect, and the rationale for achieving therapeutic levels of nutrients and nutritional supplements.

---

[*] James L. Hargrove, Department of Foods and Nutrition, The University of Georgia, Athens, GA 30602.

[†] STELLA and NetSim are copyrights of High Performance Systems, Inc., 45 Lyme Road, Suite 300, Hanover, NH 03755. The Internet address is http://www.hps-inc.com/. STELLA is an acronym for Systems Thinking, Experiential Learning Laboratory with Animation.

GENERIC STRUCTURES WITH IMPLICATIONS FOR BALANCE

Despite the complexity of living organisms, biological systems must always be simple enough to function. Once the growth phase is complete, organisms must be able to remain in balance over most of a lifespan. This simple-sounding principle implies that the intake and usage of more than 40 essential nutrients must be matched well enough to avoid deficiencies or excesses. Indeed, too much change is often an indication of disease or disorder. One important function of theory is to provide general patterns that can be understood and used in different contexts. One of the most useful organizing principles for scientists who deal with mass balance is a system comprised of individual pools that can be linked together in a series or a web. Each pool is characterized by a unique rate of formation and a proportional rate of usage. With minor adjustments, similar models can be created that represent groups of organisms, cell populations, enzymes, messenger RNAs, drugs, or nutrients in specific locations (Gallaher, 1996; Hargrove, 1998).

People who become accustomed to thinking about systems soon realize that many kinds of objects can be considered to follow simple patterns of accumulation and elimination that are very easy to represent mathematically. For example, there are probably more than 50,000 different genes in the human genome, but each gene product can be characterized by its abundance, rate of formation, and rate of degradation. Most scientists refer to this kind of thinking as compartmental analysis, but the name makes it sound more complex than it is. The elements of compartmental analysis can easily be taught to undergraduate students if they are willing to learn to use simple tools that are available for personal computers.

FEATURES OF STELLA RESEARCH SOFTWARE

Organizational Hierarchy

STELLA was designed partly to allow students to operate tutorials that illustrate dynamic aspects in any field of study. To facilitate this mode of teaching, the software operates on three levels. The topmost 'High Level Map' is meant to orient students to general aspects of a problem. Text, images, video files, diagrams, and controls over model operation can be included, as well as tables and graphs for viewing results. This level allows the user to test different scenarios without learning model details; this is comparable to operating an automobile without learning how the engine operates. A display from a High Level Map is shown in Figure 1.

From the High Level Map, clicking a down arrow takes the user to the 'Model Construction Layer.' This contains a tool bar for selecting components of the map. Each component then opens to show a dialog box that allows the user to define quantitative relationships, including any of the built-in library of mathematical functions. As an example, STELLA software was used to develop a model representing the passage of an injected isotope out of the bloodstream, through hepatic parenchyma, and into the bile. The model shown in Figure 2 was made to simulate the data that are given in the MLAB User's Guide (1997).

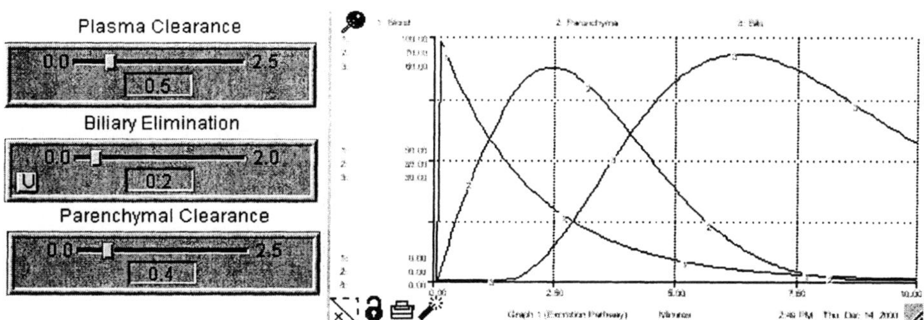

Figure 1. Controls for input and graphing output on STELLA's 'High Level Map.'

The Equations Layer

Clicking the navigation arrow on the left side of the 'Model Construction Layer' opens the 'Equations Layer,' where the set of equations that corresponds to the model is depicted. STELLA employs finite difference equations that are solved by one of three numerical methods (Euler or Runge-Kutta). The equations used in this model are shown here:

```
Bile(t) = Bile(t - dt) + (FCR - Elimination) * dt
INIT Bile = 0

INFLOWS:
FCR = DELAY(Parenchyma*Kp,1.4,0)
OUTFLOWS:
Elimination = Ke*Bile
Blood(t) = Blood(t - dt) + (Injectate - Clearance) * dt
INIT Blood = 0

INFLOWS:
Injectate = PULSE(100,0,100)
OUTFLOWS:
Clearance = Blood*Kb
Parenchyma(t) = Parenchyma(t - dt) + (Clearance - FCR) * dt
INIT Parenchyma = 0

INFLOWS:
Clearance = Blood*Kb
OUTFLOWS:
FCR = DELAY(Parenchyma*Kp,1.4,0)
Kb = 0.5
Ke = 0.003
Kp = 0.425
Bile_Data = GRAPH(TIME)
```

```
(0.00, 0.00), (1.00, 0.00), (2.00, 0.00), (3.00, 21.8), (4.00, 52.1),
(5.00, 71.8), (6.00, 84.3), (7.00, 90.9), (8.00, 96.0), (9.00, 97.3),
(10.0, 98.4)
Blood_Data = GRAPH(TIME)
(0.00, 0.00), (1.00, 0.00), (2.00, 0.00), (3.00, 23.3), (4.00, 15.1),
(5.00, 7.90), (6.00, 4.30), (7.00, 2.50), (8.00, 2.20), (9.00, 2.10),
(10.0, 0.6)
```

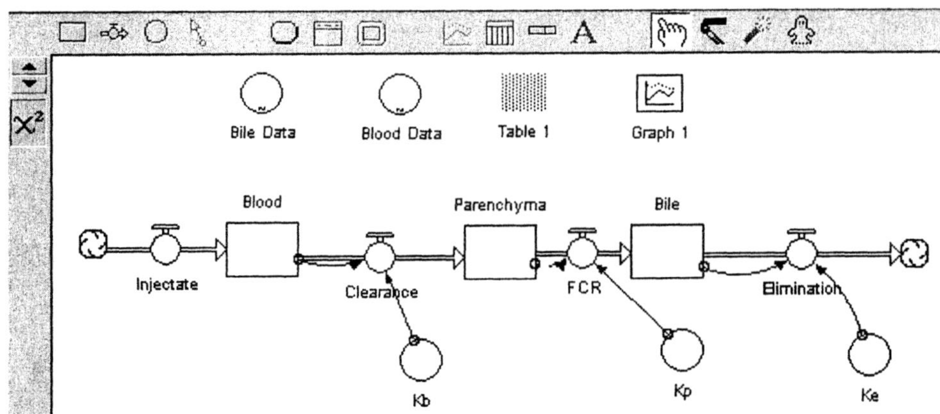

Figure 2. STELLA's 'Model Construction Layer' depicting a three-compartment model. The tool bar at the top of the model contains items (stock, flow, converter, connector) used to build the diagram and define the mathematics. Fractional clearance rates (FCR) from the blood, the hepatic parenchyma, and the bile, are indicated by k_b, k_p, and k_e, respectively.

The 'Built-In' list of mathematical functions was used to add a pulse for delivering the injected material and a delay function to retard passage of material from hepatic parenchyma into the bile. In order to enter the data to be matched in the simulation, a graphical function was opened, and the data were plotted as a function of time. The values shown in the equations resulted in the data plotted in Figure 3.

RESULTS

A bolus of 100 units of isotope was injected at the initial time. The half-life of the material in blood was set at 0.5 min^{-1}. Because a delay was noted for movement into bile, a compartment representing liver parenchyma was introduced. Despite this, a delay of 1.4 min was needed to match the data to the degree shown. Elimination from the bile was very slow; a rate of 0.003 min^{-1} was used in this example. Because STELLA does not perform parameter estimation, trial and error was used.

DISCUSSION

Computer-based presentations are becoming the standard in university classrooms. Any simulation software can be very useful in adding a dynamic capability to the teaching of principles in nutrition science. STELLA was designed to facilitate interactive teaching and learning among students of any age group. Its advantages are simplicity of use and ability to include text and images along with relatively sophisticated mathematical functions. However, it lacks the statistical capabilities of professional packages such as WinSAAM (Wastney et al., 1999). Kineticists who use software that includes methods for parameter estimation and curve fitting would note that the fit observed

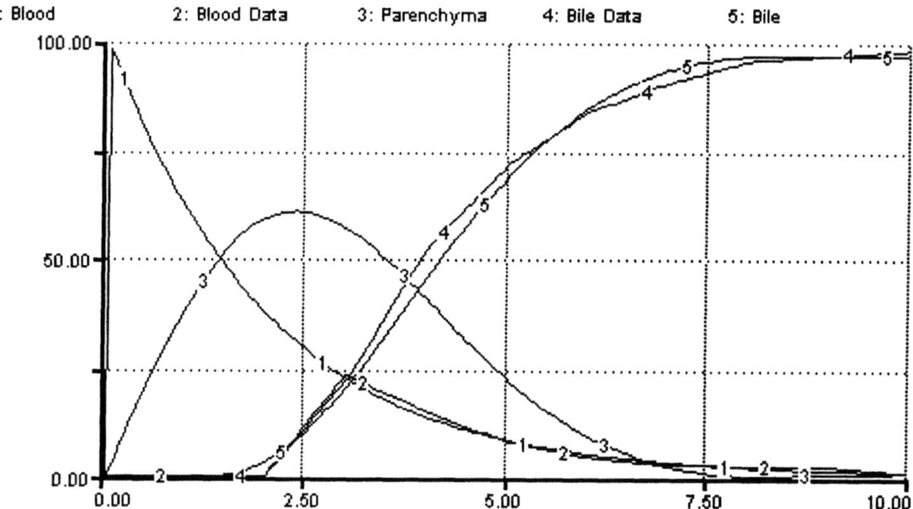

Figure 3. Results of simulation. Curves 2 and 4 plot the observed data for blood and bile, respectively; curves 1 and 5 are the corresponding simulation results. Curve 3 depicts passage of isotope through the hepatic parenchyma.

between the model outcomes and the observed data (Figure 3) is not good by professional standards. This drawback is offset by ease of use and special tools on the 'High Level Map' that make STELLA excellent for classroom presentations. Whereas advanced modeling software is typically used for research by graduate students and professionals, STELLA is used for educational purposes in undergraduate classrooms and in secondary schools. Educators who agree that interactive and enjoyable tools for hypothesis testing are useful may with to examine the STELLA software package. In the new version called NetSim, STELLA now operates from Internet browsers. This permits models to be used without requiring file transfer to local computers. It will prove interesting to observe whether simulation and mathematical modeling become important applications on the World Wide Web.

CORRESPONDENCE

Please address all correspondence to:
James L. Hargrove
Department of Foods & Nutrition
Dawson Hall
The University of Georgia
Athens, GA 30602
jhargrov@fcs.uga.edu

REFERENCES

Gallaher, E.J., 1996, Biological system dynamics: from personal discovery to universal application, *Simulation* 66: 243-57.
Hargrove, J.L., 1998, *Dynamic Modeling in the Health Sciences,* Springer, New York.
MLAB User's Guide, p. 316, 1997, Civilized Software, Inc., Bethesda.
Wastney, M.E., Patterson, B.H., Linares, O.A., Greif, P.C., and Boston R., 1999, *Investigating Biological Systems Using Modeling,* Academic Press, New York.

A REVIEW OF 'SCIENTIST' SOFTWARE

E. Paul Wileyto[*]

INTRODUCTION

'Scientist' (MicroMath Scientific Software, Salt Lake City, UT) is a program for solving systems of differential equations and for fitting systems of differential equations to data. The graphical user interface is pleasing and simple, and it does exactly what it should do: it makes it possible to use the software without reference to the manual. The model and initial conditions are entered through an equation file in which the equations look much as one would write them by hand. Parameters may be entered through the same equation file or through a parameter file. Data are entered and results are obtained in spreadsheet format; results may be easily plotted.

In this paper, I will use two demonstration problems as vehicles for describing the mechanics of using the Scientist software. This approach reflects the way in which most people would begin to use Scientist: they would start with a real, rather than a classroom, problem and stumble through without any preliminaries.

EXAMPLE 1: RUMEN PHYSIOLOGY MODEL

The first example uses the rumen physiology model presented by France et al. (1982). The model is introduced to Scientist as a model file. Shown in Figure 1 is the model file *Rumen.eqn*, which was extracted from France et al. (1982) and is set up for simulation only.

For this type of model, we have four sections. First, we declare variables, labeling them as independent variables, dependent variables, and parameters. We then specify the model. Differential equations are specified with the familiar prime (') notation. We follow model specification with statements of parameter values (preceded by the line "//Parameter Values"). Finally, we specify the initial conditions for all of the dependent variables represented by differential equations (preceded by the line "//Initial Conditions").

[*] E. Paul Wileyto, Section of Epidemiology and Public Health, School of Veterinary Medicine, University of Pennsylvania, New Bolton Center, Kennett Square, PA 19348.

```
// MicroMath Scientist Model File
IndVars: T
DepVars: V,Bn,pn,alph,Br,Pr,C,N,M,L,Mu
Params: Dv,Sv,vo,DBn,Dpn,Dalph,Kam,Dbr
Params: Kbrm,Dpr,Spr,Kprm,Dc,Lm,Mum,fc,Ym
Params: Dn,Sn,fn,Kmu,Kc,Kn,Kcn

Mu=Mum/(1+(Kc/C)+(Kn/N)+(Kcn/C/N))
L=Lm/(1+Kmu*Mu)
V'=Dv+Sv-vo
Bn'=1/V*(DBn-Bn*Dv-Bn*Sv)
pn'=1/V*(Dpn-pn*Dv-pn*Sv)
alph'=1/V*(Dalph-alph*Dv-alph*Sv)-Kam*alph*M
Br'=1/V*(Dbr-Br*Dv-Br*Sv)-Kbrm*Br*M
Pr'=1/V*(Dpr+Spr-Pr*Dv-Pr*Sv)-Kprm*Pr*M
C'=1/V*(Dc-C*Dv-C*Sv)+L*fc*M-1/Ym*Mu*fc*M+Kam*alph*M+Kbrm*Br*M
N'=1/V*(Dn+Sn-N*(Dv+Sv))+L*fn*M-Mu*fn*M+Kprm*Pr*M
M'=Mu*M-L*M-M/V*(Dv+Sv)

//Parameter values:
Sv=.0121
Dv=.0042
vo=.0163
DBn=.0673
Dpn=.0243
Dalph=.0266
Kam=1.5
M=25
Dbr=.4937
Kbrm=2
Dpr=.0729
Spr=.0121
Kprm=2
Dc=.215
Lm=0.2
Mum=5
fc=0.3
Ym=0.1
Dn=0.0422
Sn=0.0061
fn=0.5
Kmu=5.0
Kc=0.2
Kn=0.2
Kcn=0.000001

//Initial Conditions
T=0.0
V=.005
Bn=36
pn=4
alph=5
Br=30
Pr=1.5
C=7
N=2
M=26.5

***
```

Figure 1. The model file *Rumen.eqn*. See France et al. (1982) for explanation of abbreviations.

The model file is easy to write. When you create a new model file from the *File* menu, a new window arrives with a general template to aid in writing. If the general template is not to your taste, there are many sample files that may provide an easier starting point.

Under many circumstances, the next step would be to specify a parameter file listing the values of parameters. However, we have already specified the parameter values in the model file, so we do not need to write a parameter file. One would be generated automatically for us if it were needed. For example, if we were interested in estimating parameters rather than just in simulating, we would have found it necessary to create a parameter file, or one would have been created for us.

Next, we compile the model, using either the compile button or the "Compile" item in the *Model* menu. Then, we solve the model, either by pressing the solve button or by choosing "Simulation" from the *Calculate* menu. In return, the software will query for an ending value for time T. Give it a value that makes sense. In this case, I chose 3 for 3 days. You may also choose the number of intervals. When you tell the program to calculate, it will respond with a spreadsheet containing calculated values. The column headings for simulated values are the same as our names for state variables with an extension added (_CALC) to indicate that they are calculated.

You may then create a plot using the plot button. A plot from the rumen model example is shown in Figure 2, and a legend is used to indicate the correspondence between the lines and the variables they represent. Your control of the plot template will allow you to change colors to give better separation.

Throughout a Scientist session, you have many options to control the solving process and plot appearance. There are six integration techniques available for solving the model, and you also have control over the maximum number of steps, maximum step size, and tolerances. The plot template is available after solution, and it allows you to control which lines are plotted, their appearance, and the legend.

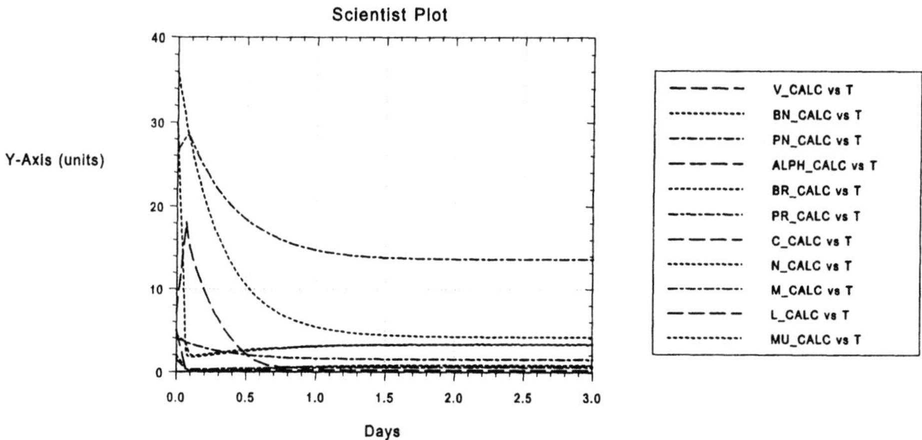

Figure 2. Plot of simulated rumen state variable.

EXAMPLE 2: KINETICS OF DRUG MOVEMENT FROM BLOOD TO BILE

The model file *Bileglow.eqn* shown in Figure 3 presents a data set from the MLAB User's Guide (1997). This model represents drug movement from the time of injection into blood and ending with uptake by bile. The model requires a delay to represent processing by the liver. Because Scientist does not have a standard means of representing a delay, we can construct one using a series of additional compartments L_i (for Limbo). We include 5 limbo compartments and use a single parameter (LB) to describe the fractional loss rate from each of them. The parameter BL represents the fractional loss rate from blood to limbo.

The model file is missing a declaration of parameter values. We have instead created a parameter file (Figure 4) from the *File* menu to contain our initial values. This is done most easily by compiling the *.eqn* file and then creating a new parameter file. If it is done after compiling, the new file will include all the declared parameters. We chose initial values for the two parameters of 0.5. The parameter set may then be saved.

For fitting, we must also supply a data set. We do this by creating a new spreadsheet from the *File* menu. If you have compiled the *.eqn* file, the new spreadsheet will contain columns for all of the compartments, including the delay compartments. You must eliminate any columns for which there are no data, or you will not be able to recover fitting statistics. However, there is no way to delete columns as we do in modern spreadsheet software. You must clear and cut and paste. Our data set is saved as *Biledat.mmd.* (Figure 5).

To start the fitting procedure, you may press the fit button, or you may select a minimization technique from the *Calculate* menu. The software will return final parameter estimates to the open parameter file (Figure 6) and return predicted values to the open spreadsheet. You may then plot in the usual way (Figure 7). Data points are usually plotted for comparison with the predicted lines, unless you choose otherwise. You may also obtain statistics on the model fit, either by pressing the statistics button (Sn) or by asking for "Statistics" from the *Calculate* menu. The statistics report for the blood/bile model is shown in Figure 8.

```
/ MicroMath Scientist Model File
IndVars: T
DepVars: L1,L2,L3,L4,L5,Blood,Bile
Params: BL,LB

Blood'=-BL*Blood
Bile'=LB*L5
L1'=BL*Blood-LB*L1
L2'=LB*L1-LB*L2
L3'=LB*L2-LB*L3
L4'=LB*L3-LB*L4
L5'=LB*L4-LB*L5

//Initial Conditions
T=0.0
Blood=100.0
Bile=0.0
L1=0.0
L2=0.0
L3=0.0
L4=0.0
L5=0.0

***
```

Figure 3. The model file *Bileglow.eqn.*

In addition to having control over selection of solvers and plotting, you also have control over the minimization technique used for fitting (4 algorithms) and over the statistics you wish to report. Both of these controls are available through the *Options* menu.

Parameter	Lower Bound	Value	Upper Bound
BL	-Infinity	0.5	Infinity
LB	-Infinity	0.5	Infinity

Figure 4. The parameter file *Bileglow.par*.

T	BLOOD	BILE
3	23.3	21.8
4	15.1	52.1
5	7.9	71.8
6	4.3	84.3
7	2.5	90.9
8	2.2	96
9	2.1	97.3

Figure 5. The spreadsheet *Biledat.mmd*.

Parameter	Lower Bound	Value	Upper Bound
BL	-Infinity	0.523077361	Infinity
LB	-Infinity	2.02614853	Infinity

Figure 6. The parameter file *Bileglow.par* after estimation.

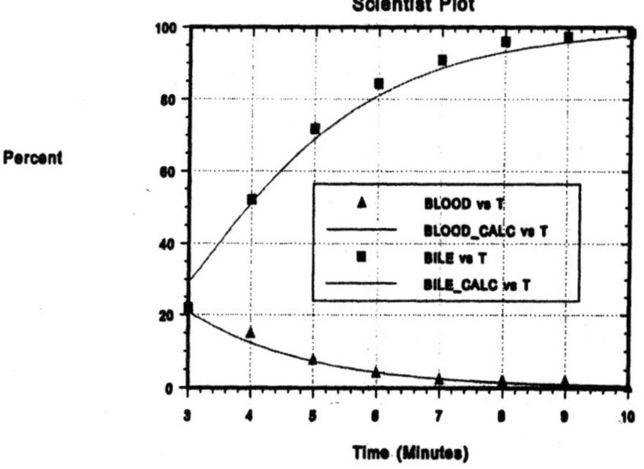

Figure 7. Plot of data and fitted model for blood to bile transfer.

```
*** MicroMath Scientist Statistics Report ***

Model File Name :       k:\scientis\bileglow.eqn
Data File Name :        k:\scientis\biledat.mmd
Param File Name :       k:\scientis\bilegl.par

Goodness-of-fit statistics for data set:      k:\scientis\biledat.mmd

    Data Column Name:     BLOOD
    Weighted                Unweighted
    Sum of squared observations :      867.660000      867.660000
    Sum of squared deviations :        16.0106983      16.0106983
    Standard deviation of data :       1.63353901      1.63353901
    R-squared :                        0.981547267     0.981547267
    Coefficient of determination :     0.964194699     0.964194699
    Correlation :                      0.996338181     0.996338181

    Data Column Name:     BILE
    Weighted                Unweighted
    Sum of squared observations :      52080.0400      52080.0400
    Sum of squared deviations :        85.8422135      85.8422135
    Standard deviation of data :       3.78246422      3.78246422
    R-squared :                        0.998351725     0.998351725
    Coefficient of determination :     0.983396716     0.983396716
    Correlation :                      0.996601506     0.996601506

    Data Set Name:        k:\scientis\biledat.mmd
     Weighted               Unweighted
    Sum of squared observations :      52947.7000      52947.7000
    Sum of squared deviations :        101.852912      101.852912
    Standard deviation of data :       2.69725935      2.69725935
    R-squared :                        0.998076349     0.998076349
    Coefficient of determination :     0.995899836     0.995899836
    Correlation :                      0.998379126     0.998379126
    Model Selection Criterion :        5.24672821      5.24672821

Confidence Intervals:
 Parameter Name :              BL
  Estimate Value =             0.523077361
  Standard Deviation =         0.0260731534
  95% Range (Univar) =         0.467156008     0.578998713
  95% Range (S-Plane) =        0.451778924     0.594375798

 Parameter Name :              LB
  Estimate Value =             2.02614853
  Standard Deviation =         0.0878513832
  95% Range (Univar) =         1.83772605      2.21457101
  95% Range (S-Plane) =        1.78591421      2.26638285

Variance-Covariance Matrix:
 0.000679809328
 -0.00168428898           0.00771786553

Correlation Matrix:
 1.00000000
 -0.735316725     1.00000000
```

Figure 8. Statistics report for fitting *Biledat.mmd.*

COMMENTS

When Scientist is working, there is nothing else like it for generating model results or fitting data quickly. The interface is used intuitively without the reference manual. However, the software is quirky and prone to crashing or locking up. It sometimes has the feel of software that was designed for in-house use, only to be released on an unprepared public. Some of the problems are machine specific. A model that can be solved or fitted without a hitch on one machine may hang up on another. Some of the problems are related to the order in which variables, parameters, or data are presented. A model that hangs up when dependent variables or parameters are presented in one order may run simply by changing the order. Plotting is an independent source of crashes. The solver is run a second time to create data for line plots. The problem may be cleared up by selecting a different solver.

Obtaining support for Scientist may be difficult as well. The program is a Windows 3.1-compatible program, and no new versions seem to be in the works. Also, the support and sales personnel do not seem to be interested in marketing the software aggressively. While using Scientist, I had minor problems and could not get through to support staff by email or telephone.

For information on purchasing Scientist software, contact
MicroMath Scientific Software
P.O. Box 71550-0550
Salt Lake City, UT 84171
801-483-3025
www.micromath.com

CORRESPONDENCE

Please address all correspondence to:
E. Paul Wileyto
Section of Epidemiology and Public Health
School of Veterinary Medicine
University of Pennsylvania
New Bolton Center
382 West Street Rd.
Kennett Square, PA 19348
epaulw@vet.upenn.edu

REFERENCES

France, J., Thornley, J.H.M., and Beever, D.E., 1982, A mathematical model of the rumen, *J. Agric. Sci. Camb.* 99:343-353.
MLAB User's Guide, p. 316, 1997, Civilized Software, Inc., Bethesda.

ACSL. *See* Advanced Continuous Simulation
   Language or software, ACSL
Adipose tissue
   daily fractional turnover, 79
Advanced Continuous Simulation Language, 225,
   243, 371
   Builder, 373, 375, 377, 381
   compiling, 377
   correlation matrix, 381
   data weighting, 375
   error structure of data, 375
   fitting, 377
   functionality, 373
   Graphic Modeler, 373, 375, 377
   input file, 374
   integration algorithms, 377
   log file, 379
   macros, 379
   Math, 373, 374
   Optimize, 373, 381
   optimizers, 377
   Sim, 373
   sub-programs, 373
   symbols, 372
   use for blood-bile model, 371
   use for rumen fermentation modeling, 371,
      381
AKA-Glucose Project, 31, 33
   chart window, 33
   control buttons, 33
   database navigation tools, 36
   database system, 33
   observation dialog window, 33
   parameter dialog window, 33
   patient dialog window, 33
   protocol dialog window, 33
Astrocyte, 317, 318, 319, 323, 324
   glutamate metabolism, 318
Atomic absorption spectroscopy, 126
Autocorrelation, 162
Balance studies, 132
Bayesian analysis, 112
Bimodality, 297
Binomial distribution, 57, 89
Bioavailability, 131

Bismuth
   modeling of, 153
Blood cell, 58
Bone, 193, 194, 195, 196, 197, 200, 202
   accretion, 202
   critical size defect, 291
   deposition, 200, 202, 203
   resorption, 200, 203
   retention, 194, 200
   wound-healing, 291
Bootstrap method, 109, 111
Brain
   astrocyte model, 322
   cell culture, 317, 321, 323
   compartmentation of, 320
   modeling of, 321
Calcium, 97, 117, 150, 194, 200
   compartmental model, 195
   kinetic study, 121
   kinetics in children, 120
   models of metabolism in children, 118
   stochastic model of, 91
Cancer, 58, 287. *See also* Tumor
   angiogenesis, 294
   catastrophe theory, 297, 298
   colon, 105
   cusp catastrophe, 297
   metastasis, 294
Carbon-13. *See* Isotope, carbon
Catastrophe theory, 297, 298
   essential features, 297
Catenary model, 97
Cattle, 267, 268
   lactation requirements, 269
   maintenance energy requirement, 269
   protein requirement, 269
   ration formulation, 267
Cell culture, 64
Cell membrane phospholipids, 80
Children
   blood sampling, 121, 123
   dietary manipulations, 124
   ethical considerations of kinetic studies, 122
   growth implications on steady state
      assumption, 122

kinetic studies, 118
   tracer dose, 122
Cholesterol, 63, 207
   blood cell metabolism, 214
   cellular traffic, 208
   degradation products, 207
   exchange, 208, 209
   LDL/VLDL metabolism, 211
   model development, 210
   related diseases, 208
   sites of synthesis, 208
   specific activity after tracer administration,
      211
   tracer, 209, 210
   tracer exchange, 209
   transformations, 207
   transport, 208
CMMB. *See* Computer-based mathematical
modeling in biology
CNCPS. *See* Cornell Net Carbohydrate and
   Protein System
Colon cancer
   Bayesian analysis, 112
   model, 107
   nonparametric model, 110
   parametric model, 108
Compartmental model, 50, 87, 88
   fibroblasts and NPC cells, 65
   homogenous, 88
   linear, 88
   of brain astrocytes, 322, 323
   rat drug metabolism, 345
   rumen, 345
Compartmental modeling, 48, 147
   calcium, 195
   estimating & adjusting fractional transfer
      coefficients, 148
   fitting, 149
   mathematical equations, 148
   minimal principle, 150
   process of, 148
   refinement of model, 327, 328
   simultaneous models of different system
      states, 150
Computer-based mathematical modeling in
biology, 21, 22, 23, 24, 33, 36
Conditional flow probability, 93
Conditional probability of transfer, 89
CONSAM, 23, 24, 324
continuous infusion method, 221
Copper
   modeling of, 153
Cornell Net Carbohydrate and Protein System,
   253, 255, 267
   absorbed energy and amino acids, 271
   applications, 282
   changes in cattle upon maturity, 279
   computing target growth rates, 277
   description, 268
   digestibility, 270

evaluation of equations through cattle
      maturity, 280
   evaluation of growth model, 275, 276
   evaluation of target weight equations, 278
   growth model, 271
   limitations, 270
   submodels, 269
   uses, 268
Correlation function, 109
Crypts, 105
Deterministic model, 49, 87, 89, 90, 91, 93
Diabetes, 2, 9
Differential equation, 51, 88, 90
Discontinuity, 297
Divergence, 298
DNA, 53, 54, 55, 56, 57
DNA adduct, 105, 106
Down syndrome, 174
Eigenvalue, 88, 98, 99
Empirical model
   generic, 48
   linear, 45
   polynomial, 46
   spline, 46
EMSA, *See* Extended Multiple Studies Analysis
Erlang distribution, 96, 98, 100
Erlang transit time, 96
Extended Multiple Studies Analysis, 196, 197
Extrinsic labeling, 132
First-order kinetics, 254
Fitting, 52, 65, 67, 162, 163, 196, 197, 225, 322,
   330
Flooding dose method, 221
Forcing function, 304, 327
Fortran, 371
Fractional catabolic rate, 304
Fractional disposal rate, 356
Fractional flow rate, 87, 88
Fractional rate of appearance, 372
Fractional rate of elimination, 372
Fractional standard deviation
   of fractional transfer coefficients, 328
Fractional synthesis rate, 221
   protein, 224, 231
Fractional transfer coefficient, 49, 162, 195, 325,
   332
   time-variant, 160, 164, 167, 168
Fractional transfer rate, 325
Functional data analysis method, 110
Gamma distribution, 96
Generalized linear modeling, 50
Global two-stage procedure, 28
Glucose, 354, 356, 360, 362
   blood regulation of, 3
   clamp, 6
   effectiveness, 5, 9
   glucose-insulin relationship, 4
   intravenous glucose tolerance test, 353, 362
   phases of response, 3
   regulation, 1

regulation, 31
Glutamate, 318
    brain metabolism of, 323, 324, 327
    metabolic regulation, 319
    metabolism by astrocytes, 318
Glutamate-glutamine cycle, 319
Hazard rate, 89, 93, 95
Heteroscedasticity, 375
High-performance liquid chromatography, 136
HPLC. *See* High-performance liquid
    chromatography
Hysteresis, 298
ICP-MS. *See* Inductively coupled plasma ion
    source MS *and* Mass spectrometry,
    inductively coupled plasma ion source
Identifiability, 119, 176, 348
Inaccessibility, 298
Inductively coupled plasma ion source MS, 142
    applications, 142
    background interference, 145
    biological sample analysis, 144
    high resolution, 145
    instrument description, 143
    limitations, 143
    noise, 144
    use in kinetic studies, 153
Initial conditions, 70, 344
Insulin
    disposition index, 13, 14
    effects on glucose production, 12
    sensitivity, 6
    signal gateway hypothesis, 13
    transendothelial transport of, 11, 12
Intrinsic labeling, 133
    atmospheric, 134
    carbon-13, 136
    copper, 133, 134
    deuterium, 133
    foliar application, 133
    hydroponic, 133
    stem injection, 133
    zinc, 133, 134
Iron
    deficiency and vitamin A absorption, 165
    deficiency and vitamin A metabolism, 162,
        165
Isomorphic model, 8
Isotope
    analysis, 142
    calcium, 194
    carbon, 134, 135
    chromium, 132
    deuterium, 133
    magnesium, 133
    zinc, 133
Isotopomer, 133, 136
Iteration, 328
JMP, 24
Kinetic homogeneity, 325
KINETICA, 91

Kolmogorov equations, 90
Lactation
    arteriovenous difference methodology, 239
    dietary manipulation and milk composition,
        244
    mammary fat synthesis, 239
    mammary lactose synthesis, 239
    mammary protein synthesis, 241
    mammary response to amino acids, 241
    mammary response to blood glucose, 240
    mammary response to dietary manipulation,
        240
    management of milk protein production, 239
    milk composition, 244
    model, 242, 246
    ratio of mammary blood flow to milk yield,
        244
    seasonal effects, 240
    temporal lactation curve following parturition,
        240
Lead
    modeling of, 152
Least squares, 46, 57, 88
Least squares fitting, 343
Likelihood ratio test, 109
Linear discriminant analysis, 58
Log likelihood function, 377, 380
Macroparameter, 89
Malthusian growth, 288
Mammillary model, 97
Markov process, 90, 91, 94, 95
Markov-chain model, 94
Mass action, 254
Mass spectrometry, 179
    abundance sensitivity, 145
    band pass-limited dynamic reaction cell, 147
    dynamic reaction cell, 146
    fast atom bombardment, 194
    hexapole collision cell, 147
    HPLC-MS, 136
    inductively coupled plasma ion source, 142,
        143
    inorganic, 141
    ion separation, 144
    mass discrimination, 145
    mass discrimination bias, 145
    plasma source, 141
    resolution, 145
    space charge effect, 145
    temporal imprecision, 145
    thermal ionization, 142
    thermal ionization magnetic sector, 126
Mathematica, 47
Mathematical model
    definition of, 287
Maximum likelihood method, 50, 225
    restricted, 109
Mean residence time, 95, 98, 100
Michaelis-Menten kinetics, 52, 246, 253
    for modeling rumen fermentation, 256

Microparameter, 89, 91
Minimal Model, 1, 7, 31
    acute insulin response, 14
    application in MLAB, 7
    assumptions, 5
    sensitivity index, 14
    validation of, 7
MINMOD, 7
MLAB, 7, 47, 52, 389
    fitting, 396
    graphics, 393, 399
    output log file, 396
    use for rumen fermentation modeling, 389
Modeless software processing, 23, 24
Modeling
    history of, 21
    process of, 45, 63
    steps of, 322
Molybdenum
    modeling of, 153
Multicolinearity, 162
Multivariate normal distribution, 58
Neurons, 317
Nickel
    modeling of, 152
Niemann-Pick C disease, 63
Nonlinear model, 332
Nonlinearity
    implications of, 333
Nonparametric, 110
nonparametric methods, 50
Non-steady state modeling, 160
Numerical integrator, 22
Osteoporosis, 193
Parallel model, 168, 178, 304, 330
    of vitamin A metabolism, 164
Parametric, 108
Particle model, 89
Partition analysis, 4
Phase-type distribution, 97
Plasma
    volume estimation in rats, 161
Poisson distribution, 57
Polyunsaturated fatty acids, 78
    chain length and exchange kinetics, 81
    effect of diet on adipose composition, 79, 82
    effect of diet on membrane composition, 78, 80
    linoleic acid, 79
Pool size, 196, 200
    in vitamin A modeling, 163
Population modeling, 28, 196, 197, 348
    continuum assumption, 288
    exponential growth model, 289
    general models, 289
    Gompertz equation, 289
    limited growth model, 289
    logistic growth model, 289
    Verhulst equation, 289
Probabilistic modeling, 87

Probability intensity coefficient, 89
Protein degradation, 221
Protein synthesis, 221
Protein turnover, 222
    leucine tracer, 223
    phenylalanine tracer, 223
    rodent dynamic model, 223
PSMS. *See* Mass spectrometry, plasma source
PUFA. *See* Polyunsaturated fatty acids
pulse dose method, 221
ration formulation models, 268
Residence time, 94, 304, 308
Retention time, 93
Retinol. *See* vitamin A
Rumen fermentation, 253, 389, 407
    degradation of fiber, 254
    France model, 390
    lag assumptions, 254, 255
    model, 381
    modeling assuming first-order kinetics, 254
    modeling assuming Michaelis-Menton kinetics, 256
    modeling assuming thermodynamic control, 258, 260
Ruminant, 389
SAAM, 22, 24, 324, 327, 343
    cholesterol modeling, 210
    forcing function, 304, 327
    terminology, 325
    use in vitamin A modeling, 159, 161
SAAM II, 23, 52, 120, 126
Scientist, 24, 407
    compile, 409
    delay, 410
    fitting, 410
    solve, 409
    statistics, 410
Sensitivity analysis, 225
    in cholesterol modeling, 214, 215, 216
Software
    ACSL, 225, 243, 371
    Continuous System Modeling Program, 389
    KINETICA, 91
    MLAB, 7, 52, 389
    moded, 24
    modeless, 24
    SAAM, 159, 161
    SAAM II, 52, 120, 126
    S-PLUS, 50, 57
    Scientist, 407
    STATA, 353
    STELLA, 401
    WinSAAM, 7, 21, 24, 63, 64, 52, 163, 180, 304
Sojourn time, 93
S-PLUS, 47, 57
Stable isotope, 119
    B-vitamers, 178
    calcium, 150
STATA, 24, 36, 353

ado files, 360
   'by' sub-group selector, 360
   'byvar' sub-group selector, 360
   command syntax, 360
   commands, 353
   demonstration of use, 353, 362
   file saving, 360
   flow control services, 360
   graphics, 354
   plot, 354
   pointers, 361
   processing machinery, 359
   releases and updates, 361
Steady state solution
   in vitamin A modeling, 163
STELLA, 23, 77, 78
   equations, 403
   example of use, 404
   mathematical functions, 404
   model construction, 402
   organizational heirarchy, 402
Stem cells, 105
Stochastic model, 89, 90, 91, 94
TCA cycle, 319, 321, 324, 330
TIMS. *See* Mass spectrometry, thermal ionization
Tracee, 197
Tracer, 87, 125, 142, 149, 194, 196, 197, 201,
   324, 325
   B-vitamers, 178
   calcium, 150
   cholesterol, 209
   dose considerations, 149
   glucose, 9
   glutamate, 323
   sucrose, 64
   vitamin A, 161, 163
   vitamin B-6, 179
   zinc, 150
Tracer:tracee ratio, 126
Transit time, 93, 94, 95, 98, 304
Tricarboxylic cycle. *See* TCA cycle
Tumor, 288
   angiogenesis, 294
   carcinogenic advantage, 288
   chemotaxis, 295
   definition of, 295
   diffusion and spheroid growth, 290
   diffusion of growth inhibitor, 290
   fractal-like boundaries, 288
   growth inhibitor-diffusion model stability, 294
   growth models, 289
   haptotaxis, 295
   outer boundary surface tension, 292
   oxygen consumption, 290
   pressure distribution, 292
   prevascular diffusion models, 289
   progression, 295, 296
   spherical symmetry assumption, 292
   strain-energy function, 294
   time-evolutionary diffusion model, 292
   vascularization, 294
Utilization rate, 304
Vitamin A
   absorption during iron deficiency, 165
   and cancer prevention, 302
   and iron kinetic studies, 162
   and TCDD kinetic studies, 166
   change in kinetics with 4-HPR, 307
   four-compartment model, 161, 163, 167
   metabolism during iron deficiency, 165
   model for iron-deficient rat, 163
   model for TCDD-treated rat, 168
   model of eye metabolism, 309
   model of prostate metabolism, 308
   non-steady state model, 163, 164, 165, 168
   pool sizes, 167
   steady state model, 161
   synthetic analog, 4-HPR, 302
   TCDD effect on homeostasis, 166
   tracer, 161, 168
   whole body metabolism, 160
Vitamin B-6, 173
   compartmental model, 176
   modeling considerations, 174
   tracer, 179
Volume of distribution, 197
Weibull distribution, 96
Weighted sum, 58
Weighting
   of data, 325
WinSAAM, 7, 21, 24, 36, 47, 52, 63, 64, 177,
   180, 343, 344, 345, 346, 348, 353
   batch window, 26, 344
   charting system, 343
   compiling, 346
   data entry, 345
   deck, 344, 345
   deck command, 346
   example of programming code, 70
   Extended Multiple Studies Analysis, 348
   file management, 26
   fitting, 346
   fractional transfer coefficients, 345
   H DAT, 345
   H PAR, 345
   iter command, 346
   iterate, 346
   logger, 26, 344
   parameter specification, 345
   plot command, 346
   population analysis, 353
   prin command, 346
   project manager, 27, 348
   project utility, 179
   QO function, 183, 70
   solv command, 346
   solving, 346
   spreadsheet window, 26, 344, 345
   statistical capability, 404
   terminal window, 25, 343, 344, 345, 346

text editor, 343, 345
time change function, 178, 180
time-change (TC) function, 71
time-change facility, 197
T-interupt, 196
use for bioavailability, 137
use for calcium modeling, 195, 196
use for vitamin B-6 modeling, 179

use in vitamin A modeling, 160, 163, 304
Zinc, 117
compartmental model, 118, 126
kinetic study, 121
kinetics in children, 125
modeling of, 151
tracer, 125